NATO ASI Series

Advanced Science Institutes Series

A series presenting the results of activities sponsored by the NATO Science Committee, which aims at the dissemination of advanced scientific and technological knowledge, with a view to strengthening links between scientific communities.

The Series is published by an international board of publishers in conjunction with the NATO Scientific Affairs Division

A Life Sciences B Physics	Plenum Publishing Corporation London and New York
C Mathematical and Physical Sciences D Behavioural and Social Sciences E Applied Sciences	Kluwer Academic Publishers Dordrecht, Boston and London
F Computer and Systems Sciences G Ecological Sciences H Cell Biology I Global Environmental Change	Springer-Verlag Berlin Heidelberg New York London Paris Tokyo Hong Kong Barcelona Budapest

PARTNERSHIP SUB-SERIES

1. Disarmament Technologies	Kluwer Academic Publishers
2. Environment	Springer-Verlag
3. High Technology	Kluwer Academic Publishers
4. Science and Technology Policy	Kluwer Academic Publishers
5. Computer Networking	Kluwer Academic Publishers

The Partnership Sub-Series incorporates activities undertaken in collaboration with NATO's Cooperation Partners, the countries of the CIS and Central and Eastern Europe, in Priority Areas of concern to those countries.

NATO-PCO DATABASE

The electronic index to the NATO ASI Series provides full bibliographical references (with keywords and/or abstracts) to about 50000 contributions from international scientists published in all sections of the NATO ASI Series. Access to the NATO-PCO DATABASE compiled by the NATO Publication Coordination Office is possible in two ways:

- via online FILE 128 (NATO-PCO DATABASE) hosted by ESRIN, Via Galileo Galilei, I-00044 Frascati, Italy.
- via CD-ROM "NATO Science & Technology Disk" with user-friendly retrieval software in English, French and German (© WTV GmbH and DATAWARE Technologies Inc. 1992).

The CD-ROM can be ordered through any member of the Board of Publishers or through NATO-PCO, Overijse, Belgium.

Series I: Global Environmental Change, Vol. 37

Springer
Berlin
Heidelberg
New York
Barcelona
Budapest
Hong Kong
London
Milan
Paris
Santa Clara
Singapore
Tokyo

Climate Change and World Food Security

Edited by

Thomas E. Downing
University of Oxford
Environmental Change Unit
1a Mansfield Road
Oxford OX1 3TB, U.K.

Springer
Published in cooperation with NATO Scientific Affairs Division

Proceedings of the NATO Advanced Research Workshop "Climate Change and World Food Security", held in Oxford, U.K., July 11–15, 1993

Cataloging-in-Publication Data applied for

Die Deutsche Bibliothek - CIP-Einheitsaufnahme

Climate change and world food security : [proceedings of the NATO Advanced Research Workshop on Climate Change and World Food Security, held in Oxford, U.K., June 1993] / ed. by Thomas E. Downing. Publ. in cooperation with NATO Scientific Affairs Division. - Berlin ; Heidelberg ; New York ; Barcelona ; Budapest ; Hong Kong ; London ; Milan ; Paris ; Santa Clara ; Singapore ; Tokyo : Springer, 1995
(NATO ASI series : Ser. I, Global environmental change ; Vol. 37)
ISBN 3-540-60562-2
NE: Downing, Thomas E. [Hrsg.]; Advanced Research Workshop on Climate Change and World Food Security <1993, Oxford>; NATO: NATO ASI series / I

ISBN 3-540-60562-2 Springer-Verlag Berlin Heidelberg New York

Printed in Germany

Typesetting: Camera ready by authors/editors
Printed on acid-free paper
SPIN: 10061715 31/3137 - 5 4 3 2 1 0

Foreword

In the last half decade since sustainable development became a serious objective, what have we achieved? Are livelihoods more secure? Are nations wealthier and more resilient? Is environmental quality being restored or maintained? These are essential questions of development. Their answers are many, varied between communities and regions, even between individuals.

Two years ago, in the aftermath of the Earth Summit and ratification of the Framework Convention on Climate Change, but before the first Conference of Parties, I participated in a panel at the inaugural Oxford Environment Conference on Climate Change and World Food Security. The panel vigorously reviewed issues of resilient development and food security.

This book is a product of the Oxford Environment Conference. It takes the essential questions of sustainability as a starting point to focus on present food security and its future prospects in the face of climate change. Why is this book important? First, I believe our goals to end hunger are under threat. We know what to do in many respects, but fail to generate the finances and political will to change the structures that thrive on poverty. Second, I believe concern about the environment has become dangerously separated from the fundamental issues of human deprivation. Third, I believe climate change is a serious threat and I am dismayed at the way nations dither over how to control greenhouse gas emissions and mechanisms to meet the challenge of adverse climate impacts. By joining the two perspectives – the present crisis of hunger and the future threat of degrading resources – *Climate Change and World Food Security* addresses both the development and environment agendas.

Who should read this book? It will be a valuable resource for professionals working in the fields of development, climate change and food security. It should also be required reading for students of these fields. More importantly, I hope it is used to support local, national and international action to apply its message and to work together in designing and building a sustainable future for our descendants. The vision for a food-secure world is captured in these papers. What must we do to achieve it?

Lord Judd
Marston, Oxfordshire

Preface

This volume of papers addresses the threat of climate change to world food security, by providing an authoritative review of climate change and surprise, contrasting trends in world food security, case studies of food and the environment, and papers on adaptive responses. Most of the papers had their inception at a North Atlantic Treaty Organisation (NATO) Advanced Research Workshop and the Oxford Environment Conference on *Climate Change and World Food Security* at the University of Oxford in July 1993, organised by Robert S. Chen, Thomas E. Downing, Robert W. Kates and Martin L. Parry.

The workshop had three objectives, each with three components. We sought to bring together three communities concerned with hunger and environmental change. *Food policy and hunger* researchers, primarily drawn from the social sciences, focus on a five to ten year time scale of the recent past and the horizon of development planning. They seek to reduce vulnerability, improve monitoring of famine and food security, and provide efficient interventions in times of crisis and as social support systems. *Agricultural system* modellers, from atmospheric, agricultural and economic sciences, seek to understand food production systems, from the plant level to world agricultural potential and trade. They provide models that integrate resources and economies with specific policy objectives. A third group focuses on *climate impact assessment*, from scenarios of future global climate change to estimates of biophysical sensitivity and economic, social and political implications. We sought to enhance our collective understanding between these three communities and to shape a common research agenda that includes food security as an essential focus, in addition to the more customary indicators of agricultural production, commodity prices and world trade.

The group of some forty invited experts from ten countries were asked to deliberate on three themes. We reviewed projections of the *incidence and distribution of hunger* over the next few decades and its dependence on economic growth and its distribution, resource constraints, and concerted action to prevent hunger or mitigate its effects. The workshop then examined the *threat of climate change*, alternative models of world agricultural potential that have simulated the impacts of climate change and the threats to food security from multiple environmental, social, economic and political change. Finally, we sought to identify practicable *adaptive strategies* that would both mitigate the effects of climate change and reduce vulnerability and food insecurity, as a contribution to the protocol on adaptation for the Framework Convention on Climate Change.

For each theme, we sought to identify *what's new*, *what's uncertain*, and *what should we do*? Over the past five years the literature on hunger and food policy and on drought and global environmental change has burgeoned. Yet, few studies integrate the social and economic dimensions of hunger and its roots in environmental change, and even fewer span the past and present to assess potential changes in the future. Given the wealth of studies and results of new research, the second question should be easily answered: how much confidence do we have in projections of future hunger and its sensitivity to climate change? Finally, the workshop sought to identify a research agenda that would support informed policies to reduce hunger, mitigate climate change and adapt to its adverse consequences.

In the intervening months since the workshop, the commissioned background papers were published in 1994 in special issues of *Food Policy* (edited by Thomas E. Downing and Martin L. Parry, Volume 19, Number 2) and *Global Environmental Change* (edited by R.S. Chen and R.W. Kates, Volume 4, Number 1). The two journal issues present papers reviewed and substantially edited in light of the workshop deliberations. This book, *Climate Change and World Food Security*, contains ten new papers and fourteen papers published in the two journals. Almost all of the papers that were previously published have been extensively revised and updated. To our knowledge, this book is the first authoritative collection of research on the prospects for climate change and food security.

The organisers of the workshop are grateful to the growing research communities who work on food security, climate change and environment; those whom we have drawn upon for reviews, advice and collegial support. This volume would not have appeared without funding for the workshop – from NATO, the Consortium for International Earth Science Information Network (CIESIN), UK Overseas Development Administration, U.S. National Oceanic and Atmospheric Administration (NOAA), and U.S. Department of Agriculture (USDA). Preparation of the manuscript was facilitated with additional funding from Nuclear Electric, Plc. Over the past year, the editor is particularly grateful to: Megan Gawith who undertook many of the editorial tasks and layout; Philippa Logan for copy-editing most of the papers; John Orr for cartography; and to Richard Macrory, Director of the Environmental Change Unit. The support of the above groups is gratefully acknowledged, while the findings reported here are those of the authors and may not necessarily reflect the views of the funding agencies.

July 1995 Robert S. Chen, Thomas E. Downing, Robert W. Kates and Martin L. Parry

CONTENTS

INTRODUCTION

Introduction: Climate Change and World Food Security

Thomas E. Downing
Environmental Change Unit, University of Oxford, Oxford, U.K.

Robert S. Chen
CIESIN, University Centre, MI 48710, U.S.

Robert W. Kates
Independent Scholar, Trenton, ME 04605, U.S.

Martin L. Parry
Department of Geography, University College, London, U.K.

1 Introduction

What is the future of hunger? Is climate change an unprecedented threat to world food security? What can be done to reduce vulnerability and promote sustainable, resilient development? The assemblage of papers in *Climate Change and World Food Security* (Downing, 1995) seeks to place the Earth and its vulnerable populations between two worlds. The *present*, in which up to a billion people are threatened by chronic undernutrition, micronutrient deficiencies and disease, is subject to drought and climatic variations. The *future,* subject to multiple environmental threats as the world warms by 1.5°C to 4.5°C by the middle or end of the next century, could see rapid improvement in food security.

The papers are organised into five parts:

2 Trends in agriculture and food security

The goals for a food-secure world encompass: the end of famine; chronic and seasonal undernutrition rates of less than 1 per cent (100 million people); virtually no deficiencies in micronutrients (e.g. iodine and vitamin A); and control of diseases such as malaria, diarrhoea and anaemia that affect nutritional status. These goals, suggested by Chen and Kates (1994, 1995, this volume), are in line with the agenda endorsed by the International Conference on Nutrition and the Bellagio Declaration (FAO and WHO, 1992). This *normative* scenario of a food-secure future is contrasted with trend projections and the 'common wisdom'. Indeed, nutritional status is

NATO ASI Series, Vol. I 37
Climate Change and World Food Security
Edited by Thomas E. Downing

improving in many parts of the world, often quite dramatically (ACC/SCN, 1992, 1993). Yet, in 1990: 15-35 million people were at risk of famine, 786 million were vulnerable to chronic undernutrition; and hundreds of millions suffered from micronutrient deficiencies and diseases. Policies to achieve a food-secure world in the next few decades include the suite of sustainable agricultural development, implementation of a social norm of food as a human right, and effective safety nets for famine prevention, emergency assistance, income support, and groups with special needs.

Crosson and Anderson (1994, 1995, this volume) consider the sustainability of the global agricultural system. Rising demand for food and fibre over the next forty years requires substantial increases in supply. What are the environmental implications in less developed countries? Unacceptable environmental effects are likely, unless knowledge (including technology) can substitute for natural resources. The required new knowledge necessitates investment in its development; past growth in knowledge did not 'just happen'. The key policy issue to achieve sustainability is increasing knowledge on economic terms to farmers. Achieving this level of investment presents political, economic and intellectual challenges as resources are allocated to people, institutions and technology.

3 Risk and global impacts of climate change on agricultural systems

Schneider (1995, this volume) reviews the current estimates of climate change: the canonical range of global temperature increases of 1.5 to 4.5°C with a doubling of greenhouse gas concentrations (measured in carbon-dioxide equivalents). This could occur by the end of the next century, in the absence of any efforts to reduce the emission of greenhouse gases (Houghton *et al.* 1990, 1992, 1995). Regional projections are highly uncertain, but suggest that the largest changes in temperature will be in more poleward latitudes and delayed warming in the Southern Hemisphere due to the thermal capacity of the oceans. Potential changes in precipitation, the key factor affecting agriculture for many vulnerable populations, is more difficult to estimate. The IPCC is now working on the second full assessment, including projections of climate change, potential impacts, and effective response strategies, to be reported in late 1995.

Estimating climate change and its impacts, however, is still an uncertain science. Average changes predicted in most assessments appear to be gradual, leading to global and continental changes that the world can cope with, albeit at some cost. On the other hand, history is replete with surprise. Schneider provides examples from the past and proposes a surprise-rich assessment paradigm. Interactions and surprises that affect food security encompass changes in demand, natural ecosystems, climate and water, crops, health, diseases and food distribution. At the regional level, each of these components includes potential surprises. Science must anticipate surprise; policy must increase its resilience and adaptability to surprise.

The effects of climate change on world agriculture are charted in two papers, reflecting comparable methodologies of global economic modelling. Fischer, Frohberg, Parry and Rosenzweig (1994, 1995, this volume) present the results of the most ambitious assessment of global climate change and international agriculture to date. Scientists in over twenty countries participated in developing climate scenarios, modelling site effects for major crops, scaling the site simulations to regional productivity changes, and modelling the world food trade system. More than 50 scenarios have now been tested. Clearly, world food production can be maintained against a broad spectrum of changes in productivity (eg. climate change) and demand (eg, population and economic growth). What is less clear, however, is the ultimate effect on regional production and food security. Developing countries appear to be less able to cope with climate change than developed countries. As a result, estimates of the population at-risk of food poverty (an economic indicator in the model) rise by 5 to 50 per cent, depending on the climatic scenario. Yet, key uncertainties dominate the pioneering analysis and span the site to global levels: carbon dioxide fertilisation; regional area suitable for agriculture (including irrigation supplies); economic comparative advantage and trade policies.

Reilly, Hohmann and Kane (1994, 1995, this volume) provide a similar analysis at the global level, using a different world trade model. Impacts differ significantly among scenarios and among countries. The total effects on welfare range from an increase of \$7 billion to a decrease of \$38 billion for three climate scenarios with CO_2 direct effects and adaptive responses. In contrast, the most pessimistic scenario results in a loss of welfare of \$250 billion. The balance of economic winners and losers is determined by the direct impact of climate change on yield, the global effect on commodity prices and national exports vs. imports. Adaptive responses must be fashioned in light of the high degree of uncertainty in national impacts and their propagation through international trade.

The future may well be one of surprise. Our understanding of climate and its interactions with oceans and the biosphere is still elementary; history is replete with lessons of unforeseen ecological change. While trends in food security are promising, the sudden disruption of armed conflict throughout the world is a salutary caution against over-optimism.

3 Vulnerability and multiple threats to sustainable agriculture

Given the (normative) vision of a food-secure world in the middle of the next century, presented by Chen and Kates, how great are the threats of population growth, environmental degradation, and climate change? What are the major impacts on agriculture and food security?

To begin with, what is vulnerability? Downing, Watts and Bohle (1995, this volume; see also Bohle *et al.*, 1994) argue that understanding the characteristics and causal structure of vulnerability to hunger and food insecurity is a precondition for grasping the potential consequences of global climate change. The conjuncture of social vulnerability, derived from human ecology, expanded entitlements and political economy, may be affected in unpredictable, or at least surprising, ways as climate change is refracted through markets, new ecological stresses, and multilateral agreements. Two pathways of resilient development, vulnerability and climate change are charted for Zimbabwe.

Crawshaw and Shaw (1995, this volume) draw upon the extensive experience of the World Food Programme to discuss regional variations and causes of vulnerability, the population at risk, and ability to cope with chronic and transitory food insecurity. Data prepared for the International Conference on Nutrition (ACC/SCN 1992) suggest that chronic undernutrition is declining, albeit slowly and less in Africa than elsewhere. The anti-hunger strategies of the poor include accumulation and diversification strategies. Traditional coping strategies may have become less effective because of social and economic transitions and because many societies have been under pressure for so long that they have exhausted their assets. The international community has not been as adaptable to the changing vulnerability of the poor – new approaches are needed to identify viable options and target assistance.

Multiple threats – environmental degradation, economic change, population growth and climate change – to regional food production are evaluated by Norse (1994, 1995, this volume). The greatest threats are sub-regional, often linked to poverty. Sub-Saharan Africa is the most threatened, due to a fragile environment, poor economic growth, rapid population growth, and prospects for adverse climate change. The greatest challenges will be in the next two to three decades, when climatic variations (possibly related to climate change) pose significant threats.

For some analysts, the dominant issue in global environmental change is the potential for unprecedented forced migration, or environmental refugees. MacGregor (1994, 1995, this volume) reviews the ongoing debate over 'environmental refugees' and 'forced migrants'. Conceptual insights and empirical studies place migrants' decisions to flee in a complex of systemic failure, of which environment is only one aspect. The economic and ecological transformations in regions that receive forced migrants are diverse, challenging host countries and international responses. Much can be achieved if constraints on the displaced populations own efforts to secure their livelihoods are removed. However, humanitarian assistance for the victims of environmental change is likely to be limited. The human right to food for the most vulnerable must be supported within their communities – legal status as refugees is not likely to be forthcoming and may not be the most suitable intervention.

Rothman and Coppock (1995, this volume) explore the questions: what are the essential components of a sustainable existence? And, how can practical scenarios of the future be developed and used? They contrast forecasting (those who blindly follow current trends) and backcasting (boldly forging new worlds and identifying plausible pathways that connect the present and the new world). A holistic perspective is required, as are regional perspectives. Unfortunately, the 2050 Project, a collaborative programme of the World Resources Institute, Sante Fe Institute and the Brookings Institute which is the basis of this innovative approach, has been scaled down.

In a wealthier, more crowded world, agriculture will have to grow at nearly 2 per cent per year to feed the world's population, allowing for changes in diet with higher incomes, but also with shifts in preferences toward more efficient nutrition. This is clearly possible, as postulated in the scenarios presented by Chen and Kates (1995, this volume), Crosson and Anderson (1995, this volume) and Fischer *et al* (1995, this volume). The environmental and economic costs of the increased agricultural demand are only acceptable if knowledge – human capital, institutional innovation and technology – keeps pace. Knowledge, rather than population growth or economic change, is seen as the key determinant of the balance between resource needs and sustainable use (Norse, 1995, this volume).

4 Local impacts and responses to global change

Food security is necessarily a local phenomenon: vulnerable individuals and households, communities under pressure, nations with large populations at risk. Regional studies in this section draw upon a wealth of local experience to chart present and future food security.

Wang and Zhao (1995, this volume) outline scenarios of climate change for China. The next 50 years could witness increases in temperature of 1.5 deg C or so and up to 5 per cent increase in precipitation, notwithstanding large regional variations. Against the prospects of climate change, agricultural development for the past 40 years shows low yields and slow development through the 1960s, a decade of medium yields and great fluctuations, and since 1980 rapid increases in yields. Potential impacts of climate change on agriculture include changes in cropping intensity as more intensive agriculture become feasible further north. At least regarding rainfed agriculture, China could benefit from climate change.

Two papers on Bangladesh summarise the depth of information and analysis available for one of the world's most vulnerable places. Karim (1995, this volume) reviews the vulnerability of the poor in Bangladesh, against a background of agroecological resources, food production systems, and drought impacts. Prospects for climate change imply quite significant increases in drought hazard – a doubling or more of land susceptible to very severe drought. With an expanding population and increased incomes, food security will depend on: rapidly revitalising agricultural production; promoting non-conventional food crops; enhancing rural agro-processing activities; and medium-term measures to restructure cropping patterns. These are feasible options, but action needs to undertaken now to cope with present and future environmental vulnerability.

A multi-sectoral vulnerability analysis is reported by Huq, Ahmed and Koudstaal (1995, this volume), drawing upon the databases described above by Karim. Regional vulnerability profiles, based on primary physical effects, impacts on natural ecosystems, socio-economic stresses and response strategies, were compared for three scenarios of climate change, three development situations, and three conditions of international water resources. Climate change and sea level rise will affect the whole of Bangladesh. The country is highly vulnerable under present conditions, although with some regional variations. The combination of high (less resilient) development and international conflict for water from India and Nepal, could lead to critical climate change and major vulnerabilities in production and food security. The vulnerability assessment methodology enables integration and relative ranking of effects. Further assessments need to focus on social groups and effective adaptive strategies.

In fragile zones, notably mountain/hill regions and dry tropical areas, food security is dynamic, sensitive to population growth and climate change. Jodha (1995, this volume) recommends a dual-purpose approach to enhance present food security and strengthen community capacity to withstand future food shortages. These fragile zones are marked by the conjuncture of environmental, economic and political threats the limit their opportunities: marginal environments sensitive to extreme events, failure of agricultural technology, lower carrying capacities, recurrent food deficits, dependence on aid, and lack of attention by policy makers. Food security in these zones is thus extremely sensitive to perturbations, of which climate change could be significant. Policies must reverse the negative processes, change their focus and context, and recognise the essence of food security.

Four papers provide comparable perspectives on southern Africa. Schulze, Kiker and Kunz (1995, this volume) outline the scientific foundations of food production in South Africa. Crop-model simulations provide 'food for thought', not off-the-shelf answers to climate change impacts and food security in southern Africa. A modest warming, of +2°C would not seriously affect national maize production, if the CO_2 'fertilisation' effect is realised in the humid zones and if precipitation is not decreased. Both of these are uncertain – solid methodologies for impact assessment are being developed but hindered by the fundamental uncertainties of atmospheric, oceanic and plant sciences.

Recent drought in southern Africa provides an analogue for insight into the effects of climate change (Magadza, 1994, 1995, this volume). The region is particularly sensitive in its water resources, wetland habitats, biodiversity, fisheries, agriculture and hydroelectricity. The cascade of effects related to climate change (including climatic extremes) could well affect food security, if not the basis of agriculture in much of the region. Long-term strategies for resource development and drought mitigation could alleviate many of the adverse impacts, but require concerted action now.

In Senegal, rice production is sensitive to water resources in the ecological-political economy (Bass *et al.*, 1995, this volume). Irrigated development of the Senegal River Basin exacerbates existing problems of landscape degradation, rural poverty and rural-to-urban migration. Adapting to climate change and alleviating poverty must understand technological or innovation bias in coping with resource fluctuations and development aims.

The effects of climate change on pastoral systems has been less well studied, and is hampered by the lack of regional climate scenarios (particularly for precipitation), models of grassland dynamics, and the complexity of pastoral resource management. Homewood (1995, this volume) sketches some fundamental features of pastoral production systems in sub-Saharan Africa and the special circumstances that make them vulnerable to climate change. Lessons learned from past interventions to raise productivity in Africa should inform future adaptive strategies in response to climate change. Despite contrasting policies and investment, food security among the Maasai of Tanzania and Kenya has not improved much since the 1930s. Adapting to adverse climate change requires buffering the production system by: maintaining access to grazing and water, retaining mobility, ensuring restocking, and fostering markets with realistic terms of trade. Efforts to increase food security, through increased production, secure individual land tenure and economic growth, can have the paradoxical effect of increasing vulnerability, particularly so for marginal pastoralists.

Variations in agricultural policy and climatic conditions have altered maize production and food security in Mexico, as reviewed by Appendini and Liverman (1994, 1995, this volume). The present food system is in transition, from goals of self-sufficiency in production to more open markets in agricultural trade, land and water. At the same time, food production has always been constrained by low rainfall and drought, and climate change may well decrease precipitation and yields. Dramatic events since the paper was written have highlighted Mexico's vulnerability. In the Spring of 1995, reservoir levels were 20 to 30 per cent of average in northern Mexico, thousands of hectares were withdrawn from agriculture. At the same time, economic policies under the North American Free Trade Agreement increased the cost of inputs, withdrew subsidies on maize and other crops, and reduced credit to peasant and small farmers. The conjuncture of an economic crisis and prolonged drought, could cause dramatic increases in food poverty and deprivation. At least for economies in transition, current adaptive strategies may not be adequate to cope with future risks.

In Asia, Africa and Latin America, national and local economies are sensitive to the present range of climatic fluctuations. Climate change could seriously reduce production of the principal food staples and the availability of water resources for irrigation. However, the real effect of climate change on food security depends on the interaction of macroeconomic policy, demographic trends and social changes. For example, land tenure in Mexico dominates access to land, water and credit, and as a consequence risk of drought and food insecurity. Economic integration in both Mexico and southern Africa provides new opportunities, but uncertain prospects for vulnerable populations.

5 Strategies to limit climate change and improve food security

The dominant argument of the papers in this part centres on lessons learned in coping with present environmental changes. However, the first paper looks at the reverse question: how much would efforts to reduce greenhouse gases affect food production and food security? Rosenberg and Scott (1994, 1995, this volume). Mitigation efforts could compete with efforts to promote food security through: higher energy costs; displacement of irrigation by hydropower; direct controls on farm emissions of methane and nitrous oxide; and competition for land for afforestation and biomass sequestration. The effects of geoengineering are less clear. Perhaps the greatest trade-off is with CO_2 – the direct fertilisation effect would be lost. The authors propose that GHG controls can be designed to minimise effects on agriculture, or in fact to increase efficiency They argue that the total costs of climate change are likely to outweigh the adverse effects of mitigation on food security.

Whereas most studies of food availability focus on food production, Bender (1994, 1995, this volume) presents an end-use analysis. Four sources of potential change in global food requirements are: changes in calorie requirements resulting from improved nutritional status and change in physical activity; efficiency changes in storage, handling, transport and consumption of food; changes in consumption of animal products; and changes in dietary structures due to increased concerns about health impacts. In the long run, there is significant scope for altering the relationship between income and demand, a critical equation in most projections of supply, demand and trade. Investment in increasing agricultural supply must be compared against other interventions in the food system that alter demand. Dietary quality need not be lost while environmental sustainability and food security can be enhanced.

Rural communities face uncertain futures, stemming from economic change, policy instability, climatic shocks, and health risks, among others. Eele (1995, this volume) reviews community vulnerability and adaptability, drawing upon African case studies. In order to survive, rural communities seek to maintain both consumption and their livelihood. *A priori*, adaptation to climate change must build upon and support existing coping strategies.

The interactions of agriculture, health and environment in food security need to be matched by a vision of research that generates new knowledge and technology to meet the challenge for sustainable, resilient, food-secure worlds. This is the challenge posed by Ruttan, Bell and Clark (1995, this volume). The 18 centres (as of 1992) in the Consultative Group on International Agricultural Research (CGIAR) have suffered from budget constraints and the difficulties of organisational maturity. The prospects for meeting future food demand are further hampered by a lack of a farm-oriented approach, weakness of the private sector in many countries, and poor performance in national research centres. Health research could learn from the experience of international agriculture: a solid national foundation and an international focus on knowledge and technology aimed at the most prevalent and debilitating diseases. Much can be done in developing countries, and even poor countries can make significant advances. Global environmental research still requires pre-natal care, in contrast to the adolescence of health systems and the challenges of maturity in international agriculture. The START initiative–a System for Analysis, Research and Training–seeks to implement a cohesive research focus at the regional level, amongst the many layers of international oversight. Ultimately, the research challenge is to 'bridge the island empires' of each discipline and sectoral focus to address global change and vulnerable regions and populations.

6 Conclusion

The following chapters, prepared by world renown researchers, were discussed in the NATO Advanced Research Workshop on *Climate Change and World Food Security* in 1993, widely reviewed, revised and now published together as an authoritative perspective on the future of hunger and its risk from climate change. What have we learned in preparing these papers? What do we believe? Certainly, the views in the papers are diverse, the future is unknowable in many aspects. Yet, we believe some insight has been gained.

6.1 How dangerous is climate change for the vulnerable?

To answer our opening question, the conclusion of the group of experts was that climate change threatens the livelihood of vulnerable populations in marginal areas already marked by serious food insecurity. At the global level, climate change over the next few decades probably will not significantly alter overall world food production or prospects for world food security. At the regional level, the extent of the threat is uncertain, depending on region-wide changes in climatic resources, agricultural and economic development and social and political efforts to promote food security. But the best existing models of national and regional agricultural impacts of future climate change project substantial negative effects on developing and low-income countries with subsequent increases in the number of hungry people.

6.2 What's new?

We are beginning to widen the scope of climate impact assessment to include a fundamental understanding of vulnerability, food security and hunger. Hunger arises from the conjunctural collapse of resources, exchange entitlement, and political systems. The threat of climate change to hunger must be understood in the context of the evolution, over the next ten to fifty years (or longer), of resources (e.g. land degradation and climate change), economies (e.g., economic growth), society (e.g., consumer preferences, equity), and political systems (e.g., civil strife, democratisation).

Interdisciplinary interaction among natural and social science disciplines is increasing and improving. Major efforts of international programmes – the International Geosphere Biosphere Programme (IGBP), Human Dimensions Programme (HDP), and International Decade for Natural

Disaster Reduction (IDNDR), for example – are beginning to reap rewards. However, differences of scale, qualitative vs. quantitative approaches, and generality vs. specificity, continue.

Similarly, the political environment for research has changed. Research on the impact of climate change is widely encouraged, although the focus on adaptation still needs selling to funding and policy organisations. Likewise, support for social science, interdisciplinary research, and data networking appears to be increasing. More pessimistic is the view that global change research has taken away from the resources available for basic science. Funding for the critical advances in agriculture and technology or monitoring and interventions in food crises may be lacking, ultimately to the detriment of effective adaptation to climate change.

6.3 What is uncertain?

Much remains uncertain, possibly unknowable. Climate modellers provide estimates of their confidence in forecasts of climate change for the next century; food systems analysts provide ranges of estimates for current rates of food poverty. Yet, we are unable to state our confidence in the interactions of climate change and food security over the next few decades, other than the broad agreement suggested above. At present, we disagree whether one billion or half a billion people are hungry, or whether climate change will increase or decrease agricultural potential at given locales. Table 1 summarises the tensions or dichotomies in social systems, their temporal and spatial components, and methods of analysis that were discussed at the conference. Possibly the most intractable issue is the integration across scales. At what point does a fundamental understanding of the individual, from nutritional requirements to social, economic and political networks, become essential for understanding future world food security and the effects of climate change?

6.4 What needs to be done?

What is required to move from 'reference' projections of the world that find hunger persisting and even increasing well into the next century to 'normative' scenarios that posit an end to hunger in the future?

The focus on hunger may require new approaches. Current hunger is a major indicator of future vulnerability to climate change. Simulation models currently employed in climate impact assessment rarely include the required dimensions to portray individual, household or community

Table 1 Tensions and dichotomies: what's uncertain, controversial, disputed?

Society:	Nature — Society
	Growth — Redistribution
	Free Trade — Self Sufficiency
	Comparative Advantage — Diversity
	Optimisation — Flexibility
	Uncertainty — Predictability
	Stability — Mobility
	Peace — Conflict
Scale:	Surprise — Trends
	Sudden — Slow
	Imminent — Distant
	Acute — Chronic
	Local — Regional — Global
	Individuals — Households — Communities — Nations — Worlds
Analysis:	Existing — New
	Top Down — Bottom Up
	Vulnerability — Adaptability
	Governance — Invisible Hands
	Preservation — Change
	Knowledge — Ignorance

Source: Downing and Parry (1994), based on notes from R.S. Chen.

food deprivation. On the other hand, methods of food security analysis are unable to cope with the evolution of vulnerability over medium to long time scales and the potential shift in resources and economies with and without the effects of climate change. However, the tools and knowledge are becoming available in both the natural and social sciences. These include: time-dependent climate scenarios, regional agricultural, and natural resource models and data, community-level case studies and models, food system supply/demand analysis, national socioeconomic models, and national and international food system models.

The climate impact assessment community now emphasises adaptation – what farmers, researchers and policy makers can do to mitigate the effects of climate change. But there is a surprisingly small amount of research at present. Too often adaptation is assumed to take place as a spontaneous, readily available, and cost-free activity. Research on adaptation can provide a necessary connection to actual social costs and significant interaction across scales, often drawing upon the extensive literature on drought, climatic variability and the use of weather forecasts. Recurrent issues include: the wide range of adaptation strategies available at different time and spatial scales, significant interactions across scales, the key role of institutions, recognition of high variability, uncertainty, and potential surprises, particularly at the local level, and possible interactions with efforts to reduce greenhouse gas emissions.

Concerted efforts to compare different methodologies and model results need to be strengthened. Methods for inter-model comparison are well established for scenarios of climate change. Crop-specific model networks have been established by the IGBP. The Food and Agriculture Organization (FAO) is beginning to evaluate alternative models of vulnerability mapping. But the wide range of models used in climate impact assessment and food system analysis need to be routinely validated and their results systematically compared. A World Data Center or network should archive complete model results and make them available for bona fide researchers.

The range of methodologies needs to be expanded, with greater integration across sectors. For example, analyses of world trade presented could be expanded to include a contextual analysis of vulnerability, the potential impacts of multiple environmental threats, and dynamic dependence of population growth and technological innovation.

Our understanding of the roots of hunger, placed in the context of human ecology, exchange entitlements and political economy, augment our efforts to abate greenhouse gas emissions and to increase resilience to climatic variations. Our concern for the future worlds of our children and their children, under the threat of global change, bolster our resolve to reduce current vulnerability to hunger. Indeed efforts to reduce hunger, to halve it within a decade and to end it in the coming century are examples of 'win-win' or 'no-regrets' policies. Reducing current vulnerability to hunger is the surest way to adapt to climate change and to create food security.

Acknowledgements

This chapter is based in part on the introduction to the special issue of *Food Policy,* Volume 19, Downing, T.E. and Parry, M.L., Climate change and world food security, pp. 99-104 (1994). Reprinted with permission from Elsevier Science Ltd, Butterworth-Heinemann Imprint, The Boulevard, Landford Lane, Kidlington OX5 1GB, U.K.

References

ACC/SCN (Administrative Committee on Coordination/Subcommittee on Nutrition) (1992 and 1993) *Second Report on the World Nutrition Situation.* Two volumes. ACC/SCN, Rome.

Appendini, K. and Liverman, D. (1994). Agricultural policy, climate change and food security in Mexico. *Food Policy*, **19** (2): 149-164.

Appendini, K. and Liverman, D. (1995). Agricultural policy and climate change in Mexico. In: Downing, T.E. (ed.) *Climate Change and World Food Security.* Springer-Verlag, Heidelberg.

Bass, B., Venema, H. and Schiller, E. (1995) Adaptation of food production to drought in the Senegal River Basin. In: Downing, T.E. (ed.) *Climate Change and World Food Security.* Springer-Verlag, Heidelberg.

Bender, W.H. (1994) An end-use analysis of global food requirements. *Food Policy,* **19**(4): 1-15.

Bender, W.H. (1995) An end use analysis of global food requirements. In: Downing, T.E. (ed.) *Climate Change and World Food Security.* Springer-Verlag, Heidelberg.

Bohle, H.G., Downing, T.E., and Watts M.J. (1994) Climate change and social vulnerability: toward a sociology and geography of food insecurity. *Global Environmental Change,* **4**(1): 37-48.

Chen, R.S. and Kates, R.W.(1994). World food security: prospects and trends. *Food Policy,* **19**(2):192-208.

Chen, R.S. and Kates, R.W. (1995). Towards a food secure world: prospects and trends. In: Downing, T.E. (ed.) *Climate Change and World Food Security.* Springer-Verlag, Heidelberg.

Chen, R.S. (ed.) *Sharing Experiences: How to End Half the World's Hunger by the Year 2000,* World Hunger Program, Brown University, Providence.

Crawshaw, B. and Shaw, J. (1995). Changing vulnerability to food insecurity and the international response: the experience of the World Food Programme. In: Downing, T.E. (ed.) *Climate Change and World Food Security.* Springer-Verlag, Heidelberg.

Crosson, P. and Anderson, J.R. (1994). Demand and supply: trends in global agriculture.*Food Policy* **19** (2): 1.5-119.

Crosson, P. and Anderson, J.R. (1995). Demand and supply: trends in global agriculture. In: Downing, T.E. (ed.) *Climate Change and World Food Security.* Springer-Verlag, Heidelberg.

Downing, T.E. (ed.) (1995) *Climate Change and World Food Security.* Springer-Verlag, Heidelberg.

Downing, T.E. and Parry, M.L. (1994) Climate change and world food security. *Food Policy* **19** (2): 99-104.

Downing, T.E., Watts, M.J. and Bohle, H.G. (1995). Climate change and food insecurity: toward a sociology and geography of vulnerability. In: Downing, T.E. (ed.) *Climate Change and World Food Security.* Springer-Verlag, Heidelberg.

Eele, G. (1995). Policy lessons from communities under pressure. In: Downing, T.E. (ed.) *Climate Change and World Food Security.* Springer-Verlag, Heidelberg.

Fischer, G., Frohberg, K, Parry, M.L. and Rosenzweig, C. (1994). Climate change and world food supply, demand and trade: who benefits, who loses? *Global Environmental Change,* **4**(1):7-23.

Fischer, G., Frohberg, K., Parry, M.L. and Rosenzweig, C. (1995). Impacts of potential climate change on global and regional food production and vulnerability. In: Downing, T.E. (ed.) *Climate Change and World Food Security.* Springer-Verlag, Heidelberg.

FAO and WHO (Food and Agriculture Organization and World Health Organization) (1992). *World Declaration and Plan of Action for Nutrition.* FAO and WHO, Rome.

Homewood, K. (1995). Pastoralist production systems and climate change. In: Downing, T.E. (ed.) *Climate Change and World Food Security.* Springer-Verlag, Heidelberg.

Houghton, J.T., Jenkins, G.J. and Ephraum, J.J. (eds.) (1990). *Climate Change: The IPCC Scientific Assessment.* Cambridge University Press, Cambridge.

Houghton, J.T., Callander, B.A. and Varney, S.K. (eds.) *Climate Change 1992: The Supplementary Report to the IPCC Scientific Assessment*, Cambridge University Press, Cambridge.

Houghton, J.T., Meira Filho, L.G., Bruce, J., Lee, H., Callander, B.A., Haites, E., Harris, N. and Maskell, K. (eds.) (1995). *Climate Change 1994. Radiative Forcing of Climate Change and An Evaluation of the IPCC 1992 Emission Scenarios.* Cambridge University Press, Cambridge.

Huq, S., Ahmed, U. and Koudstaal, R. (1995). Vulnerability of Bangladesh to climate change and sea level rise. In: Downing, T.E. (ed.) *Climate Change and World Food Security.* Springer-Verlag, Heidelberg.

Jodha, N.S. (1995). Enhancing food security in a warmer and more crowded world: factors and processes in fragile zones. In: Downing, T.E. (ed.) *Climate Change and World Food Security.* Springer-Verlag, Heidelberg.

Karim, Z. (1995). Agricultural vulnerability and poverty alleviation in Bangladesh. In: Downing, T.E. (ed.) *Climate Change and World Food Security.* Springer-Verlag, Heidelberg.

McGregor, J. (1994). Climate change and involuntary migration: implications for food security. *Food Policy*, **19** (2): 120132.

McGregor, J. (1995). Climate change and involuntary migration: implications for food security. In: Downing, T.E. (ed.) *Climate Change and World Food Security.* Springer-Verlag, Heidelberg.

Magadza, C.H.D. (1994). Climate change: some likely multiple impacts in southern Africa. *Food Policy,* **19**(2):165-191.

Magadza, C.H.D. (1995). Climate change: some likely multiple impacts in southern Africa. In: Downing, T.E. (ed.) *Climate Change and World Food Security.* Springer-Verlag, Heidelberg.

Norse, D. (1994). Multiple threats to regional food production: environment, economy, population? *Food Policy*, **19** (2): 133-148.

Norse, D. (1995). The conjunction of threats to regional food production: how serious are environment, economy, population and climate? In: Downing, T.E. (ed.) *Climate Change and World Food Security.* Springer-Verlag, Heidelberg.

Reilly, J., Hohmann, N. and Kane, S. (1994). Climate change and agricultural trade: who benefits, who loses? *Global Environmental Change,* **4**(1): 24-36.

Reilly, J., Hohmann, N. and Kane, S. (1995). Climate change and agricultural trade: who benefits, who loses? In: Downing, T.E. (ed.) *Climate Change and World Food Security.* Springer-Verlag, Heidelberg.

Rosenberg, N.J. and Scott, M.J. (1994). Implications of policies to prevent climate change for future food security. *Global Environmental Change*, **4** (1): 49-62.

Rosenberg, N.J. and Scott, M.J. (1995). Implications of policies to prevent climate change for future food security. In: Downing, T.E. (ed.) *Climate Change and World Food Security.* Springer-Verlag, Heidelberg.

Rothman, D.S. and Coppock, R. (1995). Scenarios of sustainability; the challenges of describing desirable futures. In: Downing, T.E. (ed.) *Climate Change and World Food Security.* Springer-Verlag, Heidelberg.

Ruttan, V.W., Bell, D.E. and Clark, W.C. (1995). Climate change and food security: agriculture, health and environmental research. In: Downing, T.E. (ed.) *Climate Change and World Food Security.* Springer-Verlag, Heidelberg.

Schneider, S. (1995). The future of climate: potential for interaction and surprises. In: Downing, T.E. (ed.) *Climate Change and World Food Security.* Springer-Verlag, Heidelberg.

Schulze, R.E., Kiker, G. and Kunz, R.P. (1995). Global climate change and agricultural productivity in southern Africa: thought for food and food for thought. In: Downing, T.E. (ed.) *Climate Change and World Food Security.* Springer-Verlag, Heidelberg.

Wang, F. and Zhao, Z. (1995). Climate change and agro-ecosystems in China. In: Downing, T.E. (ed.) *Climate Change and World Food Security.* Springer-Verlag, Heidelberg.

PART I

TRENDS IN GLOBAL AGRICULTURE AND FOOD SECURITY

Towards a Food-Secure World: Prospects and Trends

Robert S. Chen
Socioeconomic Data and Applications Center
Consortium for International Earth Science Information Network (CIESIN)
2250 Pierce Road
University Centre, MI, 48710, U.S.

Robert W. Kates
P.O. Box 8075
Trenton, ME, U.S.

1 Introduction

Can we envision a food-secure world of the future, a world with food enough for all and all with access to food? Drawing upon interdisciplinary research on hunger, we identify four dimensions of hunger that should be absent in a food-secure world, dimensions that go beyond conventional measures of food security defined solely by the balance between food availability and food requirements. At a minimum, a food-secure world is a world without famine, with little seasonal or chronic undernutrition, and with virtually no micronutrient deficiencies and nutrient-depleting illness. Such a world would require broad acceptance of food as a basic human right, increased food availability far in excess of increased population, extensive growth in household income, a pervasive safety net of emergency assistance, entitlements, and special needs programs, and a capability to adapt to the surprises of the future. We examine the likelihood of achieving such a food-secure world in 2060 by examining current trends and then comparing this world with two sets of scenarios of macroeconomic trends, with and without climate change. We conclude with a speculative agenda for thinking about how to attain the desired goal of food security in the world of 2060.

The standard definition of food security is 'access by all peoples at all times to enough food for an active, healthy life', but it is actually employed in different ways depending on the level of human organisation (World Bank, 1986). At a regional or national level food security tends to be equated with the adequacy of the national food balance – the sufficiency of the available food supply to meet the food requirements of national populations. The degree of national food security is highly dependent on the minimal population nutritional requirements selected (in per capita dietary calories) and usually presumes equal access by all regions and social classes to the available supply

NATO ASI Series, Vol. I 37
Climate Change and World Food Security
Edited by Thomas E. Downing

over a time period of a year or more. The available supply is determined principally by the current level of food production and releases from food stocks within a country balanced by net food trade and aid flows, uses of food for purposes other than human consumption, and the carryover of food stocks from year to year. In this context, 'world' food security is most often perceived as the ability of all nations to achieve and maintain their own national food security.

At a household level, food security tends to be equated with the sufficiency of household entitlements – that bundle of food-production resources, income available for food purchases, and gifts or assistance sufficient to meet the aggregate food requirements of all household members – as inferred from surveys of income, food expenditures, and/or consumption. The degree of household food security also depends on the minimal nutritional requirements of individuals, on the time period of analysis selected, and the presumption that all households within common income or expenditure groups have the same entitlements to food, thus ignoring non-economic vulnerabilities. The available supply of food at the household level depends on a variety of factors including food markets and prices, transportation networks and storage facilities, weather and other environmental influences, and the form of food rationing, supplementation, and other entitlement programs.

The concept of food security is least often employed at the level of individual household members. Individual food security implies an intake of food and absorption of nutrients sufficient to meet differential individual needs for activity, health, and growth and development. The level of need is determined primarily by an individual's age, gender, body size, health status, and level of physical activity. The availability of food to an individual is likely to depend on nutritional knowledge, occupation or lifestyle, and social, economic, and cultural relationships within and outside the household.

We believe that a food-secure world should provide security at all levels of human organisation. As we look to the future, it is necessary to ask not only whether the world as a whole and its national components will be able to meet aggregate food needs, but also whether all households and individuals within nations can attain and maintain food security. This adds an important social dimension to what is otherwise often perceived as purely a competition between population growth, technological development, and environmental change.

2 A food-insecure world in 1990

The easiest way to observe food security is to examine its absence – the persistence of hunger in its many guises. Four aspects of hunger stand out as indicative of the overall prevalence of hunger and its future course: famine, undernutrition, micronutrient deficiencies, and nutrient-depleting illness. We begin with a review of the status of these four aspects circa 1990 and assess current trends (see Table 1, Chen 1990b; Millman *et al.*, 1991).

For most people hunger is synonymous with famine. Famine is frequently associated with large and persistent food shortages, although we now know that people can suffer famine because of their inability to have access to food even when food is available (Sen, 1981). Indeed most recent and ongoing famines (e.g., in Angola, Bosnia, Mozambique, Somalia, and the Sudan) stem primarily from the use of hunger as a weapon in war through the appropriation, interdiction, or destruction of civilian food supplies (Messer, 1990).

Despite the common perception that famine is now rampant, the number of people affected by famine as compared to other, less acute forms of hunger, is relatively small. Indeed, the trend in famines is clearly downward even though there are no good time series of famine mortality, morbidity, or even occurrence. To demonstrate this trend, we have used a simple indicator, the total population residing in countries where famine was reported in The New York Times, and we have constructed such a measure from 1950 onwards (Figure 1) (Kates *et al.*, 1993). Using this measure, the population affected directly or indirectly by famine in their country of residence has declined greatly since the end of World War II, reflecting a lessening of the prevalence of famine and a major shift in famine incidence from populous Asia to less-populated Africa. Population in countries with reported famine peaked in the period 1957-63 at a yearly average of almost 788 million, then declined to a yearly average of 264 million in 1978-84. This decline continued throughout the period 1985-91, when the population of famine-plagued countries averaged 141 million. In 1992, The New York Times reported famine in 12 countries, which have a combined population of 157 million (Table 1).

Of course, these estimates based on national-level population surely overestimate the number of people directly subjected to famine, since famine conditions are almost always localised within countries, even in Somalia. At the same time, important pockets of famine may be readily overlooked by major news media. Based on an examination of actual reports of famine during the past three decades, we estimate that no more than 15-35 million people have been at risk of death due to famine in any given year.

There are two widely used measures of undernutrition. The first is a measure of household access or 'food poverty', which is defined the number of individuals living in households whose access to food as measured by income, expenditures, or consumption is insufficient to provide a dietary

Table 1 Recent estimates of hunger

Dimension of Hunger/Food Security	Population Affected (millions)	(%)	Year	Source
Famine (population at risk)	157	0.3	1992	WHP
Undernutrition (chronic and seasonal)				
FAO food poverty (1.2 BMR)	477	9	1990	WHP
Updated FAO food poverty (1.54 BMR)	786	20	1988-90	FAO
Child malnutrition (weight below -2 s.d.)	184	34	1990	ACC/SCN
Micronutrient deficiencies				
Iron deficiency (women 15-49)	370	42	1980s	ACC/SCN
Iodine deficiency	211	5.6	1980s	ACC/SCN
Vitamin A deficiency (children <5)	14	2.8	1980s	ACC/SCN
Nutrient-depleting illness				
Diarrhoea, measles, malaria (deaths of children <5)	6.5	0.8	~1990	UNICEF
Parasites (infected population)[a]				
Giant roundworm	785-1,300	15-25	1980s	World Bank
Hookworm	700-900	13-17	1980s	World Bank
Whipworm	500-750	10-14	1980s	World Bank

[a] Includes those people expected to have multiple infections.

Sources: WHP (World Hunger Program): Uvin (1993); FAO (Food and Agriculture Organisation of the United Nations): 1992; ACC/SCN (Advisory Committee on Coordination – Subcommittee on Nutrition): ACC/SCN, 1992; UNICEF (United Nations Children's Fund): Grant (1991); World Bank: Warren, *et al.* (1993).

intake adequate for growth, activity, and good health. The United Nations Administrative Committee on Coordination-Subcommittee on Nutrition (ACC/SCN) estimates that, in 1990, 786 million people lived in households too poor to obtain the food they needed for health and modest levels of activity (ACC/SCN, 1992).[1]

A second measure of undernutrition is based on actual body measurements (anthropometry), which are closely linked to nutritional status. Sample surveys of children in developing countries indicate that in 1990, 34 per cent of children under the age of five – about 184 million children – had weights that fell two standard deviations or more below the mean of the international reference population. This represents a decrease from the 1975 level of 42 per cent but due to rapid population growth the absolute number of undernourished children increased (from 168 million in 1975). Regionally, this indicator suggests that undernutrition has increased or remained static in Africa, decreased slowly in South Asia and Latin America, and declined rapidly in East and Southeast Asia.

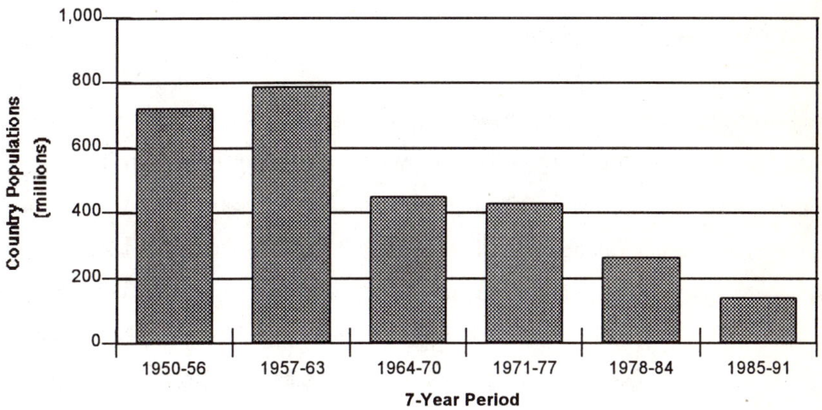

Figure 1 Famines. Average total population of countries where famine was reported in *The New York Times*, 1950-91.

[1] The ACC/SCN estimates utilised a methodology updated by FAO which allows for a more consistent cutoff point for minimum average energy intake (<1.54 times the Basal Metabolic Rate at all ages) over the course of a year. This methodology produces higher estimates of food poverty than the earlier FAO methodology utilized by the IIASA study by Fischer *et al.* (1994). See Uvin (1994) for more detail.

Even when diets are adequate from the viewpoint of total food calories, they may be inadequate with respect to important micronutrients such as iron, iodine, and vitamin A. These micronutrient deficiencies affect hundreds of millions of people worldwide (Table 1) and can lead to significant morbidity, long-term impairment, and death. For example, about one billion people live in areas where diets tend to lack iodine, and more than 200 million people actually suffer from goitre. This condition is characterised by its distinctive swelling of the thyroid glands and often leads to lethargy and reduced mental function. Severe iodine deficiency in pregnant women can increase rates of miscarriage, stillbirth, congenital defects, foetal brain damage, low birth weight, and infant death. Severe deficiency in infants can cause brain damage and increased infant mortality. Iodine deficiency is the most prevalent in Asia, but pockets of deficiency exist in Africa, the Andes, and even Europe.

Vitamin A deficiency poses a threat to nearly 200 million children of pre-school age as well as to hundreds of million older children and pregnant women. An estimated 14 million children under five have clinical signs of vitamin A deficiency (xeropthalmia) and perhaps 500,000 young children become partially or totally blind each year as a result. Many of these children subsequently die, in part because of the frequent association of this deficiency with diarrhoea, respiratory disease, and other infections.

Anaemia stemming from inadequate intake of iron or low intestinal absorption of iron affects hundreds of millions of people, especially adult women and children under 13. Iron-deficiency anaemia can affect learning and work performance and may reduce resistance to infection. For pregnant women anaemia can be especially serious, as it increases the risk of haemorrhages or even congestive heart failure during labour and delivery. Other micronutrient deficiency diseases of concern include pellagra (lack of niacin), scurvy (vitamin C), and beriberi (thiamine). Though much more rare than in the past, outbreaks of these diseases have occurred in recent years in a number of different refugee and famine situations (CDC, 1992; Chen, 1990).

Even if food availability is adequate to meet nutritional needs, several types of illness may lower the absorption of nutrients within the body as well as affect food intake. These include diarrhoea, measles, and malaria and intestinal parasites such as giant roundworm, hookworm, and whipworm. The United Nations Children's Fund (UNICEF) estimates that diarrhoeal disease is the major cause of death for 4 million children under five annually (Grant, 1991). Acute diarrhoea typically results in short-term reductions in solid food intake, malabsorption of nutrients, and endogenous nutrient loss. Persistent diarrhoea (more than 2 weeks) can lead to severe growth faltering and other signs of nutritional deficiency (Tomkins and Watkins, 1989).

Measles is commonly associated with weight loss and growth faltering, poor food intake, malabsorption, and vitamin A deficiency. It also appears to suppress the immune system, allowing pneumonia, diarrhoea, croup, and other infections to take hold with consequent effects on nutritional status. UNICEF estimates that measles is the major cause of death for some 1.5 million young children each year and a contributing factor in many other child deaths (Grant, 1991).

Malaria kills an estimated 1 million children annually. Its most common nutritional consequence is anaemia and growth faltering. Other effects include low birth weight, neonatal anaemia, and immune suppression. Intestinal parasites plague more than one billion people worldwide (Table 1). They typically cause weight loss and anaemia and some species may also contribute to poor food intake, malabsorption, endogenous nutrient loss, and vitamin A deficiency. Deworming efforts have in many instances led to measurable improvements in nutritional status.

3 A warmer, more crowded, connected, and diverse world in 2060

The world of 2060 is less than a lifetime away and is likely to be experienced by up to half of those who today might be graduating from secondary school. We have chosen to examine the prospects of a food-secure world using such a long time frame because current scientific estimates project that by 2060 two major changes in the global situation will be fully evident. In the view of most demographers, the great demographic transition from a developing world with high births and high deaths to a world with low births and low deaths will be virtually complete by 2060. And most atmospheric scientists expect to see by 2060 at least a doubling of greenhouse gases accompanied by significant changes in the world's climate.

It is difficult to sketch such a world in great detail but it may be helpful to contemplate an oceanographic metaphor. We might think of change over the next 67 years as currents, tides, and surges, undertows, and rip tides. The currents are the long-term trends, the tides are the cyclical swings, and the surges, undertows, and reptiles are the surprises.

We believe that the world of 2060 will be warmer and more crowded, more connected but more diverse. Environmental change, population growth, and increasing connectedness and diversity are powerful trends as deep-running as the great ocean currents, seemingly set in place with little possibility of reversal, although clearly subject to slower or more rapid movement. Unless there is some fatal flaw in present-day scientific understanding, we are already deeply committed to a warmer world. Recent scientific assessments suggest that a doubling of atmospheric carbon dioxide concentrations and other increases in trace gases will lead to an average temperature rise of perhaps 1.5-4.5°C with the potential for significant spatial and temporal variation from this mean (Houghton *et al.*, 1992, 1994; see also Schneider, 1995, this volume). Changes in precipitation, cloudiness, humidity, sea level, weather extremes, and other aspects of the environment could also result. In combination with other human-induced changes in local and regional environments, we foresee a world environment in 2060 significantly different from that of today.

The world of 2060 will have a population of *at least* 10 billion people, perhaps many more. Expectations of future population growth have steadily increased as present trends toward lower fertility rates have slowed. By 2060 the proportion of population living in what is now considered the developing world will have grown from 77 per cent in 1990 to more than 85 per cent. More than half of the world's population will be living in urban areas, and the largest cities will likely grow into the tens of millions.

The world of 2060 will also be much more closely connected by ties of communication, economic production and consumption, interlinked technologies, and migration. Within the next few years, at least two and probably three major common-market, free-trade blocs are likely to emerge with associated blocs in the south. Flows of goods, capital, and technology will most certainly be accompanied by flows of people. New information technologies and mass communication techniques will continue the trend towards faster, more intensive, and more inclusive flows of information, overcoming many different geographic, temporal, linguistic, cultural, and political barriers. It is not even clear that in 2060 countries will still be relevant units for calculating large scale food security.

Paradoxically, such connectedness will not necessarily homogenise the world's people beyond the ever present Coca-Cola, but may well increase the diversity of both people and things. Goods, information, and people are generally drawn to places of existing wealth or opportunity, which can make such places more diverse. And strong countercurrents that emphasise ethnic, national, and religious distinctiveness may create eddies and whirlpools where differing currents mix and clash.

In contrast to the foregoing long-term trends, there are the more familiar cycles or tides that are superimposed on the great underlying currents. Many of these are short term such as the current recession of the business cycle or El Niño phenomenon that affects the Pacific Ocean and environs at irregular intervals. Then there are decades-long swings. For example, in the great democracies, there may be recurrent swings towards the left or the right of the political spectrum. Over larger time scales of 50 years, there are the so-called Kondratief cycles, based on the introduction of clusters of new technologies (e.g., textiles, steam, steel, petrochemicals, and electronics).

We are not at all confident where the world of 2060 will fall in the ebb and flow of these great tides. But we expect that such recent trends as those towards democracy or market economies which are often regarded as deep currents of change may well emerge as recurrent tides. We will probably not return to a bipolar world in the near future, but U.S. dominance is already proving elusive. New configurations of economics and politics are already evolving. The infatuation with the market will diminish in some countries as the high social costs of inequity begin to outweigh the benefits of overall growth. What will replace the market is not certain. Some of the chaos, gridlock, and corruption of emerging democracies may encourage the re-emergence of authoritarian institutions or worse. Younger leaders now taking the place of the Cold War generation in positions of influence will in turn give way to a younger generation moulded by the beginning of a new century and new millennium.

Finally, we are least confident about the undertows, reptiles, and storm surges that batter our conventional expectations, other than the wisdom of surely expecting surprises. National boundaries that had seemed immutable for decades have been swept away in a matter of months. These changes in turn have had repercussions around the world that may take years if not decades to play out in full. Coalitions and alliances formed in reaction to specific crises create new channels of communication and action that can deepen into new norms of human behaviour and interaction. Hidden or ignored undercurrents like the spread of AIDS can eat away at the foundations of society, increasing the potential for disaster and surprise.

4 Towards 2060: A food-secure world

Charting the confluence of these diverse phenomena and their implications for building a food-secure world during the next 6 to 7 decades is at the least a monumental task. Instead, we propose four outcomes that might characterise such a world and speculate about how they might be attained. In the food-secure world of 2060, famine will be an historic occurrence of the past, less than 1 per cent of the world's population would experience extremes of seasonal or chronic undernutrition, endemic micronutrient deficiencies would virtually disappear, and nutrient-depleting illness would be prevented or controlled.

To move towards such a food-secure world requires a global implementation of food as a basic human right. Food supplies would have to grow at rates as great as in the previous half century. International and regional welfare systems will need to be institutionalised to provide for famine prevention, emergency assistance, maintenance and supplementation of entitlements, and special programs addressed to the distinctive needs of women, children, and other groups vulnerable to hunger. And in a highly connected world of 2060, there should be much interest in learning how to cope with surprise, to achieve resilience by maintaining flexibility.

A food-secure world would be one in which food as a human right would be the norm of social behaviour, an expectation upheld and enforced by all and for all. The rudiments for the international recognition of a human right to food currently exists in the form of the Universal Declaration of Human Rights, the International Covenant on Economic, Social, and Cultural Rights, and in conflict situations, the 1977 protocols to the Geneva Conventions of 1949. Yet in 1984, the foremost legal expert on rights to food could write:

> Few human rights have been endorsed with such frequency, unanimity, or urgency as the right to food, yet probably no other human right has been as comprehensively and systematically violated on such a wide scale in recent decades (Alston, 1984, p. 162).

Since then there has been a slow but marked improvement on the leading edge in the creation of such a social norm – the provision of humanitarian assistance in cases of widespread disaster caused by natural or technological hazards or war. Indeed, preventing or mitigating famine is the oldest form of collective human assistance and may have been one of the bases for the early organisation of urban society. The latest development in this long evolution has been the gradual emergence of a human right to food by civilians in zones of armed conflict. Simply stated, there is growing agreement that no nation or governmental authority has the right to starve its own people. Major precedents for the enforcement of such a right, even within the borders of a civil conflict or against the wishes of host governments, have been made in Bosnia, Northern Iraq, and Somalia. But humanitarian assistance becomes increasingly complex as the cold war definitions of self-interest wither away and new agreements as to common interest have yet to emerge. The transition to the case where all nations and peoples live up to their responsibilities to ensure that everyone is adequately fed will be long and difficult. Nevertheless, though war and conflict may themselves still persist, we are optimistic that by the middle of the next century the use of food as a weapon will be eliminated.

A food-secure world has food enough for all. As we come to the close of the century and consider a future of doubled or perhaps tripled population, the classic Malthusian concern rises again. In the last two years there has been a sudden spate of books and papers asking the question 'Will there be enough food to feed the population of tomorrow?' (see Bongaarts, 1994; Crosson and Anderson, 1994, 1995, this volume; Ehrlich *et al.*, 1993; Mitchell and Ingco, 1993; Pinstrup-Andersen, 1994; Smil, 1994).

While there is widespread interest in the overall adequacy of a future food supply, there is much less current interest. It is widely believed that there is actually plenty of food in the world and that hunger results mainly from its maldistribution. This is certainly true for a world universally content with a nutritionally adequate, but rather basic vegetarian diet. Distributed according to nutritional need, the vegetarian food supply plus the production from naturally grazed animals could support as much as 120 per cent of the world's current population. But for a world whose diet contains a modest amount of animal products derived from cereal grains, then there is only enough food produced at present to meet fully the nutritional needs of about three-quarters of the world's population. And to feed people with a healthy but animal-rich industrialised nation diet, there would be only enough food for a little more than half of the world's population.

In a more recent analysis, Smil (1994), drawing on work by Bender (1994, 1995, this volume), has estimated that current consumption, if distributed according to need, comes close to meeting most nutritional needs, requiring perhaps a 10 per cent increase (200 calories) in current calorie availability (Table 2). This could easily be met from the large quantities of grain that go into the feeding of animals. These figures do not necessarily imply that people cause hunger by eating cereal-based animal products or that there is or will be a global food shortage. Economists rightly point out that if poor countries and peoples had greater purchasing power then more food would be produced – the world has much unused capacity for producing food. And without such purchasing power, cereal products not used in animal feed would still not be available to poor people unless given away as food aid. Studies of the expenditures of poor people also show that with increased income, most poor people want to spend some of that income for a diverse diet that includes animal products except where restricted by religious preference.

All of the foregoing suggests that in a world of somewhat more than doubled population, food availability will have to increase three to fourfold in order to meet both nutritional food requirements and the modest expectations of improved diets while using current food production and consumption techniques. Between 1934 and 1989, food production measured in terms of food calories grew an average of 2.1 per cent per annum with per capita growth of about 0.6 per cent. For a fourfold increase in food production between 1990 and 2060 growth will have to be at a rate of 2 per cent per year, or about the historic average.

How likely are we to achieve such increases in the next century? Some recent trends have raised questions as to the world's ability to maintain the historic rate of growth in food production. For example, Ehrlich *et al.* (1993) offer eleven 'biophysical' constraints to maintaining the growth in food production:

- losses of farmland
- limits to freshwater supplies
- erosion and degradation of soils
- biological limits to yields
- diminishing returns from fertilisers
- problems associated with chemical control of pests
- declining genetic diversity
- increased ultraviolet radiation
- air pollutants
- climate change and sea-level rise
- a general decline in the free services of natural ecosystems.

These biophysical constraints are in addition to the many socio-economic constraints. For example, in the short term, Pinstrup-Andersen (1994) reflecting mainstream agricultural economic thinking, cites the dependence of increases in food production on economic policies that would:

- Complete structural adjustment and economic reforms.
- Remove subsidies, trade and market distortions in developed and developing countries.
- Enhance access by the poor to land, capital, and technology.
- Expand investment in rural infrastructure, health, education, and agricultural research and technology.

Table 2 Estimated global per capita averages (kcal/day) of food harvests, availability, losses, consumption, and requirements in 1990

	Global average	Subtotals
Edible crop harvests	4,600	
Cereals		3,500
Tubers		300
Pulses		100
Vegetables, fruits, nuts		150
Oils		350
Sugar		200
Less animal feed	1,700	
Grain		1,300
Grain milling residues		300
Other crops and processing residues		100
Less apparent postharvest losses	600	
Available plant food	2,300	
Available animal food	400	
Available food (FAO)	2,700	
Less apparent restaurant, institution, and household losses	700	
Food Consumption	2,000	
Food Requirements		
Healthy, active smaller children (2.6 per cent over consumption)	2,050	
Healthy, active, taller children; active adults universal development (10 per cent over consumption)	2,200	
Healthy, active, taller children; active adults but current development (19.3 per cent over consumption)	2,360	

Source: (Smil, 1994: p. 266)

- Facilitate sustainability in agricultural production.
- Reverse the decline in international assistance to agriculture.

As alternatives to these specific biophysical and socio-economic constraints, agricultural scientists are quick to point out at least four major opportunities for increased production:

- The unrealised potential to increase yields from the application of current techniques and technologies.
- The possibilities provided by expected advances in biotechnology.
- The development of organic and sustainable agriculture techniques.
- The opportunity to increase efficiency in the end use of food.

While the increases in yield potential from improved seeds, fertiliser, pesticides, and water has slowed in both industrialised and the developing countries of Asia, considerable opportunities exist for yield increases by the application of readily available technology. Yields of cereals in Asia and Latin America are twice those of most African countries. It is argued that even modest increases in seed quality, fertiliser, pesticide and water use, encouraged by appropriate policies, prices and markets, could lead to a rapid increase in yields. For example, Smil (1994) argues that it would be possible by 2050 to increase by 35 per cent current food availability with no additional inputs just by improving the efficiency of production through the use of better agronomic practices, increased uptake of available nitrogen, and more efficient use of the available water (Table 3, line 1). All told, an additional equivalent of three quarters of current production could be had, he claims, from very conservative assumptions as to agricultural progress between now and 2050 (Table 3).

Beyond the intensive utilisation of what is already available, new possibilities are provided by the promise of a biotechnological revolution just getting under way. A new set of molecular, cellular, and whole-plant biology techniques are tailoring plants used in agriculture to respond more efficiently to nutrients and moisture, to tolerate heat and drought, and to resist pests. For example, pest-resistant seeds for legumes widely used in developing countries has just been genetically engineered to provide post-harvest protection against weevils (Schmidt, 1994). However, the first biotechnology products in the marketplace will probably benefit farmers in rich nations.

Agricultural biotechnology is unlikely to contribute to the alleviation of hunger in developing nations during the coming decade. Continued progress into the next century will depend greatly on

Table 3 Conservative estimates of additional harvest achievable by the year 2050 in harvest equivalents (tons unmilled grain/per year)

Agronomic Measures	Million metric tons
Intensification of existing cropping practices (raise yields 35 per cent)	700
Extension of cultivated land (20 per cent increase compared to 1990)	500
Cultivation of idle land	75
High-efficiency irrigation	100
No beef production with intensively grown feed	50
Irrigation with saline waters	25
Agrisilviculture, aquaculture, and new crops	50
Total	1,500

Source: (Smil, 1994: p. 280)

improved capabilities on the part of developing country scientists and farmers (Messer and Heywood, 1990).

Even if food production could be trebled or quadrupled, could it be sustained under the additional burdens of heightened soil and water loss, pesticide and fertiliser use, and changing climates that will accompany such increase? There is much interest in developing a 'sustainable' agriculture of the future. But there is also much division in the directions such a sustainable agriculture might take, roughly divided between those who would encourage 'natural' or organic methods, often built on indigenous knowledge, and those that would apply technology, including biotechnology, to developing farming techniques that minimise negative environmental impacts while maintaining high productivity.

Our research suggests two important questions for an uncertain future. First, we must ask whether more food can be made available from existing agricultural production. The present global food system is not very efficient in transforming raw agricultural products into useable food (Chen, 1990). Bender (1994, 1995, this volume) has shown that over 50 per cent of the available food is wasted, most of it in richer countries and most of it in waste at 'end uses' – food sold retail and used in restaurants, institutions, and households. The least wasteful systems found in developing countries can keep such waste to 30 per cent. An estimated 62 per cent of the world's population lives in countries where waste is over the 30 per cent best-practice threshold. Diets rich in fat from animal sources are harmful and are wasteful as well because of the grain required to feed animals.

Thus Bender estimates that an amount almost equal to a quarter of the world's food consumption could be created by using the current best practice in limiting waste and adopting the minimal healthy diet of 30 per cent or less fats from animal sources.

Second, we need to examine alternatives for increasing food production in the light of their potential impacts on the global environment. For example, it may be necessary to weigh the benefits of bringing more cropland, irrigation, and rangeland into production and of using more fertilisers against their contributions of methane and nitrous oxide emissions and loss of biomass. Similarly, measures to prevent or delay global environmental change that affect energy use, water management, and forest practices could have important near-term effects on agricultural sustainability (Chen, 1990a; Rosenberg and Scott, 1994, 1995, this volume). Notably, the success of these measures in actually reducing or spreading out environmental changes is by no means guaranteed.

Household income affects food security directly, providing the major source of entitlements for the food required by the household, income that is currently inadequate in households with a total population of more than 700 million. Indirectly, household incomes in the aggregate affect the size of the available food supply by virtue of its effects on encouraging greater production. This suggests that in a world of somewhat more than doubled population, in order to meet the food requirements and modest expectations of improved diets of the poorest households, household income will have to increase four to six-fold in order to meet the food requirements and modest expectations of improved diets of the poorest households.

This large range in needed income arises from the great inequities found in poor countries. While the gap between the amount of wealth commanded by the upper and lower fifths of society diminishes with rising national income from 15-20:1 in poor countries to the 5-7:1 found in wealthy countries , the process is slow and inequitable. For the poor, 'trickle down' is not very efficient. And even increasing income is not enough. Although it is true that for the poorest households up to 85 per cent of income is spent on nutrition, it appears that in general a 10 per cent increase in income increases dietary calories only by 4-6 per cent (Marek, 1992). This occurs because some of the additional income is spent on non-food needs and some on greater and more costly food varieties. Thus for a food-secure world, even raising most households above the poverty level would still require a pervasive safety net.

5 Safety net

To achieve a food-secure world will require a pervasive safety net just as such a welfare system is required in the wealthiest of countries today. Famine can be abolished because it is systematically anticipated, monitored, and prevented. Natural and technological disasters will continue to require emergency assistance. Cycles of crop production, income, illness, and family creation and loss will continue to generate the need for entitlement maintenance and supplementation. The special needs of women and children will have to be met by responsive maternal and child care systems. Just how such a safety net would be created and supported in 2060 cannot be foreseen, as the very units of social organisation required for implementation are in doubt. But the outlines of such systems can be foreseen, both in the prototypes now operating within many developing countries and the advanced systems in place in the wealthiest countries.

In 1992-93, a major drought affected southern Africa, reducing crop production by as much as 50 per cent in some countries. Yet there was no famine. Indeed, the remarkable achievement in preventing famine in Botswana, Malawi, Mozambique, South Africa, Zambia, and Zimbabwe was further evidence for the maturity of national and international efforts to end famine deaths. These efforts, beginning in the 1980s and designed to cope with drought, flood, war, and famine, had led to great improvement in the global system for providing emergency food aid. International early-warning systems co-ordinated by the United Nations Food and Agriculture Organisation were established in 1975, and national early warning systems are in operation in many countries. Efforts to improve understanding of the underlying vulnerability of particular groups to famine and to co-ordinate development, response, and relief efforts based on this understanding hold promise of more timely and effective interventions before famine conditions spread. The importance of providing income and not just food to prevent famine is increasingly recognised and the use of emergency works programmes has increased in both Asia and Africa.

Thus it is now possible to anticipate widespread food shortages and to provide food quickly almost anywhere in the world. It is increasingly possible to identify vulnerable groups and famine potential even where national food stocks may be sufficient (Downing, 1991; Babu and Quinn, 1994; Bohle *et al.*, 1994; Downing *et al.*, 1995, this volume). It is also increasingly possible to rally international concern when justified even in the face of national denial. It is only in zones of armed conflict that the major obstacles to eliminating deaths due to famine remain.

In zones of armed conflict, the rudiments for international protection of civilian rights to food are provided by the 1977 protocols to the Geneva Conventions of 1949 that prohibit the starvation of civilians as a means of combat. Recent interventions in conflicts in Angola, Bosnia, the Kurdish regions of Turkey, Iraq, and Iran, Liberia, Mozambique, Rwanda, Somalia and the Sudan have had mixed success in alleviating mass starvation but do provide a series of precedents in which humanitarian concerns can overcome the barriers posed by national sovereignty if necessary to prevent famine. But they also provide a series of cautions as to how difficult it is to intervene in these renewals of political, ethnic, religious, and even clan-based rivalries and civil wars; how narrow is the line between humanitarian assistance, peacekeeping, and enforced peacemaking; and how easily the commitment of even the greatest of powers can be diverted by armed resistance. Nonetheless, it is clear that in this painful search for a new post-cold war standard of international responsibility, there is also the opportunity for a major breakthrough in the standards of international conduct. Indeed, hunger cannot be ended without it.

The international community has distributed some 10-11 million metric tons (mmt) of food aid annually in recent years to developing countries (excluding Russia and Eastern Europe). Of this total, some 2-4 mmt usually go to emergency situations, and an additional 1-2 mmt help to feed a growing number of refugees and other displaced peoples. Improved coordination between donor and recipient countries and international agencies is making it possible to place emergency stocks nearer to where they may be most useful and to deliver them when and where needed. The overall availability of food aid in the future will depend on a range of factors, including national agricultural and trade policies, world food markets, and international use of food aid as a tool for humanitarian assistance, socio-economic development, and foreign policy. However, it is worth noting that current levels of food aid, which have remained relatively stagnant in recent years, are far lower than the projected 20-60 mmt needed in this decade both to provide emergency assistance and to stabilise food prices in developing countries (BOSTID, 1989; Bender, 1994, 1995, this volume; Chen, 1990b).

Many problems remain in emergency assistance efforts, especially when 'human' disasters like war and political oppression strike. In developing countries, local capacity to assist those in need in times of disaster is often limited not by concern but by resources. International responses to such disasters are frequently haphazard, delayed, and at times counterproductive, as evidenced by the varied assistance given to refugees and returnees during the Persian Gulf War (U.S. Committee for Refugees, 1991; Minear *et al.*, 1992; Chen, 1992) and in the tragedy in Rwanda. Nevertheless, there appears to be a slow but steady improvement in the logistical, institutional, and legal capabilities to deal with complex humanitarian situations. Whether this trend will be reversed by the burgeoning number of refugees and other displaced persons around the world and the unwillingness of developed nations to accommodate them remains to be seen.

Various systems to subsidise or distribute food or their income equivalents are a least as old as the Roman empire. Maintaining a general entitlement to food can be quite costly, whether done as a direct subsidy or gift or whether by mandating an artificially low consumer price, which if not directly costly to governments, can discourage needed food production. But food or income maintenance can be quite effective when targeted to the truly needy. Especially effective are those that utilise existing markets to distribute food through food stamps, ration shops, or specially subsidised foods. Supplementing rural food entitlements is more difficult since it is harder to reach rural households on a sustained basis with food-welfare programs.

Notwithstanding, even in rural areas, there has been considerable success in providing work opportunities during drought and flood crises that provide some income in return for labour to construct needed agricultural infrastructure and to restore environmental resources. This has been harder to do in Africa where low densities of population and smaller numbers of technicians make the organisation of such projects more difficult. Nonetheless they have succeeded in Botswana and elsewhere.

Creating diverse sources of income can also help maintain the entitlements of the poor by making them less vulnerable to post-harvest seasonal hunger or the threat of natural disaster imposed by the sole reliance on agriculture. In encouraging such diversity, community and women's banking programs that provide self-sustaining sources of credit to start small businesses or to produce handicrafts and services have proven very effective.

Food-poor households that raise their own food have to cope with the deterioration of their natural resources, the loss of crucial access to common resources, and restriction to all but the most ecologically marginal land. For the self-provisioning population, maintaining access to the natural resource base and the inputs needed for agriculture, herding, or fishing is becoming increasingly difficult in the face of growing population and increased competition for land. There are important opportunities for redistribution to smallholders of land that is little used and a variety of low-cost techniques that have demonstrated ability to sustain productivity, provide fuelwood, limit soil erosion, and increase food and income.

Nonetheless, agricultural smallholders are still at the mercy of 'too perfect' markets as well as the imperfect ones economists frequently complain about. In such markets, what smallholders have to sell is at low prices in highly competitive markets and what they have to buy is at high prices from relatively restricted ones. It seems likely that the developing countries will have to follow the path of the developed countries in providing subsidies and basic entitlements to their rural populations in order to stem the flow of population into urban areas and retain viable agricultural economies.

Immunisation and low-cost treatment of diarrhoea, malaria, and measles promise to reduce the impact of disease on nutrition for most young children even as early as the end of this century. But such systems need to be maintained and even wealthy countries such as the U.S. find growing numbers of children without immunisation in the absence of continuous programs focused on the poor and the needy. Sustained breastfeeding, expanded supplemental feeding, and growth monitoring, in combination with limiting the effects of childhood illness and undernutrition, could eliminate the common forms of childhood wasting and stunting. The prevalence of breastfeeding is stable or even increasing in many developing countries, possibly as a result of continuing efforts to encourage and maintain breastfeeding. Anaemia especially among pregnant women is especially troublesome as it requires regular iron supplementation. All of these activities – breaking the disease-nutrition nexus and meeting the special dietary needs of women and children – can be handled well within the context of effective low-cost systems of maternal and child health care supported by adequate nutrition education.

Reducing the burdens imposed by intestinal parasites on two billion or more people will not be an easy task. Ongoing clinical trials of deworming drugs suggest that a near-term public health solution is possible at least for those most severely affected. In the long run, however, improvements in sanitation and safe drinking water are critical. The pace of improvements will likely depend on financial, institutional, and technological developments in the rapidly growing cities, densely populated watersheds, and environmentally threatened rural areas of the future.

5.1 Adaptations to surprise

To achieve a food-secure world by 2060, societies will have successfully adapted to a series of unexpected surprises. These will have made the task easier and harder, and certainly different than it might have been. The surprises themselves are by definition unpredictable, but the mechanisms are easily imaginable. For example, one can readily conceive of a new disease outbreak that would seriously disrupt plant or animal production at the same time that new breakthroughs in biotechnology begin to provide substantial improvements in yields and nutrition. New and unexpected sources of conflict may well interfere with the flow of food, people, and caring, while social and religious movements enhance the development of global responsibility. Optimistic outlooks in some regions will undoubtedly give way to pessimism but perhaps a few of today's 'basket cases' will confound the conventional wisdom.

To achieve food security in the world of 2060 will require a capacity to deal with surprise – to take advantage of surprising opportunities and to maintain social and technical flexibility to cope with surprising adversity. Such a capacity could entail the maintenance of large food stocks and other emergency response resources to cope with fluctuations in food production due to a changing climate or the weather extremes associated with a volcanic eruption of the magnitude of the Tambora eruption in 1715. It will certainly require support for the research and development infrastructure to provide continuing improvements in food production in the face of a variable environment and ongoing population change. And though the details are hard to foresee, it will need to include a safety net to ensure that the most vulnerable people and regions are not caught unprepared.

6 Scenarios for 2060

Our scenario of a food-secure world in 2060 thus incorporates a multidimensional, normative view of future food security based upon trends and prospects in agriculture, economy, technology, and social development. A world without famine must not only have plenty of food, but also a widespread recognition of the human right to food and effective mechanisms to prevent famine and alleviate its root causes. A world with relatively low levels of undernutrition must not only be more wealthy, but also willing and able to provide food entitlements and to poor and vulnerable groups as needed. A world with virtually no micronutrient deficiencies and nutrient-depleting illness must not only have more diverse diets, but also the income to support widespread access to adequate sanitation, safe water, public health and primary care services, and nutritional and health education.

How does our normative scenario compare with other views of future food security? There are at least two other research communities that have attempted to address long-term food security issues: first, economists, agriculturalists, and others concerned with the future supply and demand for food in relationship to population growth, technological change, and self-sufficiency and trade; and second, those involved in assessing climate change and its implications for agricultural systems. Thus it is instructive to examine future scenarios of 2060 based on (1) 'conventional' expectations of future food supply and demand and trends in agricultural and economic systems; and (2) a world affected by a changing climate.

Fortunately, we are able to draw on the recent work of Fischer *et al.* (1994, 1995, this volume) utilising the Basic Linked System (BLS) of the International Institute for Applied Systems Analysis (IIASA). The BLS is a general equilibrium system of agricultural policy models that are linked through trade, world market prices, and financial flows. It presently consists of a mix of some 35 different country and country-group models each designed to simulate the overall national or regional economy with particular attention to the agricultural sector. These individual models allow for evolutionary changes in capital investment and labour flows within and between the agricultural and non-agricultural sectors and for technical progress (though not induced innovation).

The BLS was used to generate a 'reference' scenario for the year 2060 by integrating the model in annual time steps with population growth, labour force participation, and technological change as the main exogenous inputs. Growth in Gross Domestic Production (GDP) is computed endogenously. The reference scenario projects that GDP will increase by a factor of 4.4 over 1980 levels by 2060. Total cereal production increases by a factor of 2.25 over this period, barely keeping up with population growth. Assuming no change in income distribution, the BLS projects that the number of undernourished (computed according to the older FAO methodology) will rise from 501 million in 1980 to 641 million in 2060. This represents a significant decline in the proportion of undernourished, from 23 per cent to 9 per cent. Since it is primarily an economic model system, the BLS makes no projections about famine, micronutrient deficiencies, or nutrient-depleting illness. It says nothing explicit about recognition of the human right to food, safety nets, and adaptability to surprise.

A second set of scenarios were created to test the sensitivity of the BLS to a changing climate. Synthetic time series of temperature and precipitation for individual GCM grid squares were generated by combining the results of three different general circulation model (GCM) simulations under conditions of doubled atmospheric concentrations of carbon dioxide (CO_2) with observed climatic data. These time series were then used as inputs into crop models for wheat, maize, rice, and soybeans applied to more than 100 sites in 18 countries. The crop models take into account the effects of higher CO_2 concentrations on plant growth and water use and allow changes in management practices, e.g., in response to a changing climate, to be tested. The resulting yield changes for three sets of GCM runs and four different cases (with and without the CO_2 fertilisation effect and with 2 levels of adaptation) were aggregated regionally and nationally and used as inputs into the BLS.

In all cases, climate change leads to net declines in cereal production and agricultural GDP in the developing world. Modest increases in production in the developed world in most instances do not fully counterbalance the developing world declines, except at high levels of adaptation in the less dramatic GCM scenarios. World market prices increase at the same time that developing countries are forced to import more cereals. As a result, the estimated number of undernourished people is projected to increase in 11 of the 12 climate change scenarios examined.

Tables 4 and 5 summarise the results of the BLS scenario experiments in comparison with our scenario of a food-secure world. These illustrate that there is a large gap between a 'desirable' food-secure scenario and the BLS scenarios with respect to the prevalence of undernutrition and future growth in food availability and income. About 500 million more people, or an additional five per cent of the world population in 2060, suffer from severe undernutrition in the BLS Reference Scenario compared with the food-secure scenario. Under the most adverse climate changes considered, the number of undernourished could increase drastically to more than two billion, or 20 per cent of the 2060 population. These higher numbers stem in part from the much lower growth in food availability and income projected to occur in the BLS scenarios.

It is useful to review the differing assumptions embedded in these scenarios (Table 6). Both sets of scenarios utilise similar projections of population. Both allow for the persistence of poverty, but in the food-secure scenario, there is room for changes in income distribution whereas in the BLS scenarios income distribution is fixed. Both foresee greater global economic integration, but only the food-secure scenario allows for dietary change. By intent, the BLS Reference Scenario posits no drastic changes and the climate change scenarios allow only for changes in climate and carbon dioxide levels. Both sets of scenarios allow for technological change, but in the BLS scenarios this is represented as a slow but steady growth in yields whereas the food-secure scenario may well require substantial technological improvements in most developing countries.

Of course, we must reiterate here that the food-secure scenario is purely normative and speculative, i.e., it addresses the question of what targets might have to be met if food security is to be achieved and maintained by 2060. The BLS scenarios project from the present in an effort to assess plausible future conditions with and without climate change. Comparison of these two

Table 4 Scenarios of food security in 2060

Dimension of Hunger	Food-Secure Scenario	BLS Reference Scenario	12 Climate Change Scenarios
Famine	None	?	?
Undernutrition (chronic and seasonal)			
FAO food poverty (1.2 BMR)	<100 million	641 million	629-2087 million
Micronutrient deficiencies	Virtually none	?	?
Nutrient-depleting illness	Controlled	?	?

Table 5 Requirements for achieving scenario levels of food security, 1990 to 2060

Requirements	Food-Secure Scenario	BLS Reference Scenario	12 Climate Change Scenarios
Food as a basic human right, not a weapon of war	Recognised by all	?	?
Sustainable food availability	3-4 x higher	1.9 x higher	1.5-2.1 x higher
Income growth	4-6 x higher	3.3 x higher	
Safety net	Pervasive	?	?
Ability to cope with surprise	High	?	?

Source: Fischer *et al.* (1994) for BLS and climate change scenarios.

approaches illustrate the potential difference in magnitude between 'business as usual' and what we pose as a desirable outcome. This comparison thus sets an agenda for thinking about the feasibility and multi-dimensionality of a future food-secure world.

7 Conclusion: an agenda for future food security

In our efforts to assess world food security in the next century, it will be important to look beyond changes in the world's capacity to produce food given the risk of possible climate changes. We argue here for a broader notion of food security, one that considers food availability and access at a range of levels in ways that support the activity, health, and development of all human beings.

Such an expanded notion of food security raises the standards by which we can judge success in achieving food security but it also offers multiple pathways for achieving it. Not only can we attempt to increase the overall supply of food, but we can and should address questions of income distribution, dietary needs, and food losses and waste. Efforts to promote individual food security through better health and nutrition education will complement efforts to ensure greater entitlements to food at individual, household, and regional levels. Strengthening emergency response capabilities and other aspects of societal safety nets will help to reduce pockets of food insecurity as well as to improve overall resilience in the face of change and surprise.

There is little doubt that the overall 'delivered' supply of food must increase to meet the combined needs of both population growth and improved diets. In the long run, such growth need only match that of the past 50 years, if population 'only' doubles during the next 67 years. Some of this

Table 6 Major assumptions in achieving scenario level of food security.

Key Assumptions	**Food-Secure Scenario**	**BLS Reference Scenario**	**12 Climate Change Scenarios**
Population increase	~ 2 x higher	~ 2 x higher	~ 2 x higher
Dietary change	More diverse	?	?
Income distribution	Changeable	Fixed	Fixed
Relative poverty	Persists	Persists	Persists
Economic integration	Increases	Increases	Increases
Surprises	Continue	No drastic changes	Mean climate change only

growth could come from more efficient utilisation of food production. Still, ongoing debates about the sustainability of the historic growth rates do call into question the feasibility of such continued growth, especially in the face of possible climate change. Were this level of growth not to be achieved, attainment of a truly food-secure world would certainly become much more difficult.

And even with the greater levels of food availability and access herein proposed – increases of 3-4 times in food production and 4-6 in income – achieving food security will require major strengthening of the human right to food, a greatly expanded safety net, and an increasing capability to adapt to surprise. Fortunately, the rudiments for value changes, social institutions, and increased flexibility can already be found. With nurturing (and perhaps a dose of optimism) it seems possible that these could serve as the basis for building the food-secure world of the future. Many of those concerned about climate change assume that a changing climate will make the task of achieving greater food security much more difficult. Changes in temperature, precipitation, cloudiness, sea level, and other aspects of climate, if they do occur, seem likely to disrupt growth in food production and income. Measures to mitigate or adapt to climate change may also inhibit production growth or divert income from efforts to alleviate poverty or food insecurity.

But we also can conceive of ways in which the confluence of efforts to adapt to climate change and the effort to achieve food security can be mutually reinforcing in beneficial ways. First, those who presently have the least food security are concentrated in South Asia and Africa, where some scenarios of regional climate change suggest that there could be improvements in growing conditions. It may be possible for these areas to take advantage of better growing conditions as they intensively develop their agricultural infrastructure in coming decades. Second, the global nature of the climate change threat reinforces inclinations towards international responsibility and solidarity. Indeed, bargaining between developed and developing nations on potential preventive and adaptive measures could highlight the achievement of food security as a relatively low-cost objective with high payoff in terms of enhancing international co-operation and building adaptive capabilities. Finally, achieving food security for both developed and developing countries in an uncertain and changing climate will likely require the establishment of responsive policies, programs, and institutions that could benefit the rich and poor alike.

Our vision of a food-secure world in the year 2060 is clearly based on a high degree of speculation – and perhaps unfounded optimism. Nevertheless, we argue that such a normative vision can help to widen perspectives on the interactions between future food security and global climate change in ways that can encourage creative thinking about how to transform new possibilities – and threats – into beneficial opportunities. In today's food insecure world, darkened by the spectre of unprecedented anthropogenic changes in the environment, such creative thinking may be our best defence.

Acknowledgement

The preparation of this paper was supported in part by CIESIN through NASA Grant NAGW-2901. The opinions, conclusions, and recommendations are those of the authors and do not necessarily represent the views of CIESIN or NASA.

This paper is a revised and updated version that appeared in *Food Policy* Volume 19, Chen, R.S. and Kates, R.W., World food security: prospects and trends, pp.192-208 (1994). Reprinted with permission from Elsevier Science Ltd, Butterworth-Heinemann Imprint, The Boulevard, Landford Lane, Kidlington OX5 1GB, U.K.. The paper also draws on material from Kates, R.W., Ending hunger: 1999 and beyond, a paper presented at the International Conference on Overcoming Hunger: The 1990s and Beyond, held at Mahidol University, Salaya, Thailand, November 6-11, 1994.

References

ACC/SCN (Advisory Committee on Coordination - Subcommittee on Nutrition) (1992). *Second Report on the World Nutrition Situation Vol I: Global and Regional Results.* ACC/SCN, Geneva.

Alston, P. (1984). International law and the right to food. In: Eide, A. *et al.* (eds). *Food as a Human Right*, United Nations University, Tokyo. p 162.

Babu, S.C. and Quinn, V. (1994) Household Food Security and Nutrition Monitoring: The African Experience. Special Issue. *Food Policy*, **19**(3): 211-343.

Bender, W. H. (1990). Food aid and hunger. The Hunger Report: 1990. Robert S. Chen, (gen. ed) HR-90-1, Alan Shawn Feinstein World Hunger Program, Brown University, Providence RI, 1990, pp 49-70.

Bender, W. H. (1994). An end use analysis of global food requirements. *Food Policy*, **19**(4): 381-395.

Bender, W. H. (1994). An end use analysis of global food requirements. In: Downing, T.E. (ed.) *Climate Change and World Food Security.* Springer-Verlag, Heidelberg.

Board on Science and Technology for International Development (1989). *Food Aid Projections for the Decade of the 1990s.* National Academy Press, Washington DC.

Bohle, H.G., Downing, T.E. and Watts, M.J. (1994). Climate change and social vulnerability. *Global Environmental Change*, **4**(1): 37-48.

Bongaarts, J. (1994). Can the growing human population feed itself? *Scientific American,* **270** (3): 36-42.

Brown, L. and Kane, H. (1994). *Full House: Reassessing the Earth's Population Carrying Capacity.* W. W. Norton, New York.

Centers for Disease Control (1992). Famine-affected, refugee, and displaced populations: recommendations for public health issues. *Morbidity and Mortality Weekly Report.* **41** No. RR-13: pp 1-76.

Chen, R.S. (1990). Global agriculture, environment, and hunger: past present, and future links. *Environmental Impact Assessment Review,* **10** (4): 335-358.

Chen, R.S. (1992). Hunger among refugees and other people displaced across borders. *Hunger 1993: Third Annual Report on the State of World Hunger – Uprooted People, Bread for the World.* Institute on Hunger and Development, Washington DC: pp 15-36.

Chen, R.S. (ed). (1990). *Refugees and hunger. The Hunger Report: 1990.* Alan Shawn Feinstein World Hunger Program, Brown University, Providence: pp 49-70.

Crosson, P. and Anderson, J. R. (1994). Demand and supply: trends in global agriculture. *Food Policy,* **19** (2): 105-119.

Crosson, P. and Anderson, J. R. (1994). Demand and supply: trends in global agriculture. In: Downing, T.E. (ed.) *Climate Change and World Food Security.* Springer-Verlag, Heidelberg.

Downing, T. E. (1991). *Assessing Socio-economic Vulnerability to Famine.* Brown University, Alan Shawn Feinstein World Hunger Program, Providence.

Ehrlich, P.R., Ehrlich, A. and Daily, G.C. (1993). Food security, population, and environment. *Population and Development Review,* **19**(1): 1-32.

FAO (Food and Agriculture Organization of the United Nations) (1992). *FAO Statistical Analysis Service, World Food Supplies and Prevalence of Chronic Undernutrition in Developing Regions As Assessed in 1992.* ESS/MISC/1/92, Statistical Analysis Service, Statistics Division, Economic and Social Policy Department, FAO, Rome, 1992.

Fischer, G., Frohberg, K., Parry, M.L. and Rosenzweig, C. (1994). Climate change and world food supply, demand and trade: who benefits, who loses? *Global Environmental Change,* **4** (1): pp. 7-23.

Fischer, G., Frohberg, K., Parry, M.L. and Rosenzweig, C. (1995). Impacts of potential climate change on global and regional food production and vulnerability. In: Downing, T.E. (ed.) *Climate Change and World Food Security.* Springer-Verlag, Heidelberg.

Grant, J.P. (1991) *The State of the World's Children 1991.* United Nations Children's Fund, New York.

Houghton, J. T., Callander, B.A. and Varney, S.K. (eds.) (1992). *Climate Change 1992: The Supplementary Report to the IPCC Scientific Assessment.* Cambridge University Press, Cambridge.

Kates, R.W, Chen, R.S., Downing, T.E., Kasperson, J.E., Messer, E. and Millman, S.R. (1988). *The Hunger Report: 1988.* Alan Shawn Feinstein World Hunger Program, Brown University, Providence.

Kates, R.W. (1993). Ending deaths from famine: the opportunity in Somalia. *New England Journal of Medicine,* **328** (14): 1055-1057.

Marek, T. (1992). Ending Malnutrition: Why Increasing Income Is Not Enough, Population, Health, and Nutrition Division,Technical Working Paper No 5, Africa Technical Department, The World Bank, Washington DC.

Messer, E. (1990). Food wars: hunger as a weapon of war. In: Chen, R.S. (ed.) *The Hunger Report: 1990.* Alan Shawn Feinstein World Hunger Program, Brown University. pp 27-35.

Messer, E. and Heywood, P. (1990). Trying technology: neither sure nor soon. *Food Policy,* **15** (4): 336-345.

Millman, S.R. with Chen, R.S., Emlen, J., Haarmann, V., Kasperson, J.X. and Messer, E. (1991). *The Hunger Report: Update 1991.* World Hunger Program, Brown University, Rhode Island.

Minear, L., Chelliah, U.B.P., Crisp, J. MacKinlay, J. and Weiss, T.G. (1992). *United Nations Coordination of the International Humanitarian Response to the Gulf Crisis 1990-1992*. Occasional Paper 13, Institute for International Studies, Brown University, Providence.

Mitchell, D.O. and Ingco, M.D. (1993). *The World Food Outlook*. International Economics Department, World Bank, Washington.

Pinstrup-Andersen, P. (1994). *World Food Trends and Future Food Security*. Washington, DC, IFPRI.

Rosenberg, N. J. and Scott, M. J. (1994). Implications of policies to prevent climate change for future food security. *Global Environmental Change,* **4** (1): 49-62.

Rosenberg, N. J. and Scott, M. J. (1995). Implications of policies to prevent climate change for future food security. In: Downing, T.E. (ed.). *Climate Change and World food Security*. Springer-Verlag, Heidelberg.

Schmidt, K. (1994). Genetic engineering yields first pest-resistant seeds. *Science,* **265**: 739.

Sen, A.K. (1981). *Poverty and Famines: An Essay on Entitlement and Deprivation*. Clarendon Press, Oxford.

Smil, V. (1994). How many people can the earth feed?. *Population and Development Review,* **20**(2): 255-292.

Tomkins, A. and Watkins, F. (1989). *Malnutrition and Infection: A Review*. ACC/SCN State-of-the-Art Series, Nutrition Policy Discussion Paper No. 5, United Nations Administrative Committee on Coordination-Subcommittee on Nutrition, Geneva.

Uvin, Peter (1994). The state of world hunger. In: P. Uvin (ed.) *The Hunger Report: 1993*. Yverdon, Switzerland.

U.S. Committee for Refugees (1991). *Mass Exodus: Iraqi Refugees in Iran, Issue Brief,* U.S. Committee for Refugees, Washington DC.

Warren, K.S., Bundy, D.A.P., Anderson, R.M., Davis, A.R., Henderson, D.A., Jamison, D.T., Rescott, N. and Senft, A. (1993). Helminth infections. In: D. T. Jamison and W. H. Mosley *Disease Control Priorities in Developing Countries,* The World Bank, Washington DC.

World Bank (1986). *Poverty and Hunger: Issues and Options for Food Security in Developing Countries.* The World Bank, Washington DC.

Demand and Supply: Trends in Global Agriculture

Pierre Crosson
Resources for the Future
1616 P St. NW
Washington, DC 20036, U.S.

Jock R. Anderson
World Bank
1818 H St. NW
Washington DC 20433, U.S.

1 Introduction

This paper is addressed to the broad question: can the global agricultural system satisfy rising demand for food and fibre to the year 2030 at 'acceptable' economic and environmental costs? Costs are understood to include those borne by people living between now and 2030, as well as those imposed on subsequent generations by management of the system between the 1990s and 2030. The paper is thus about the sustainability of the global agricultural system in the sense that one's judgement of whether the system is sustainable would likely depend on whether the answer to the question about 'acceptable' costs is yes or no.

We choose 2030 as the target year because it is far enough into the future to expect sustainability issues to arise, but not so far as to make speculation fruitless because of the high uncertainties involved.

The emphasis on economic and environmental costs in the question addressed indicate that the paper is focused primarily on the supply side of the global agricultural system. However, the economic and environmental costs of increasing supply depend importantly on the rate of increase in global demand for food and fibre. Other things being equal, holding economic and environmental costs within 'acceptable' limits confronts more problems at higher rates of demand growth than at lower rates. More specifically, the pressure of agriculture on the natural resource base and environment is greater with higher than with lower rates of global demand growth.

NATO ASI Series, Vol. I 37
Climate Change and World Food Security
Edited by Thomas E. Downing

Consequently, the capacity of the agricultural system to respond satisfactorily to rising demand – the sustainability of the system – cannot be assessed without some notion of the likely growth in global demand for food and fibre.

The supply response to increasing demand, and the consequent behaviour of economic and environmental costs, will depend on the quantity and quality of the resources that, on a global scale, can be mobilised for agricultural production. We concentrate here on five categories of resources: land, water, plant genetic resources for crop breeding, climate resources, and knowledge about agricultural production embodied in people, institutions and technology. At any given time, these resources may be combined in relationships of both complementarity and substitutability, depending on technical and economic conditions. Over the past 50 years, however, knowledge has increasingly substituted for the other resources, as it has proved far easier to increase the supply of knowledge than the supply of resources.

All present evidence, including that marshalled in this report, indicates that this trend in the relationship between knowledge and the other resources will continue into the indefinite future. Indeed, the critical question with respect to agricultural sustainability is what, over the long term are the limits, if any, of substitution of knowledge for the other resources employed in agricultural production? This view of the sustainability issue determines the structure of the paper.

2 The demand scenario

To assess the sustainability of the global agricultural system, we need some notion of the demand pressure that might be brought on the system. We need a demand scenario.

The scenario sketched here deals with future global demand for grains. Grains are estimated to account presently for roughly one-half of global consumption of food energy; and, as animal feed, grains also account for a rising share of global protein consumption. The demand for animal products is assumed to be reflected primarily in the demand for feedgrains. However, the resource and environmental impacts of forest clearing to grow crops or graze animals are considered in the supply-side assessment.

Projections of the growth of food demand over several decades typically are based on projections of population growth and per capita income. United Nations and World Bank projections of population coincide in showing that, from 1990 to 2030, some 90 per cent of global population growth will be in the less developed countries (LDCs). Moreover, it is only in those countries that increases in per capita income add much to demand for food measured at the farm gate. Most people in the more developed countries are sufficiently well fed that additions to per capita income induce little if any increase in food demand.

The combination of these population and per capita income factors is such that, in the demand scenario sketched here, over 90 per cent of the growth in food demand occurs in the LDCs. On a global scale, demand doubles from the late 1980s to 2030. In the LDCs, demand grows 2.7 times. In the LDCs the growth rates for wheat and rice decline sharply from 1988/89 to 2030 compared with the rates in 1979/81 to 1988/89. The growth rate for coarse grain, however, almost doubles (from 1.7 per cent to 3.2 per cent). This pattern of change in demand for the three types of grain reflects experience in South Korea and other parts of East Asia over the last couple of decades (Mitchell, 1991). The region generally enjoyed rapid urbanisation and fast rising per capita income, especially in South Korea, associated with a rapid increase in demand for wheat, and subsequently for coarse grain, in relation to demand for rice. It is argued (Mitchell, 1991) that a shift to a more urban, higher income society induces increased demand for a higher quality diet, particularly with respect to animal protein, and the observed shift in patterns of demand for grain reflect this.

In the rest of this paper, the demand scenario sketched above is taken as the challenge confronting the global agricultural system over the next 40 years. Can the system increase the supply of food in response to the demand scenario at economic and environmental costs consistent with the aspirations of people all around the world, especially in the LDCs, for a rising standard of welfare? That is the broad question addressed. Because almost all of the projected increase in grain consumption is in the LDCs, the focus is mainly on them.

3 The supply response

The global agricultural system must somehow mobilise the social capital necessary to satisfy the demand scenario to economic and environmental costs consistent with sustainability. The critical components of the social capital are natural resources – land, water, plant and animal genetic resources, and climate resources – and human-made resources – knowledge embodied in people, technology and institutions.

In the study on which this paper is based (Crosson and Anderson, 1992), we analysed in some depth the potential future supplies of the four categories of natural resources. Estimates of potential supplies of land and water for irrigation based on physical characteristics of soil, terrain and climate suggest considerable potential for additional supplies of both resources. We concluded, however, that the economic and environmental costs of realising much of the potential would be unacceptably high; consequently, additional supplies of these resources would contribute little to meeting the demand scenario.

Our review also satisfied us that the international and national systems for protecting and distributing plant genetic material work reasonably well, so improvements in management of the systems would not contribute much to increasing supplies of the material. With respect to climate resources, we concluded that if the climate changes, as is generally expected, the effect would more likely be to reduce the supply of climate resources than to increase them.

We concluded that, *with no increase in the supply of knowledge resources*, the global agricultural system could not meet the demand scenario at acceptable economic and environmental costs. If demand is met at acceptable costs it will only be because the supply of knowledge resources is increased enough to do the job. More knowledge is the key to achieving a sustainable agricultural system.

Space limitations preclude the development of this line of argument here. Instead we seek to give the flavour of the argument by presenting the principal elements of the discussion of land resources, of the effect on economic and environmental costs of trying to meet the demand scenario without an increase in the supply of knowledge resources, and of issues involved in expanding the supply of knowledge on the requisite scale.

4 The land resource

Land supply has both a quantitative dimension and a qualitative dimension. With respect to quantity, the Food and Agricultural Organisation (FAO) estimates that in the 1980s there were about 1.5 billion hectares of land in crops with a roughly equal amount now in forest, pasture and range with high-to-medium-to-low potential for conversion to crop production. As a first approximation, the estimate of potential cropland suggests that most of the increased demand in the global scenario could be met by bringing the potential land into crop production.

The first approximation, however, is misleading. There are several reasons. One is the geographic distribution of the potential cropland. The land is quite unevenly distributed among LDCs, 45 per cent of it being in Africa and 49 per cent in South America. In principle, the better endowed countries of Latin America and Africa could export food to land-short Asian countries. However, for domestic, political and national security reasons, Asian countries are not likely to regard a hectare of uncultivated cropland in Africa or South America as equivalent to a hectare within their own borders.

An additional constraint on the supply of potential cropland is that most of such land in South America is in the humid tropics, most of it now in forest. Compared with presently cultivated land, the potential land is far from domestic and foreign markets, and is poorly connected by road, rail and air to those markets. In Africa, because of a poorly developed road and rail system, this lack of transport infrastructure is perhaps even more of a constraint to opening new land to crop production than in South America.

Much of the potential cropland in South America and Africa could be exploited only by clearing land now in forest. This would involve some opportunity cost measured by the value of forestry services which conversion would foreclose. Conversion of range to crop production also would exact an economic opportunity cost in terms of lost animal production and the complementarity that this brings to the cropping phase of rotations. It is clear that the estimates of potential cropland do not take these opportunity costs into account.

Tropical forest clearing could also incur environmental costs in losses of biological diversity that clearing entails and of additions of carbon to the atmosphere, contributing to global warming.

These costs are not generally reflected in markets, so no one knows their present or potential future magnitude. Nonetheless, the costs are believed to be high by influential members of the world community, and widely publicised efforts are under way to persuade the Brazilian and other governments in tropical areas to slow if not halt tropical forest clearing. So far these efforts have met with little success, but there is every reason to believe they will continue. The pressure to control deforestation will mount and likely will have more effect. In this case, the realisable cropland potential of South America, and possibly Africa, would be considerably less than the FAO numbers suggest, other things being equal.

Considerations of land quality would also limit the land supply response. There are two aspects to the land quality constraint. One is that the potential cropland is inherently less productive in crop production than land already in that use. This should not be surprising. The better soils would naturally be brought into crop production first. It follows that a hectare of potential cropland is not the equivalent in productivity of a hectare of land already in crop production.

The other aspect of land quality – land degradation – is more important than the quality difference between present and potential cropland because it affects both land already in production as well as potential cropland. Moreover, there is more disagreement about the extent of land degradation, disagreement with much significance for policies to achieve a sustainable agricultural system.

A review of the literature on land degradation reveals both a deep concern about the consequences of degradation for the sustainability of the global agricultural system and a recognition of how little is reliably known about present rates of degradation and their long-term implications for sustainability. The concern dates at least from the United Nation's Desertification Conference held in Nairobi in 1977. And a follow-up study designed to estimate changes five years after the conference (Mabbutt, 1984) concluded that, by 1982, land degradation (called desertification in the paper) annually reduced to zero the productivity of some 20 million hectares, that another 6 million hectares annually were converted to 'wasteland' and that the number of people living in 'desertified' areas increased by 35 per cent from 1977 to 1982.

Nelson (1988) in a study for the World Bank reviewed the evidence on land degradation, including that presented in Mabbutt's 1982 paper, and found it often ambiguous and contradictory.

Some studies about the advance of deserts in Africa and India were contradicted by subsequent studies which showed advances in periods of drought followed by retreat when the rains returned.

Nelson's paper illustrates the theme that little is known about the extent and severity of land degradation. Dregne (1988, p. 679) notes that estimates of land degradation, including his own, are based on '...little data and much informed opinion'. Writing specifically of soil erosion and its productivity effects Dregne (1988, p. 680) asserts that 'there is an abysmal lack of knowledge of where water and wind erosion have adversely affected crop yields'. He notes that the equations for calculating wind and water erosion were developed for use in the temperate zone (more specifically, the American midwest) and their accuracy for the tropics is uncertain.

El-Swaify *et al.* (1982, p. 1), authors of the most comprehensive published study of soil erosion in LDCs, assert that '....there is little or no documentation of the extent, impact or causes of erosion....' in tropical environments. Lal and Okigbo (1990) share these several views about the lack of reliable land degradation data. They note, with respect to Nigeria, that most of the evidence on soil degradation is from analysis of soil sampled from experiments at research stations. 'There is little, if any, research done on farmers' fields to provide concrete and reliable quantitative information on the rate, extent and distribution of soil degradation' (Lal and Okigbo, 1990, p. 7).[1]

The research on land degradation indicates that the most important sources of degradation by far are erosion by wind and water. Stocking (1984), after systematically reviewing this research, concluded that almost all of it had been done in the United States. Stocking's conclusion still is valid.

Before the late 1970s, all of the research in the U.S. on erosion-productivity relationships was conducted on small experimental plots. The small scale of these experiments and the highly

---

[1] Oldeman *et al.* (1991) prepared a map with an accompanying explanatory text, showing the state of human-induced degradation of the world's soils. The total degraded area is assessed as 1964 million ha, 15 per cent of the total mapped area of 13,013 million ha. Water erosion accounts for 56 per cent of the 1964 million degraded hectares, wind erosion for 28 per cent, chemical degradation for 12 per cent and physical degradation for 4 per cent. For reasons given in Crosson and Anderson (1992, pp. 33-34), the significance of these estimates for the global supply of cropland is not clear. However, more research to advance the initiative started by Oldeman *et al.* should have great value in adding to knowledge in this important but poorly understood area.

specific soil and climatic conditions under which they were undertaken made it impossible to extrapolate the results to national-scale estimates. Surveys undertaken by the U.S. Soil Conservation Service in 1977 and 1982 provided the kind of data needed to make such estimates. Three models were developed to do this, the Productivity Index (PI) model (Larson *et al.*, 1983; Pierce *et al.*, 1984), Erosion Productivity Impact Calculator (EPIC) (Alt, Osborn and Colaccio, 1989) and a third model developed at Resources for the Future (RFF) (Crosson, 1986). PI and EPIC simulate the responses of crop growth to changes in soil characteristics – pH, water-holding capacity, nutrient supply, bulk density, etc. – under the impact of soil erosion. The RFF model is a regression type which accounts for intercounty differences in crop yields among a sample of counties in the Midwest as a function of intercounty differences in soil erosion, soil depth, fertiliser application, and other inputs.

Studies with these three models indicated that continuation of 1982 rates of cropland erosion in the United States for 100 years would reduce crop yields in a range from 3 per cent (Alt *et al.*, 1989) to 5-10 per cent (Crosson, 1986). Two features of this research stand out. One is the close similarity of the results, particularly in view of the quite different modelling approaches taken. (EPIC is a far more complex model than PI.) The second notable feature is the smallness of the erosion effect on soil productivity, even after 100 years at rates which the U.S. Soil Conservation Service officially regards as too high on close to one-half of the nation's cropland.

The findings that soil erosion in the United States almost surely is not a long-term threat to the supply of agricultural land does not mean that the erosion threat is similarly small in the LDCs. Authorities on characteristics of the tropical soils found extensively in the LDCs agree that those soils generally are more susceptible to erosion than temperate zone soil and that a given amount of erosion on LDC soils is likely to induce a greater decline in soil productivity (Stocking, 1984; Lal, 1984). Consequently we do not infer that because erosion is not a serious productivity threat in the United States it is not also in the LDCs. Unfortunately, the present ignorance about soil erosion and its productivity effects in the LDCs precludes firm judgements about this important issue.

In summary, in the LDCs, the amount of land now in forest, pasture and range that could be converted to crop production at acceptable economic and environmental costs, within the existing knowledge regime, is almost surely far less than the 2.7-fold increase in crop demand in the demand scenario.

5 Economic and environmental costs

Over the next 40 years, what would be the effect on economic and environmental costs of global agricultural production of seeking to meet the demand scenario within the existing knowledge regime? Another way of phrasing the question is this: holding the supply of knowledge resources fixed, what would be the economic and environmental costs of meeting the demand scenario by increasing supplies of land, water and genetic resources, assuming that the supply of climate resources, on a global scale, is unchanged?

5.1 Economic costs

No attempt is made to estimate economic costs quantitatively. This would be a task quite beyond the scope of this paper, even if the data and models for estimating such costs were available, which they are not. Work done by Binswanger (1989) and others is suggestive, however, of the possible impact of the demand scenario on economic costs if the supply response of farmers was limited only to bringing more land, water and other resources into production within the existing knowledge regime.

In brief, these studies suggest that under these supply circumstances, farm level prices would have to rise 500-1000 per cent to induce the necessary supply response to the demand scenario in the LDCs. The world community would surely consider such economic cost increases to be inconsistent with any reasonable concept of sustainability.

5.2 Environmental costs

Environmental costs of agricultural production are unpriced and are paid by someone other than the farmer who generates them. The lack of prices for environmental costs creates a major problem in estimating the importance of the costs. The problem is well illustrated by the case of pesticides. Pesticides are designed to kill things: weeds, insects, diseases and fungi. Over the past four decades, they have done this very well and, in the process, made an enormous contribution not only to increased world food production but also human health, for example, through control of the malaria-carrying mosquito. But pesticides often affect unintended as well as intended targets and can, therefore, be a threat to human health, wildlife and whole ecological systems. It is these unintended damages that constitute environmental costs of pesticides. But nowhere are there comprehensive estimates of the monetary value of the costs that provide a basis for judging their importance relative to other costs of agricultural production, or their trend over time.

Although the present levels and trends in environmental costs of pesticides are not known for any country, it seems reasonably clear that, with the supply of knowledge resources fixed, the demand scenario for 2030 would induce farmers around the world, particularly in LDCs, to increase the application of pesticides, both in total and per hectare. A doubling of global food demand and a 2.7-times increase in LDCs would seem certain to stimulate this response, in the absence of new knowledge about pest management. In this event, the probability of rising environmental costs of pesticides over the next 40 years would appear high.

Sediment carried in runoff from farmers' fields may also impose significant environmental costs. Eckholm (1976) published evidence of substantial damages from sedimentation in countries all around the world, but especially in LDCs. He noted increased flooding in the Indus plain of Pakistan, in parts of India, Thailand, the Philippines, Indonesia, Malaysia, and in the Cauca Valley of Colombia, and he attributed it to rising streambed levels because of sediment from eroded land upstream. Eckholm also noted that accelerated siltation of the Mangla reservoir in Pakistan has reduced the expected life of the facility by one-third to one-half, and that reservoirs in other Asian countries also were threatened.

Eckholm's evidence was anecdotal and did not provide the basis for a comprehensive estimate of the costs of sediment damage in either the more developed countries or the LDCs. Subsequent studies have not filled this knowledge gap, although they appear to support Eckholm's argument that costs of sediment damage are high, particularly in LDCs. The India Department of the World Bank (1991a), for example, cited a study of eight reservoirs in India, which found that rates of siltation are substantially higher than design rates, and that on average actual reservoir life will be only 35 per cent of the design life unless siltation rates are reduced. Similar accounts of sediment damage are reported in other studies (e.g. World Bank, 1992, with respect to China).

As in the case of pesticides it appears that farmers' responses to the demand scenario would increase sediment costs. With the supply of knowledge resources fixed, farmers inevitably would respond to the growth of demand by converting forest and range-land to crops. Historically this has increased erosion and sedimentation. Although more sedimentation does not necessarily imply higher environmental costs, the scale of the response to the demand scenario would almost surely force a rise in costs.

Tropical deforestation also would entail losses of plant animal genetic resources. Although no one knows the present and potential value of these resources, it is widely believed that the value is high. For example Wilson (1989, p. 108) refers to these resources as '...a potential source for immense untapped material wealth in the form of food, medicine and other commercially important substances'.

Without attempting quantitative estimates of losses of biological diversity, McNeely *et al.*. (1990, p. 41) cite Wilson (1988) as the authority for an assessment that '...by many indications, the world is already experiencing extinction rates of greater scale and impact than at any previous time in the earth's history'.

The poor quality of the data on tropical deforestation and on whatever losses of biological diversity this might entail, does not permit confident judgement about trends in the environmental costs of diversity losses. With respect to the future, McNeely *et al.*. (1990) believe that rising world population and economic activity are likely to increase the rate of deforestation. We believe this is the implication also of the demand scenario.

5.3 Conclusion on economic and environmental costs

The discussion of the two kinds of costs points unequivocally to the conclusion that, with the supply of knowledge resources fixed, the demand scenario would imply rising economic and environmental costs of agricultural production, probably in all countries with an active agricultural sector, and with virtual certainty in the LDCs. On a global scale and in most countries, the supplies of land and water would not be increased enough to accommodate the demand scenario at acceptable economic costs; and the drive to increase production inevitably would result in increased use of agricultural chemicals and land clearing that would force equally unacceptable increases in environmental costs.

6 The knowledge resource

The inadequacy of present and potential supplies of natural resources for agriculture implies that, if the demand scenario is to be met at acceptable economic and environmental costs, it will only be because the supply of knowledge is increased enough to do the job. Institutional improvements, for example, establishing clear and enforceable property rights in land and water resources, could increase the productivity of land and water resources somewhat, but would not basically change the prospect for rising economic and environmental costs, given the demand scenario.

This argument for investing in knowledge rests on two propositions:

- that the elasticity of the supply of knowledge relevant to agricultural production is high relative to the elasticity of supply of natural resources; and
- that across a wide range of situations the elasticity of substitution of knowledge for natural resources is high.

No effort is made here to demonstrate the validity of the two propositions, beyond noting that they are consistent with the experience of global agriculture over the past 40 or 50 years. That experience was characterised by an unprecedented increase in global agricultural output accompanied by a massive expansion of knowledge embodied in people, technology and institutions and, by comparison, a modest expansion in the quantities of land and water devoted to agricultural production.

The discussion in this section is focused on the conditions necessary to expand knowledge pertinent to agricultural production on the requisite scale, at the requisite time and of the requisite sort. Most attention is given to increasing knowledge embodied in technology and management practices, not because human capital and institutions are less important, but because the processes for directly developing new technology and management practices are in the 'front line' of the knowledge battle. They depend, of course, on institutional effectiveness and, especially in the longer term, on human capital of the requisite quality. The roles of both education and institution-building in development are crucial.

7 The importance of yield-increasing technology

Rice. Rice is predominantly an Asian phenomenon, and is expected to remain so into the foreseeable future. Asia currently accounts for roughly 90 per cent of global production and consumption of rice, and the International Economics Department of the World Bank/UNDP (1990) expects this to hold also into the next century. Future increases in rice production, therefore, will depend overwhelmingly on increasing production in Asia.

The land constraint is, and will probably remain, particularly binding in Asia. Consequently, we assume that the entire growth of demand for rice will have to be met by increasing yields. How much of an increase is that? In the demand scenario, LDC demand for rice rises an average of 1.7 per cent annually from 1988/89 to 2030. The rate of increase in the first 15 years is about twice the rate over the rest of the period.

The question for rice technology therefore is: can an average annual yield increase of 1.7 per cent be achieved over the next 40 years at economic and environmental costs that Asian societies would find acceptable? Experience in the 1980s suggest a cautiously optimistic answer to the question. Global rice yields increased by 2.4 per cent annually from 1978/82-1985/89 and production rose by 11 per cent (U.S. Department of Agriculture, 1990). FAO data indicate that consumption of nitrogen fertilisers in Asia increased from 19.5 metric tons in 1979/80 to 33.4 metric tons in 1988/89. Much of this increase was surely applied to rice land. Use of pesticides in Asia also increased sharply in the 1980s, and much of this also must have been applied to rice. No firm conclusion is possible about the environmental costs of these increases in intensity of use of fertiliser and pesticides in Asia, except that the costs doubtless rose.

The yield and economic cost experiences of Asian rice production in the 1980s support a tentative judgement that the region can increase rice yields and production over the next several decades at the requisite rates and on socially acceptable terms, bearing in mind the uncertainty about recent and possible future environmental costs. This judgement needs further consideration, however, because there now is widespread concern among rice specialists that the recent generally favourable yield experience may not be sustainable. The concern is based on evidence of growing environmental difficulties and stagnant or even declining yields in some areas of Asia (e.g. Pingali, 1991) and, more basically, on the fact that the yield *potential* of rice varieties has not increased since the 1960s, when the International Rice Research Institute (IRRI) released the first modern varieties for general use (Ruttan, 1991; Pingali, 1991). The rapid increase in rice yield since the 1960s was achieved by exploiting that initial potential, which, in turn, came mainly through increasing the rice harvest index, i.e. the ratio of grain to total biomass.

With allowance for these concerns, Anderson and Herdt (1989) are cautiously optimistic that if rice research receives funding not less than at present, average yields can continue to increase relative to potential yield and that, over the longer term, some increase in potential can also be achieved.

Meeting the yield challenge to rice will require that Asian agricultural research institutions, including the International Rice Research Institute (IRRI), have a clear and broad perspective on the longer-term problems facing the Asian rice economy and that they receive financial support from the international donor community and from Asian governments sufficient to do the job. Recent experience in this regard is mixed and is anything but uniform across Asia (Pardey, *et al.*, 1991, pp. 226-43, p. 416). It must be observed, however, that most Asian governments have done better than their counterparts in other regions in protecting and sustaining their research infrastructure. Since rice is overwhelmingly an Asian crop, the support for research is grounds for encouragement that the future challenge can be met.

Wheat and coarse grains. Unlike rice, production and consumption of wheat and coarse grains are widely scattered all around the world. Although some regional shifts in production of these crops are likely over the next several decades, especially when account is taken of possible changes in regional climates, and of the probability of agricultural reforms in the Commonwealth of Independent States (CIS), production and consumption of the crops are not likely to be less geographically dispersed than they are now.

Over the past couple of decades, technological advance in LDC wheat production, as measured by the rate of yield increase, has been remarkable, no doubt reflecting in part the fact that the modern wheat varieties are particularly well suited to irrigated production – some 42 per cent of LDC wheat area (CIMMYT, 1989, p. 1). The achievement in yield performance suggests a well-developed capacity to develop higher yielding varieties of wheat.

Taking account of potential cropland in Latin America and Africa, Crosson and Anderson (1992) concluded that an average annual increase of 1.9 per cent in wheat yields over the next 40 years would be sufficient to meet the demand scenario for wheat, and that this should be quite achievable. Implicit in this relatively optimistic long-term view is a judgement that there will be considerable contributions both from biotechnology-based innovations and through improved crop management as farmers become more skilled in realising the potential that already exits for wheat-yield improvements. Needless to say, all such advances depend crucially on the sustained and strong support for investments in national and international research extension systems.

One reason for giving explicit attention here to coarse grains is that the demand scenario indicates a faster rate of increase in global demand for coarse grains over the period to 2030 than for wheat. This pattern is based heavily on Asian experience showing that, with rising per capita income and urbanisation, demand for coarse grains as animal feed grows relative to demand for wheat and rice. The demand scenario assumes that this pattern will be increasingly evident not only in Asia but in Latin America and Africa as well.

The faster expected growth in LDC demand for coarse grains and the so far slower rate of advance in coarse-grain technology suggest that, in considering prospects for developing higher yielding technologies, major attention must be given to coarse grains. Most coarse grain production in LDC is under rainfed conditions, although comprehensive data are lacking. Yields of coarse grains generally have not increased as rapidly as wheat yields in the LDCs over the past few decades. The reasons for this are not obvious, but Anderson *et al.* (1988, p. 24), focusing on maize, by far the dominant coarse grain, offered some explanations. One is that maize is grown under highly diverse conditions, so that individual varieties have to be adapted to narrow circumstances. By contrast, lowland rice and irrigated wheat are grown under much more homogeneous natural conditions. A related reason is that maize is grown all around the less-developed world unlike rice, mostly an Asian crop, and most wheat, which, among the LDCs, is grown mainly in West Asia, the Middle East and North Africa. A consequence of the greater geographical dispersion of maize production is that international maize researchers have had to establish and maintain effective working relationships with a large number of national agricultural research institutions. These institutions are diverse in needs and situation, yet often slender in resources compared with many of those focused on wheat and rice. The maize research establishment thus is in some senses inherently more cumbersome and difficult to manage than that for wheat and rice.

Researchers will not have an easy task if LDCs are to succeed in meeting future demand for coarse grains at acceptable economic and environmental costs. In the demand scenario, LDC consumption of coarse grains will grow at an average annual rate of 2.8 per cent between the late 1980s and 2030, a three-fold increase. Some of the increase – perhaps one per cent per year – could probably be met by bringing more land under coarse grains in Africa and Latin America, but most of the burden will have to be borne by rising yields.

8 Farming-systems resource-management approaches

The research described in the preceding section is designed to increase yields by developing improved cultivars of plants – the kind of research that was fundamental in the Green Revolution in wheat and rice production. There is a role also for research designed to achieve yield increases by modifying the physical environment in which plants grow. This latter kind of research can be called farming systems research (FSR) because it focuses on the farm operations as a whole.

Farming systems research covers an enormous range of farming and research activities including work that is increasingly designated as resource management research (RMR). We offer two illustrative examples of research especially appropriate for addressing coarse-grain yields under rainfed conditions. One concerns measures to increase conservation of soil moisture. The second describes research designed to increase yields on the acid infertile soils that characterise much of the tropics.

Conserving soil moisture. Yudelman and Hillel (1988) assert that, in India and elsewhere in the less-developed world, engineering approaches to soil conservation have proved uneconomical for many farmers. As a consequence, researchers, and at least some farmers, have turned to alternative techniques to conserve soil moisture (Greenfeld, 1988, p. 146). Low-tillage, or minimum-tillage, techniques that leave much of the previous crops' residue on the soil surface as mulch and leave the soil in rougher aggregates that conventional tillage are being investigated, and adopted by some farmers (Yudelman and Hillel, 1988).

In India, results of pilot studies from watershed development projects indicate that ploughing on the contour combined with vegetative bundling offers 'scope for significantly improved water retention' (World Bank, 1991b, p. 51). The use of votive grass in the bunding is argued to be especially promising. The techniques can both reduce the susceptibility of crops to drought and lengthen the growing season by holding soil moisture longer than do current practices. Coupled with shorter and more water-stress tolerant varieties of cereals, pulses and other crops, these techniques have potential for 'significant yield improvement' (World Bank, 1991b, p. 52). Moreover, the technology 'is both easily replicable at the farm level and involves very modest investment costs, the core of the new technology only involving the very simple and low-cost practices of contour ploughing and vegetative bunding, coupled with short-season varieties and some fertiliser' (p. 52).

The evidence described here indicates that there are a number of practices that, if the economics are right, farmers can draw on to increase soil moisture, with favourable impacts on crop yields in areas where water is the principal limiting factor in crop growth. Much of the coarse grain production in LDCs is grown in such areas.

Although the technical feasibility of these water-conserving practices is well established, the fact that the practices have not been generally adopted by farmers is prima facie evidence that they are not generally economical. Research to overcome the economic barriers may have high payoff in higher coarse grain yields, especially in dry areas.

Continuous cropping on acid infertile soils. Traditional farming practices on the acid, infertile soils characteristic of much of the tropics result in low yields and severe loss of soil productivity after two or three years of cultivation. Research over the past 10 or 15 years suggests that, with adoption of new management practices, these soils can be much more productive on a sustainable basis. Buringh and Dudal (1987, p.33) assert that:

> 'Various experiments have clearly shown that many deep tropical soils can be permanently cultivated if an appropriate management is applied: the main problem of these soils is not irreversible hardening or laterite formation, as is often believed, but the aluminum toxicity that can be corrected by liming. A poor nutrient status is corrected by appropriate fertilization.'

Olson writes that several examples exist worldwide of situations where low productivity brush or sandhills have been converted to productive land by the introduction of fertilisers 'and other appropriate agronomic practices' (Olson, 1987, p. 216). Olson does not say so, but some of the research to which he referred likely was that done by a team headed by Pedro Sanchez, a soil scientist who worked at North Carolina State University. Reporting on that research, Sanchez *et al.*. (1982) stated that they had developed technologies that permit continuous production of annual crops on some of the acid infertile soils in the Amazon Basin. Studies in the region show that three grain crops can be produced annually with appropriate use of fertiliser. In the eight and a half years preceding the report, 21 crops were harvested in the same field with annual average production of 7.8 t/ha of grain. (In the American Midwest annual yields of maize currently average about 7.5 t/ha). Sanchez *et al.*. reported further that, under this regime of continuous cropping, soil properties improved. A number of farmers in the region were induced to try the production system on parts of their land. The first eight who participated obtained yields comparable to those obtained at the research station. Other more long-term field results elaborating the findings of this group are emerging (e.g. Sanchez *et al.*, 1990). The results provide new and substantive data underpinning an optimistic assessment of effective and sustainable cropping of tropical soils.

9 Conclusion

The main theme of this paper is that, if the demand scenario is met at acceptable economic and environmental costs, it will be because the supply of knowledge that farmers are able and willing to use increases enough to compensate for the insufficient supplies of land, water and other natural resources. In this argument, knowledge is the key resource, and increasing the supply of it on economical terms to farmers is the key policy issue in achieving global sustainable agriculture. The argument rests on the notion that the supply elasticity of knowledge is greater than it is for land, water, genetic and climate resources, and that the elasticity of substitution of knowledge for natural resources generally is high. Experience over the past 50 years provides empirical support for the notion. It would be a mistake, however, to assume from this experience that the supply of knowledge will be as readily increased in the future as it was in the past. The past growth of knowledge did not 'just happen.' It was the result of a deliberate commitment of resources by public and private institutions and by farmers. It cannot be assumed that the future commitment of resources will be on the requisite scale or that the payoff to the resources will be as high as in the past. Whether the incentives of private institutions will be sufficiently strong and the political will and perception of need in public institutions sufficiently well-developed is problematical. And if the knowledge needed should involve major scientific breakthroughs, for example, if expanding crop yields at the needed rate should require substantial increases in photosynthetic efficiency, then even a large increase in commitment of research resources may not be enough to do the job.

Finally, it is not obvious how investments to expand the supply of knowledge should be distributed among human capital, institutional innovation and new technology. The discussion in the previous section concentrated on issues in development of new technology, but noted that knowledge embodied in people and institutions is also critically important. There is no reason to believe that the marginal return to these three kinds of investment are the same. Indeed, a major question in devising a strategy to expand the supply of knowledge is how to allocate the necessary investments among people, institutions and new technology. We seek no answers to the question here. Our point instead is to emphasise that the issue of the scale of the needed investment in knowledge should be addressed simultaneously with the issue of the optimal distribution of the investment among the three forms in which the knowledge will be embodied. Clearly, the problem of expanding the supply of knowledge on a scale and in a way required for a sustainable response to the demand scenario will present political, economi and intellectual challenges of the first order.

Acknowledgement

This paper is a revised version that first appeared in *Food Policy* Volume 19, Crosson, P. and Anderson, J.R., Demand and supply: trends in global agriculture, pp 105-119 (1994). Reprinted with permission from Elsevier Science Ltd, Butterworth-Heinemann Imprint, The Boulevard, Landford Lane, Kidlington OX5 1GB, U.K.

References

Alt, K., Osborn, C. and Colaccio, D. (1989). Soil erosion: what effect in agricultural productivity, *Agriculture Information Bulletin* No. 556, U.S. Department of Agriculture, Economic Research Service, Washington, DC.

Anderson, J.R., Herdt, R.W. and Scobie, G.M. (1988). *Science and Food: The CGIAR and Its Partners.* World Bank, Washington, DC.

Anderson, J.R. and Herdt, R.W. (1989). The impact of new technology on foodgrain productivity in the 21st century. In: Maunder, A and Valdés, A. (eds.) *Agriculture and Governments in an Interdependent World.* Proceedings of the 20th International Conference of Agricultural Economists, Dartmouth for IAAE, Aldershot, pp. 683-694.

Binswanger, H. (1989). *The Policy Response of Agriculture*. Proceedings of World Bank Annual Conference on Development Economics. Washington, DC. pp. 231-58.

Buringh, P. and Dudal R. (1987). Agricultural land use in time and space. In: Wolman, M.G. and Fournier, F.G.A. (eds.) *Land Transformation in Agriculture*. Wiley, Chichester. pp. 9-43.

CIMMYT (1989). *1987-88 CIMMYT World Wheat Facts and Trends. The Wheat Revolution Revisited: Recent Trends and Future Challenges.* CIMMYT, Mexico.

Crosson, P. (1986). Soil erosion and policy issues. In: Phipps, T., Crosson, P. and Price, K. (eds.) *Agriculture and the Environment.* Research for the Future, Washington, DC.

Crosson, P. and Anderson, J.R. (1992). *Resources and Global Food Prospects: Supply and Demand for Cereals to 2030.* World Bank Technical Paper No. 184, Washington, DC.

Dregne, H.E. (1988). Desertification of drylands. In: Unger, P., Sneed, T., Jordan, W. and Jensen, R. (eds.) *Challenges in Dryland Agriculture: A Global Perspective.* Proceedings of the International Conference on Dryland Farming. Texas Agricultural Experiment Station, Amarillo/Bushland. pp. 610-12.

Eckholm, E. (1976). *Losing Ground: Environmental Stress and World Food Prospects.* Norton, New York.

El-Swaify, S. Dangler, E. and Armstrong, C. (1992). *Soil Erosion by Water in the Tropics.* University of Hawaii, Honolulu.

Greenfeld, J.C. (1988). Water conservation: fundamental to rainfed agriculture in developing countries. In: Unger, P., Sneed, T., Jordan, W. and Jensen, R. (eds.) *Challenges in Dryland Agriculture: A Global Perspective.* Proceedings of the International Conference on Dryland Farming, Texas Agricultural Experiment Station, Amarillo/Bushland, pp. 145-6.

Hawkes, J. (1985). *Plant Genetic Resources: The Impact of the International Agricultural Research Centers.* CGIAR Study Paper No. 3, World Bank, Washington, DC.

Lal, R. (1984). Productivity assessment of tropical soils and the effects of erosion. In: Rijsberman, F. and Wolman, M. (eds.) *Quantification of the Effect of Erosion on Soil Productivity in an International Context.* Delft Hydraulics Laboratory, The Netherlands.

Lal, R. and Okigbo, B. (1990). *Assessment of Soil Degradation in the Southern States of Nigeria.* Environment Department Working Paper No. 39, World Bank, Washington, DC.

Larson, W., Pierce, F. and Dowdy, R. (1983). The threat of soil productivity to long-term crop production. *Science,* **2319** (4584): 458-65.

Mabbutt, J. (1984). A new global assessment of the status and trends of desertification. *Environmental Conservation,* **11**(2): 103-13.

McNeely, J.A., Miller, K.R., Rid, W.V., Mittermeir, R.A. and Werner, T.B. (1990). *Conserving the World's Biological Diversity.* International Union for the Conservation of Nature and Natural Resources/World Resources Institute/Conservation International/World Wildlife Fund/ World Bank. Gland, Switzerland.

Mitchell, D. (1991). *Changing Patterns of Food and Raw Materials Consumption in Asia/Pacific Region: Implications for Primary Commodities.* International Economics Department, World Bank, Washington, DC. (Mimeo.)

Nelson, R. (1988). *Dryland Management: The Land Degradation Problem.* Environment Department Working Paper No. 8, World Bank, Washington, DC.

Oldeman, L.R., Hakkeling, R.T.A. and Sombroek, W.G. (1991). *World Map of the Status of Human-Induced Soil Degradation: An Explanatory Note.* International Soil Reference and Information Centre, Wageningen, and United Nations Environment Programme, Nairobi. Second edition.

Olson, R.A. (1987). The use of fertilisers and soil amendments. In: Wolman, M.G. and Fournier, F.G. (eds.) *Land Transformation in Agriculture.* John Wiley, Chichester.

Pardey, P.G., Roseboom, J. and Anderson, J.R. (1991). Regional perspectives on national agricultural research. In: Pardey, P.G., Roseboom J. and Anderson, J.R. (eds.) *Agricultural*

Research Policy: International Quantitative Perspectives. Cambridge University Press, Cambridge.

Parry, M.L. (1990). *Climate Change and World Agriculture.* Earthscan, London.

Peeters, J. and Williams, J. (1984). Towards better use of gene-banks with special reference to information. *Plant Genetic Resource News,* **60**: 22-32.

Pierce, F., Dowdy, R. Larson, W. and Graham, W. (1984). Soil productivity in the Corn Belt: an assessment of erosion's long-term effect. *Journal of Soil and Water Conservation,* **39**(2): 131-36.

Pingali, P. (1991). *Technological Prospects for Reversing the Declining Trend in Asia's Rice Productivity.* Paper presented at the World Bank Conference on Agricultural Technology, Current Policy Issues for the International Community and the World Bank, Airlie House, Virginia, October 21-23.

Plucknett, D.L. and Smith, N.J.H. (1988). *Plant Quarantine and the International Transfer of Germplasm.* CGIAR Study Paper No. 25, World Bank, Washington, DC.

Plucknett, D., Smith, N., Williams, J. and Anishetty, N. (1987). *Gene Banks and the World's Food.* Princeton University Press, Princeton.

Ruttan, V. (1991). Challenges to agricultural research in the 21st century. In: Pardey, P.G., Roseboom J. and Anderson, J.R. (eds.) *Agricultural Research Policy: International Quantitative Perspectives.* Cambridge University Press, Cambridge.

Sanchez, P.A., Bandy, D.E., Villachica, J.H. and Nicholaides, J.J. (1982). Amazon Basin soils: management for continuous crop production. *Science,* **216**: 821-7.

Sanchez, P.A., Palm, C.A. and Smyth, T. (1990). Approaches to mitigate tropical deforestation by sustainable soil management practices. In: Scharpenseel, H.W., Schomarker, M. and A. Ayoub (eds.) *Soils on a Warmer Earth: Developments in Soil Science.*

Stocking, M. (1984). *Erosion and Soil Productivity: A Review.* Consultants Working Paper No. 1, AGLS, FAO, Rome.

U.S. Department of Agriculture (1990). *World Grain Situation and Outlook.* Foreign Agricultural Service FG 3-90 (March). Washington, DC.

Wilson, E.G. (1988). The current state of biological diversity. In: Wilson, E.O. and Peters, F.M. (eds.) *Biodiversity.* National Academy Press, Washington, DC.

Wilson, E.O. (1989). Threats to biodiversity. *Scientific American,* **261**(3): 108-16.

World Bank (1991a). *India-Irrigation Sector Review.* Vol. I, Main Report, Report No. 9518-IN. World Bank, Washington, DC.

World Bank (1991b). *India-Irrigation Sector Review.* Vol. II, Supplementary Analysis, Report No. 9518-IN. World Bank, Washington, DC.

World Bank (1992). *China: Environmental Strategy Paper.* Vol. I, Main Report, Report No. 9669-CHA. World Bank, Washington, DC.

World Bank/UNDP (1990). *Irrigation and Drainage Research: A Proposal.* Agriculture and Rural Development. World Bank, Washington, DC.

Yudelman, M. and Hillel, D. (1988). Prospects for technological change in dryland agriculture: an economic perspective. In: Unger, P., Sneed, T., Jordan, W. and Jensen, R. (eds.) *Challenges in Dryland Agriculture: A Global Perspective.* Proceedings of the International Conference on Dryland Farming. Texas Agricultural Experiment Station, Amarillo/Bushland, pp. 591-7.

PART II

RISK AND GLOBAL IMPACTS OF CLIMATE CHANGE ON AGRICULTURAL SYSTEMS

The Future of Climate: Potential for Interaction and Surprises

Stephen H. Schneider
Department of Biological Sciences
Stanford University
Stanford CA 94305, U.S.

1 Introduction

Surprises, by the very definition of the word, deal with unforeseen aspects of an issue. In the context of global warming, nearly all assessments recognise a wide range of uncertainty owing to a variety of processes that are not yet fully quantifiable, such as the feedback effects of clouds, soil organic matter changes, photosynthetic enhancement, changes in species composition or populations and alterations in the extent of snow and ice fields with climate changes. These feedback mechanisms could amplify or damp by up to several times projections made using preliminary calculations that neglect their proper interaction. Such assessments try to estimate these uncertainties and typically have suggested that if carbon dioxide (or its radiative equivalent in methane, chlorofluorocarbons, nitrous oxides etc.) were to double and be held fixed indefinitely (i.e. an 'equilibrium' experiment), then globally-averaged surface air temperatures would rise by about 1.5-4.5°C (Table 1) (National Academy of Sciences, 1978, 1987; Houghton *et al.*, 1990, 1992). This canonical range reflects the best estimate of the subjective judgements (i.e. intuition) of broad cross-sections of knowledgeable scientists across a wide range of institutions over the uncertainties associated with these feedback processes. They are not surprises, strictly speaking, as variation within this range is anticipated, even though we cannot assign more than subjective or intuitive probability as to whether the actual climatic sensitivity to equilibrium CO_2 doubling would be nearer the higher or lower end of the range. Scientific intuition and some calculations have even suggested a modest chance (perhaps 10 per cent) for temperature warming above the upper 4.5°C limit or below the 1.5°C lower limit (Jager, 1987) – and this would constitute a 'surprise', as most of the knowledgeable scientific community consider it much more probable that the actual sensitivity will turn out within the canonical 1.5-4.5°C range.

NATO ASI Series, Vol. I 37
Climate Change and World Food Security
Edited by Thomas E. Downing

Table 1 Executive summary from the Intergovernmental Panel on Climate Change 1990

1. All models show substantial changes in climate when CO_2 concentrations are doubled, even though the changes vary from model to model on a sub-continental scale.

2. The main equilibrium changes in climate due to doubling CO_2 deduced from models are given below. The number of *'s indicates the degree of confidence determined subjectively from the amount of agreement between models, our understanding of the model results and our confidence in the representation of the relevant process in the model. Five *'s indicate virtual certainties, one * indicates low confidence.

Temperature:

- ***** the lower atmosphere and Earth's surface warm;
- ***** the stratosphere cools;
- *** near the Earth's surface, the global average warming lies between +1.5°C and +4.5°C, with a 'best guess' of 2.5°C;
- *** the surface warming at high latitides is greater than the global average in winter but smaller than in summer. (In time dependent simulations with a deep ocean, there is little warming over the high latitude southern ocean);
- *** the surface warming and its seasonal variation are least in the tropics.

Precipitation:

- **** the global average increases (as does that of evaporation), the larger the warming the larger the increase;
- *** increases at high latitudes throughout the year;
- *** increases globally by 3 to 15% (as does evaporation);
- ** increases in mid-latitudes in winter;
- ** the zonal mean value increases in the tropics although there are areas of decrease. Shifts in the main tropical rain bands differ from model to model, so there is little consistency between models in simulated regional changes;
- ** changes little in subtropical arid areas.

Soil moisture:

- *** increases in high latitudes in winter;
- ** decreases over northern mid-latitude continents in summer.

Snow and sea-ice:

- **** the area of sea-ice and seasonal snow-cover diminish.

The results from models become less reliable at smaller scales, so predictions for smaller than continental regions should be treated with great caution. The continents warm more than the ocean. Temperature increases in southern Europe and central North America are greater than the global mean and are accompanied by reduced precipitation and soil moisture in summer. The Asian summer monsoon intensifies.

3. Changes in the day-to-day variability of weather are uncertain. However, episodes of high temperature will soon become more frequent simply due to an increase in the mean temperature. There is some evidence of a general increase in convective precipitation.

4. The direct effect of deforestation on global mean climate is small. The indirect effects (through changes in the CO_2 sink) may be more important. However, tropical deforestation may lead to substantial local effects, including a reduction of about 20 per cent in precipitation.

5. Improved predictions of global climate changes require better treatment of processes affecting the distribution and properties of cloud, ocean-atmosphere interaction, convection, sea-ice and transfer of heat and moisture from the land surface. Increased model resolution will allow more realistic predictions of global-scale changes, and some improvement in the prediction of regional climate change.

Source: Houghton *et al.* (1990).

By definition, true global change surprises constitute currently unimagined outcomes from the global change 'experiment' (Revelle and Suess, 1957) that we are now performing on earth. Let us pause briefly and consider more formally what defines a 'surprise'. In statistics, probabilities are assigned to various outcomes, and such probabilities can be objective (e.g. the odds of flipping an unloaded coin) or subjective (based on prior information that leads an expert to offer a subjective probability that some outcome will occur). 'Risk' is classically defined as probability multiplied by the consequence of some known outcome, and 'uncertainty' typically refers to situations where an event or process is factually or hypothetically known, but the probabilities that it will occur are unknown or highly subjective. Surprise, then, is reserved for situations in which even the event or process is not known. But this formal definition of surprises is not very useful. Therefore, following Holling, a workshop on global change surprises adopted a working definition of what is most commonly viewed as surprise: '*Surprise occurs when perceived reality departs qualitatively from expectations.*' (Holling, 1986; Schneider and Turner, 1994). With this working definition, we need not worry whether someone, somewhere has imagined or modeled some unconventional outcomes, let alone assigned them probabilities, but can concentrate on methods for reducing already known areas of uncertainties (often by applications of 'normal science' as commonly practised) or generating new processes for imagining novel outcomes (best accomplished by improving the process of interdisciplinary exchanges).

Over the past two centuries, there has been a documented 25 per cent increase in carbon dioxide, a 150 per cent increase in methane, the introduction of unnatural chemicals such as chlorofluorocarbons, injection of sulphur oxides, major disturbances to land surfaces, and alterations in precipitation chemistry – together they are popularly known as global change (National Academy of Sciences, 1988). That humans have created these global change 'forcings' is widely accepted, although the precise quantitative degree to which various human activities have created each of these alterations is uncertain. The response of environmental systems to the global change forcings is subject to great uncertainties and the potential for substantial surprises. As argued later, the likelihood and severity of surprises will increase the faster and harder the environmental system is forced to change.

Against this backdrop of potential anthropogenic influence is a rich geologic record filled with evidence of paleoclimatic and paleoecological changes due to natural causes. One hundred million years ago, dinosaurs moved about in vegetation that today could not withstand sustained freezing, and such vegetation dominated a planet with surface temperatures perhaps 10°C warmer than present, with much warmer poles and no permanent polar ice. That world saw reduced diversity of habitat types relative to those associated with the wide range of surface temperatures now on earth. The mid-Cretaceous earth had much higher sea levels and perhaps contained substantial increases in the amount of carbon dioxide relative to the present (Barron and Hecht, 1985; Berner, 1991). Since the beginning of the current geologic epoch (Cenozoic) at the end of the Cretaceous period about 65 million years ago, there has been an uneven decline in global temperatures, a decrease in sea level, an increase in land fraction, isolation of the Antarctic at the South Pole, formation of permanent polar ice, the differentiation of tropical, temperate and polar climatic zones, along with attendant diversity of habitats and isolation and evolution of certain species associated with each differentiated biogeographic zone.

In the past several million years, many periodic expansions and contractions of polar ice have taken place – so-called glacial/inter-glacial cycles. The most recent 160,000 years (see Figure 1) have seen two inter-glacial cycles typically lasting 10,000-15,000 years, preceded by intense cold periods followed by abrupt terminations in which temperatures rise by typically 5°C, sea level increases by 100 meters or more, and inter-glacial flora (and fauna) habitats become established into a relatively static inter-glacial climate – at least over the most recent 10,000 years of the Holocene inter-glacial period. The typical sustained, globally-averaged rates of surface temperature change from an ice age to an inter-glacial period is on the order of a degree Celsius per millennium. That is, as Figure 1 shows, it took about 10,000 years – from about 15,000 to 5,000 years before the present – for the earth to warm up by 5°C, for sea levels to rise some 100-120 meters, for CO_2 to increase by some 40 per cent, for methane to increase by 100 per cent, for farily stable biomes to become established and for the relatively stable climate in which human civilisation evolved to set up. Similar rates and magnitudes can be observed in Figure 1 for the previous glacial/inter-glacial transition about 140,000 years ago.

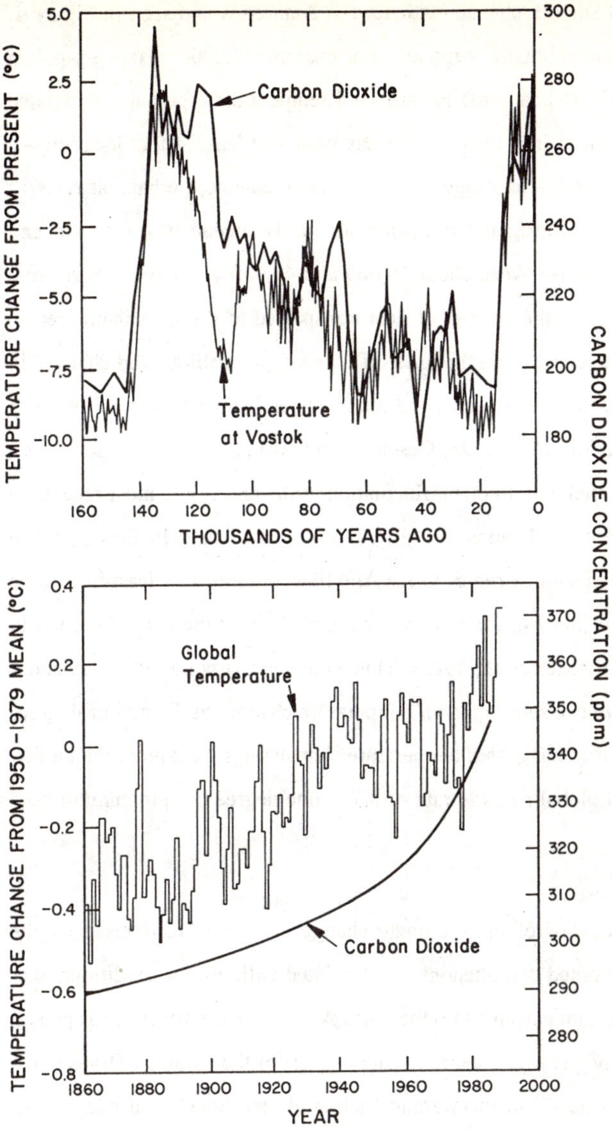

Figure 1 Carbon dioxide and temperature are very closely correlated over the past 160,000 years (top) and, to a lesser extent, over the past 100 years (bottom). The long-term record, based on evidence from Antarctica and interpreted by the glaciology group at Grenoble, France, shows how the local temperature and atmospheric carbon dioxide rose nearly in step as an ice age ended about 130,000 years ago, fell shortly after the onset of a new glacial period and rose again as the ice retreated about 10,000 years ago. The recent temperature record shows a noisy global warming trend, as traced by workers at the Climatic Research Unit at the University of East Anglia. Whether the accompanying build-up of carbon dioxide in the atmosphere caused the approximately half-degree Celsius warming is hotly debated.
Source: Schneider (1989).

But against this background of smooth change evidenced at Antarctica and seen in Figure 1, the geologic record has some paleoclimatic surprises. For example, 75,000 years ago, pollen in north-west Europe suggests a radical shift toward cold temperatures, perhaps associated with the eruption of Mount Toba in Indonesia, but in any case – at least at that location – a dramatic change in ecological conditions suggests rapid climate change, perhaps at rates of degrees per century. Moreover, during the transition to our Holocene warm period and during the recent late glacial times (between about 10,000 and 40,000 years ago), there were many episodes of abrupt changes in the amount of dust transported to the Greenland icecap, in the amount of carbon dioxide (changes on the order of 60 parts per million) and changes in at least North Atlantic regional climates that were much faster than 1°C per millennium (Woillard and Mook, 1982; Taylor *et al.*, 1993; Oeschger and Arquit, 1991; Johnsen *et al.*, 1992). Also, there was a celebrated and dramatic fluctuation at the end of the most recent ice age in which warm weather flora and fauna had just become established in Europe when about 10,800 years ago there was a quick return to Ice Age-like conditions, at least in Europe and to some extent downstream into Eurasia with weaker signals over the rest of the world (Berger and Labeyrie, 1987; Broecker *et al.*, 1985). This Younger Dryas period lasted about several hundred years, and may have seen *regional* temperature changes at the rate of degrees per century, although it is doubtful that *global* temperature change rates were more than a few times higher than the sustained global average rate of about one degree per millennium seen in Figure 1.

Thus, such natural surprise scenarios of rapid climate change have been recognised by the climatic community, which has noted that phenomena associated with glacial conditions (e.g. cutting-off of the warm Gulf Stream currents into the north Atlantic near northern Europe and consequent alteration in the conveyer belt oceanic circulation) or for the Younger Dryas (such as a meltwater spike of fresh water from the waning Laurentide Ice Sheet) can occur in as little as 100 years. The change in oceanic circulation has, in turn, led to speculations of an analogous, major 'greenhouse surprise,' should the redistribution of regional climatic conditions associated with greenhouse gas increases or other anthropogenic effects lead to a condition where the north Atlantic would no longer continue to evaporate more water than runs into it from rivers or rains on it. This is the condition that makes the North Atlantic salty enough to create sinking waters at high latitudes in winter, which invites northward penetration of the Gulf Stream and liberates to the atmosphere the great amounts of heat needed to keep wintertime Europe much warmer than, say, Hudson Bay at the same latitude. Broecker and Denton (1989), for example, have speculated on the possibility of such surprises, although neither they nor anyone else are able to assign confident probabilities to such events, not because these probabilities are believed to be minuscule, but because our understanding of the phenomena is insufficient and it is difficult to do more than speculate on this potential climatic surprise.

Most impact assessments in which agricultural, hydrological, or ecological consequences of hypothesised global warming are given almost always rely on smooth climatic change, even if the scenarios do include high, medium and low sensitivity cases to account for feedback uncertainty (National Academy of Sciences, 1992). The uncertainty typically accounted for is the conventional kind resulting from lack of knowledge of the quantitative aspects of known feedback mechanisms such as clouds, not from a range of dimly imaginable surprise outcomes such as flip-flops in ocean currents. Thus, our current range of estimates of effects cannot be considered definitive, for, given our ignorance of the whole climate system and the rates at which human activities are forcing it to change, there is always a significant chance of surprises that might be discovered as global change scenarios play themselves out. Such surprises could be positive as well as negative, but also could be very large relative to standard smooth varying scenarios and resultant assessments. It is in that spirit of trying to see into the shadows of the tails of the distribution of plausible consequences of global changes that these ideas are offered here.

Finally, we must confront the following philosophical dilemma: some scientists have argued that even the very suggestion in public of surprises with potentially severe negative consequences could leverage the popular and political debate over the impacts of global warming and possible policy responses by numbing people into paralyses, on the one hand, or, on the other hand, scaring them into draconian actions that might exceed those that are efficient given the uncertainty associated with global change. Indeed, those are real risks, but there are also risks of self-imposed silence. Society constantly asks its experts to forewarn us of 1 in 100 or even 1 in 10,000 possibilities if they carry potential dread outcome or potentially highly beneficial outcome – such as health risks from chemicals or risks to national or international security on the one hand or slight chances of some new technique of radical medicine saving lives on the other hand. It is my strong belief that the range of surprises we might discover in the greenhouse context carry such impact potential, both routine and uncertain, and thus as long as the 'warning labels' are clearly understood by the readership so they know that what they are reading here is based either on very low probability cases or cases whose probabilities are not currently assessable (although possibly not small), then this work can be interpreted without distortion. To me, it is ethically unacceptable for a small number of scientists or other experts to keep such information or ideas within the community, for withholding that information would be substituting our values for those of society at large in terms of deciding how to react to such information. Likewise, no health official could responsibly refuse to warn society about potential risks even with probabilities as low as 1 in 100 to 1 in 10,000. Society, not experts, must decide how to take risks. However, the tremendously speculative nature of the surprises offered here must be understood by all who think about this topic, as must the potentially serious consequences be noted by those who wish to risk global changes without hedges to slow down global change forcings or improve adaptation by anticipatory action to the prospect of global changes.

2 Traditional 'surprise-free' paradigm: CO_2-doubling equilibrium scenarios

The present paradigm for climatic impact assessment is to specify scenarios of growth in human population, affluence and associated technologies (e.g. Houghton *et al.*, 1990, 1992). These, in turn, produce changes to the land surface or chemical composition of the atmosphere and waters, which, in turn, can modify climate or other environmentally important services essential to food production, pest control etc. (Ehrlich *et al.*, 1977). Uncertainty is always taken into account in most responsible assessments, but usually in the form of assuming two or three different climate model scenarios. That is, using two or three general circulation models (GCMs) in which CO_2 is doubled in the model, the model is allowed to reach a new equilibrium climate, the climate is then compared with the present climate and the difference from the perturbed (i.e. 2 x CO_2) climate considered to be that which the future world would be forced to experience. This equilibrium, CO_2-doubled climate change scenario is then fed into crop models, hydrologic models, ecological models, sea-level rise models, hurricane intensification models and a variety of other assessment tools to estimate the distribution of impacts (Smith and Tirpak, 1988; Pearman, 1988; National Academy of Sciences, 1992). Economic or other decision analytic techniques are used to estimate costs and benefits, distribution of 'winners and losers' or the implications of alternative policy actions. Indeed, the results of Rosenzweig *et al.* (1993) used just such an approach in which three different CO_2-doubled general circulation models are used to evaluate impacts on crop yields, with globally averaged yields decreasing in the range of 5-10 per cent if only climate change is considered.

Another traditional part of the assessment paradigm is to recognise that there could be mitigating factors in addition to climate change, such as the potential for CO_2 fertilisation, whereby enhanced photosynthesis from the increased concentration of CO_2 in the atmosphere would restore some of the crop yields lost to increased moisture stress that most of the CO_2-doubled GCMs predict. This CO_2 fertilisation is believed to make up a substantial fraction of the yield lost (Martin *et al.*, 1989). Furthermore, others have argued that farmers do not passively sit back and allow climate to change their livelihood and would adapt to changes as they occur, thereby mitigating potential negative effects (Rosenberg, 1993). There are few 'dumb farmers', these analysts argue, as agriculture has shown itself to be a highly evolving, technologically sophisticated and resilient economic sector. Of course, agriculture is highly vulnerable to climatic shocks such as the heatwaves in 1988, which reduced corn yields in the United States by about one-third. But if such weather were to recur continuously in a globally warm world, then, at least in developed countries such as the United States, it has been asserted that on-farm adaptations and CO_2 fertilisation could reduce the yield losses to only a few percentage points rather than tens of percents (Waggoner, 1992; Rosenberg and Scott, 1994, 1995, this volume).

The assumptions associated with the standard paradigm of global change climate impact assessment, although recognising a wide range of uncertainty, are, however, essentially surprise-free. Rather than postulating low (or uncertain) probability cases in which little climate change on the one hand or catastrophic surprises on the other hand might occur, and then multiplying that lower probability by the much larger potential costs or benefits, most analysts use a few standard general circulation model CO_2-doubling scenarios to 'cover the uncertainty', rather than take extremely negative or relatively negligible climatic change cases. However, I believe that a strategic approach – i.e. one that considers a very wide range of probabilities and consequences – is more appropriate, given the high plausibility of surprises, even if we have little capacity to anticipate specific details now. For example, two critical assumptions of the standard assessment paradigm are that climate extremes – drought, floods, hurricanes etc. – either remain unchanged or will change with the mean change in climate according to unchanged variability distributions. However, as Mearns *et al.* (1984) have shown, changes in the daily temperature variance or the autocorrelation of daily weather extremes can either significantly reduce or dramatically enhance the vulnerability of agriculture, ecosystems or other climate-extreme-sensitive components of the environment to global warming. However, how such variability measures might change as the climatic mean changes is at yet highly uncertain (Rind *et al.*, 1989; Mearns *et al.*, 1990). Nevertheless, the possibility that variability will change with climate change is not at all remote, even though it is not possible with current techniques to ascertain how that might occur.

Another assumption of the standard assessment paradigm is that 'nature' is either constant or irrelevant in cost-benefit calculations (Daly, 1991). For example, ecological services, whereby pest control is maintained, are assumed constant or of no value in most assessment calculations. Yet, as will be argued shortly, should climatic change occur in the middle to upper range of that typically projected, it is highly likely that communities of species will be disassembled, and the probability that significant alterations to accustomed patterns of pests and weeds seems virtually certain. Some argue that pests, should their patterns be altered, can be simply controlled by pesticides and herbicides. What is not considered in the standard paradigm, but should be, is the consideration of a surprise scenario such as a change in public consciousness over the value of nature that would reject pesticide/herbicide application as an automatic on-farm response to global changes.

3 A new assessment paradigm: interactions and surprises

Although it has long been known that understanding and projection of the impact of human activities on climate and the converse involve detailed interactive studies of social, physical, biological and chemical systems, the multidisciplinary nature of the analysis is exceedingly difficult to undertake in practice (Chen, 1981; Schneider, 1988). Of course, there have been attempts to develop multidisciplinary assessments, which spawned the movement towards an international geosphere/biosphere program (IGBP, 1991) and the Global Change program (Malone and Roederer, 1984). These national and international efforts certainly suggest a major recognition of the need to integrate physical, biological and social factors. The NASA

Earth Systems Science approach produced the well-known 'wiring diagram' of Bretherton (Bretherton, 1986). This example focused on the interactive natural systems, with social systems drafted only as small side ovals – but at least they were recognised. However, as Schneider (1987) noted, it is not clear whether such a diagram, although in principle including all known important processes in the natural system, reflects any chance for reliable predictability, given some human actions that cause global change forcings. Perhaps the best that we will ever be able to do in practice is to evaluate the interactions of several sub-components of these interacting sub-systems and then later on look at several different sets of possible interactions. After that, we would be in a better position to make a heuristic judgement on appropriate mitigation or adaptation policies that is more quantitatively informed than presently achievable.

Rather than speculate further here on whether there is any useful system-wide predictability in such a wiring diagram, let us instead turn to our own simplified system diagram (Figure 2), which stresses physical, biological and social interactions as they might affect food security. We will then choose several examples at various points in this simple picture (simple only in the sense that it is simpler than reality although still very complex to analyse). These examples give an opportunity to identify plausible areas of surprises and interactions. Indeed, interactions among physical, biological and social systems may prove to be the biggest surprise of all for adaptability of human systems to climate change.

Figure 2 is my suggested economic/environment/food security assessment flowchart. It is designed to approximate the way in which one might model the food security system as affected by climate change, rather than as a description of complex reality. In addition, although interconnected systems in reality do not flow linearly from the social sphere to food security as this flowchart suggests, if small enough time steps are taken, and the process is repeated many times, then the practical necessity of proceeding linearly eventually will cause smaller errors.

In Figure 2, the upper circle represents, in essence, the 'demand side', which leads to human activities that are responsible for the global change forcings that have been documented – greenhouse gas increases, increase in sulphur oxide emissions, habitat fragmentation etc. Population, technology and affluence combine at the global scale to provide scenarios for

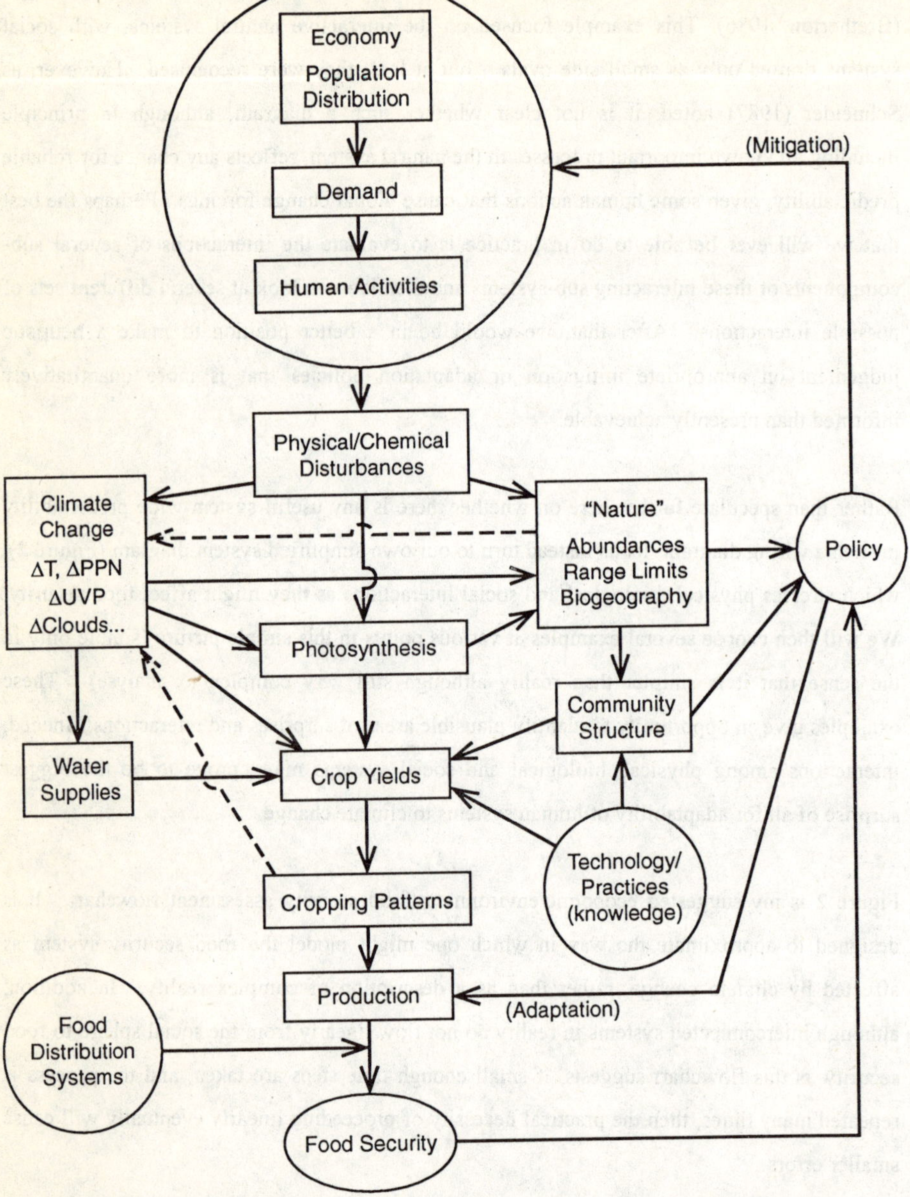

Figure 2 System flowchart for diagnosing interactions between climatic change factors and food security.

future potential physical and chemical disturbances. Based on these human behavioural assumptions, physical and chemical disturbances can then be supplied to the next level of analysis, such as models of climate change – typically in the form of general circulation models – to determine changes in temperature, precipitation, ultraviolet light (if ozone depletion is involved), a change in clouds etc. In reality, an intermediate step is needed to determine the way in which the natural and human disturbed carbon cycle operates, as the carbon dioxide and methane injected are actively transformed by atmospheric, oceanic and biospheric chemical processes, as well as by the direct disturbances to physical and biological sources and sinks (e.g. deforestation). Changes in decomposition rates of soil organic matter arising from feedbacks involving increases in temperature on increased soil microbial activity is a good candidate for surprise, as is potential photosynthetic enhancement of crops, forests and other biota. The latter has already been postulated by a wide range of people as a 'positive surprise' (Idso, 1991; Rosenberg and Scott, 1994). That is why there is a direct arrow on Figure 2 from the disturbances to photosynthesis. There is also a direct arrow from photosynthesis to Nature, as the effect of carbon dioxide enhancement of photosynthesis is not likely to force equal enhancements on all vegetative species (agricultural or natural). It seems virtually certain that natural biological systems will primarily respond to CO_2 enhancement by conferring competitive advantage on some species at the expense of others. Using the phrase 'winners and losers' has no traditional economic meaning in this context, for the conservation of nature is hard to define in human value terms – if dollars are the sole measure of worth. The only statement that can be made with some assurance is that, the larger and more rapidly the environmental system is forced to change, the more communities of species will be altered by differential CO_2 photosynthetic enhancement of species within natural systems. This will likely lead to surprises, which currently are neither calculated nor valued, positively or negatively, in any of the assessments of climate change, but are debated philosophically in the typical dichotomy of economists versus environmentalists (Myers and Simon, 1994; Daly, 1991; Nordhaus, 1994).

It is already clear that changes in climate would change the abundances and range limits of many species (COHMAP members, 1988; Davis, 1988; Root, 1992; McDonald and Brown, 1992; Peters and Lovejoy, 1992). As Root and Schneider (1993; 1995) have noted, however, this combination of differential photosynthetic enhancement along with the differential response of various species to climatic changes will undoubtedly lead to reorganisations of biological communities that are difficult to foresee, with potential implications for evapotranspiration which might in turn affect soil moisture and run-off, as well as implications for pest management, biodiversity preservation and conservation of nature. For example, as shown in Figure 3 (using the standard assessment paradigm of three CO_2-doubled, equilibrium general circulation model runs to drive an ecological model), a change in climate predicted by these three GCMs would dramatically alter the pattern of ticks in North America, substantially reducing the populations in hot southern zones of these types of ticks, and dramatically increasing the population in northern zones, such as Nova Scotia (Smith and Tirpak, 1988, p. 544). This has implications not only for human health but also for natural communities, as insects are also food for birds, which depend on vegetation for nesting, which in turn serve as prey for animals such as foxes, which together are part of the community structure of ecosystems.

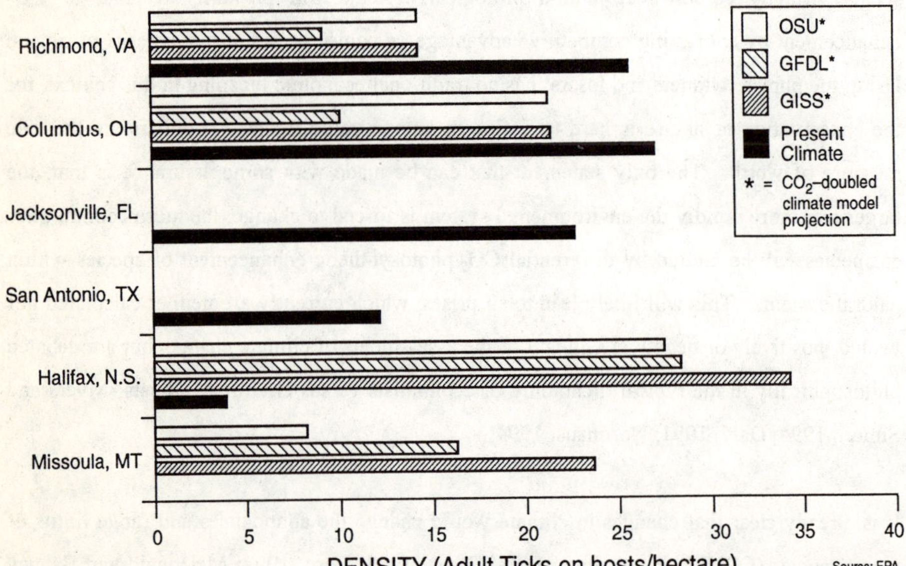

Figure 3 Simulated tick densities for selected cities under various scenarios of global warming. Source: Smith and Tirpak (1988).

It is typically assumed that individual species may respond to climate change by shifting their range limit or abundances. However, most studies concentrate on individual species at the scale of a field plot the size of a tennis court (Kareiva and Andersen, 1988). How to scale up such studies to larger geographic and community levels is not yet based on general theory (Levin, 1992; Root and Schneider, 1995; Ehleringer and Field, 1993). It can be inferred from such studies that the migration rates for specific species of vegetation, mammals, insects, reptiles or birds are all different, yet the interactions among these species may have unique properties (i.e. communities), which suggests that climate change could disrupt the community structure.

The case of ticks, as they also have implications for the health of humans and animals, suggests another category of surprises, for which there already has been considerable analysis: human health. Several studies have already suggested that the direct effects of increased heating on cardiovascular disease (Figure 4) could lead to increased mortality rates in already warm climates for those who are vulnerable (Weihe and Mertens, 1991). However, it is usually noted that adaptations, ranging from simple acclimatisation to proper choice of clothing, nutrition, diet, housing etc., can substantially mitigate high vulnerability to each stress. There is potential for 'surprise' impact on human health if adaptations are not properly anticipated. More importantly, as Weihe and Mertens have noted, is the potential not just for redistribution of ticks, but for much more serious disease factors associated with climate. Their summary (see Table 2) is an example of their serious concern for expanding the range of malaria or altering the habitats of other potential disease vectors. Dobson and Carper (1992) have also suggested that significant redistribution of tsetse fly zones could spare some currently infested areas while threatening vast other tracts not currently threatened. The potential of surprise to human and animal health with a combination of heat-stress, altered disease patterns and their interactions with malnourishment is a synergism that must not be neglected in future impact assessments. This possibility should come as no surprise, even though reliable estimates of interactions or of specific scenarios of alternate outcomes are not easy to derive.

In referring back to Figure 2, the food security oval appears below the box labelled crop yields. It is affected by photosynthesis, and water supplies as affected by climate (as well as human adaptation such as nutrient fertilisation or groundwater pumping). Yields are also subject to influence from changes to community structure on the 'Nature' side of the box in Figure 2 in which pests, soil organic matter, or soil microbes should be included in the analysis. It is useful to focus on climate change/food security impact assessment studies such as Rosenzweig *et al.* (1993). Most such analyses of climatic impacts are based on the static forecast of an equilibrium CO_2-doubled GCM run, although the standard range of feedback uncertainties are explicitly accounted for through the choice of different GCMs. (Discussed later is the need to add a dynamical transient scenario of time and space evolving climatic changes, and the tremendous increase in surprise potential that such an increased dimension of analysis will imply.)

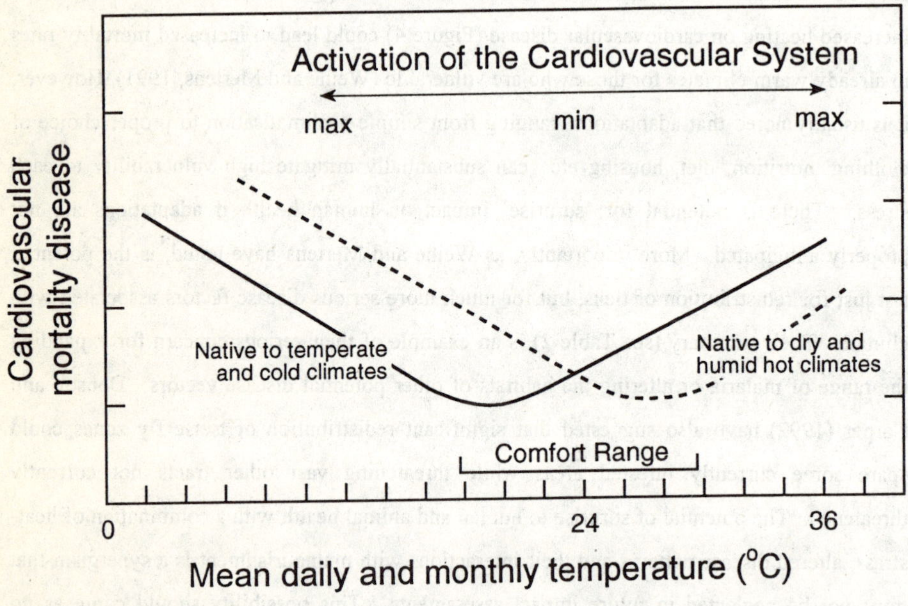

Figure 4 Relationship between CVD mortality rate and adaptation to native climates. The cardiovascular system is activated and strained under both hot and cold conditions, with a minimum of strain at comfort temperatures. People native to temperate and cold climates prefer and tolerate lower comfort temperatures; people native to dry and humid hot climates prefer and tolerate higher comfort temperatures. On the global scale, comfort temperatures range from about 17 to 30°C. Cold-native people show a higher vulnerability to heat, heat-native people to cold. Source: Weihe and Mertens (1991).

Table 2 Examples of major vector-borne tropical diseases

Disease	Causative agent	Vector	Population at risk† (millions)	Prevalence of infection (millions)	Present distri-bution	Climate change impact‡
Malaria	*Plasmodium* spp. (protozoa)	*Anopheles* mosquitos	2,100	270	Tropics/ subtropics	+++
Lymphatic filariasis	Nematode worms: family filaridae	Various mosquito species	900	90	Tropics/ subtropics	++
Onchocerciasis (river blindness)	*Onchocerca volvulus* (filaridae)	*Simulium* blackflies	90	17.6	Africa/ Latin America	+
Schistosomiasis	*Schistosoma*	Water snails	600	200	Tropics/ subtropics	++
African Trypanosomiasis (sleeping sickness)	Trypanosom spp. (protozoa)	Tsetse flies	50	25,000 new cases per year	Tropical Africa	+
Leishmaniases	Leishmanias es spp. (protozoa)	Sandflies	350	12	Asia/ S.Europe/Af rica/ S. America	?
Arbivoral Diseases						
Dengue fever	Dengue virus	Aedes mosquitos		(30-60 million infections/ yr)	Tropics/ subtropics	++
Yellow fever	Y.F. virus	Various mosquito species		>3,000 deaths in 1986-88	Africa/ The Americas	+
Japanese encephalitis	J.E. virus	Various mosquito species			Asia/ Pacific islands	+
Other arbivoral diseases	+90 other viruses	Anthropods, mosquitos and ticks			Asia/ Pacific islands Cosmo-politan	+

† Based on a world population estimated at 4.8 million (1989).

‡ Chance of spreading owing to climate change: +++ = highly likely; ++ = very likely; + = likely; ? = unknown.

Source: Weihe and Mertens (1991).

To return to Figure 2, crop yields then are translated to cropping patterns based on land use, technology and knowledge which leads to overall production. Then, depending on economic and political factors that influence food distribution systems, food security can be ascertained. Currently, it is estimated that between half a billion and a billion people are food-insecure due to dietary restrictions, political strife, poverty, maldistribution and the normal variability of the weather resulting in droughts, floods, or weather-related pest attacks (Chen and Kates, 1994). A debate surrounds the possibility that a world of doubled current population by the mid-21st century could not easily become food-secure (Daily and Ehrlich, 1992; Crosson and Anderson, 1992).

It is sufficient for the purpose of this paper to note that the complex chain of interactions cited leads from human activities to global change forcing, to a climatic change response, to photosynthetic response, to community structure changes, to crop yield changes, to production changes – which, depending on adaptations involving knowledge, technology and investment, could modify crop yield changes that otherwise would have occurred. The degree of modification of crop yields from adaptive responses cannot be definitely determined in advance, because of uncertainties (and surprises) in future climatic changes, ecological changes and social and technological factors. Nevertheless, adaptation responses have been estimated, although all analyses suggest that there could be a wide distribution of impacts. Rosenzweig *et al.* (1993) suggest that countries with access to technology and economic resources are less likely to have their crop yields negatively impacted by GCM-produced 2xCO_2 climate scenarios, but that those already with marginal food security problems are more likely to be negatively impacted and less likely to have effective adaptive responses. That is partly because richer, more northerly countries would benefit from longer growing seasons and increased CO_2 fertilisation (although they acknowledge that yield reduction from heat-stress-induced drought is unquestionably a serious potential threat), whereas tropical countries would not have the differential advantage of a shortened frost period, may suffer further problems with heat stress, do not often have the capital or technological infrastructures for effective adaptations and have many more pests to cope with than those countries with wintertime frosts to interrupt pest cycles. This analysis, if valid, could further complicate the gap between 'have' and 'have not' countries, adding to the possibility of environmentally displaced people and increasing differences in living standards, which could cause further disruption to global stability (Myers, 1993).

All this potential for change naturally leads to the policy circle in Figure 2, which could result in mitigation strategies that cause modified economy, population distribution, demand and thus human activities, thereby altering the physical and chemical disturbances which, one hopes, could reduce threats to nature or food security. Alternatively, policy response might also be to enhance adaptive capacity, shown by the adaptations arrow on Figure 2, which is drawn directly into the food technology/practices and food security part of the systems diagram. Whether currently-envisioned adaptations could easily be implemented given the fluctuating nature of climate is questioned later.

One more potential case of interactions must not be neglected: a change in the paradigm of the 'value of Nature'. In Figure 2, arrows are drawn from the 'Nature' box directly to policy. In the Victorian era it was common for ladies' hats, in the name of high fashion, to contain dozens of stuffed carcasses of birds. Today, a person wearing such a hat would be viewed by almost all 'civilised' people as odious. During the intervening years a dramatic shift of values took place, and nearly all cultures now agree that the wanton destruction of nature, certainly for the purpose of style (and some even think for economic growth) is inappropriate. Indeed, it may not be such a surprise if a continuation and extension of those values into the 21st century would make it difficult for farmers trying to adapt to new pests associated with changed biotic community structure caused by altered climate and photosynthetic enhancement to engage in indiscriminate use of pesticide and herbicide chemical controls against these new pests, given the threats that those chemicals might pose to biodiversity or animal or human health. One of the major surprises may be growing political opposition to expanded use of conventional techniques of adaptation, if these standard farming practices degrade soil, cause erosion, pollute waters or waste energy – particularly if the energy sources used also inject greenhouse gases into the atmosphere. Thus, in this category of interactions we cannot exclude a surprise in the form of the continuation of the public's desire to value Nature more explicitly rather than treat it as an unpriced externality, as is typical in most of today's economics (National Academy of Sciences, 1992; Daly, 1991). Thus, the adaptability of future agricultural systems may require different techniques from those currently envisioned and so the alleviation of most of the yield reductions from climatic change scenarios that Rosenzweig *et al.* (1993) or MINK (Rosenberg, 1993) estimate from current types of adaptations may not be as easily attainable as currently assumed.

But not all surprises need be negative. Climate changes could open new possibilities for growing new kinds of crops. Or, on the social side, new technologies may very well yield new positive adaptations or mitigations, such as a breakthrough in the already rapidly decreasing price of windmills and wind energy (Johansson *et al.*, 1993). Farmers may already find it cost-effective, particularly in windy places such as North Dakota, to install windmills on high poles in their grazing or cropping zones, not only to provide electricity for irrigation pumps that would provide a large measure of resilience against future climate change, but also as a cash crop, selling their electric power to national or international grids – which itself may require legal adaptation on the part of some societies to modify power regulations so as to require the grid network to buy power from such independent, distributed producers. Thus, surprises need not be negative, but not to anticipate the negative ones simply because some could be beneficial would be equivalent to not anticipating negative threats to security or health and thus forgoing personal insurance or defence investments. A wide range of potential outcomes and subjective probabilities should be considered by society in choosing how to face a future with increasing changes to habitats, land surfaces and atmospheric and oceanic chemical composition.

4 Transient effects, surprises from non-uniform forcing

In addition to potential surprises such as shifts in the basic structure of ocean currents, or an increase in the intensity of storms or the length of the hurricane season, there are two other categories of climatic 'surprises' that should be mentioned: (i) transient effects and (ii) effects from heterogeneous forcing. 'Surprises' is in quotation marks, despite the working definition at the outset, to remind us that both these effects are known to be plausible and potentially significant, but the extent to which they will substantially alter climate patterns produced in current three-dimensional climatic models is still speculative. Nevertheless, they could be sufficiently different from most current analyses, so it is worth elaborating on these possibilities.

4.1 Transient effects

It has long been known that the oceans, with their large heat capacities, could delay the realisation of climate warming or cooling from any radiative forcing (i.e. any change in the heat budget of the surface-atmospheric system due to changes in the energy output of the sun, greenhouse gases, scattering of sunlight back to space from fossil fuels, aerosols arising from volcanic eruptions, deforestation or other land use effects or from changes to the water cycling capacity or reflectivity of the surface) (SMIC, 1971; Hasselmann, 1979; Hansen *et al.*, 1981; Thompson and Schneider, 1979). However, Schneider and Thompson (1981) argued that the differential heat capacity of the center of continents (equivalent to only a meter or less of water), the center of tropical oceans (at which only a shallow 75 m or so mixed layer is in close contact with the atmosphere) and high latitude oceans (in which the ocean mixes to extreme depths of thousands of metres) causes significantly different thermal response times to radiative heating disturbance – weeks, years or centuries respectively. Thus, even if most current general circulation models (which do not include adequate representation of deep oceans and the mixing processes within the oceans both vertically and horizontally) could provide reasonable estimates of the distribution of regional climatic effects in equilibrium (including such phenomena as increased dryness in mid-latitudes, enhanced monsoon rainfall in south Asia or polar amplification of warming [Houghton *et al.*, 1990, 1992]), Schneider and Thompson (1981) argued that it is not at all clear that, during the century or so transition towards that equilibrium, the world would experience a pattern of regional climatic change similar to that in equilibrium. In other words, if in 2025 AD, when the combined effect of greenhouse gases is equivalent to, say, a 50 per cent increase in current greenhouse gas concentrations, the world would not simply experience a distribution of regional climatic effects one-half of that in, say 2050 AD, when the equivalent of CO_2 doubling would occur. Rather, the differential heat capacities could cause climatic anomalies to unfold in regional transient patterns that would be completely different to the pattern of equilibrium. This is because the rates of warming in the middle of continents, in the middle of oceans and from the tropics to the equator are such that the surface temperature gradients across them evolve over a century or two towards their equilibrium value differential. As alterations to these surface gradients in temperature help to determine regional climatic anomalies, the earth could experience transient climatic change that would not fit the patterns currently seen in most GCM analyses in which CO_2 is doubled in equilibrium experiments – even if the GCMs were reliable equilibrium simulators.

Although this possibility should not be viewed as a surprise, it calls into question the ease with which some analysts cite adaptation as an ameliorative factor for preserving agricultural crop yields in the face of projected climate change. Although all farmers might be aware of the need to adapt to projected climate change, it may be difficult for them to adapt without knowing what that change was to be. A time-evolving, transient climate-change scenario could unfold in which the regional climatic anomalies of the 2010s would be different from those of the 2030s. Under such conditions, adapting to recent experience may be counterproductive for ameliorating future impact. Some recent coupled atmosphere/ocean general circulation models have suggested precisely this conclusion – that there may be substantial surprises in the regional distribution of time-evolving climatic anomalies (Manabe and Stouffer, 1994; Washington and Meehl,1989) to which agriculture, hydrological systems and nature will have to adapt. A great deal more research is needed to develop an earth system model (ESM) in which atmosphere, oceans, land surface processes, chemical composition, biota and ice must be coupled to help answer more definitively the potential surprises that may be associated with this transient phenomenon. At present, however, it must be considered as plausible and a significant probability that substantial deviations from the already uncertain range of regional climatic projections using current equilibrium 2 x CO_2 GCMs must be anticipated, and that adaptation to the pattern of equilibrium climate change may not be a practical ameliorative for yield losses due to climate change for the next century.

4.2 Regional forcing

It has long been known that increasing the quantities of greenhouse gases in the atmosphere causes a heating of the surface-tropospheric system of several watts per square meter (for equivalent CO_2 doubling) (Ramanathan, 1981) which, through a number of climatic feedback mechanisms, leads to the canonical 1.5-4.5°C equilibrium warming estimate for CO_2 doubling. A few adherents have suggested positive feedback surprises that could cause warming greater than 4.5°C (e.g. liberation of polar zone methane clathrates; MacDonald, 1990) on the one hand, or strong negative feedbacks (Lindzen, 1990; Schneider, 1990, 1991; Elsaesser, 1991) on the other hand. It has also long been known (SMIC, 1971) that emissions of sulphur dioxide (principally associated with high sulphur coal and oil burning) could lead to further chemical conversions in the atmosphere into sulphate aerosol particles with the potential for cooling the earth. This old debate has recently been rekindled with the suggestion that human industrial production of sulphate aerosols has partially compensated for anticipated global warming, at least over 30 per cent or so of the Northern Hemisphere in which these aerosols are principally generated – the United States, Europe, Eurasia, China and downwind from these areas. Charlson *et al.* (1991) and Kerr (1992) summarise some of the debate. However, the net effect on global average forcing from sulphate aerosols as is typically computed in model experiments trying to detect a greenhouse signal is equivalent to subtracting about one-third of the equivalent greenhouse effect on a global averaged basis (IPCC, 1995; Schlesinger and Ramankutty, 1992; Kelly and Wigley, 1992; Charlson *et al.*, 1991).

The agriculturally and ecologically important component of this problem is that impacts in these sectors typically depend on alterations to regional climates – another potential for surprise. As greenhouse gases are relatively uniformly distributed around the earth, one might expect that they would heat the earth's surface layer relatively uniformly. This, it turns out, is not true, as Kiehl and Briegleb (1993) have shown. That is, in regions where there is a lot of high altitude cloudiness, these clouds carry very strong greenhouse properties, masking the effect of either aerosols or greenhouse gases resident in the atmosphere in the air column below them. Thus, there would be a factor of several-fold difference in the greenhouse heating potential between the humid and the dry parts of the tropics, with the dry parts of the tropics having a heating of up to 4 W/m^2 from greenhouse gas increases since the Industrial Revolution, whereas the humid parts would be heated by only about 1-2 W/m^2. This heterogeneous picture is further complicated if one adds a map of potential human-induced sulphate aerosol cooling to the Kiehl and Briegleb map of greenhouse gas forcing, as these authors have also done. This combined greenhouse gas/aerosol map shows regions in which heating is even negative (over China, central Europe and parts of the eastern USA, associated with the reflection of sunlight in the cloud-free parts of the atmosphere in which sulphate aerosols are generated).

In addition to this complication is the potential for another 'surprise' – the effect of sulphate aerosols in increasing the brightness of clouds in relatively unpolluted places, such as far downwind of the continents in the oceanic areas. The potential effect has been known since 1971 (see Twomey's results in the SMIC Report, 1971, Fig. 8.9) but is not subject to credible quantifiable estimation (IPCC, 1995). The implication for the purposes of this article on food security and climatic surprises is not simply the potential global average contribution of aerosols to the uncertainty associated with the global average rates of radiative forcing from human activities (i.e. global warming from greenhouse gases minus some degree of global cooling from sulphate aerosol enhancement), but the heterogeneous nature of that radiative forcing itself.

It is already known that heterogeneous forcings such as unusually warm or cold anomalies of sea surface temperature in the oceans can create significant atmospheric anomalies above them, which then, through dynamical interactions in the atmosphere known as teleconnections, can extend anomalous effects thousands of kilometers up- and downstream. El Niño-Southern Oscillation (ENSO) is a prime example of such teleconnections associated with regional oceanic heating or cooling patterns. Thus, there will be a highly heterogeneous radiative global change forcing of the atmosphere through the combination of greenhouse gases, cloudiness distribution and sulphate aerosols (to say nothing of biomass burning and land-use changes in the tropics which both change the surface albedo and put soot and other aerosols in the atmosphere). Not only does this imply the need for global correction to the anthropogenic greenhouse effect per se, but it could also create regional climatic anomalies and downstream teleconnections (Schneider, 1994). Recent calculations with transient models driven by CO_2 and sulphate aerosol forcings show patterns of climate changes much closer to observations than previous runs (IPCC, 1995).

Furthermore, the sulphate effect on cloud albedo saturates with an increasing concentration of particles, so the greenhouse-sulphate interactions not only could be regionally heterogeneous, but also could change in relative importance with time. Thus, to provide advice on appropriate adaptive techniques to farmers, one would need to have some sense of whether some period of anomalous climate that farmers may have experienced is the kind of anomalous climate they will continue to experience in the future – an expectation that cannot reliably be assumed at this stage given the significant potential for both transient effects and heterogeneous forcing and associated teleconnection surprises.

5 Geoengineering: cheap technical fix or nasty surprise?

One more aspect of the Kiehl-Briegleb results is connected with the issue of geoengineering – the deliberate attempt to alter the earth's radiation balance or biogeochemical cycles to offset inadvertent climatic modifications such as those that might result from an equivalent doubling of CO_2 (National Academy of Sciences, 1992; Schelling, 1983). Although I, and others, long ago argued that geoengineering schemes are fraught with a high degree of technical uncertainty and thus are ripe for both environmental and political surprises (Kellogg and Schneider, 1974; Schneider and Mesirow, 1976), the Kiehl-Briegleb results imply that it would not be possible to offset exactly the climatic effects of increased greenhouse gases by, for example, spreading aerosol particles in the stratosphere, as those particles would spread evenly around latitude zones, reflecting sunlight in zonal patterns, whereas the infrared radiative forcing for greenhouse gases is regionally heterogeneous owing to the non-uniform distribution of water vapour and high clouds. In other words, even if one could geoengineer an exact, globally averaged cancellation of the inadvertent greenhouse radiative heating with a deliberate injection of stratospheric particle cooling, patchy regional heating/cooling patterns could still have significant regional climatic impacts (Schneider, 1994). Global temperature change is not a sufficient measure of climate change for impact assessment purposes – regional effects are essential in the analysis.

6 The latest Greenland ice cap surprise

Recently, a major paleoclimatic surprise was dug up in the Greenland ice sheet by an international team of European scientists, the Greenland Ice-core Project members (GRIP, 1993). They found that, in addition to the fairly rapid climatic oscillations previously uncovered in Greenland ice cores during glacial periods or glacial-interglacial transitions (e.g. the Younger Dryas), during the previous interglacial, the Eemian between about 135,000 and 120,000 years ago, the climate may not have been as stable or continuously warm (about 2°C warmer than the present Holocene interglacial) as previous Antarctic ice cores had suggested (e.g. Figure 1). At the GRIP site they found that Eemian temperatures were (like Antarctica) often about 2°C warmer than the Holocene, but (unlike Antarctica) this stable warmth occurred for only part of the record. In Greenland several intermittent local climatic shifts of up to 5°C seemed to have occurred in a century or less (comparable – locally, at least – with the rates of global climatic change projected into the 21st century from human activities).

Before these results can be confirmed as global events, measurements of greenhouse gas shifts are required. Confirmation from a neighbouring coring effort from US scientists is also lacking, as preliminary results from these two sites disagree in detail, suggesting that the Eemian cycles may not be climatic phenomena. However, some Atlantic fossil plankton records and pollen records in the Alps suggest simultaneous, large fluctuations in the local ecological conditions – presumably from the rapid climatic fluctuations. Regardless of whether these are connected, the decision of the GRIP team to use past records to anticipate climatic surprises are relevant for our concerns.

7 Discussion

The advent or prospect of surprises for events such as those either of low probability or probability that is not currently assessable suggests the obvious need for flexibility in managing all systems that depend on environmental factors that are influenced by human activities, not least of which is climate change. I believe that impact analysts need to give more attention to the emerging climatic impact paradigm of strategic analysis of a spectrum of probabilities and outcomes. Typically, impact assessments, as mentioned earlier, tend to focus on 'best guesses' (i.e. IPCC's best guess of 2-3°C warming in the middle of the 21st century), with uncertainty often accounted for by the selection of two or three equilibrium, CO_2-doubled model scenarios, or perhaps even a low, middle and high warming scenario.

I understand, even applaud, such analyses as important early steps in the attempt to integrate physical, biological and social components of environmental impact assessment. But I also think the alternative paradigm needs to begin to creep into analytic activities in the form of strategic assessments. By strategic assessment I mean not only analysing central tendencies and equilibrium changes, but also looking at transients and a distribution of outcomes including some significant, positive and negative surprise scenarios, and attempting to assign to each of these scenarios some probability or range of probabilities – even though such probabilities will necessarily be subjective. The assignment of an intuitive or subjective probability should not be viewed as pejorative, for most purported objective probabilities, such as the significance level to which a global climate signal can be attributed to human activities (e.g. Solow and Broadus, 1989) are not really objective, as the assignment of formal statistical significance requires two fundamentally *subjective* assumptions: (i) what is the long-term (i.e. low-frequency) natural climatic variability, and (ii) what have been the time-evolving concentrations of both natural and anthropogenic forcings (such as greenhouse gas build-ups or changes in the solar output or aerosols). Thus, rather than delude ourselves into a false sense of objective analytic security based on some arbitrary (typically 99 per cent) statistical confidence limit, it is better to recognise that most probabilities assigned to the likelihood of specific environmental consequences in the global change area are based on intuition about the plausibility of underlying assumptions. One hopes that intuition is based on the integrated expertise of people from wide-ranging fields familiar with many aspects of the problems and their interconnections (Morgan and Keith, 1995).

Nevertheless, a strategic approach in which even relatively low probability or unassignable probability events are analyzed for their potential consequences would seem prudent, as is done by nearly all defence establishments or in medical practice where relatively low or unknown probability scenarios with significant potential impact outcomes are analyzed and sometimes major public or private investments made to hedge against or accelerate the onset of such scenarios. Due to the vast array of potential surprises from physical, biological or social scientific sub-components, and the kinds of interactions described earlier, it seems that a risk management approach (Morgan, 1993) in which multiple scenarios of different probability are analyzed may be the only reasonable way for impact assessment and the sensitivity of such assessments to alternative policy actions to proceed.

We need to re-examine with considerable scepticism how easy adaptation might prove to be in the real world of transient climatic and ecological changes that will be driven by heterogeneous, radiative global change forcings. For example, the best way to adapt may be to have a credible forecast of time-evolving changes which could guide farmers, water-supply planners, conservation biologists, dike- and dam-builders etc. A credible forecast of global changes would provide the capacity not only to mitigate potential negative consequences, but perhaps could even lead some to take advantage of positive opportunities thereby increasing global welfare – let alone any values for the externality of altered nature. However, I believe the predictability of the earth system will depend on the rate at which we force it to change (Thompson and Schneider, 1982). That is, the slower the build-up of greenhouse gases, SO_2, land uses or other global changes that alter the radiative and chemical balances of the earth-atmosphere system, the closer to an equilibrium approximation each calculated response will be. A world forced to change slowly will more closely approximate a world of a sequence of equilibrium snapshots. This would eliminate to a considerable degree the need to couple very high resolution models of atmosphere, ocean, land surfaces, biosphere etc. into a complex ESM which would be needed to make credible projections for a world in which the transient changes are very fast. Thus the time evolution of regional changes would be very different from a series of equilibrium snapshots.

7.1 Who is credible?

This discussion of subjective or intuitive judgements to estimate uncertain probabilities for some specified outcomes raises a fundamental question: 'What constitutes a *credible* subjective judgement?' The typical media debate that puts 'end of the world' Cassandras against 'nothing will go wrong' Pollyannas is not the model to choose for credible opinions! It may sell newspapers, but it is a distortion of reality. It is better to turn to the many national and international assessment teams consisting of a spectrum of advocates and researchers, such as IPCC or the US National Research Council. More recently, decision analysts have turned to sampling surveys to assess subjective probability judgements of scientists deemed knowledgeable by the analysts (Morgan and Keith, 1995). Not surprisingly, 15 out of 16 scientists surveyed by Morgan and Keith (1995) expressed opinions very close to those of IPCC or National Academy of Sciences (NAS) reports. However, one scientist, often prominently featured in media reports as a 'contrarian' included to 'balance' conventional wisdom, estimated climatic sensitivity to CO_2 doubling an order of a magnitude smaller than the other 15. Interestingly, the 15 'standard paradigm' scientists offered broad probability distributions that included a 5-20 per cent chance of both negligible and catastrophic future human-induced climatic change, whereas the famous contrarian assigned less than a two per cent subjective likelihood of climate change greater than 1°C if CO_2 doubled. Although reasonable people may disagree with who should be included in such surveys, the belief that, despite all the acknowledged uncertainties, only a two per cent chance of detectable climatic change is likely is simply not a credible scientific judgement, in my opinion. Unfortunately, choosing who should be surveyed in subjective probability studies is itself subjective.

7.2 Adaptation

One further complication discussed earlier needs to be reconsidered: natural variability. It is already clear that natural variability in the climate, including extremes of heat, dry, wet, cold etc., can cause havoc with food production systems, and is a principal cause of the year-to-year variations in crop yields and food security in many countries. As the year-to-year natural climatic variability is likely to be larger than even a decade's worth of projected anthropogenic climate change, how are we to convince farmers to adapt to hypothetical forecasts when their recent experience may (by random chance) have seen fluctuations in the opposite direction to the change in mean conditions projected by climate models? Thus, the credibility of easy adaptation to climate model scenarios may raise another key concern over how adaptable farmers will be in the face of both natural climate variability and surprises, rather than standard smooth projections of climate change. The prospect of natural fluctuations combined with time-evolving, regional changes different from standard paradigm forecasts could make the ameliorative effects of adaptation assumed in current assessments highly questionable. Thus, the triple effect of natural variability, uncertainty due to transient evolution and heterogeneous radiative forcing, to say nothing of the potential interactions with ecological or human health aspects, render any total integrated assessment highly speculative. The value of such anthropogenic techniques is to demonstrate the sensitivity of the system's response to various policy options, not the creation of a credible 'optimal' policy.

Nevertheless, the world must examine a plausible range of events with some assigned (even if subjective) probabilities. That task is what impact assessors must do to help to provide some insight into the potential impacts of human activities on climate, society and nature (e.g. Nordhaus, 1994). However, this analyst has long believed, given the wide range of possible outcomes and the considerable difficulty in assigning credible probabilities to each of these outcomes, that slowing down the rate at which human activities are likely to force global changes would certainly seem to be a prudent hedge against the prospect of some of the more serious outcomes (Schneider, 1990). The fact that there is a great deal of inefficiency in energy (Lovins and Lovins, 1992; Office of Technology Assessment, 1991) and agricultural systems suggests that already there are many cost-effective ways in which we can alter human activities that force global changes, and thus slow down the rate at which we force the system to change. Slowing down the rate of global change forcing would also reduce the unpredictability associated with rapid transients as well as the lack of credibility of climate forecasts that would also be associated with such unpredictability (the latter of which diminishes the capacity for adaptability).

8 Concluding remarks

There are obvious first priorities for action: (i) a high degree of policy flexibility, and adaptive management and societal resilience, (ii) focusing research on a range of outcomes and estimates of probabilities, and (iii) a 'tie-in' strategy in which policies that help to reduce global change forcing would also provide other benefits (for example, reducing threats to biodiversity, ozone layer reduction, acid deposition, or investments to eliminate technical inefficiency that reduces economic efficiency). Tables 3 and 4 (Schneider and Turner, 1994) suggest ways to better anticipate and adapt to global change surprises. Meanwhile, despite the bright prospect for some degree of cost-effective short-term mitigation, there will almost certainly be continued global change forcings. Thus, significant global changes can be anticipated with a high degree of probability, even if the specifics are speculative and require considerable efforts to construct multiple scenarios and strategic analyses. The high

Table 3 Improving the anticipation of scientific surprise

1. Encourage and integrate the role of synthesis and synthesizers - appreciating 'putting the puzzle together' and searching for connections across problem domains, disciplines and perspectives.
2. Focus a larger fraction of the research effort on 'outlier' outcomes (e.g. applying methods to sample the opinion of a broad range of knowledgeable experts as to the likelihood of a wide range of imagineable outcomes).
3. Support work at (and across) the edges of conceptual and problem areas.
4. Promote process- as well as product-oriented research and encourage multiple disciplines and communities to communicate and integrate their knowledge about global-change problems.
5. Ensure the following attributes of research discourse and funding that have been insufficiently appreciated to date:
   * sceptical welcoming of advocacy science/scientists and of the airing and professional evaluation of unconventional views; and
   * encouraging overlap without cloning research domains, approaches and institutions.
6. Backcast scenarios from posited future states and/or reconstruct past scenarios in alternative ways to identify events or processes that might happen, but be cognisant that diffusion processes usually are not reversible.
7. Encourage the 'resilience paradigms' as well as the 'optimising paradigms' to inform society and offer alternative strategies for dealing with global change and its impacts.

Source: Schneider and Turner (1994).

probability that some global changes will be unfolding suggests the urgent need for focus on adaptation, which is made more viable through mitigation – slowing down the rate of disturbance to the system.

But future adaptation also may be made more viable through some kinds of current actions which we might label 'anticipatory adaptations' (Schneider and Thompson, 1985). These insurance-like activities could include making a dam 10 per cent higher than it otherwise would have been made for the purposes of flood insurance, or a bridge a meter higher than it otherwise would have been. Such extra expenses may be minimal if added to current construction costs, but very expensive if added later (National Academy of Sciences, 1992). The same principle applies to super-insulating houses and application of stricter

Table 4 Increasing resilience and adaptability in surprise

1. Diversifying economic productive systems: the tendency towards increased economic specialisation carries the risk of vulnerability to controls (e.g. markets or absentee landlords) well beyond the local area which can have both positive and negative impacts on local resilience to environmental perturbations.
2. Avoidance of technological monocultures: reliance on a single technology, such as the extensive use of a single crop variety, may be vulnerable to environmental or other perturbations with negative impacts on the economy. (In this case, long-term impacts might follow without a back-up system, such as a 'crop variety bank'.)
3. Strengthening 'entitlements' or access to resources broadly defined: providing robust safety nets to respond to unforeseen events is a critical part of resilience.
4. Adaptive management systems: organisational theory suggests that different management systems have different capacities for dealing with surprise; those characterised by openness, participation of all parties, and flexibility may be more adaptive than those characterised by command-and-control (e.g. IPCC embedded into the Conference of Parties - climate convention - by means of rolling re-assessments).
5. Disaster coping systems: improving designs of early-warning, monitoring and altering systems, and strengthening the capablity of private and public sectors to respond rapidly to potential disasters should be encouraged.
6. Organisational memory and social learning: measures that improve memory and the ability to learn from surprises improve overall resilience and reduce vulnerability to surprise.
7. Keep public/decision-makers ever cognisant of the subjective nature of surprise assessment and of the need for reflexive and flexible ways of thinking and learning about the subject (as opposed to command approaches).

Source: Schneider and Turner (1994).

building code standards for energy efficiency, or higher standards for efficient machinery in manufacturing processes. Other forms of anticipatory adaptation might be non-market investments in less environmentally impacting energy systems such as making solar, wind or biomass energy more cost-effective or less land-consumptive, or nuclear energy less costly or threatening on a variety of grounds. If such investments are viewed as strategic 'insurance premiums' at the scale of nations – the minimal level at which most global change forcings will need to be controlled – rather than by individuals and firms, it may be possible to have some rational anticipation that will make adaptation easier over time. If those investments also make sense through the reduction of multiple threats or through other multiple benefits (the tie-in strategy) (Kellogg and Schware, 1981; Schneider and Thompson, 1985), then it would seem that these are candidates for the highest priority investments.

Developing countries will not be in the same financial position as developed countries to make anticipatory environmental investments, nor do they necessarily have ready access to the most efficient technologies. Dealing with planetary-scale problems will ultimately require planetary-scale management. Working out a way for nations to cede a small degree of national sovereignty to allow planetary-scale management of the earth so that all would benefit through the preservation of our joint environmental commons may ultimately become the best positive surprise that we could envision. At present, such a surprise is hardly imaginable. But, as awareness grows over the potential benefits of planetary-scale management of planetary-scale problems, what is currently unimaginable may change as rapidly as the unimaginable collapse of the Berlin Wall may have seemed in 1984. What is hardest to imagine is an alternative solution to make the world of humans more resilient and that of Nature less disrupted as the 21st century struggles through rapid transitions toward a more politically and economically stable world with warmer temperatures, fewer wildlands, more people and a host of barely foreseeable interactions and surprises.

References

Barron, E.J. and Hecht, A.D. (eds.) (1985). *Historical and Paleoclimatic Analysis and Modeling.* Wiley, New York.

Berger, W.H. and Labeyrie, L. (eds.) (1987). *Abrupt Climatic Change.* NATO/NSF Workshop, Reidel, Dordrecht.

Berner, R.A. (1991). A model for atmospheric CO_2 over phanerozoic time. *American Journal of Science,* **291**: 339-376.

Bretherton, F. (1986). *Earth System Science.* National Aeronautics and Space Administration,

Washington, DC. pp 24-25.

Broecker, W. S. and Denton, G. H. (1989). The role of ocean-atmosphere reorganizations in glacial cycles. *Geochim. Cosmochin. Acta,* **53**: 2465-2501.

Broecker, W.S., Peteet, D. and Rind, D. (1985). Does the ocean-atmosphere system have more than one stable mode of operation? *Nature,* **365**: 21-26.

Charlson, R.J., Langner, J., Rodhe, H., Leovy, C.B., and Warren, S.G. (1991). Perturbation of the northern hemisphere radiative balance by backscattering from anthropogenic sulfate aerosols. *Tellus,* **43**: 152-163.

Chen, R.S. (1981). Interdisciplinary research and integration: The case of CO_2 and climate. *Climatic Change,* **3** (4): 429-448.

Chen, R.S. and Kates, R.W. (1994). Climate change and world food security. *Global Environmental Change,* **4** (1): 3-7.

COHMAP members (1988). Climatic change of the last 18,000 years: observations and model simulations. *Science,* **241**: 1043-1052.

Crosson, P. and Anderson, R.A. (1992). *Resources and Global Food Prospects, Supply and Demand for Cereals to 2030.* World Bank Technical Paper Number 184. World Bank, Washington, DC.

Daily, G. and Ehrlich, P. (1992). Population sustainability and earth carrying capacity. *BioScience,* **42** (10): 761-771.

Daly, H.E. (1991). Ecological economics and sustainable development. In: Rossi, C. and Tiezzi, E. (eds.) *Ecological Physical Chemistry, Proceedings of an International Workshop.* Elsevier, Amsterdam. pp. 185-201.

Davis, M.B. (1988). Ecological systems and dynamics. In: *Toward an Understanding of Global Change.* National Academy Press, Washington, DC.

Dobson, A. and Carper, R. (1992). Global warming and potential changes in host-parasite and disease-vector relationships. In: Peters, R. and Lovejoy, T. (eds.) *Global Warming and Biological Diversity.* Yale University Press, New Haven, Connecticut. pp. 201-214.

Ehleringer, J.P. and Field, C.B. (eds.) (1993). *Scaling Physiological Processes: Leaf to Globe.* Academic Press, New York.

Ehrlich, P., Ehrlich, A. and Holdren, J. (1977). *Ecoscience: Population, Resources, Environment.* W. H. Freeman Press, San Francisco, California.

Elsaesser, H.W. (1991). Comment on 'The global warming debate heats up: an analysis and perspective by S. H. Schneider'. *Bulletin of the American Meteorological Society,* **72**: 1009.

GRIP (Greenland Ice-Core Project) members (1993). Climate instability during the last interglacial period recorded in the GRIP ice core. *Nature,* **364**: 203-207.

Hansen, J., Johnson, D., Lacis, A., Lebedeff, S., Lee, P., Rind, D. and Russell, G. (1981). Climate impact of increasing atmospheric carbon dioxide. *Science,* **213**: 957-966.

Hasselmann, K. (1979). On the problem of multiple time scales in climate modeling. In: Bach, W., Pankrath, J. and Kellogg, W. (eds.) *Man's Impact on Climate.* Elsevier, New York. pp. 43-55.

Holling, C.S. (1986). The resilience of terrestrial ecosystems: local surprise and global change. In: Clark, W.C. and Munn, R.E. (eds.) *Sustainable Development of the Biosphere.* Cambridge University Press, Cambridge.

Houghton, J.T., Callender, B.A. and Varney, S.K. (eds.) (1992). *Climate Change 1992: The Supplementary Report to the IPCC Scientific Assessment.* Cambridge University Press, Cambridge.

Houghton, J.T., Jenkins, G.J. and Ephraum, J.J. (eds.) (1990). *Climate Change: The IPCC Scientific Assessment.* Cambridge University Press, Cambridge.

Idso, S.B. (1991). The aerial fertilization effect of CO_2 and its implications for global carbon cycling and maximum greenhouse warming. *Bulletin of the American Meteorological Society,* **72**(7): 962.

IPCC (Intergovernmental Panel on Climate Change) (1995). *Second Assessment of Working Group I.* Cambridge University Press, Cambridge.

IBGP (International Geosphere-Biosphere Programme) (1991). *A Study of Global Change (IGBP).* IGBP Press, Stockholm.

Jager, J. (1987). Developing policies for responding to climatic change. A summary of the discussion and recommendations of the workshops held in Villach 28 September to 2 October 1987. WCIP-1, WMO/TD-No. 225.

Johansson, T.B., Kelly, H., Reddy, A.K.N. and Williams, R.H. (eds.) (1993). *Renewable Energy: Sources for Fuels and Electricity.* Island Press, Washington, DC.

Johnsen, S.J., Clausen, H.B., Dansgaard, W., Fuhrer, K., Gundestrup, N., Hammer, C.U., Iversen, P., Jouzel, J., Stauffer, B. and Steffensen, J. P. (1992). Irregular glacial interstadials recorded in a new Greenland ice core. *Nature,* **359**: 311-313.

Kareiva, P. and Andersen, M. (1988). Spatial aspects of species interactions. In: Hastings, A. (ed.) *Community Ecology: Workshop held at Davis, California, April 1986.* Springer, New York. pp. 35-50.

Kellogg, W.W. and Schneider, S.H. (1974). Climate stabilization: for better or worse? *Science*, **186**: 1163-1172.

Kellogg, W.W. and Schware, R. (1981). *Climatic Change and Society: Consequences of Increasing Atmospheric Carbon Dioxide.* Westview Press, Boulder, Colorado.

Kelly, P.M. and Wigley, T.M.L. (1992). Solar cycle length, greenhouse forcing and global climate. *Nature,* **360**: 328-330.

Kerr, R.A. (1992). Pollutant haze cools the greenhouse. *Science*, **255**: 682-683.

Kiehl, J. and Briegleb, B.P. (1993). The relative roles of sulfate aerosols and greenhouse gases in climate forcing. *Science,* **260**: 311-314.

Levin, S.A. (1992). The problem of pattern and scale in ecology. *Ecology,* **73**: 1943-1967.

Lindzen, R.S. (1990). Some coolness concerning global warming. *Bulletin of the American Meterological Society,* **77**: 288-299.

Lovins, A.B. and Lovins, L.H. (1992). Profitability stabilizing global climate. *Climatic Change,* **22**: 89-94.

MacDonald, G.J. (1990). Role of methane clathrates in past and future climates. *Climatic Change,* **16** (3): 247-282.

Malone, T.F. and Roederer, J.G. (eds.) (1984). *Global Change.* ICSU Press, New York.

Manabe, S. and Stouffer, R.J. (1994). Multiple century response of a coupled ocean-atmosphere model to an increase of atmospheric carbon dioxide. *Journal of Climate,* **7**: 5-23.

Martin, P., Rosenberg, N. and McKenney, M.S. (1989). Sensitivity of evapotranspiration in a wheat field, a forest, and a grassland to changes in climate and direct effects of carbon dioxide. *Climatic Change,* **14** (2): 117-152.

McDonald, K.A. and Brown, J.H. (1992). Using montane mammals to model extinctions due to global change. *Conservation Biology,* **6** (3): 409-415.

Mearns, L.O., Katz, R.W. and Schneider, S.H. (1984).Changes in the probabilities of extreme high temperature events with changes in global mean temperature. *Journal of Climate and Applied Meteorology,* **23**: 1601-1613.

Mearns, L.O., Schneider, S.H., Thompson, S.L. and McDaniel, L.R. (1990). Analysis of climate variability in general circulation models: comparison with observations and changes in variability in $2xCO_2$ experiments. *Journal of Geophysical Research,* **95**:

20469-20490.

Morgan, G. and Keith, D. (1995). Subjective judgements by climate experts. *Environmental Science and Technology.* (In press.)

Morgan, M.G. (1993). Risk analysis and management. *Scientific American*, July: 32-41.

Myers, N. (1993). Environmental refugees in a globally warmed world. *BioScience*, **43**(11): 752-761.

Myers, N. and Simon, J.L. (1994). *Scarcity or Abundance? A Debate on the Environment.* W.W. Norton, New York.

National Academy of Sciences (1978). *International Perspectives on the Study of Climate and Society.* National Academy Press, Washington, DC.

National Academy of Sciences (1987). *Current Issues in Atmospheric Change.* National Academy Press, Washington, DC.

National Academy of Sciences (1988). *Toward an Understanding of Global Change: Initial Priorities for US Contributions to the International Geosphere-Biosphere Programme.* National Academy Press, Washington, DC.

National Academy of Sciences (1992). *Policy Implications of Greenhouse Warming: Mitigation, Adaptation, and the Science Base.* National Academy Press, Washington, DC.

Nordhaus, W. (1994). Expert opinion on climatic change. *American Scientist,* **82**.

Oeschger, H. and Arquit, A. (1991). In: Bradley, R.S. (ed.) *Global Changes of the Past.* UCAR/Office for Interdisciplinary Earth Studies, Boulder, Colorado.

Office of Technology Assessment (1991). *Changing by Degrees, Steps to Reduce Greenhouse Gases.* US Government Printing Office, Washington, DC.

Pearman, G.I. (ed.) (1988). *Greenhouse: Planning for Climate Change.* CSIRO, Melbourne, Australia.

Peters, R. and Lovejoy, T. (eds.) (1992). *Global Warming and Biological Diversity.* Yale University Press, New Haven, Connecticut.

Ramanathan, V. (1981). The role of ocean-atmosphere interactions in the CO_2 climate problem. *Journal of Atmospheric Sciences,* **38**: 918-930.

Revelle, R. and Suess, H. (1957). Carbon dioxide exchange between atmosphere and ocean and the question of an increase of atmospheric CO_2 during past and present decades. *Tellus,* **9**: 18-27.

Rind, D., Goldberg, R. and Ruedy, R. (1989). Change in climate variability in the 21st century. *Climatic Change,* **14**: 5-37.

Root, T.L. (1992). Effects of global climate change on North American birds and their communities. In: Kingsolver, J., Kareiva, P. and Huey, P. (eds.) *Biotic Interactions and Global Change.* Sinauer Associates, Sunderland, Massachusetts. pp. 280-292.

Root, T.L. and Schneider, S.H. (1993). Can large-scale climatic models be linked with multiscale ecological studies? *Conservation Biology*, **7**: 256-270.

Root, T.L. and Schneider, S.H. (1995). Ecology and Climate: Research Strategies and Implications. *Science,* **269**: 331-341.

Rosenberg, N.J. (ed.) (1993). Towards an integrated impact assessment of climate change: the Mink Study. *Climatic Change,* **24**(1-2). (Special Issue).

Rosenberg, N.J. and Scott, M.J. (1994). Implications of policies to prevent climate change for future food security. *Global Environmental Change,* **4**(1): 49-62.

Rosenberg, N.J. and Scott, M.J. (1995). Implications of policies to prevent climate change for future food security. In: Downing, T.E. (ed.) *Climate Change and World Food Security.* Springer-Verlag, Heidelberg.

Rosenzweig, C., Parry, M., Fischer, G. and Frohberg, K. (1993). Climate change and world food supply, demand and trade: who benefits, who loses? *Global Environmental Change,* **4**(1): 7-23.

Schelling, T.C. (1983). Climatic change: implications for welfare and policy. In: *Changing Climate, Report of the Carbon Dioxide Assessment Committee.* National Academy Press, Washington, DC. pp. 449-482.

Schlesinger, M.E. and Ramankutty, N. (1992). Implications for global warming of intercycle solar irradiance variations. *Nature,* **360**: 330-333.

Schneider, S.H. (1987). An international program on global change: can it endure? *Climatic Change,* **10**: 211-218.

Schneider, S.H. (1988). The whole earth dialogue. *Issues in Science and Technology,* **4**(3): 93-99.

Schneider, S.H. (1989). The changing climate. *Scientific American,* **261**: 70-79.

Schneider, S.H. (1990). *Global Warming: Are We Entering the Greenhouse Century?* Vintage Books, New York.

Schneider, S.H. (1990). The global warming debate heats up: an analysis and perspective. *Bulletin of the American Meteorological Society,* **71**: 1292-1304.

Schneider, S.H. (1991). Response to Hugh Elsaesser. *Bulletin of the American Meteorological Society,* **72**: 1009 -1011.

Schneider, S.H. (1994). Detecting climatic change signals: are there any 'fingerprints'? *Science,* **263**: 341-347.

Schneider, S.H. and Mesirow, L.E. (1976). *The Genesis Strategy: Climate and Global Survival.* Plenum Publishing, New York.

Schneider, S.H. and Thompson, S.L. (1981). Atmospheric CO_2 and climate: importance of the transient response. *Journal of Geophysical Research,* **86**: 3135-3147.

Schneider, S.H. and Thompson, S.L. (1985). Future changes in the global atmosphere. In: Repetto, R. (ed.) *The Global Possible: Resources, Development and the New Century.* Yale University Press, New Haven, Connecticut. pp. 397-430.

Schneider, S.H. and Turner, B.L. (1994). *Report of the 1994 Aspen Global Change Institute Summer Session on Surprise and Global Environmental Change.* Aspen Global Change Institute, Aspen, Colorado.

SMIC Report (1971). *Inadvertent Climate Modification: Report of the Study of Man's Impact on Climate.* MIT Press, Cambridge, Massachusetts. p. 308.

Smith, J.B. and Tirpak, D. (eds.) (1988). *The Potential Effects of Global Climate Change on the United States: Draft Report to Congress.* Vols. 1 and 2. US Environmental Protection Agency, Office of Policy Planning, and Evaluation, Office of Research and Development. US Government Printing Office, Washington, DC.

Solow, A.R. and Broadus, J.M. (1989). On the detection of greenhouse warming. *Climatic Change,* **15** (3): 449-454.

Taylor, K.C., Lamorey, G.W., Doyle, G.A., Alley, R.B., Grootes, Mayewski, P.A., White, J.W.C. and Barlow, L.K. (1993). The 'flickering switch' of late Pleistocene climate change. *Nature,* **361**: 432-436.

Thompson, S.L. and Schneider, S.H. (1979). A seasonal zonal energy balance climate model with an interactive lower layer. *Journal of Geophysical Research,* **84**: 2401-2414.

Thompson, S.L. and Schneider, S.H. (1982). CO_2 and climate: the importance of realistic geography in estimating the transient response. *Science,* **217**: 1031-1033.

Waggoner, P.E. (chairman) (1992). *Council for Agricultural Science and Technology, Preparing U.S. Agriculture for Global Climate Change.* Task Force Report 119. Ames, Iowa. p. 96.

Washington, W.M. and Meehl, G.A. (1989). Climate sensitivity due to increased CO_2: experiments with a coupled atmosphere and ocean general circulation model. *Climate Dynamics,* **4**: 1-38.

Weihe, W.H. and Mertens, R. (1991). Human well-being, diseases and climate. In: Jager, J. and Ferguson, H.L. (eds.) *Climate Change: Science, Impacts and Policy, Proceedings of the Second World Climate Conference.* Cambridge University Press, Cambridge. pp. 345-359.

Woillard, G.M. and Mook, W.G. (1982). Carbon-14 dates at grande pile: correlation of land and sea chronologies. *Science,* **215**: 159-161.

Impacts of Potential Climate Change on Global and Regional Food Production and Vulnerability

Günther Fischer
International Institute for Applied Systems Analysis, Laxenburg, Austria.

Klaus Frohberg
Institut für Agrarpolitik, Rheinische Friedrich-Wilhelms-Universität, Germany.

Martin L. Parry
Department of Geography, University College London, UK.

Cynthia Rosenzweig
Goddard Institute for Space Studies, U.S.

1 Introduction

Since the late 1950s, global agricultural output has increased on a scale unprecedented in human history. Much of the productivity increase is attributed to the development of high yielding crop varieties, intensive use of inorganic fertilisers and pesticides, irrigation expansion and capital intensive farm management. In the 1970s the euphoria surrounding the Green Revolution was questioned in the wake of the energy crisis and growing awareness of long-term environmental consequences. Soil erosion, ground water contamination, soil compaction and decline of natural soil fertility, and destruction of traditional social systems, led to a reappraisal of what were still considered to be the most advanced agricultural production techniques. Agricultural research has since expanded its scope to include sustainable and resource-efficient cropping systems and farm management practices.

Since the beginning of the 1980s yet another threat to agriculture has caused concern and attracted much attention. Many climatologists predict significant global warming in the coming decades due to increasing atmospheric carbon dioxide and other trace gases. Changes in climate will alter agricultural potential in various agro-ecological regions of the world. An increase in the atmospheric concentration of carbon dioxide will result in an increase in potential agricultural productivity and enhance the efficiency of water use by various crops. The effects of global warming will also tend to expand the agro-ecological potential poleward

NATO ASI Series, Vol. I 37
Climate Change and World Food Security
Edited by Thomas E. Downing

and into higher altitudes. These positive aspects, however, may be constrained and outweighed by altered temperature conditions, precipitation, and evaporation patterns.

The magnitude and geographical distribution of such climate-induced changes may affect our ability to expand food production as required to feed a population of more than 10 billion people projected for the middle of the next century. Climate change could have far-reaching effects on patterns of trade among nations, development, and food security. To help prepare for this uncertain but challenging future, this study examined the potential effects of climate change on crop yields and world food supply, and trade.

Although much agriculture is totally dependent on weather and climate, there is considerable uncertainty about the potential impact of climate change. One incontestable fact is the rising concentration of carbon dioxide (CO_2) in the earth's atmosphere. An additional certainty is the soundness of the basic greenhouse theory: the composition of the gas mix in the atmosphere strongly affects the planet's temperature. Based on that theory, scenarios suggest that a doubling of atmospheric CO_2 concentrations could produce a 1.5 to 4.5° C increase in global temperature in fifty years (FAO, 1994) or by the end of the next century (Houghton *et al.*, 1992, 1995).

Beyond what is known about greenhouse gases and the climate system, however, lie great uncertainties: How much warming will occur, at what rate, and according to what geographical and seasonal pattern? What secondary processes will the warming trend induce, and what might be the physical and biological impacts of such processes? Will some areas benefit while other areas suffer? And, if such damages are unavoidable, what can be done to adapt or modify our systems so as to minimise or overcome them? These are important and complex questions, and research teams are only beginning to understand them and to develop robust methods for analysis.

In 1989 the US Environmental Protection Agency (EPA) commissioned a three year study on the effects of climate change on world food supply. The study was jointly managed by the Goddard Institute for Space Studies (GISS) and the Environmental Change Unit (ECU) in collaboration with the International Institute for Applied Systems Analysis (IIASA) and involved about fifty scientists worldwide.

Recent research has focused on regional and national assessments of the potential effects of climate change on agriculture. For the most part this work has treated each region or nation in isolation, without relation to changes in production elsewhere and without paying attention to climate change effects on the world market and their feedbacks. Assessments of potential impacts have been achieved in national studies completed in the United States (Adams *et al.*, 1990 and 1993; Smith and Tirpak, 1989), Australia (Pearman, 1988), and the United Kingdom (UK Department of the Environment, 1991). Regional studies have been conducted in high-latitude and semi-arid agricultural areas (Parry *et al.*, 1988). These regional and national studies have been summarised in the Intergovernmental Panel on Climate Change (IPCC) Working Group II Report (Tegart *et al.*, 1990). The present study was an attempt to arrive at an integrated global assessment of the potential effects of climate change on agriculture and the world food system.

The complexity of the problem at hand requires an analysis that takes into account the relevant physical and economic relationships that govern the world food system. IIASA's research has provided a framework for analysing the world food system, viewing national agricultural systems as embedded in national economies which in turn interact with each other at international level. This paper provides a brief description of the study approach and analysis of results obtained from a set of simulation runs carried out with the Basic Linked System of National Agricultural Models (BLS) (Fischer *et al.*, 1988) constructed by the Food and Agriculture Program at the International Institute for Applied Systems Analysis (IIASA). The study involved a four step procedure:

- Selection of climate change scenarios.
- Estimation of site-specific potential changes in crop yields.
- Aggregation of site crop-modelling results to estimates of potential national or regional changes in productivity.
- Dynamic simulation of climate change yield impacts on the world food system.

Changes in crop productivity were estimated for 2060, based on site-specific models. These were extrapolated to area-commodity changes, and used as input in the BLS model of world food supply and trade.

2 Climate change scenarios

Scenarios of climate change were developed in order to estimate their effects on crop yields and food trade. A climate change scenario is defined as a physically consistent set of changes in meteorological variables, based on generally accepted projections of atmospheric concentrations of CO_2 and other greenhouse gases. The range of scenarios analyzed is intended to capture the range of possible effects. The scenarios for this study were created by changing observed data on current climate (1951-80) according to the results of doubled CO_2 simulations of three equilibrium general circulation models (GCMs) (Table 1):

GISS: Goddard Institute for Space Studies (Hansen *et al.*, 1983),
GFDL: Geophysical Fluid Dynamics Laboratory (Manabe and Wetherald, 1987), and
UKMO: United Kingdom Meteorological Office (Wilson and Mitchell, 1987).

The temperature changes of these GCM scenarios (4.0-5.2°C) are at or near the upper end of the range (1.5-4.5°C) projected for doubled CO_2 warming by the IPCC (Houghton *et al.*, 1990, 1992, 1995). The GISS and GFDL scenarios, however, are near the mean temperature change (3.8°C) of recent doubled- CO_2 experiments documented for atmospheric GCMs with a seasonal cycle and a mixed-layer ocean (Houghton *et al*, 1992, 1995).

Mean monthly changes in climate variables from the appropriate gridbox were applied to observed daily climate records to create climate change scenarios for sites used in the crop model assessment. GCMs currently provide the most advanced means of predicting the potential future climatic consequences of increasing radiatively active trace gases. They have been shown to simulate current temperatures reasonably well, but do not reproduce current precipitation very accurately; and their ability to reproduce current climate varies considerably from region to region. Of special importance for agricultural climate change

Table 1 Climate change scenarios

GCM	Resolution lat × long	CO_2 ppm	Change in average global temperature, °C	Change in average global precipitation, %
GISS	7.83°×10°	630	4.2	11
GFDL	4.4°×7.5°	600	4.0	8
UKMO	5.0°×7.5°	640	5.2	15

impacts, there is a notable lack of consensus among GCMs in prediction of regional soil moisture changes (Kellogg and Zhao, 1988). Furthermore, GCMs have so far not been able to produce reliable projections of changes in climate variability, such as alterations in the frequencies of drought and storms, even though these could affect crop yields significantly.

For the crop modelling part of this study, climate changes from doubled CO_2 GCM simulations are utilised with an associated level of 555 ppm CO_2; assuming these conditions will occur in 2060. This approach, which mixes equilibrium climate and transient CO_2 projections, was appropriate at the time of the EPA study given the lack of transient climate change simulations for impact studies and the need for a dynamic time course for the economic model. Rates of future emissions of trace gases and the point in time when their effects will be fully realised are not certain.

The level of CO_2 is important when estimating potential impacts on crops, because crop growth and water use benefit from increased concentrations of CO_2 (Cure and Acock, 1986). A CO_2 level of 555 ppm was associated with the effective doubled CO_2 climate projections for use in the crop modelling simulations. This was based on the GISS GCM transient trace gas scenario A described in Hansen *et al.* (1988), in which the simulated climate had warmed to the effective doubled CO_2 level of about 4°C by 2060. This level assumes that non-CO_2 trace gases contribute approximately 15 per cent of the change in radiative forcing from 300 to 600 ppm.

Impacts were assessed for the period 1990 up to year 2060, with population growth and technology trends projected to that year. Simulations of a reference scenario without climate change were compared with scenarios of climate change (outlined below).

3 Modelling site-specific crop yields

Crop models and a decision support system developed by the International Benchmark Sites Network for Agrotechnology Transfer (IBSNAT, 1989) were used to estimate how climate change and increasing levels of carbon dioxide may alter yields of major crops at 112 sites in 18 countries. The IBSNAT models simulate crop growth and yield formation as influenced by genetics, climate, soils and management practices. Models used were for wheat, maize, rice, and soybean. The sites represent both major and minor production areas at low, mid, and high latitudes. The crop models used were CERES-Wheat (Ritchie and Otter, 1985; Godwin *et al.*, 1989), CERES-Maize (Jones and Kiniry, 1986; Ritchie *et al.*, 1989), CERES-Rice (Godwin *et al.*, 1993) and SOYGRO (Jones *et al.*, 1989).

The IBSNAT models are comprised of parameterisations of important physiological processes responsible for plant growth and development, evapotranspiration, and partitioning of photosynthate to produce economic yield. The simplified functions enable prediction of growth of crops as influenced by the major factors that affect yields, i.e., genetics, climate (daily solar radiation, maximum and minimum temperatures, and precipitation), soils, and management practices. This type of dynamic process crop growth model is considered to be a significant advance over traditional regression-based methods. The models include a soil moisture balance sub-model so that they can be used to predict both rainfed and irrigated crop yields. The cereal models simulate the effects of nitrogen fertiliser on crop growth, and these were studied in several countries in the context of climatic change. For the most part, however, the results of this study assume optimum nutrient levels.

The IBSNAT models were selected for use in this study because they have been validated over a wide range of environments (e.g., Otter-Nacke *et al.*, 1986). They are better suited for large-area studies in which crop growing and soil conditions differ greatly than more detailed physiological models that have not been as widely tested. The validation of the crop models over different environments also improves the ability to estimate effects of changes in climate. Furthermore, because management practices, such as the choice of varieties, the planting date, fertiliser application, and irrigation, may be varied in the models, they permit experiments that simulate adaptation by farmers to climate change.

Plant species vary in their response to CO_2 in part because of differing photosynthetic mechanisms. C_3 plants use some of the solar energy they absorb in photorespiration. In this process, which occurs only in light, a considerable fraction of the carbon initially reduced from CO_2 and fixed into carbohydrates is deoxidised to CO_2. C_3 species tend to respond readily to increased CO_2 levels because photorespiration is suppressed in these conditions. Important crop plants with a C_3 photosynthetic pathway are wheat, rice, and soybean. In C_4

plants, on the other hand, CO_2 is trapped inside the leaf and then concentrated in the cells which carry out photosynthesis. These plants are more efficient photosynthetically than C_3 plants under present CO_2 levels, but are likely to be less responsive to CO_2 enrichment. C_4 plants of economic importance include maize, sorghum, millet, and sugarcane. In a CO_2-enriched atmosphere, such plants may be more vulnerable to increased competition from C_3 weeds.

Another important physiological effect of CO_2 enrichment is the closure of stomata, the small openings in leaf surfaces through which CO_2 is absorbed and water vapour released. Accordingly, a rise in atmospheric CO_2 may reduce transpiration even while promoting photosynthesis. This dual effect may improve water-use efficiency. Thus increased CO_2 without other climatic stresses can increase yield and reduce water use per unit of biomass.

The crop models used in this study account for the physiological effects of increased atmospheric CO_2 concentrations on crop growth and water use (Peart *et al.*, 1989; Hendrey *et al.*, 1993). Ratios were calculated between measured daily photosynthesis and evapotranspiration rates for a canopy exposed to high CO_2 values, based on published results (Allen *et al.*, 1987; Cure and Acock, 1986; and Kimball, 1983), and the ratios were applied to the appropriate variable in the crop models on a daily basis. The photosynthesis ratios (555 ppm CO_2/330 ppm CO_2) for soybean, wheat, rice, and maize were 1.21, 1.17, 1.17, and 1.06, respectively. Changes in stomatal resistance were set at 49.7/34.4 s/m for C_3 crops and at 87.4/55.8 s/m for C_4 crops, based on experimental results by Rogers *et al.* (1983).

This promising picture of improved food production under higher atmospheric CO_2 is modified by other factors and by uncertainty about the validity of extrapolations. First, most of our understanding of the positive effects on crops relies on short-term and controlled studies at the individual plant level. Extrapolation and generalisation to large-scale field conditions or to long-term global food production are still uncertain. Also, since crop responses to climate change are site-specific and species-dependent, the knowledge of one kind of grouping or plants (e.g., annuals) may have little relevance to other species or groupings (e.g., perennials). Second, under conditions of limited soil nutrients, low solar radiation (e.g., through enhanced cloud cover) or higher temperature stress, higher CO_2 may do little to improve yields; in much of the world such stress conditions are the rule rather than the exception (FAO, 1994). Third, plants growing in experimental settings are often subject to fewer environmental stresses and less competition from weeds and pests than are likely to be encountered in farmers' fields. Given these uncertainties, the CO_2 responses may be overestimated; they are clearly an area requiring further research.

Crop modelling simulation experiments were performed for baseline climate (1951-80), arbitrary sensitivity tests (+2° and +4°C increase in temperature and +/-20% precipitation), and GCM doubled CO_2 climate change scenarios with and without the physiological effects of CO_2. This involved the following tasks:

- For the countries studied, geographical boundaries were defined for the major crop production regions; agricultural systems (e.g., rainfed and/or irrigated production, number of crops grown per year) were described, and data on regional and national rainfed and irrigated production of major crops were gathered.
- Observed climate data for representative sites within these regions were obtained for the baseline period (1951-80), or for as many years of daily data as were available; the soil, crop variety, and management inputs necessary to run the crop models at the selected sites were specified.
- The crop models were validated with data from field trials where possible.
- The crop models were run with baseline data, arbitrary sensitivity tests, and GCM climate change scenarios, with and without direct effects of CO_2 on crop growth. Rainfed and/or irrigated simulations were carried out as appropriate to current growing practices.
- Alterations in farm-level agricultural practices that would lessen any adverse consequences of climate change were identified and evaluated, by simulating irrigated production and other adaptation responses, e.g., shifts in planting date and substitution of crop varieties.

In each country, agricultural scientists used the crop models to test adaptive responses to the worst climate change scenario (this was usually, but not always, the UKMO scenario). These adaptations included change in planting date, change of cultivar, irrigation, fertiliser, and change of crop. Irrigation simulations in the crop models assumed automatic irrigation to field capacity when plant available water dropped to 50 per cent, and an irrigation efficiency of 100 per cent. Not all adaptation possibilities were simulated at every site and country: the choice of adaptations was made by the participating scientists, based on their knowledge of current agricultural systems. Thus, the adaptation simulations were not comprehensive because not all possible combinations of farmer responses were tested at every site. Spatial analyses of crop, climatic, and soil resources are needed to test fully the possibilities for crop substitution. Neither the availability of water supplies for irrigation nor the costs of adaptation were considered in this study; these are both critical needs for further research.

Global warming and CO_2 enrichment can have positive or negative impacts. Simulated yield increases in the mid- and high-latitudes are caused primarily by:

- *Positive physiological effects of* CO_2. At sites with cooler initial temperature regimes, increased photosynthesis more than compensated for the shortening of the growing period caused by warming.
- *Lengthened growing season and amelioration of cold temperature effects on growth.* At some sites near the high-latitude boundaries of current agricultural production, increased temperatures extended the frost-free growing season and provided regimes more conducive to greater crop productivity.

The primary causes of decreases in simulated yields are:

- *Shortening of the growing period.* Higher temperatures during the growing season speed annual crops through their development (especially grain-filling stage), allowing less grain to be produced. This occurred at all sites except those with the coolest growing-season temperatures in Canada and the former USSR.
- *Decrease in water availability.* This is due to a combination of increases in evapotranspiration rates in the warmer climate, enhanced losses of soil moisture and, in some cases, a projected decrease in precipitation in the climate change scenarios.
- *Poor vernalisation.* Vernalisation is the requirement of some temperate cereal crops, e.g., winter wheat, for a period of low winter temperatures to initiate or accelerate the flowering process. Low vernalisation results in low flower bud initiation and ultimately reduced yields. Decreases in winter wheat yields at some sites in Canada and the former USSR were caused by lack of vernalisation.

With the direct effects of CO_2 and precipitation held at current levels, average crop yields weighted by national production show a positive response to +2°C warming and negative response to +4°C. Wheat and soybean yields increase 10-15 per cent, and maize and rice yields increase about 8 per cent with a +2°C temperature rise. Yields of all four crops turn negative at +4°C, indicating a threshold of the compensation of direct CO_2 effects for temperature increases between 2 and 4°C, as simulated in the IBSNAT crop models. Rice and soybean are most negatively affected at +4°C. These averaged results, however, mask differences among countries. For example the effects of latitude are such that in Canada, a +2°C temperature increase with no precipitation change results in wheat yield increases (with direct effects of CO_2 taken into account), while the same changes in Pakistan result in average wheat yield decreases of about 12 per cent. In general, a 20 per cent increase in precipitation improved the simulated yields of the crops tested, and a 20 per cent decrease lowered yields of all crops.

4 Aggregation of site crop modelling results to national and regional productivity

Data on crop yield changes expected for different scenarios of climate change had to be compiled for all the crop sectors and all the geographical groupings represented in the Basic Linked System. Crop model results for wheat, rice, maize and soybean from the selected sites were aggregated by weighting regional yield changes, based on current production, to estimate changes in national yields. The regional yield estimates represent the current mix of rainfed and irrigated production, the current crop varieties, nitrogen management and soils. Production data were gathered by scientists participating in the study and from the Food and Agriculture Organisation (FAO) and U.S. Department of Agriculture (Crop Production Statistical Division and International Service) (USDA).

Changes in national yields of other crops and commodity groups and of regions not simulated with crop models, were estimated based on similarities to modelled crops and growing conditions and published and unpublished climate change impact studies. Estimates were made of yield changes for the three GCM scenarios with and without direct effects of CO_2. The yield changes with the direct effects of CO_2 were based on the mean responses to CO_2 for the different crops in the crop model simulations.

The spatial aggregation appears to be reasonably robust for the major commodity producing regions, such as the U.S. and Argentina. However, sites in most developing countries, particularly those with precarious food balances, were insufficient to capture the diversity of agricultural conditions.

5 Climate change impacts on crop yields and adaptive responses

The differences among countries in simulated crop yield responses to climate change without the direct effects of CO_2 are primarily related to differences in current growing conditions. Higher temperatures tend to shorten the growing period at all locations tested. At low latitudes, however, crops are currently grown at higher temperatures, produce lower yields, and are nearer the limits of temperature tolerances for heat and water stress. Warming at low latitudes thus results in accelerated growing periods, more severe heat and water stress, and greater yield decreases than at higher latitudes. In many mid- and high-latitude areas where current temperature regimes are cooler, increased temperatures, do not significantly increase stress levels although grain-filling periods are shorter and yields are potentially decreased. At some sites near the high-latitude boundaries of current agricultural production, increased temperatures can benefit crops otherwise limited by cold temperatures and short growing seasons, although the extent of soil suitable for expanded agricultural production in these regions was not studied explicitly. The potential for expansion of cultivated land is embedded in the BLS world food trade model and is reflected in shifts in production (see below).

The estimated changes in crop yields, without farm-level adaptation varies. Global wheat yield changes, weighted by national production, are negative without the physiological effects of CO_2 but are positive with direct CO_2 effects. Maize yield is most negatively affected, reflecting its greater production in low-latitude areas where simulated yield decreases are greater. Maize production declines most with direct CO_2 effects, probably due to its lower response to the physiological effects of CO_2 on crop growth. Simulated soybean yields are most reduced without the direct effects of CO_2, but are least affected in the less severe GISS and GFDL climate change scenarios when direct CO_2 effects are simulated. Soybean responds positively to increased CO_2, but is the crop most affected by the high temperatures of the UKMO scenario.

Without adaptation, the GISS and GFDL climate change scenarios produced yield changes ranging from +30 to -30 per cent. Effects under the GISS scenario are, in general, more adverse than under the GFDL scenario to crop yields in parts of Asia and South America, while effects under the GFDL scenario result in more negative yields in the USA and Africa and less positive results in former USSR. The UKMO climate change scenario, which has the greatest global warming (5.2°C), causes average national crop yields to decline almost everywhere.

The adaptation studies, conducted by the scientists participating in the project suggest that ease of adaptation to climate change varies with crop, site, and adaptation technique. For example, at present many Mexican producers can only afford to use small doses of nitrogen fertiliser at planting; if more fertiliser becomes available to more farmers some of the yield reductions under the climate change scenarios might be offset. However, given the current economic and environmental constraints in countries such as Mexico, a future with unlimited water and nutrients is unlikely (Liverman *et al.*, 1994; Appendini and Liverman, 1994; Appendini and Liverman 1995, this volume). In contrast, switching from spring to winter wheat at the modelling sites in the former USSR produces a favourable response (Menzhulin *et al.*, 1992), suggesting that agricultural productivity may be enhanced there with a relatively easy shift to winter wheat.

Adaptation Level 1, simulating minor changes to existing agricultural systems, partially compensated for the climate change scenarios, particularly in developing countries. For the GISS and the GFDL scenarios, adaptation implying major changes to current agricultural systems (Adaptation Level 2) compensated almost fully for the negative climate change impacts. With the high level of global warming as projected by the UKMO climate change scenario, neither Level 1 nor Level 2 adaptation fully overcomes the negative climate change effects on crop yields in most countries, even when the direct CO_2 effects are taken into account.

6 Modelling the world food system

The world food system consists of many actors, some powerful and others dependent. Historically, food and fibre have been produced in increasingly efficient ways. Producers and consumers interact through national and international markets. While there is a trend towards internationalisation in the world food system, only approximately 15 per cent of world cereal production currently crosses national borders (Fischer *et al.*, 1990). National governments shape the system by imposing regulations and by investments in agricultural research, infrastructure improvements, and education. Although the system does not guarantee stability nor equity, it has generated long-term real declines in prices of major food staples.

The Basic Linked System of National Agricultural Policy Models (BLS) consists of some thirty-five national and/or regional models: eighteen national models, two models for regions with close economic cooperation (EC and Eastern Europe + former USSR[1]), fourteen aggregate models of country groupings, and a small component that accounts for statistical discrepancies and imbalances during the historical period. The individual models are linked together by means of a world market module. A detailed description of the entire system is provided in Fischer *et al.* (1988). Earlier results (not on climate change) obtained with the system are discussed in Parikh *et al.* (1988) and in Fischer *et al.* (1990).

The BLS is a general equilibrium model system. This necessitates that all economic activities are represented in the model. Financial flows as well as commodity flows within a country and at the international level are consistent in the sense that they balance. Whatever is produced will be demanded, either for human consumption, feed or intermediate input; it might be traded or put into storage. Consistency of financial flows is imposed at the level of the economic agents in the model (individual income groups, governments, etc.), at the national as well as the international level. This implies that total expenditures cannot exceed total income from economic activities and from abroad, in form of financial transfers, minus savings. On a global scale, not more can be spent than what is earned.

The country models are linked through trade, world market prices and financial flows. The system is solved in annual increments, simultaneously for all countries. It is assumed that supply does not adjust instantaneously to new economic conditions. Only supply that will be marketed in the following year is affected by possible changes in the economic environment. A first round of exports from all the countries is calculated for an initial set of world prices, and international market clearance is checked for each commodity. World prices are then revised, using an optimising algorithm, and again transmitted to the national models. Next, these generate new domestic equilibria and adjust net exports. This process is repeated until the world markets are cleared in all commodities. At each stage of the iteration the domestic

[1] The political changes as well as changes in national boundaries of the very recent past are not captured in the BLS, although the model formulation has been adjusted, away from centrally planned economies to more market-oriented behaviour.

markets are in equilibrium. Since these steps are taken on a year-by-year basis, a recursive dynamic simulation results.

The concept of an economic agent who decides on production and disappearance is the basis on which the BLS is built. Producers maximise returns to primary factors they are endowed with in their production activities. Consumers are assumed to maximise utility. And governments follow prescribed objectives in their policy setting within the constraints of balancing expenditures with the revenues generated through taxes, tariffs or other means, and international transfers.

Although the BLS contains different types of models, all adhere to some common specifications. The models contain two main sectors: agriculture and non-agriculture. Agriculture produces nine aggregated commodities, all non-agricultural activities are combined into one single aggregate. Production is critically dependent on the availability of the primary production factors – land, labour and capital. The first is used only in the agricultural sector, while the latter two are determinants of output in both the agricultural and the non-agricultural sectors.

An upper bound on land available for cropping and for use as pasture is determined by the availability of land resources as well as economic conditions; e.g., by the economic returns to land. The physical resource limits in developing countries were derived from an FAO assessment of potential arable land (FAO/UNFPA/IIASA, 1983, and FAO, 1988). The responsiveness of how much land can be cultivated due to changing economic conditions is rather low since time and investment are needed to bring new land into cultivation.

For agricultural commodities, acreage or animal numbers and yield are determined separately. Yield is represented as a function of fertiliser application (crops) or feeding intensity (livestock). Technological development is assumed to be largely determined by exogenous factors. Technical progress is included in the models as biological technical progress in the yield functions of both crops and livestock. Rates of technical progress were estimated from historical data and, in general, show a decline over time. Mechanical technical progress is part of the function determining the level of harvested crop area and livestock husbandry. Induced (endogenous) technical progress is not considered for any of these cases or for non-agricultural production.

Several factors cause consumers and producers to react differently over time to political changes and altered economic and technological conditions. While it is mainly the formation of taste and habit, and changing prices and incomes that alter consumers' responses, producers are most affected by their past investment decisions, by technical innovations, or – as in this study – changes in productivity due to climate change. The ability of producers to respond to changes in economic conditions is relatively limited in the short-run. It increases in the medium to long term as investment allocation changes accumulate over time.

Simulating with the BLS produces a variety of variables. At the world market level these include prices, net exports, global production and consumption. At the country level the information generated varies, depending on the model specification, including, in general, producer and retail prices, level of production, use of primary production factors (land, labour and capital), intermediate input use (feed, fertiliser, and other chemicals), level of human consumption, stocks and net trade, gross domestic product and investment by sector, population number and labour force, welfare measures such as equivalent income, and the level of policy measures as determined by the government (e.g., taxes, tariffs).

7 Reference projections without climate change

This section presents the reference projections generated with the model system without taking into account possible climate change and increases in atmospheric CO_2 concentrations. This indicates the range of projections that can be simulated from a relatively narrow set of assumptions. The following section compares these reference projections to scenarios dealing specifically with climate change. Note that the scenarios used in this study are considered long-term projections, not forecasts. The considerable length of the projection period makes it impossible to avoid judgement concerning some of the exogenous variables in the system (see Chen and Kates, 1994, 1995, this volume).

The standard reference scenario we describe, termed scenario REF-M, mostly reflects a 'business as usual' assumption, in the sense that no radical shifts in technological or political trends are assumed. This particular long-term projection has been designed to serve as a point of departure in studying potential impacts of climate induced yield and productivity changes in agriculture.

Another reference projection, scenario REF-H, assumes faster economic growth than the standard reference scenario. In the BLS, growth dynamics can be influenced by adjusting the rate of investment in the economy, and through altering assumed technological progress in production functions. In scenario REF-H, growth of GNP is especially higher for Asian countries compared to scenario REF-M. Similarly, a lower growth simulation run, scenario REF-L, has been created by curtailing investment in favour of consumption.

Additional reference projections with the BLS were based on the standard reference scenario REF-M. Two projections deal with the sensitivity of the world food systems with respect to land development and availability of agricultural inputs. In scenario REF-MA, expansion of arable land after year 2000 is constrained to half the expansion in scenario REF-M. Another simulation run, scenario REF-MF, limited the use of fertilisers by implementing a tax on fertilisers of 50 per cent in developed countries and of 33 per cent in developing countries, respectively. In addition, a ceiling on fertiliser use per hectare was enforced. Both these scenarios were simulated to test the impact of policy measures geared towards reducing greenhouse gas emissions from agriculture.

Finally, reference projection REF-MP uses the economic assumptions of the standard reference run REF-M but alters the demographic projections from medium population growth to low population growth.

All of the simulations were carried out on a yearly basis from 1980 to 2060. The amount of information generated is fairly detailed and only a small fraction can be presented here. We concentrate on the results at the global and regional level – the thirty five geographic units of the BLS were aggregated into ten regions. Emphasis is given to population and economic growth, prices, and changes in the structure of production and trade patterns. These developments are the outcome of simultaneous processes and decisions and are only discussed in sequence for clarity.

As a global system of linked national and regional models, the BLS contains only a few variables that are fully exogenous. Most important among these are population growth and technical progress.

Population growth rates were obtained from the United Nations (UN) medium variant in all simulations except for scenario REF-MP where the low variant was used (UN, 1989). Since the UN projected national population levels only up to the year 2025, the remainder of the projection period was covered by growth rates compiled from long-term population projections of the World Bank (1990). Growth rates have been compiled for each country or country group in the BLS. They are presented in Table 2. Labour participation rates are taken from International Labour Organisation (ILO) projections. This implies that the trajectory of total labour force in a country is exogenously prescribed and does not change between scenarios using the same population projections. However, the allocation of total labour force between agriculture and non-agricultural sectors is endogenous and responds to relative prices and incomes differentials.

During the last two decades population, at the global level, has been growing at an average rate of about 1.8 per cent per annum, from 3.6 billion in 1970 to about 5.3 billion people in 1990. In the BLS standard reference projection, scenario REF-M, the average annual population growth rate is projected to decline gradually from 1.7 per cent per annum during 1980 to 2000 to around 0.5 per cent per annum for the period 2040 to 2060. This would bring global population numbers to 6.1 billion in year 2000 and about 10.3 billion in year 2060. Most of the increase in population occurs in developing countries, increasing their share in world population from around 73 per cent in 1980 to 78 per cent in year 2000 and some 86 per cent in year 2060, i.e., almost 9 billion people in developing countries of which approximately 5 billion are projected for Asia, and 2.2 billion for Africa. With lower population growth, such as in scenario REF-MP, population reaches 7.2 billion in year 2020, and some 8.6 billion in year 2060, about 17 per cent less than in the medium variant of scenario REF-M.

Table 2 Projected population numbers and average annual growth, 1980 - 2060

Region	1980		2000		2020		2060		1980-2000	2000-2020	2020-2040	2040-2060
	mill.	% of total	mill.	% of total	mill.	% of total	mill.	% of total	% p.a.	% p.a.	% p.a.	% p.a.
Medium variant												
World	4378		6125		7883		10315		1.7	1.3	0.8	0.5
Developed	1186	27.1	1340	21.9	1445	18.3	1470	14.3	0.6	0.4	0.1	-0.0
NAM	810	18.5	915	14.9	980	12.4	970	9.4	0.6	0.4	0.1	-0.1
WEU+ODE	428	9.8	472	7.7	501	6.4	505	4.9	0.5	0.3	0.1	-0.1
EEU+FSU	374	8.5	424	6.9	465	5.9	500	4.8	0.6	0.5	0.3	0.1
PAO	134	3.1	151	2.5	156	2.0	145	1.4	0.6	0.2	-0.1	-0.2
Developing	3193	72.9	4786	78.1	6437	81.7	8843	85.7	2.1	1.5	1.0	0.6
AFR	412	9.4	757	12.4	1274	16.2	2240	21.7	3.1	2.6	1.7	1.1
LAM	351	8.0	522	8.5	695	8.8	878	8.5	2.0	1.4	0.9	0.3
WAS	188	4.3	325	5.3	485	6.2	777	7.5	2.8	2.0	1.4	1.0
SAS	898	20.5	1396	22.8	1889	24.0	2541	24.6	2.2	1.5	0.9	0.6
CPA	1086	24.8	1419	23.2	1637	20.8	1843	17.9	1.4	0.7	0.4	0.2
PAS	258	5.9	366	6.0	458	5.8	564	5.5	1.8	1.1	0.7	0.4
Low variant												
World	4380		5968		7215		8565		1.6	1.0	0.5	0.3
Developed	1185	27.1	1319	22.1	1364	18.9	1348	15.7	0.5	0.2	-0.0	-0.0
Developing	3195	72.9	4649	77.9	5851	81.1	7217	84.3	1.9	1.2	0.6	0.4

Note: See Appendix for region names.

Technical progress enters the representation of the production processes at different levels in a country model. Time is used in most cases as a proxy variable. The limitation of this approach is that technical progress does not vary among different scenarios. Hence, the model system does not allow for induced technical progress according to economic conditions. This limitation should be taken into account when interpreting the results, especially in scenarios where large adjustments are required.

The BLS reference projections used in this study assume that the policy regimes in the various countries continue along past trends without drastic changes. Protectionist policy measures in agriculture, however, are assumed to be lowered to half the observed historical levels of 1980. The transition to such reduced protection of the agriculture sectors is implemented between the beginning of the simulation period and year 2020. Thereafter the respective policy settings are kept constant.

Growth of the world economy is a rather important attribute of a scenario. Agriculture in many developing countries is still a major source of foreign exchange earnings. But it is also highly dependent for its performance on the development of the overall economy. This dependence is particularly strong over the longer term due to the impact of income changes on demand. Furthermore, movements of production factors between agriculture and other sectors of the economy as well as technological developments are important for the performance of agriculture. It is projected that the contribution of agriculture to total GDP will in all regions gradually decline over time. For the standard reference scenario REF-M, in year 2060, the share of agriculture in total GDP is estimated to be only about 40 per cent of its present level in developed countries, and about half the present level in developing countries.

Growth rates in most of the national models of the BLS are endogenously determined based on essentially three elements: (i) capital accumulation through investment and depreciation, related to a savings function that depends on lagged GDP levels as well as balance of trade and financial aid flows, (ii) dynamics of labour force as a result of demographic changes, and (iii) exogenous technical progress. Table 3 presents some indicators of economic development derived from the simulation results of the reference projections.

Table 3 Economic growth indicators for different reference projections (average annual per cent change)

	REF-L			REF-M			REF-MP			REF-H		
	World	**Deve-loped**	**Deve-loping**	**World**	**Deve-loped**	**Deve-loping**	**World**	**Deve-loped**	**Deve-loping**	**World**	**Deve-loped**	**Deve-loping**
GDP												
1980-2020	2.2	2.0	3.1	2.4	2.2	3.3	2.3	2.1	3.1	2.7	2.2	4.0
1980-2060	1.7	1.5	2.3	1.8	1.6	2.4	1.7	1.6	2.3	2.2	1.7	3.2
GDP per capita												
1980-2020	0.8	1.5	1.3	1.0	1.7	1.5	1.1	1.8	1.6	1.3	1.8	2.2
1980-2060	0.7	1.3	1.0	0.8	1.4	1.1	0.9	1.4	1.2	1.2	1.5	1.9
Agriculture												
1980-2020	1.5	0.7	2.0	1.5	0.7	2.0	1.4	0.7	1.9	1.6	0.7	2.1
1980-2060	1.2	0.5	1.5	1.2	0.5	1.6	1.1	0.4	1.4	1.2	0.5	1.6
Food Production												
1980-2020	1.5	1.0	1.8	1.5	1.0	1.9	1.4	0.9	1.7	1.6	1.0	2.0
1980-2060	1.1	0.7	1.4	1.1	0.7	1.4	1.0	0.6	1.2	1.2	0.7	1.5
Population												
1980-2020	1.5	0.5	1.8	1.5	0.5	1.8	1.3	0.4	1.5	1.5	0.5	1.8
1980-2060	1.1	0.3	1.3	1.1	0.3	1.3	0.8	0.2	1.0	1.1	0.3	1.3

All regions, developed and developing alike, show a declining rate of GDP growth over time, which is consistent with historical developments. Declining rates of population growth and growth of the labour force and an assumed general slowdown in productivity growth contribute to declining GDP growth rates. In the standard reference scenario (REF-M), GDP at the global level increases at an average 2.4 per cent annually during 1980 to 2020. Economic growth declines to about 1.5 per cent per annum during 2020 to 2040. Overall, global GDP increases 4.4 times during the simulation period compared to a 2.4 fold increase in population numbers, resulting in an average annual increase in GDP per capita of 1.4 and 1.1 per cent in developed and developing countries, respectively, over the 80 year period from 1980 to 2060. It should be noted, however, that increases in per capita indicators are higher at the regional and national level compared to global figures owing to an aggregation effect induced by the demographic development, giving increasingly higher weights to poorer developing countries. For instance, in scenario REF-H per capita GDP increases by 1.5 and 1.4 per cent per annum in developed and developing countries, respectively, whereas at aggregate global level the average annual increase is 1.2 per cent.

With faster economic growth, especially in developing countries, such as assumed in scenario REF-H, world GDP grows almost 5.6 times between year 1980 and 2060. In comparison, the lower growth projection REF-L results in a 3.9 fold increase in GDP. This demonstrates that even with relatively minor variations on the basic economic assumptions the resulting output in year 2060 because of the length of the simulation period, varies by some 40 per cent.

The standard reference projection (REF-M) presents the perspective of a world in which the effective demand for food grows substantially owing to higher incomes and larger populations. Technological progress and economic development assumed in the reference scenario meet this increase in demand at somewhat decreasing world market prices for agricultural products, consistent with historical trends. As shown in Table 3, food production (measured in terms of net food energy, i.e., production less feed, seed and waste) as well as overall agriculture growth is projected to exceed population growth throughout the simulation period in all regions. Table 4 shows global production of agricultural commodities in the standard reference scenario (REF-M) and in the higher income scenario (REF-H).

For agricultural commodities, average annual growth rates of production, and hence effective demand, during the period 1980 to 2060 range from 0.9 to 1.3 per cent per annum implying a two to threefold increase compared to 1980 levels. Gross agricultural production[2] increases on average 1.2 per cent per annum, i.e., by 2060 it reaches about 2.6 times the 1980 level. This compares favorably to the projected average population increase of less than 1.1 per cent annually during this 80 year period.

Global trade in the reference scenario increases somewhat faster than global agricultural production. For cereals, the share of net exports in global production is estimated to increase from 13 per cent in 1980 to 15 per cent in 2060, with wheat and coarse grains showing an almost threefold and rice a fourfold increase in trade levels. In general, the share of global trade in global production of commodity aggregates increases gradually over time. Increasing demand in developing countries, due to rising incomes and growing populations, leads to a deterioration in the level of agricultural self-sufficiency for this group of countries, which

Table 4 Global agricultural production in the BLS reference projections

	Scenario REF-M					Scenario REF-H			
	1980	**2000**	**2020**	**2040**	**2060**	**2000**	**2020**	**2040**	**2060**
Wheat	441	603	742	861	958	658	811	911	1056
Rice*	249	367	480	586	659	415	545	661	749
Coarse grains	741	1022	1289	1506	1669	1065	1349	1587	1772
Bovine + ovine meat	65	83	105	123	136	84	107	125	139
Dairy	470	613	750	877	997	616	758	893	1021
Other animal production	17	25	33	41	48	25	34	42	49
Protein Feed	36	52	64	76	85	52	65	77	87
Other Food	225	326	433	538	629	326	436	545	640
Non-Food	26	34	41	47	52	34	41	48	53
Agriculture	370	522	676	821	942	533	696	848	977

Note: *Production is in milled rice equivalent; conversion factor from paddy is 0.667.

---

2 Gross agricultural production, labelled Agriculture in Table 4 is calculated at constant 1970 world market prices.

changes from a net surplus of about 3 per cent in 1979/81 into a 1 per cent deficit by 2060 caused primarily by increasing deficits in cereals, meat and milk.

Besides their significance in international trade, cereals continue to be of major importance in human consumption. Table 5 shows regional differences in the shares of different commodity groups in total food energy intake. Although cereals become somewhat less important as a direct energy source in human diets in all regions, the share at global level stays above 50 per cent throughout the simulation period. The share of livestock products goes up in all regions, but on average – as a consequence of changing regional weights in the global total – remains at about thirteen per cent. The balance goes to other food commodities, like vegetable oils, sugar, fruits, vegetables, pulses and root crops. In developed countries, cereals and other commodities account for just over one third each, livestock products for a little more than a quarter of food energy intake. The BLS food demand projections for developing countries suggest that well over half the energy intake will continue to originate from cereals, about one third from other crop commodities and about ten per cent from meat and milk.

Global cereal production in 1979/81 is estimated to amount to 1.4 billion tons (rice is included in milled form). In the standard reference scenario (REF-M) production increases to about 2.5 billion tons by 2020 and some 3.3 billion tons by 2060. Depending on scenario assumptions in the suite of scenarios REF-L to REF-H, projected world cereal production

Table 5 Projected share of human energy intake by commodity group (in per cent)

	World	Developed	Developing	Africa	Latin America	South East Asia	West Asia
1980							
Cereals	55	39	64	60	42	69	65
Livestock	14	26	8	7	16	6	11
Other	31	35	28	33	42	25	24
2060							
Cereals	51	34	55	55	41	58	54
Livestock	13	27	10	8	16	9	12
Other	36	39	35	37	43	33	34

Note: Food energy from livestock commodities includes milk and milk products (e.g., cheese and butter), meat and edible offals, eggs and fish.

in 2060 varies between 2.9 to 3.6 billion tons, i.e., the width of the band around the projected level of the standard reference scenario REF-M is about 20 per cent.

Table 6 shows regional levels of cereal production for different reference projections, Table 7 lists regional cereal trade. The share of developed countries in global production of cereals is projected to steadily decline in all these scenarios, from 56 per cent in 1980 to between 41 to 46 per cent by 2060. Over the same period, the share of developed countries in global demand of cereals declines from 53 per cent in 1980 to between 32 to 35 per cent in 2060 implying an increased net trade of cereals to developing countries.

To evaluate the impact of alternative scenarios on the poor in different countries, it was necessary to generate a consistent indicator of food poverty in the BLS[3]. Country-wise estimates of the number of undernourished persons have been made by the FAO (1984,

Table 6 Cereal production in the BLS reference projections (million tons)

	REF-M			REF-MP	REF-MA	REF-MF	REF-L	REF-H
	1980	**2020**	**2060**			**2060**		
World	1460	2510	3286	2930	3199	3200	3214	3576
Developed	822	1229	1449	1348	1395	1353	1428	1474
NAM	317	575	702	624	626	621	687	709
WEU + ODE	198	245	275	264	289	267	276	272
EEU + FSU	275	357	410	413	418	419	395	433
PAO	32	52	63	47	62	46	69	60
Developing	638	1282	1836	1582	1804	1847	1787	2102
AFR	52	148	277	239	258	286	265	286
LAM	84	199	274	211	279	290	274	273
WAS	52	103	152	130	147	160	146	153
SAS	147	325	470	412	479	469	448	528
CPA	260	407	530	488	507	506	518	728
PAS	43	100	133	102	135	137	135	134

Note: See Appendix for region names.

[3] We use the term food poverty, following the FAO methodology of the lack of food or income to achieve a dietary intake above 1.4 times the basal metabolic rate. Earlier results of the BLS reported populations 'at risk of hunger'. However, in this volume of papers, hunger implies the full range of deprivation, including micro-nutrients and a wide range of causes, such as armed conflict (see, for example, in this volume, Appendini and Liverman, 1995, Chen and Kates, 1995, Downing *et al.*, 1995, Homewood, 1995, Huq *et al.*, 1995, Jodha, 1995, Karim, 1995 and Norse, 1995).

Table 7 Net exports of cereals in the BLS reference projections (million tons)

	REF-M			REF-MP	REF-MA	REF-MF	REF-L	REF-H
	1980	**2020**	**2060**			**2060**		
World	182	369	504	423	484	466	509	520
Developed	56	189	305	238	307	257	298	322
NAM	97	247	356	288	318	300	345	365
WEU + ODE	14	-35	-30	-21	-15	-29	-36	-23
EEU + FSU	-46	-23	-25	-24	-8	-11	-24	-27
PAO	-9	0	5	-6	14	-3	14	8
Developing	-56	-189	-305	-238	-307	-257	-298	-322
AFR	-17	-71	-128	-93	-136	-107	-135	-126
LAM	-6	5	13	-10	19	31	22	4
WAS	-15	-75	-129	-90	-132	-119	-121	-135
SAS	-7	-26	-13	-8	4	-2	-19	12
CPA	-11	-27	-44	-26	-60	-61	-47	-65
PAS	0	5	-3	-9	-2	1	2	-10

Note: See Appendix for region names.

1987). To introduce the FAO method in a reduced form in the BLS, suitable for use in the simulation models, a cross-country regression was estimated, expressing the share of people at risk of food poverty by a measure of food energy availability relative to nutritional requirements. Food availability, in turn, depends on income and price levels.

In the standard reference scenario (REF-M) the incidence of food poverty decreases markedly from the estimated 23 per cent of the population in developing countries (excluding China) in 1980 to some 9 per cent in 2060[4]. Yet, despite this remarkable improvement the estimated number of people at risk increases somewhat, from about 500 million in 1980 to almost 720 million in 2020, and some 640 million in 2060. The projected number of undernourished people in developing countries is shown in Table 8. No estimate was attempted for developed regions. Of course, the projected number of people at risk of food poverty is sensitive to the scenario assumption – ranging from less than 400 million under a low population run (REF-MP), about 500 million when assuming faster economic development (REF-H) and about 760 million under the lower growth scenario (REF-L).

4 Subsequent to this research, the FAO has estimated (FAO, 1993) that the number of undernourished in developing countries amounted to 941 million people in 1979/81 and to 843 million people in 1988/90, based on a threshold food energy level of 1.54 times the basal metabolic rate (BMR).

Policy measures can specifically address poverty and hunger even under a lower growth economic scenario (see Chen and Kates, 1995, this volume). For the sake of comparability of scenarios, however, such specific measures have not been considered in this climate change impact study.

While the estimates of food poverty show an improvement in relative and absolute terms in Asian countries, the African continent experiences a mixed outcome mainly owing to large increases in population. The estimated share of people at risk of food poverty in Africa declines from 28 per cent in 1980 to 18 per cent by 2060. However, the number of food poor is projected to increase more than threefold, from about 120 million people in 1980 to 415 million in 2060, thereby making Africa the region with the largest number of undernourished. With lower population growth (REF-MP), the estimated per centage of people at risk is reduced by more than 25 per cent; the higher economic development (REF-H) reduces the number by some 140 million (approximately 22 per cent). On the other hand, lower economic growth (REF-L) increases the estimate of food poverty by almost 20 per cent. Again, it is important to note that even from a relatively narrow set of assumptions, the resulting estimates of the undernourished in developing countries varies significantly.

Table 8 Food poverty in reference projections (millions)

	REF-M			REF-MP	REF-MA	REF-MF	REF-L	REF-H
	1980	**2020**	**2060**			**2060**		
Developing	501	715	641	395	727	722	757	498
AFR	116	291	415	305	447	446	441	375
LAM	36	39	24	13	34	33	32	20
WAS	28	55	72	46	84	85	78	68
SAS	265	319	128	30	160	157	202	35
CPA*	26	33	0	0	0	0	0	0
PAS	30	6	2	0	3	2	4	0

Note: * estimate does not include China
See Appendix for region names

8 The world food system and climate change yield impacts

Evaluation of the potential impact of climate change on production and trade of agricultural commodities, in particular on food staples, is carried out by comparing the results of the climate change scenarios to reference projections of the world food system without climatic change. The climate change impact scenarios relate to:

- Three GCM simulations of doubled CO_2.
- With and without the physiological effects of increased CO_2.
- Timing of climate change and its impacts.
- No responses and two levels of farm adaptation to mitigate yield impacts.
- Policy changes that affect both the reference run and climate change experiments, e.g., population growth, trade policies, economic growth, and policies to reduce emission of greenhouse gases from agriculture, such as limitation of arable land expansion, rice cultivation, or use of chemical fertilisers.

Well over 70 experiments have been simulated. Results from some twenty scenarios, are reported:

- Simulations without considering physiological effects of increased atmospheric CO_2 concentrations on crop growth and yield.
- Simulations with physiological effects of 555 ppm CO_2 on crop growth and yield.
- Simulations with physiological effects of CO_2, and adaptations at the farm-level that would not involve any major changes in agricultural practices to mitigate negative yield impacts, Adaptation Level 1.
- Simulations with the physiological effects of CO_2, and adaptations at the farm-level that would involve major changes in agricultural practices, Adaptation Level 2.
- Simulations with the physiological effects of CO_2, but with temperature and precipitation changes projected by the GISS GCM for 2030. This scenario was termed the low-end scenario (LE) and was used to test the consequences of a lower sensitivity of the climate system to increasing atmospheric greenhouse gas concentrations.
- Simulations with physiological effects of CO_2 and farm-level adaptation for a transient impact assessment.

Data on crop yield changes estimated for different scenarios of climate change were compiled for thirty-four countries or major regions of the world as modelled in the BLS. Most models included in the BLS distinguish yield and acreage functions. The yield response functions of major crops use the level of fertiliser application and a term related to technology as explanatory variables. While technical progress is exogenous, the level of fertiliser application is derived from optimality conditions, i.e., by comparing the marginal value product of fertiliser to its price. Yield variations caused by climate change were introduced into the yield response functions by means of a multiplicative factor impacting upon the relevant parameters in the mathematical representation. This implies that both average and marginal fertiliser productivity are affected by the imposed yield changes.[5]

Country and crop-specific information on changes in the area suitable for crop cultivation due to climate change was not available[6]. Therefore, the land allocation is indirectly influenced through the implied changes in overall performance of the agricultural sector as well as changing comparative advantage of the competing crop production activities.

Exogenous variables, population growth and technical progress, were left at the levels specified in the respective reference projections. No specific adjustment policies to counteract altered performance of agriculture have been assumed beyond the farm-level adaptations specified above. In other words, in the scenarios presented here, adaptation suggested in the crop modelling experiments takes place and actors adjust to changing economic conditions. Beyond that no specific policy measures to combat negative climate impacts on agriculture are considered. The adjustment processes taking place in the different scenarios are the outcome of the imposed yield changes triggering changes in national production levels and costs, leading to changes of agricultural prices in the international and national markets; this in turn affects investment allocation and labour migration between sectors as well as reallocation of resources within agriculture.

Time is an important aspect in this assessment: the yield changes due to climate change start occurring in 1990, gradually reaching their full impact in 2060.

---

5 Alternative schemes for introducing yield changes are conceivable, e.g., with an additive term in the response functions rather than a multiplier. More empirical knowledge with regard to the effect of climate-induced yield changes on marginal productivity is needed to select the most appropriate implementation.

6 An assessment of agro-ecological zones (AEZ) under altered climatic conditions, as currently being conducted for several countries, could provide such information. The BLS is equipped to handle explicit area constraints in the resource allocation module of the agricultural production component.

What distortion would an exogenous change in agricultural productivity imply for the world food system? This distortion has been termed a *static* climate change yield impact, as it measures the hypothetical effect of yield changes, without adjustments by the economic system over time. It refers to a state of the system that is *not* in equilibrium, in the sense of meeting commodity balances. As such it is only of theoretical interest, but helps understanding and quantifying the nature and magnitude of adjustments taking place due to changing economic conditions.

To obtain for any particular year τ an estimate of the static climate change yield impact, $\Lambda(\tau)$, for any particular scenario, we apply the exogenously provided crop-wise percentage yield changes, $\lambda_i^j(\tau)$ to the yield and production levels as observed in a BLS reference projection in year τ. For cereals these impacts can be added up without weighting. To arrive at static impact estimates for other groups of crops and the entire sector, world market prices of year τ as simulated in the respective reference projection are used. In mathematical notation,

$$\Lambda_R(\tau) = \{\sum_{j\in R}\sum_{i\in C} P_{i\tau}^W \cdot Q_{i\tau}^j \cdot \lambda_i^j(\tau)\} / \{\sum_{j\in R}\sum_{i\in C} P_{i\tau}^W \cdot Q_{i\tau}^j\}$$

where

$\Lambda_R(\tau)$	static climate change yield impact on region R in year τ.
$\lambda_i^j(\tau)$	static climate change yield impact for crop i, in country j, in year τ.
$P_{i\tau}^W$	world market price of commodity i in year τ of BLS reference projection.
$Q_{i\tau}^j$	production of commodity i, in country j, in year τ of BLS reference projection.

Table 9 shows static climate change yield impacts estimated for the global and regional level. The estimates of static climate change yield impacts, without assuming direct physiological effects of 555 ppm CO_2 on crop growth and yields, represent a fairly pessimistic outlook, with decreases in crop productivity on the order of 20 to 30 per cent. It may be noted that such an assumption is not regarded as very probable.

When direct physiological effects of CO_2 on yields are included, the magnitude and even direction of the aggregate static impact at world level varies between GCM climate scenarios and with assumptions regarding farm-level adaptation. In all cases, the most negative effects are obtained in scenarios using the UKMO climate change estimates. Scenarios derived from GISS estimates show only small negative effects or even gains at the global level.

Table 9 Climate Change Yield Impacts in 2060 (% change)

	GISS			GFDL			UKMO		
	Cereals	**Other Crops**	**All Crops**	**Cereals**	**Other Crops**	**All Crops**	**Cereals**	**Other Crops**	**All Crops**
World									
Without phys. effect of CO_2	-22.1	-21.8	-22.4	-25.4	-24.4	-25.0	-33.6	-33.0	-33.5
With phys. effect of CO_2	-5.1	2.8	-0.1	-9.0	0.3	-2.8	-18.2	-8.9	-12.2
Adaptation Level 1	-1.7	2.8	0.9	-5.5	0.3	-1.7	-12.9	-8.3	-10.1
Adaptation Level 2	1.4	4.6	3.2	-1.1	2.3	1.0	-6.1	-3.3	-4.4
Developed									
Without phys. effect of CO_2	-13.9	-6.1	-10.3	-21.3	-15.3	-18.6	-30.4	-27.1	-28.9
With phys. effect of CO_2	2.6	18.6	10.6	-5.1	9.6	2.1	-15.8	-3.2	-9.8
Adaptation Level 1	7.8	18.6	13.1	0.1	9.9	5.0	-6.7	-0.1	-3.6
Adaptation Level 2	7.8	18.7	13.1	3.3	9.9	6.4	-2.8	1.4	-0.8
Developing									
Without phys. effect of CO_2	-28.5	-25.3	-26.5	-28.6	-26.4	-27.1	-36.2	-34.3	-35.1
With phys. effect of CO_2	-11.2	-0.7	-3.7	-12.0	-1.8	-4.5	-20.1	-10.2	-13.0
Adaptation Level 1	-9.2	-0.7	-3.2	-10.0	-1.8	-3.9	-17.8	-10.1	-12.3
Adaptation Level 2	-3.6	1.4	-0.1	-4.5	0.6	-0.8	-8.7	-4.3	-5.6

Note: See Appendix for region names.

Table 10 Static climate change yield impacts with Adaptation Level 1 in 2060 (%).

	GISS			GFDL			UKMO			LE		
	Cereals	Other crops	All crops	Cereals	Other crops	All crops	Cereals	Other crops	All crops	Cereals	Other crops	All crops
World	-1.7	2.8	0.9	-5.5	0.3	-1.7	-12.9	-8.3	-10.1	5.7	11.9	9.6
Developed	7.8	18.6	13.1	0.1	9.9	5.0	-6.7	-0.1	-3.6	11.5	19.5	15.6
NAM	2.7	12.6	5.9	-3.8	3.6	-0.7	-10.8	-9.8	-10.1	6.6	22.0	12.1
WEU + ODE	6.2	13.5	10.3	4.3	9.6	7.1	2.7	7.3	5.1	18.7	22.2	20.5
EEU + FSU	17.7	25.9	22.8	2.6	13.5	8.6	-8.3	-0.3	-4.0	14.0	16.9	15.9
PAO	8.3	13.9	11.0	8.5	10.5	9.1	7.4	8.0	7.4	17.6	17.2	17.0
Developing	-9.2	-0.7	-3.2	-10.0	-1.8	-3.9	-17.8	-10.1	-12.3	1.1	10.3	7.6
AFR	-20.6	0.8	-3.0	-24.0	-4.7	-8.1	-25.6	-7.0	-10.3	-4.7	7.4	5.4
LAM	-16.7	-6.1	-8.7	-14.5	0.2	-3.2	-22.7	-15.8	-17.7	3.1	16.6	13.8
WAS	-12.2	-4.6	-6.5	-17.4	-9.2	-11.1	-22.5	-15.0	-16.9	8.8	13.5	12.3
SAS	-9.8	-4.0	-6.7	-10.7	-6.6	-8.4	-28.8	-26.1	-26.6	1.8	7.8	5.0
CPA	3.3	9.4	6.9	1.4	7.2	5.0	-0.8	4.7	2.6	1.9	13.2	9.7
PAS	-14.9	-11.1	-11.4	-5.6	-3.7	-3.2	-14.8	-5.1	-8.1	6.1	-3.2	-3.6

Note: See Appendix for region names.

However, as demonstrated by Table 9, the impacts are geographically quite unevenly distributed. At the aggregate level, developed countries experience, in all but the UKMO scenarios, an increase in productivity. In contrast, developing regions suffer a loss in productivity in all estimates presented here, except for the low-end scenario estimates which fully take into account CO_2 physiological effects but assume only modest climate sensitivity to increased atmospheric greenhouse gas concentrations. Table 10 shows continental level results of scenarios assuming direct physiological effects of increased atmospheric CO_2 concentrations and, where applicable, some farm-level adaptation (Adaptation Level 1). Under GISS and GFDL climate projections, crop productivity in developed regions benefits substantially, especially in the former Soviet Union (EEU+FSU) and in Pacific OECD countries (PAO) which include Australia, Japan and New Zealand. Impacts on developing regions are all negative, except for the group of Centrally Planned Asia (CPA) including China. Under the GISS and UKMO scenarios countries in Central and South America are most affected; the GFDL estimates are worst for West Asia (WAS), South Asia (SAS), and Africa (AFR). Only the UKMO scenario results in major negative impacts for both cereals and other crops. Static impacts derived for the low-end climate scenario are mostly positive, amounting to an increase of some 10 per cent globally. Note, however, that the increase in developed regions is estimated to be more than twice the gain in developing countries.

The above calculations discuss an effect that would result if climate-induced yield changes were to occur without economic adjustment, overnight so to say. In the scenario assumptions, however, yield productivity changes are introduced gradually from 1990 to 2060 to reach their full impact only after a 70 year period. In scenarios with shortfalls in food production caused by climate change yield impacts, market imbalances push international prices upwards and provide incentives for reallocation of capital and human resources. At the same time, consumers react to price changes and adjust their patterns of consumption. Let us first discuss the simulated changes on the world market.

Table 11 contains changes in world market prices for cereals and an overall crop price index as simulated in the respective climate change yield impact scenarios relative to the standard reference projection (REF-M). When direct physiological effects of CO_2 on plant growth and yields are not included, major increases in world market prices result – four to ninefold

Table 11 Change in world market prices in 2060 (per cent)

	Cereals				All Crops			
	GISS	GFDL	UKMO	LE	GISS	GFDL	UKMO	LE
Without CO_2	306	356	818	81	234	270	592	70
With CO_2	24	33	145	-21	8	17	90	-25
Adaptation Level 1	13	22	98	N.S.	2	10	67	N.S.
Adaptation Level 2	-4	2	36	N.S.	-8	-3	25	N.S.

Note: N.S. – not simulated

increases of cereal prices depending on GCM scenario. Apart from the scientific evidence of the beneficial physiological effects of elevated CO_2 levels, we note that such increases would call for strong public reactions and policy measures to mitigate the negative yield impacts. Hence, the outcome for scenarios without the physiological effects of CO_2 on yields do not seem convincing. They are mentioned here for the sake of completeness.

When physiological effects of CO_2 on yields are included in the assessment, then cereal prices increase in the order of 24 to 145 per cent relative to the standard reference projection. The index of crop prices increases by 8 to 90 per cent, depending on the climate change scenario. Changes caused by the GISS and GFDL scenarios are rather modest, an increase of about 25 to 33 per cent in cereal prices, and less than 20 per cent increase in overall crop prices. Only the yield impacts derived under a climate change as projected by the UKMO scenario produce large agricultural price increases. On the other hand, under a low-end climate scenario, where impacts are dominated by the positive physiological effects of CO_2, major price decreases occur.

Price changes are further reduced when farm-level adaptation is considered. The crop price index rises less than 10 per cent in both GISS and GFDL simulation runs. The UKMO projection, however, still produces a two-thirds increase in crop prices. With adaptation measures involving major changes in agricultural practices, adaptation level 2, prices would fall below the reference run levels in the GISS and GFDL scenarios. Note that the assumptions underlying adaptation level 2, requiring some major investments, are hardly consistent with such an economic development, and the stipulated adaptations would often not be economically viable. In such cases low-cost adaptation measures, adaptation level 1, appear to be more relevant.

Looking at the results in the final year of the simulations, 2060, may not tell the full story. To capture potential impacts in earlier decades we have also undertaken simulations based on the GISS Transient scenario A. This set of climate experiments is not a dynamic simulation using transient CO_2 concentration levels, but consists of separate GCM runs with increased levels of atmospheric CO_2. The CO_2 concentrations in this GISS experiment were 405 ppm, 460 ppm and 530 ppm, respectively in 2010, 2030 and 2050. Crop modelling experiments were conducted separately for these three time slices. Yield impacts incorporated in the yield response functions of the BLS were built up gradually as a function of time between 1990-2010, 2010-2030 and 2030-2050, reaching the modelled impacts in 2010, 2030 and 2050 respectively. Beyond 2050 the yield change multiplier was extrapolated extending the trend of the period 2030-2050 to 2060. Table 12 summarises the price changes relative to the standard reference scenario in 2010, 2030 and 2050. Note that in the GISS Transient run positive crop impacts dominate for about half the simulation period. The crop price index falls initially (around year 2010) as much as 10 per cent below the reference level; then, as negative impacts in developing countries increase and beneficial impacts in developed countries flatten, prices return to and exceed the price index of the reference scenario.

The BLS reference projections result in different production levels and agricultural prices. It is of interest to compare the sensitivity of relative prices under different development paths tothe magnitude of climate change-induced price distortions. Table 13 shows per cent changes in overall crop and cereal prices for different reference projections relative to

Table 12 Change in world market prices in GISS Transient Scenario A (per cent)

	Cereals			All Crops		
	2010	**2030**	**2050**	**2010**	**2030**	**2050**
With CO_2	-0	13	15	-6	6	6
Adaptation level 1	-5	8	9	-8	4	3
Adaptation level 2	-11	-2	-5	-12	-3	-6

Table 13 Change in world market prices under different reference projections (per cent)

	Cereals		All Crops	
	2020	**2060**	**2020**	**2060**
REF-L	7	13	8	10
REF-MA	24	37	18	30
REF-MF	18	32	14	23
REF-H	-3	-8	-3	-4
REF-P	-21	-24	-18	-18

the BLS standard reference projection scenario (REF-M). Deviations of cereal prices range from -24 to 37 per cent, and from -18 to 30 per cent for a price index of all crops. Note that lower population growth (REF- MP) or faster economic development (REF-H) are projected to lower agricultural prices whereas slower economic growth (REF-L) and additional constraints on agricultural production (REF-MA and REF-MF) would increase crop prices. Furthermore, these negative and positive price effects are larger than the range of price changes derived under GISS and GFDL simulation runs when physiological effects of CO_2 are considered (as in Table 11).

Tables 14 and 15 highlight the estimated *dynamic* impacts of climate change on agriculture resulting after 70 years of simulation with economic adjustment. According to these calculations, and with direct physiological effects of 555 ppm CO_2 on crop yields, the impact on global agriculture GDP would be between -2 and +1 per cent in all but the UKMO scenarios where it ranges between -2 and -6 per cent. Developed countries are likely to experience some increase in agricultural output. On the other hand, developing countries are projected to suffer a production loss in all of the scenarios. Table 16 lists the simulated regional impacts considering physiological effects of CO_2 on crop growth and some farm-level adaptation (Adaptation Level 1). It also includes results from the low-end scenario. Among developed regions, simulated positive impacts on agriculture output are largest for Europe (WEU+ODE), former Soviet Union (EEU+FSU) and Pacific OECD (PAO). Dynamic impacts in developing regions are mostly negative except for Centrally Planned Asia (CPA) which benefits in all these scenarios.

Table 14 Impact of climate change with economic adjustment in 2060 (per cent)

	Cereals Production % change			Crop Production % change			GDP Agriculture % change		
	GISS	**GFDL**	**UKMO**	**GISS**	**GFDL**	**UKMO**	**GISS**	**GFDL**	**UKMO**
World									
Without phys. effect of CO_2	-10.9	-12.1	-19.6	-11.5	-12.8	-18.0	-10.2	-11.7	-16.4
With phys. effect of CO_2	-1.2	-2.8	-7.6	-0.5	-1.7	-6.4	-0.4	-1.8	-5.4
Adaptation Level 1	0.0	-1.6	-5.2	0.2	-1.0	-5.0	0.2	-1.2	-4.4
Adaptation Level 2	1.1	-0.1	-2.4	1.1	0.2	-2.3	1.0	0.0	-2.0
Developed									
Without phys. effect of CO_2	-3.9	-10.1	-23.9	3.8	-5.5	-12.7	1.1	-6.2	-12.5
With phys. effect of CO_2	11.3	5.2	-3.6	15.6	7.6	-0.9	11.6	5.1	-1.9
Adaptation Level 1	14.2	7.9	3.8	17.6	9.1	4.0	13.3	6.5	1.8
Adaptation Level 2	11.0	3.0	1.8	15.1	8.6	2.2	11.8	6.5	1.3
Developing									
Without phys. effect of CO_2	-16.2	-13.7	-16.3	-16.6	-12.8	-19.8	-13.9	-13.5	-17.7
With phys. effect of CO_2	-11.0	-9.2	-10.9	-5.8	-4.9	-8.2	-4.4	-4.0	-6.6
Adaptation Level 1	-11.2	-9.2	-12.5	-5.6	-4.4	-8.1	-4.1	-3.7	-6.4
Adaptation Level 2	-6.6	-5.6	-5.8	-3.6	-2.7	-3.9	-2.6	-2.2	-3.1

Table 15 Dynamic impact of climate change with Adaptation Level 1 in 2060 (per cent changes)

	Cereal Production				Crop Production				GDP Agriculture			
	GISS	GFDL	UKMO	LE	GISS	GFDL	UKMO	LE	GISS	GFDL	UKMO	LE
World	0.0	-1.7	-5.3	2.5	0.2	-1.0	-5.0	3.4	0.2	-1.2	-4.4	2.5
Developed	14.2	7.9	3.8	10.6	17.6	9.1	4.0	11.0	13.3	6.5	1.8	8.7
NAM	10.6	5.9	-5.2	4.1	9.3	4.8	-3.2	2.0	7.5	3.2	-3.0	0.8
WEU + ODE	6.5	7.7	12.2	14.7	10.7	6.7	11.7	14.5	8.0	5.2	7.9	10.9
EEU + FSU	24.6	7.6	6.0	19.9	30.7	12.7	-1.3	20.1	26.8	11.2	-2.7	17.3
PAO	19.6	31.7	53.2	4.2	16.3	24.0	52.0	-1.1	4.0	5.5	12.1	0.2
Developing	-11.2	-9.2	-12.5	-3.9	-5.6	-4.4	-8.1	0.9	-4.1	-3.7	-6.4	0.5
AFR	-23.7	-24.5	-16.2	-8.2	-4.1	-9.3	1.4	-3.3	-2.2	-7.6	1.4	-2.8
LAM	-25.0	-17.8	-14.5	-10.3	-13.8	-2.6	-11.1	4.0	-10.9	-3.2	-7.9	1.3
WAS	-13.6	-17.0	-18.6	2.8	-7.9	-11.5	-12.4	6.3	-5.6	-8.7	-9.5	5.8
SAS	-11.9	-8.5	-26.8	-3.5	-7.8	-7.2	-25.1	2.0	-6.2	-5.2	-20.0	2.2
CPA	4.1	2.6	2.1	1.6	3.8	2.5	1.7	3.0	3.3	2.1	1.4	1.9
PAS	-12.3	0.3	-1.6	-12.8	-14.4	-2.3	-2.3	-11.3	-12.8	-3.1	-3.6	-9.1

Note: See Appendix for region names

Table 16 Changes in production under different reference projections in 2060 (per cent)

	Cereal Production				Crop Production				GDP Agriculture			
	REF-L	REF-H	REF-MA	REF-MF	REF-L	REF-H	REF-MA	REF-M	REF-L	REF-H	REF-MA	REF-MF
World	-2.2	8.8	-2.7	-2.6	-2.3	4.9	-2.4	-1.9	-2.4	3.7	-1.7	-1.3
Developed	-1.5	1.7	-3.8	-6.7	-1.2	1.2	-3.9	-5.7	-0.8	0.8	-3.1	-3.1
NAM	-2.0	1.0	-10.8	-11.5	-1.0	0.2	-9.1	-9.1	-2.1	0.9	-6.9	-5.7
WEU + ODE	0.5	-0.8	5.2	-2.9	4.1	-1.7	2.7	-1.5	5.5	-2.6	1.5	0.3
EEU + FSU	-3.7	5.5	1.8	2.2	-4.4	5.2	-4.5	-3.0	-4.3	5.4	-5.0	-3.6
PAO	9.6	-4.2	-0.8	-27.3	-2.8	-6.4	7.9	-15.6	-3.4	-4.3	0.8	-6.0
Developing	-2.7	14.5	-1.8	0.6	-2.7	6.1	-1.8	-0.6	-2.9	4.7	-1.2	-0.7
AFR	-4.2	3.4	-7.0	3.3	-1.8	4.0	-1.8	-2.1	-1.9	4.0	-0.5	-2.3
LAM	0.0	-0.3	1.8	5.6	-2.4	2.8	1.2	3.8	-3.7	4.0	1.8	2.7
WAS	-4.1	0.3	-3.8	5.2	-0.9	1.3	1.6	2.9	-2.0	2.6	1.9	2.7
SAS	-4.8	12.3	1.9	-0.2	-4.3	10.7	-1.2	0.2	-3.7	8.5	-1.6	-0.3
CPA	-2.3	37.3	-4.5	-4.7	-2.8	8.0	-5.6	-4.4	-2.5	2.9	-4.0	-2.9
PAS	1.7	0.7	1.9	3.3	-2.5	3.6	-1.8	0.2	-3.5	3.8	-2.1	-0.2

Note: See Appendix for region names

It is important to note that these changes in comparative advantage between developed and developing regions are likely to accentuate the magnitude of the impacts suggested by the agronomic assessment, i.e., winners are likely to gain more, losers lose even more than projected without economic adjustment. In fact, we can distinguish two prototypical situations in these scenario results. First, when global supply is only marginally affected and hence with only little impact on prices, then the shift in relative productivity from developing to developed regions dominates the adjustment process. For instance, in the GISS and GFDL scenarios with farm-level adaptation, agriculture production shifts somewhat from developing to developed countries taking account of the differences in projected yield changes. Second, only in the UKMO scenario the supply gap is so substantial that massive price increases result; these in turn provide production incentives to both regions to recover more than half of the production forgone due to climate change (according to the static crop model estimates).

Net imports of cereals into developing countries increase under all scenarios. The change in cereal imports, relative to the standard reference projection, is largely determined by the magnitude of the estimated yield changes, the change in relative productivity in developing and developed regions, the change in world market prices and changes in real incomes of consumers in developing countries. For example, under the GISS climate change scenarios, productivity is altered in favour of developed countries with relatively small changes in incomes and prices, resulting in pronounced increases of net cereal imports into developing countries.

With less agricultural production in developing countries and higher prices on international markets the estimated number of people at risk of food poverty is likely to increase. This occurs in all but two scenarios. The largest increase is to be expected from the UKMO scenario without physiological effects of CO_2, the smallest change, a decline of 2 per cent, occurred in the GISS scenario considering physiological effects of increased CO_2 and adaptation level 2. Finally, in the low-end scenario the estimated number of people at risk of food poverty decreases by more than 10 per cent. Table 17 summarises these estimates.

Table 17 Impact of climate change on people at risk of food poverty in 2060

	million people				% change			
Developing*	GISS	GFDL	UKMO	LE	GISS	GFDL	UKMO	LE
Without CO_2	721	801	1446	265	112	125	225	41
With CO_2	63	108	369	-84	10	17	58	-13
Adaptation Level 1	38	87	300	N.S.	6	14	47	N.S.
Adaptation Level 2	-12	18	119	N.S.	-2	3	19	N.S.

Note: * excludes China; N.S. - not simulated

9 Conclusions

The impact of climate change on agriculture and the world food system has been evaluated with a system of linked national models and several scenarios of climate-induced yield changes have been analysed in this paper. It must be realised that the ability to estimate climate change yield impacts on world food supply, demand and trade is surrounded by large uncertainties regarding important elements, such as the magnitude and spatial characteristics of climate change, the range and efficiency of adaptation possibilities, the long term aspects of technological change and agricultural productivity, and even future demographic trends.

Despite progress made in enhancing the spatial resolution of GCMs and increasing the complexity of these models through coupling of atmosphere and ocean modules, the resulting climate projections are still considered fairly uncertain, especially with regard to changes in regional precipitation patterns. Transient climate projections depend on trajectories of greenhouse gases in the atmosphere which in turn depend on future levels of emissions. Our understanding of the many interacting processes affecting biogeochemical cycles is still incomplete. Further, the levels and kinds of future human activities are uncertain. Long-term projections of national economies are highly speculative. This translates into uncertainties both of future demand and supply of agricultural products as well as uncertainties regarding prospects for future technological development and capital investment, both key ingredients of productivity in agriculture. Finally, uncertainty concerning the use of crop models stems from the paucity of data and validation for tropical regions and crops; the lack of effects of pests and diseases; and inadequate representation of the physiological effects of CO_2.

In spite of these caveats, there appear to be a number of robust conclusions emerging from the study. The effects of changes in climate on crop yields are likely to vary greatly from region to region across the globe. The results of these scenarios indicate that the effects on crop yields in mid and high latitude regions appear to be positive or less adverse than those in low latitude regions, provided the potentially beneficial direct physiological effects of CO_2 on crop growth can be realised. At an aggregate level, given the range of climate scenarios used in the study, world food supply does not appear to be seriously threatened by global climate change in the period to the middle of the next century. In the near-term, positive effects from increased atmospheric CO_2 levels are even likely to enhance agricultural productivity.

Globally, the simulated effects of climate change are small compared to the production increases required to feed a doubling of world population. Even moderate variations in scenario assumptions about future economic growth and policies generate a wider range of outcomes than those resulting from simulated climate change impacts on agriculture. Yet, the ability of the world food system to absorb negative yield impacts decreases with the magnitude of the impact. Economic adaptation can largely compensate for moderate yield changes as in the GISS and GFDL scenarios but not changes such as in the UKMO scenario.

In all climate change yield impact scenarios productivity of agriculture shifts in favour of developed countries. Economic feedback mechanisms are likely to emphasise and accentuate the uneven distribution of climate change impacts across the world, resulting in net gains for developed countries in all but the UKMO scenarios and a noticeable loss to developing countries. As a result, net imports of cereals into developing countries increase in all climate change yield impact scenarios compared to trade in the reference scenario.

Including direct physiological effects of CO_2 on crop yields, world cereal production is estimated to decrease between 1 and 7 per cent, depending on the GCM climate scenario. Assuming adaptation to climate change at farm-level, cereal production would still be reduced by up to 5 per cent. The largest negative changes would occur in developing countries, averaging around -10 per cent.

This loss of production in developing countries in the majority of simulation runs, together with rising agricultural prices, is likely to increase the number of people at risk of hunger in the order of 5 to 50 per cent, depending on the GCM scenario. Again, these results fall well within the range of outcomes generated without considering climate change by varying basic scenario assumptions such as the rate of growth of the non-agricultural sector.

It must be noted that the indicator of food poverty in the BLS essentially compares food energy consumed to estimated requirements. This assumes that no major changes in income distribution will occur, except in some of the national models (e.g. India) where class-specific income and consumption are included. Nevertheless, we may conclude that both rural and urban poor are likely to be negatively affected by the impacts of climate change. Rural poor will find it more difficult to secure sufficient access to resources under reduced land productivity projected for many developing countries relative to the reference simulations. Poor urban consumers will be affected by reduced domestic food production as well as higher world market prices, both of which under simulated conditions translate into higher food prices to consumers. Although these conclusions appear to be fairly robust it must be recognised that there is much controversy about how to adequately define food poverty; further, many studies have shown that well-targeted programmes could be relatively inexpensive and effective in combating food poverty in both rural and urban environments.

From a political and social standpoint, the results of the study indicate the potential for a decrease in food security in developing countries. The analysis suggests that the worst situation would arise from a scenario of severe climate change, low economic growth, continuing large population increases, and little farm-level adaptation. In order to minimise possible adverse consequences, the way forward is to encourage the agricultural sector to continue developing crop breeding and management programmes for heat and drought conditions, in combination with measures taken to preserve the environment, to more efficiently use resources, and to slow the growth of the human population. The latter step would also be consistent with efforts to slow emissions of greenhouse gases, and thus the rate and eventual magnitude of global climate change.

Acknowledgements

This paper is a revised and extended version that first appeared in *Global Environmental Change* Volume 4, G. Fischer, K. Frohberg, M.L. Parry' and C. Rosenzweig, Climate change and world food supply, demand and trade: who benefits, who loses? pp7-23. Reprinted with permission from Elsevier Science Ltd, Butterworth-Heinemann Imprint, The Boulevard, Landford Lane, Kidlington OX5 1GB, U.K.

Appendix

Code	Region
NAM	North America
WEU + ODE	Western Europe and other developed market economies
EEU + FSU	Eastern Europe and former Soviet Union
PAO	Pacific OECD countries
AFR	Africa
LAM	Latin America
WEA	West Asia
SEA	South and South-east Asia
CPA	Centrally planned Asia
PAS	Pacific Asian countries

References

Adams, R.M., Fleming, R.A., Chang, C., McCarl, B.A. and Rosenzweig, C. (1993). *A Reassessment of the Economic Effects of Global Climate Change on U.S. Agriculture*. Oregon State University, Corvallis.

Adams, R.M., Rosenzweig, C., Peart, R.M., Ritchie, J.T., McCarl, B.A., Glyer, J.D., Curry, R.B., Jones, J.W., Boote, K.J. and Allen, L.H. Jr. (1990). Global climate change in US agriculture. *Nature,* **345**(6272): 219-224.

Allen, L.H., Jr., Boote, K.J., Jones, J.W., Jones, P.H., Valle, R.R., Acock, B., Rogers, H.H., and Dahlman, R.C.(1987). Response of vegetation to rising carbon dioxide: Photosynthesis, biomass and seed yield of soybean. *Global Biogeochemical Cycles* **1**:1-14.

Appendini, K. and Liverman, D. (1994). Agricultural policy, climate change and food security in Mexico. *Food Policy,* **19**(2): 149-164.

Appendini, K. and Liverman, D. (1995). Agricultural policy and climate change in Mexico. In: Downing, T.E. (ed.) *Climate Change and World Food Security*. Springer-Verlag, Heidelberg.

Chen, R.S. and Kates, R.W. (1994). World Food Security: Prospects and Trends. *Food Policy,* **19**(2): 192-208.

Chen and Kates (1995). Towards a Food Secure World: Prospects and Trends. In: Downing, T.E. (ed.) *Climate Change and World Food Security*. Springer-Verlag, Heidelberg.

Cure, J.D. and Acock, B.(1986). Crop responses to carbon dioxide doubling: A literature survey. *Agriculture and Forest Meteorology,* **38**: 127-145.

FAO/UNFPA/IIASA (1983). *Potential Population Supporting Capacity of Lands in the Developing World.* FAO, Rome.

FAO (Food and Agriculture Organisation) (1984). *Fourth World Food Survey*, FAO, Rome.

FAO (Food and Agriculture Organisation) (1987). *Fifth World Food Survey*, FAO, Rome.

FAO (Food and Agriculture Organisation) (1988). *World Agriculture Toward 2000*, FAO, Rome.

FAO (Food and Agriculture Organisation) (1993). *Agriculture: Towards 2010*, FAO, Rome.

FAO (Food and Agriculture Organisation) (1994). *Global Climate Change and Agricultural Production. Highlights from and expert consultation.* FAO, Rome.

Fischer G., Frohberg K., Keyzer M.A., and Parikh K.S. (1988). *Linked National Models: A Tool for International Policy Analysis.* Kluwer Academic Publishers, Netherlands.

Fischer G., Frohberg K., Keyzer M.A., Parikh K.S., and Tims W. (1990). *Hunger - Beyond the Reach of the Invisible Hand.* IIASA, Laxenburg.

Godwin, D., Ritchie, J.T., Singh, U. and Hunt, L.(1989). *A User's Guide to CERES-Wheat - V2.10.* International Fertilizer Development Center, Muscle Shoals.

Godwin, D., Singh, U., Ritchie, J.T. and Alocilja, E.C.(1993). *A User's Guide to CERES - Rice*. International Fertilizer Development Center, Muscle Shoals.

Hansen, J., Russell, G., Rind, D., Stone, P., Lacis, A., Lebedeff, S., Ruedy, R. and Travis, L.(1983). Efficient three-dimensional global models for climate studies: Models I and II. *Monthly Weather Review,* **111**(4): 609-662.

Hansen, J., Fung, I., Lacis, A., Rind, D., Russell, G., Lebedeff, S., Ruedy, R. and Stone, P. (1988). Global climate changes as forecast by the GISS 3-D model. *Journal of Geophysical Research,* **93**(D8): 9341-9364.

Hendry, G.R., Lewis, K.F. and Nagy, J.(1993). Free air carbon dioxide enrichment: development, progress, results. In: J. Rozema, H. Lambers, S.C. van de Geijn, and M.L. Cambridge (eds). *CO_2 and Biosphere*. Kluwer Academic Publishers, Dordrecht, pp. 17-31.

International Benchmark Sites Network for Agrotechnology Transfer (IBSNAT) Project (1989). *Decision Support System for Agrotechnology Transfer Version 2.1 (DSSAT V2.1)*. Department of Agronomy and Soil Science. College of Tropical Agriculture and Human Resources. University of Hawaii, Honolulu, H.I.

Houghton, J.T., Callander, B.A. and Varney, S.K. (eds.) (1992). *Climate Change 1992. The Supplementary Report to the IPCC Scientific Assessment.* Intergovernmental Panel on Climate Change. Cambridge University Press, Cambridge. 200 pp.

Houghton, J.T., Jenkins, G.J. and Ephraums, J.J. (eds.) (1990). *Climate Change: The IPCC Scientific Assessment*. Cambridge University Press. Cambridge.

Houghton, J.T., Meira Filho, L.G., Bruce, J., Lee, H., Callander, B.A., Haites, E., Harris, N. and Maskell, K. (eds.) (1995). *Climate Change 1994. Radiative Forcing of Climate Change and An Evaluation of the IPCC 1992 Emission Scenarios.* Cambridge University Press, Cambridge.

Kellogg, W.W. and Zhao, Z.-C.(1988). Sensitivity of soil moisture to doubling of carbon dioxide in climate model experiments. Part 1. *North America. Journal of Climate,* **1**: 348-366.

Kimball, B.A. (1983). Carbon dixoide and agricultural yield. An assemblage and analysis of 430 prior observations. *Agronomy Journal,* **75**: 779-788.

Jones, C.A. and Kiniry, J.R.(1986). *CERES - Maize: A Simulation Model of Maize Growth and Development*. Texas A&M Press, College Station.

Jones, J.W., Boote, K.J., Hoogenboom, G., Jagtap, S.S. and Wilkerson, G.G.(1989). *SOYGRO V5.42: Soybean Crop Growth Simulation Model. User's Guide*. Department of

Agricultural Engineering and Department of Agronomy. University of Florida, Gainesville.

Liverman, D., Dilley, M., O'Brien, K. and Menchaca, L.(1994). *Possible Impacts of Climate Change on Maize Yields in Mexico. Implications of Climate Change for International Agriculture: Crop Modeling Study.* U.S. Environmental Protection Agency, EPA 230B-94-003, Washington DC.

Manabe, S. and Wetherald, R.T. (1987). Large-scale changes in soil wetness induced by an increase in CO_2. *Journal of Atmospheric Science,* **44:** 1211-1235.

Menzhulin, G.V., Koval, L.A. and Badenko, A.L.(1994). *Potential Effects of Global Warming and Carbon Dioxide on Wheat Production in the Former Soviet Union. Implications of Climate Change for International Agriculture: Crop Modeling Study.* USEPA EPA 230B-94-003, Washington DC.

Otter-Nacke, S., Godwin, D.C. and Richie, J.T.(1986). *Testing and Validating the CERES-Wheat Model in Diverse Environments. AgGRISTARS YM-15-00407.* Johnson Space Center No. 20244, Houston.

Parikh K.S., Fischer, G., Frohberg, K., and Gulbrandsen, O. (1988). *Toward Free Trade in Agriculture.* Martinus Nijhoff, The Hague.

Parry, M.L., Carter, T.R. and Konijn, N.T.(eds.) (1988). *The impact of climate variations on agriculture. Vol. 1 Assessments in cool temperate and cold regions. Vol. 2 Assessments in semi-arid regions.* Kluwer, Dordrecht.

Pearman, G. (1988). *Greenhouse: Planning for Climate Change.* CSIRO, Canberra.

Peart, R.M., Jones, J.W., Curry, R.B., Boote, K. and Allen, L.H. Jr. (1989). Impact of climate change on crop yield in the southeastern U.S.A. In: J.B. Smith and D.A. Tirpak (eds.), *The Potential Effects of Global Climate Change on the United States.* U.S. Environmental Protection Agency, Washington DC.

Ritchie, J.T. and Otter S. (1985). Description and performance of CERES-Wheat: A user-oriented wheat yield model. In: W.O. Willis (ed.) *ARS Wheat Yield Project.* Department of Agriculture, Agricultural Research Service, ARS-38, Washington D.C.

Ritchie, J.T., Singh, U., Godwin, D. and Hunt, L.(1989). *A User's Guide to CERES-Maize - V 2.10.* International Fertilizer Development Center, Muscle Shoals.

Rogers, H.H., Bingham, G.E., Cure, J.D., Smith, J.M. and Surano, K.A.(1983). Responses of selected plant species to elevated carbon dioxide in the field. *Journal of Environmental Quality,* **12**: 569-574.

Smith, J.B. and Tirpak, D.A.(eds.) (1989). *The Potential Effects of Global Climate Change on the United States. Report to Congress.* EPA-230-05-89-050. U.S. Environmental Protection Agency, Washington D.C.

Tegart, W.J. McG., G.W. Sheldon, and D.C. Griffiths (eds.) (1990) *Climate Change: The IPCC Impacts Assessment.* Australian Government Publishing Service, Canberra.

UK Department of the Environment (1991). *The Potential Effects of Climate Change in the United Kingdom.* Climate Change Impacts Review Group. HMSO, London.

UN (1989). *World Population Prospects 1988.* United Nations, New York.

Wilson, C.A. and Mitchell, J.F.B. (1987). A doubled CO_2 climate sensitivity experiment with a global climate model including a simple ocean. *Journal of Geophysical Research,* **92** (13): 315-343.

World Bank (1990). *World Population Projections 1989-90 Edition.* Johns Hopkins University Press, Baltimore.

Climate Change and Agricultural Trade: Who Benefits, Who Loses?

John Reilly and Neil Hohmann
Economic Research Service
U.S. Department of Agriculture
1301 New York Ave, NW
Washington, DC, 20005, U.S.

Sally Kane
Council for Economic Adivsers
317 Old Executive Office Building
Washington, DC, 20235, U.S.

1 Introduction

The net economic effect of climate change on the agriculture of any country will depend as much on its role in agricultural trade as on the impacts of the changed climate on crop yields. To examine the central role of trade, the balance between the gains and losses of agricultural producers and consumers, and the implications for agricultural policy, we employ a model of world food markets under conditions of simulated climate change and crop yields.

Early attempts to evaluate the economic impact of climate change on agriculture included studies by Parry *et al.* (1988), Adams *et al.* (1990), Arthur (1988) and Mooney and Arthur (1990) (Sonka, 1991, provides a complete review). Kane *et al.* (1992), Tobey *et al.* (1992) and Reilly and Hohmann (1993) have emphasised the importance of trade in determining the net economic effect on a country. In this paper, we present economic model simulation results based on the significant effort of Rosenzweig *et al.* (1993), (reported in Fischer *et al.*, this volume). Our specific focus differs from Fischer *et al.* (1994, 1995, this volume) in that we attempt to compute the economic losses or gains for different countries and country groups using the static world policy simulation model (SWOPSIM), which is directly comparable with the earlier work of Kane *et al.* (1992) and Tobey *et al.* (1992). The broader purpose of the paper is to consider what these findings imply for agricultural policy. Two features of the results stand out. (i) Economic impacts of climate change must be considered in the context of global effects recognising the role of international trade. (ii) Despite the considerable advance in understanding, the implications of climate change for world agriculture, and even more so for individual countries or regions, is highly uncertain in both

NATO ASI Series, Vol. I 37
Climate Change and World Food Security
Edited by Thomas E. Downing

magnitude and direction of effect. Considerations of national or international responses that might improve adaptive response to climate or compensate potential losers must be considered in the light of these two features.

Key uncertainties persist, even in the face of considerable progress in climate and agronomic modelling. Limitations in climate forecasts make it difficult to translat forecasted expected changes in the climate system into agronomic and economic impacts. The derived impacts on the agronomic and economic systems are most useful for understanding the climate-agronomic-economic system linkages and broad patterns of consequences. Agronomic uncertainties include, for example, considerations regarding the biophysical response of plants to climate or to carbon dioxide and the interactions of insect and plant pests. Work is continuing to investigate how best to extrapolate from crop studies that model the individual plant response to countries or groups of countries as represented in economic models.

Many questions remain on the socioeconomic adjustment to climate and agronomic changes. The response of farmers, consumers, and private and public investment is important in determining the combined adjustment that could take place and thus helps set the bounds on the disruption costs associated with climate change. Questions not yet well addressed include: will individual farmers change planting dates, try new cultivars and crops, and adopt new tillage techniques readily as climate changes? Or will they have difficulty detecting climate change and finding successful adaptations against the noisy background of climate variability? Will public and private sector investments (agricultural research and development, irrigation facilities, transportation infrastructure) respond effectively to changing opportunities?[1]

With all the scientific and methodological questions mentioned above, debate ensues about the usefulness of impact studies. It is, in our view, critical to translate climate impacts into impacts of more direct relevance to society, such as agricultural production effects, to strengthen policy analysis, and to develop and improve impact estimation methodologies. However, taken as predictions, the value of impact studies can be no better than the climate predictions on which they are based.

---

1 For more discussion on economic methods and models applied to agriculture and climate change, see Mendelsohn and Reilly (1993) and Sonka (1991). For a thorough discussion of climate change analysis, see Kates *et al.* (1985).

2 SWOPSIM model structure

We used SWOPSIM of world food markets to simulate economic effects of climate change. SWOPSIM describes world agricultural markets with supply and demand equations specified by matrices of own and cross-price elasticities.[2] The model contains 20 agricultural commodities, including eight crops, four model/livestock, four dairy product, two protein meal, and two oil product categories. The base year for the model was 1989; it was constructed to treat 33 countries/regions separately. The country coverage is globally comprehensive. Crop coverage includes all grains, oilseeds, sugar crops, and tobacco. The model also treats livestock and protein meal and vegetable oil that are indirectly affected by climate change because directly affected crops are inputs into the production of these commodities. Fruits, vegetables, and root and tuber crops are not included.

The basic model structure is a follows. For each country/region *i* and commodity *j* in the model, constant elasticity demand and supply functions are specified. For country *i* and commodity *j*, quantity demanded (*QD*) and supplied (*QS*) are given as:

$$QDj = d_{ij}\,(1 + sd_{ij}).\ \pi_{k=1}^{n}\, CP_{ik}^{\alpha x} \tag{1}$$

and

$$QSj = s_{ij}\,(1 + ss_{ij}).\ \pi_{k=1}^{n}\, PP_{ik}^{\beta x} \tag{2}$$

where CP_{ik} and PP_{ik} are domestic prices facing consumers and producers of commodity k=1,...,n. For *k=j*, the α_x and β_x are the own, uncompensated demand and supply price elasticities and for *k=j* they are cross-price elasticities. The d_{ij} and s_{ij} are base quantities and the sd_{ij} and ss_{ij} are demand and supply shift parameters. A climate impact scenario involves specifying ss_{ij} for all countries *i* and all commodities *j* affected by climate change. In the simulations reported her, climate is assumed directly to affect only crop production. Secondary products such as livestock, oils and protein meals are affected primarily because crops such as coarse grains and oilseeds are used in producing these secondary products, and secondarily by changes in demand brought on by price changes. Percentage yield changes developed from crop response models are used as a

2 This section is based on Roningen *et al.* (1991).

measure of ss_{ij}. Domestic consumer and producer prices reflect world prices; government interventions in production, PSW_{ij} (domestic producer, export and import subsidies); and government intervention in consumption, CSW_{ij}. Because of these interventions, prices in any one country may differ from the world price and the consumer price for a country may differ from the producer price. The basic price linkage equations are:

$$PP_{ij} = pp_{ij} + w_{ij}\, WP_{ij}^{\gamma ig} + PSW_{ij} + TP_{ij} + NW_{ij} \tag{3}$$

and

$$CP_{ij} = cp_{ij} + PP_{ij}\, CSW_{ij} + PSW_{ij} \tag{4}$$

The w_{ij} are either 0 (non-traded commodities) or 1 (traded commodities) and NW_{ij} are domestic prices (solved within the model). The NW_{ij} are zero when $w_{ij} = 1$. The PP_{ij} and cp_{ij} are constants determined by the base data and the TP_{ij} are trade (export and import) interventions. The γij are price transmission elasticities and reflect additional government interventions that limit the transmission of world price changes to domestic prices.

Market equilibrium is characterised by a set of prices where excess world demand (world demand minus world supply) equals zero for all commodities. A numerical solution is obtained in the model through iteration. Once a solution is obtained, producer and consumer losses (or gains) are measured as the integral of the supply and demand functions from the initial price to the new equilibrium price.[3] Existing government intervention generates revenues (or tax expenditures) and changing commodity production and prices changes these revenues creating a third source of economic loss (or gain). Part of this change results from the price transmission elasticity which implicitly generates government revenue or tax expenditure. Changes in government revenues or tax expenditures are borne by taxpayers who are also consumers and/or producers. The final distributional effects of climate change depend, therefore, on how the government revenue/expenditure change is distributed among taxpayers but SWOPSIM does not make any assumptions with regard to how these revenue changes would be distributed.

---

[3] Economic losses (or gains) are measured in terms of Marshallian economic welfare that include changes in consumer surplus, producer surplus, and government payments. Consequently, a change in consumer or producer surplus is the amount of money that the consumer or producer would have to be paid (or have taken away) to be as well off as before the change occurred. (see Tobey *et al.*, 1992, and Roningen *et al.*, 1991).

Some particular caveats apply to this model structure. SWOPSIM is a static, partial-equilibrium model that does not capture agricultural interactions with other economic sectors. Nor does it explicitly capture the cost of adjustments. The partial-equilibrium framework is likely to capture adequately the consequences of climate change for the agricultural sector – work by Kokoski and Smith shows that the climate change effects of fairly large, single-sector impacts, can be adequately measured in a partial-equilibrium setting.[4] As a static model, however, SWOPSIM does not consider changes in climate or the resource base over time. The model essentially evaluates the counter factual hypothesis: what would have been the case had the climate been different in 1989? Because the parameters of the SWOPSIM model are set to represent medium-term to long-term supply and demand relationships, the results should not be interpreted as if climate suddenly changed in 1989. The counter factual setting might more reasonably be described as one where the climate changes over the 1970s and producers and consumers then have about a decade to adjust to the changes that occur.

3 Basic results based on general circulation model (GCM) simulations

For purposes of simulating the economic effects of climate change on world agriculture, the yield effects estimated be Rosenzweig *et al.* (1993), were used as inputs into the SWOPSIM agricultural trade model. For purposes of our simulations, we did not use the more extensive adaptations, labelled level 2, because at least some of these responses appeared to be costly and were inconsistent with the implicit assumption in our work that the supply losses did not involve costly adaptations. In fact, the level 2 adaptations were mixed: for example, large shifts in planting dates seem minor adjustments when one considers that the shift would occur over the course of 100 years (a one-month shift is only about eight hours per year). Similarly, the change from winter to spring wheat does not appear to be a difficult or costly change given that farmers shift among crops from season to season without particular difficulty. The most costly adaptation among those in level 2, however, is irrigation which may be further limited by water supply, requiring substantial changes in investments in inputs, and therefore exceeding assumptions embedded in shifts in supply.

---

4 This may hold less well in developing countries where agriculture is a larger share of the economy.

The principal results of the simulations are:

- For the three general circulation models (GCMs), net annual economic change in world agriculture without adaptation strategies and without (with) the positive effects of enhanced fertilisation from increased carbon dioxide in the atmosphere in billions of 1989 US dollars were: the Goddard Institute for Space Studies (GISS) model, -\$115.5 (-\$0.1); the Geophysical Fluid Dynamics Laboratory (GFDL) model, -\$148.6, (-\$17.0); the United Kingdom Meteorological Office (UKMO) high-resolution model, -\$248.1 (-\$61.2). That is, under the GISS climate the positive effects of CO_2 fertilisation offset all but \$0.1 billion of the losses due to climate change alone. Without the carbon dioxide fertilisation effect, losses are substantial. These global losses are from 0.5 to 1.5 per cent of global gross national product (GNP) in 1989.[5]
- For the three GCMs, the net annual global economic changes with the CO_2 fertilisation were: GISS, +\$7.0; GFDL, -\$6.1; UKMO, -\$37.6. That is, the adaptations considered further reduced losses by \$7 to \$25 billion. Note that for the GISS case there were net global benefits. In comparison, carbon dioxide fertilisation reduced losses on the order of \$115 to \$190 billion.
- Even under the GISS climate with CO_2 fertilisation and adaptation where the net annual economic effect for the world is positive, all three developing country income class groups suffered economic losses (Table 1). But, individual developing countries experienced economic impacts different from those indicated by these aggregate results. For example, China, included among the less than \$500 per capita countries, experienced small losses or net economic gains in most scenarios reported in Table 1 (See Figures 2-4).
- The commodity price changes are closely linked to economic losses or gains (Table 2). In the cases with CO_2 fertilisation and adaptation, prices generally fall for wheat, soybeans, cotton, and tobacco under the GISS and GFDL climates but rise for rice, sugar and maize. Under the UKMO scenario, prices rise for most commodities. Without CO_2 fertilisation and adaptation

---

5 Economic effects are compared with GNP because GNP represents total money income and provides an indication of the agronomic and economy-wide ability to compensate for economic losses. GNP (which is sometimes used as an indicator of economic well-being) is not equivalent to welfare as measured in SWOPSIM. This means, for example, that welfare losses as measured by SWOPSIM are not bounded by agriculture's share of GNP or even by total GNP. Quite obviously if no food were produced in the world then all people would starve and the welfare losses would include both the monetary value of the agronomic and economy plus non-monetary welfare. It is somewhat non-sensical to speculate about the total value of the welfare loss in such a case – clearly it would be much larger than GNP. We gratefully acknowledge comments from Klaus Frohberg and Pierre Crosson urging clarification on this point.

the price rises as much as 620 per cent (rice, UKMO). Crops generally grown in warmer regions (e.g. rice, sugar cane) show the largest increase.

The above general results are similar to the Rosenzweig *et al.* (1993) study conclusions and flow directly from the yield estimates. The contribution of the SWOPSIM results are to provide the net economic effects for individual countries and by producer and consumer groups within countries. As we have argued elsewhere (Tobey *et al.*, 1992), the net economic effect on a country depends jointly on the country's status as a net exporter or importer and whether the yield change was negative or positive.

Table 1 Welfare effects by country group (millions of 1989 US dollars)

	With CO_2 and adaptation			**With CO_2 no adaptation**			**No CO_2 no adaptation**		
Region	**GISS**	**GFDL**	**UKMO**	**GISS**	**GFDL**	**UKMO**	**GISS**	**GFDL**	**UKMO**
Developing	-210	-2573	-14588	-2070	-5322	-19827	-56692	-66110	-121083
\$500-\$2000/ capita	-429	-2927	-10669	-1797	-5135	-15010	-26171	-27839	-48095
>\$2000/capita	-603	-534	-1021	-818	-878	-328	-6661	-4351	-3870
E. Europe/ USSR	2423	-125	-4875	1885	-2048	-10959	-12494	-28854	-57471
OECD	5822	25	-6470	2674	-3644	-15101	-13453	-21485	-17606
Total	7003	-6135	-37623	-126	-17028	-61225	-115471	-148640	-248124

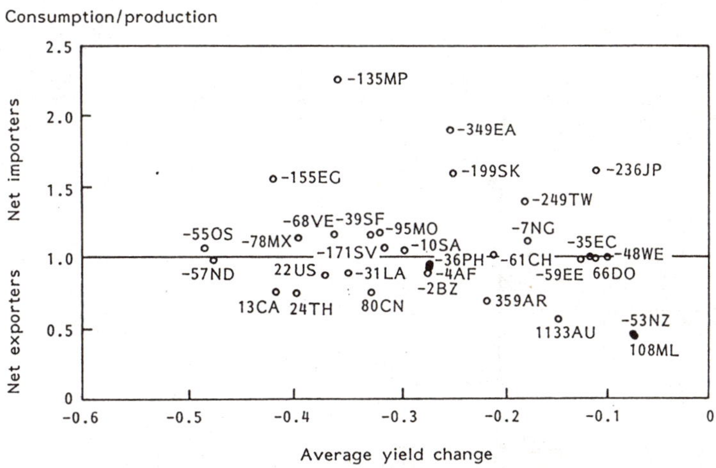

Figure 1 Per capita welfare change: UKMO, no CO_2 fertilisation, no adaptation (1989 \$ per capita)

The specific relationship suggested by Tobey *et al.* (1992) was that exporting countries could gain if world prices rose. They observed that in some cases exporting countries could gain if they suffered yield losses because producer gains were substantial enough to make up for the country's consumer losses from higher prices. The opposite was observed if global climate change generally enhanced production worldwide and world prices fell. In such a circumstance, agricultural exporting countries could suffer economic losses even if agricultural yields improved in the country. The basic economic reasoning for this result is that agricultural demand is generally

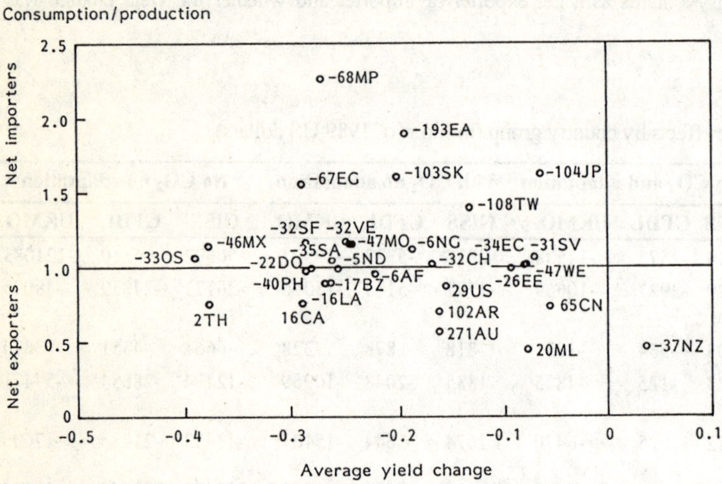

Figure 2 Per capita welfare change: GISS, no CO_2 fertilisation, no adaptation (1989 $per capita)

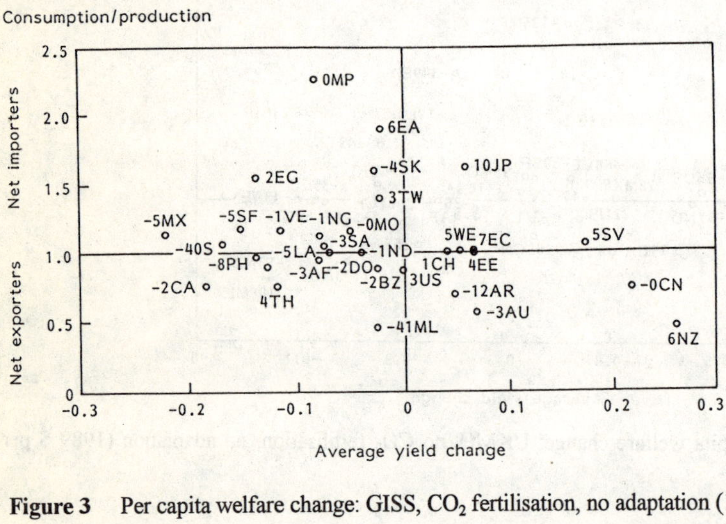

Figure 3 Per capita welfare change: GISS, CO_2 fertilisation, no adaptation (1989 $per capita)

observed to be insensitive to price changes. Thus, production restrictions will generally increase revenues to producers. When climate change is deleterious, it acts as a production restriction. In a closed economy (as is the world as a whole), the net effect of such reductions will be negative because consumer losses will outweigh producer benefits. But, as some countries are large exporters while others are large importers, individual country effect can be positive.

Figures 1-4 plot the average yield change for a country against the ratio of consumption for several of the cases investigated. (See Table 3 for a definition of country codes). As such, these figures display the SWOPSIM results in a way that test the relationship suggested by Tobey *et al.* (1992).

Table 2 Percentage price change from base resulting from climate change

Beef	-0.39	0.98	2068	0.74	2.19	4.82	5.30	7.17	10.30
Pork	-1.75	2.79	9.27	1.38	6.62	16.33	19.31	25.98	37.38
Lamb	-0.51	-0.02	-0.33	-0.14	0.14	0.41	-1.21	-0.17	0.96
Poultry meat	-1.52	2.95	9.22	1.84	6.88	16.43	19.14	25.74	37.72
Poultry eggs	-1.60	2033	7.86	1.00	5.58	13.96	16.46	22.56	33.50
Milk-whole	0.00	0.00	0.00	0.00	0.00	0.00	0.00	0.00	0.00
Butter	-0.05	-0.97	-2.72	-0.56	-1.94	-3.79	-5.77	-6.53	-7.97
Cheese	-0.15	0.10	0.36	0.04	0.28	0.75	0.67	1.08	1.70
Milk powder	-0.17	0.72	2.06	0.40	1.55	3.28	4.25	5.35	7.18
Wheat	-21.84	2.18	49.70	-17.83	20.41	88.20	130.48	207.18	351.58
Maize	1.30	19.59	44.21	24.35	43.80	91.66	98.55	137.94	219.41
Sorghum	-6.72	12.79	42.35	1.02	27.19	74.10	95.55	141.77	235.64
Rice	24.15	22.84	78.09	34.01	41.17	109.12	359.66	371.59	618.18
Soybeans	-20.26	-7.15	28.31	-17.14	-3.66	63.42	73.74	102.60	248.94
Soybeans meal	-5.51	3.49	19.14	0.45	10.22	37.22	42.15	57.40	98.26
Soybean oil	-18.57	-10.50	12.92	-19.04	1.21	27.76	38.14	50.48	109.95
Groundnuts	-22.76	-11.96	23.48	-21.38	-8.90	36.19	111.93	156.65	289.13
Groundnut cake	-7.27	1.05	17.44	-2.71	6.80	30.15	48.66	66.35	105.38
Groundnut oil	-12.43	-6.97	9.51	-12.22	-6.19	14.31	39.78	51.78	83.49
Cotton	-22.22	-14.23	26.61	-21.32	-12.09	42.47	131.75	164.26	393.41
Sugar	14.48	20.10	78.15	16.30	25.99	87.29	179.24	196.52	359.49
Tobacco	-42.02	-32.89	-5.39	-26.43	-13.90	28.11	222.32	296.29	550.78

The yield change is the exogenously determined supply shift that served as an input to SWOPSIM. The average yield change for a country is the value-weighted average of all crops. The consumption/production ratio provides a normalised measure of import/export status. When the ratio is above 1.0 the country is a net importer. When the ratio is below 1.0 the country is a net exporter. The ratio was developed from the base data: i.e. prior to climate change impacts. The plotted numbers are per capita economic losses for the country.

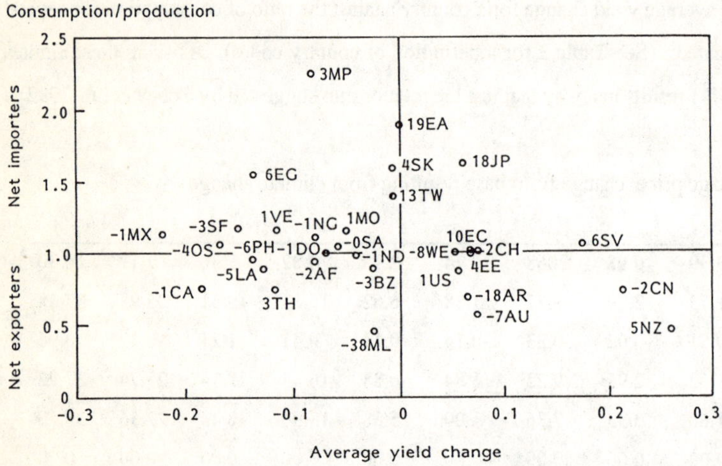

Figure 4 Per capita welfare change: GISS, CO_2 fertilisation, adaptation (1989 $per capita).

Table 3 Country/region coverage in SWOPSIM

US	United States	NG	Nigeria
CN	Canada	AF	Other sub-Saharan Africa
EC	European Community	EG	Egypt
WE	Other Western Europe	MP	Middle East and North Africa - oil exporters
JP	Japan	MO	Other Middle East and North Africa
AU	Australia	ND	India
NZ	New Zealand	OS	Other South Asia
SF	South Africa	DO	Indonesia
EE	Eastern Europe	TH	Thailand
SV	Former Soviet Union	ML	Malaysia
CH	People's Republic of China	PH	Philippines
MX	Mexico	SA	Other South-east Asia
CA	Central America and Carribbean	SK	South Korea
BZ	Brazil	TW	Taiwan
AR	Argentina	EA	Other East Asia
VE	Venezuela	RW	Rest-of-world
LA	Other Latin America		

The figures provide a reasonably compact presentation of key inputs and results. Note that for the cases where yield changes induced by climate change are not moderated by the CO_2 fertilisation effect or adaptation, yields are uniformly negative under the UKMO (Figure 1) case and, for all but New Zealand, are negative for the GISS case. The value-weighted yield changes for countries ranged from -5 to -50 per cent in the UKMO case and from +5 to -40 per cent in the GISS case. Thus, it is not surprising that the net global effect of these cases shows fairly severe economic losses. Once CO_2 fertilisation is considered, the yield losses are far smaller and some countries experience yield gains: for the GISS case (Figure 3) the yield changes range from +25 to -25 per cent.

The plots basically confirm the relationship suggested by Tobey *et al.* (1992).[6] The UKMO case with no fertilisation and no adaptation (Figure 1) provides some of the strongest support for the relationship because world crop prices rise strongly and consistently across crops. Price increases range from about 200 per cent for maize to over 600 per cent for rice (see Table 2). Hence, exporting countries generally gain substantially despite experiencing yield losses.

The GISS case with both CO_2 fertilisation and adaptation (Figure 4) demonstrates how these effects are reversed when the net effect of climate change on world agriculture is positive. In this case, commodity prices generally fall on the order of 20-25 per cent (Table 2). The exceptions are rice, with a price increase of nearly 25 per cent and maize, with a price increase of about 2 per cent. Net importing countries (e.g. Egypt, Japan) with yield increases (the top left quadrant, Figure 4) benefit from lower commodity prices and better yields. These gains are only partly offset by producer losses due to lower prices. Negative economic effects are generally experienced by exporting countries even though some experience positive yield effects – i.e. the lower two quadrants. As the same time, some importing countries show net benefits despite yield losses because consumers benefit from lower prices and this effect dominate producer losses – i.e. the top right quadrant. As might be expected, the GISS case with fertilisation but without

---

6 The one notable outlier in all cases in the lower right is New Zealand. New Zealand yield changes are generally the most favourable, but the country generally experience net welfare losses. The principal reason is that New Zealand's agricultural exports are dominated by lamb, the price of which does not rise substantially. Hence producer gains are somewhat limited but consumer losses are substantial because of rising world prices for other commodities. This result may depend on the omission of grazing effects in existing estimates of climate impacts.

adaptation (Figure 3) shows the weakest pattern because the net economic effect on the world is near zero and crop prices increase for some crops but decrease for other crops.

The per capita economic changes are also of interest. For the cases where the CO_2 fertilisation effect and adaptation are included, the per capita economic changes are fairly small. For these three scenarios, losses never exceed \$50 (1989 dollars) per capita. Economic gains are as much as \$134 per capita. In all the scenarios, the large per capita economic *changes* are experienced by exporting countries and (except for the GISS with CO_2 fertilisation and adaptation) the *largest changes* are *gains*. This should not be a particularly surprising result: a country's exposure to the risk of agricultural climate change depends on how important agriculture is in the economy. There is wide variation in agricultural production per capita but relatively less variation among countries in consumption (of basic commodities) per capita.[7] Thus the greater than average exposure to climate change risk (on a per capita basis) occurs for those countries that are very large producers. The risk for this group, however, is generally the risk of significant *increases* in income if climate change generally causes a deterioration of global agricultural conditions. Big losses for exporting countries occur when climate change leads to generally lower world commodity prices. Thus, the largest and most concentrated financial losses from climate change would likely occur to agricultural exporters if climate change were generally beneficial. Observers of commodity markets generally should not be surprised by such a finding. The most severe economic distress tends to occur for basic commodity exporters at times when prices are depressed.

In the most severe cases, those without adaptation or CO_2 fertilisation, the per capita losses for commodity importers are quite high for some countries. In the UKMO case, losses in a number of countries are on the order of \$100-\$350 per capita. These countries/regions include Japan, East Asia (Hong Kong, Macao, North Korea), South Korea, Taiwan, the Middle East Oil Producers, and Egypt. For developed countries, such losses may not provide substantial problems. However, if these effects were experienced in the poorer developing countries where per capita incomes are on the order of \$500-\$1500, serious malnutrition and starvation could result. These large losses occur primarily for large importing countries (a consumption/ production ratio of greater than 1.5) when world prices rise substantially. In this scenario, Australia benefits by over \$1100 per capita. These scenarios indicate the potential for significant change in comparative advantage and

---

7 Household consumption expenditures on food items vary between developed and undeveloped countries but this is in part due to post-farm gate processing which is not directly affected by climate.

redistribution of wealth among countries not unlike that which occurred when petroleum price rose in 1970s. These cases, however, probably should be considered extreme bounds on potential world price increases, because at least some CO_2 fertilisation effect and some adaptation are likely. Moreover, by the time these climate changes occur per capita income may have improved in developing countries such that they can better afford these magnitudes of commodity price increases. Although rising incomes would not necessarily limit the magnitude of the wealth transfer, they limit the extent to which income losses result in starvation because wealthier consumers can choose food at the expense of other goods whereas subsistence consumers have few such choices.

4 Some policy implications

This paper began by asserting that the following two principal realities must shape any consideration of policy in anticipation of climate change impacts on agriculture: (i) international trade and (ii) uncertainty. The numerical simulations provide indications of the possible types of shifts and changes that could occur. Because even the direction of change is not certain, specific policy conclusion, such as increasing stockpiles of food or targeting specific countries for assistance, are not particularly robust. The broadest conclusion of the results is that as long as there is fair confidence that CO_2 fertilisation is likely to operate more or less as represented in the crop yield of Rosenzweig *et al.* (1993), global agriculture does not appear particularly threatened by climate change. Thus, one might cautiously conclude that the basic capacity of the world to produce food does not appear to be undermined. Marginalisation of production conditions because of political turmoil or economic restructuring could intensify the climate change effects. Broader uncertainties, such as surprise and discontinuity in transient climate scenarios, remained unexpected.

Although the above conclusions do not provide a solid numerical foundation for specific policy directions, the existence of considerable uncertainty and the role of trade suggest several policy directions.

First, trade is able to play a significant moderating role, allowing farmers in countries less severely affected by climate change to profit by selling production to consumers in the more severely affected regions. In this way, markets act to pool the risk of locally severe effects. Even with highly uncertain scenarios, regionally differentiated effects are very likely. Can markets continue to play this risk-pooling role? One scenario that could arise that would prevent markets from operating as effective risk-pooling mechanisms would be that countries would intervene to prevent adjustment in trade flows. Why would they do this? Producer groups are relatively well organised political forces. A growing loss of comparative advantage suggesting that a country's agricultural resources should shift to other activities could easily generate pressure for import restrictions or domestic subsidies to maintain the domestic agricultural industry. Sometimes the protection of domestic agriculture is cast as supporting national food security. To maintain a national 'comfort level' in food security, whether that level is 120 per cent, 100 per cent, 80 per cent or 25 per cent of domestic consumption, trade restrictions would have to increase if the country's basic ability to produce is degrading relative to the rest of the world. Whether couched as food security or domestic industry protection, one policy implication resulting from significant changes in regional comparative advantage could be increased pressure for trade barriers, even though this would lead to larger economic losses overall. These pressures depend on relative – not absolute – changes in productivity, and hence could operate equally strongly even when worldwide productive capacity is increasing.

Second, the changing and uncertain comparative advantage among countries suggests that one should, if some type of national agricultural climate adjustment policy is considered, view the policy in terms of strategic trade policies. Is it a good strategic investment based on the ability to compete in the future or is it costly support that will prolong an industry that will only grow less competitive? If deteriorating agricultural conditions were predicted for a nation, a reasonable strategy might be to expand research into improved crops and cropping techniques, develop irrigation resources, and expand cropped area. If, however, these adaptations represent public-funded responses that occur while the rest of world's agriculture is doing quite well, this country would be throwing good money after bad. A better use of the funds would be to develop new skills among the rural population so that they might not labour indefinitely without hope of overcoming the worsening relative agricultural position. With export earnings from new industries the country could import food for less than it would cost to produce it domestically. If the same climate conditions face the country, but the rest of the world is more severely affected by climate change, the conclusion is reversed. The country can not only produce food more cheaply itself but may find itself a growing and profitable agricultural export industry. In a broad context, cultural values and environmental policy goals need to be considered.

Third, given the uncertainties, flexibility must be a key consideration, but what does flexibility mean and what does it cost? Flexibility and diversity are frequently linked but are not necessarily the same concept. One can imagine a very diverse agricultural sector but one with little flexibility. Rather than producing only wheat and rice, one might produce maize, rice, wheat, and groundnut as well. Thus, if conditions deteriorate for rice, only one-fifth of the crop would be at risk rather than one-half of the crop. Flexibility, on the other hand, might involve assuring that all farmers could grow both rice and wheat. Hence, if conditions deteriorate for rice, one need lose very little because everyone could grow wheat instead. However, because agricultural conditions for different crops are closely correlated (few grow well under extreme drought), neither diversity nor flexibility within agriculture alone is likely to be particularly effective. Being able to grow both maize and wheat or growing maize, rice, wheat, and groundnuts is not an effective risk strategy if none of the crops do well. Thus, access to markets (both within countries and among countries) seems to provide a more likely opportunity to avoid local crop failure (due to both short-term drought conditions and long-term climatic trends). Development of other employment opportunities (unrelated to agriculture) and improved education and training probably provide more effective diversification and flexibility to insure farmers against financial risk. Furthermore, these human capital investments are consistent with any overall development programme.

Fourth, subsistence farms and farming communities merit special policy consideration because the risks are starvation and malnutrition rather than financial loss. By definition, such groups are unaffected by commodity markets, either domestic or global. As the subsistence farm is both the supplier and demander of the products, the implicit price of commodities is fully determined within the farm. It must bear the full risk of local bad years (food storage is the private insurance mechanism). Subsistence agriculture is also, by definition, near the edge. If climatic resources and other resources, such as skill levels and financial capital, were better, the amount of food produced would exceed the amount needed by the household and the farm could enter the commercial markets and thus financial savings could allow the farmer to take advantage of markets in times of food shortage. On the other hand, any worsening of climate conditions means that the household must use whatever resources it has at hand (largely its own labour) more intensively or face hunger, malnutrition and starvation. These same groups are subject to the current vagaries of climate, which include cycles of drought. Although the risks are greater for such groups, the general direction of policy is probably no different from that in the general case discussed above; i.e. policies that improve skill levels and integrate these groups into markets.

General education and training will allow these peoples to become better farmers (or potentially enter a manufacturing workforce). With better incomes and improved skills, they will be in a better position to deal with climate changes.[8]

Fifth, the modelling results suggest a fairly consistent picture of negative effects in the developing countries in the aggregate. What does this mean for adaptive policy responses for individual developing countries? Unfortunately, very little. Even across the three scenarios that one might consider 'reasonable' projections rather than thought experiments (i.e. the three climate scenarios with fertilisation and modest adaptation), the direction of effect varies for individual developing countries and among developing countries. For example, the 'other sub-Saharan' region shows net economic losses in all three scenarios, but under the GISS scenario consumers benefit from lower food prices while producers suffer from the same lower prices plus degraded climatic conditions. Further, the results probably create more uniformity of effect than one might actually expect because (i) the climate models themselves are coarsely resolved; (ii) the Rosenzweig *et al.* (1993) approach of extrapolating from a relatively few sites tends to create uniformity that may not occur; and (iii) the SWOPSIM model aggregates countries. Thus, again the watchword appears to be uncertainty.

Sixth, can adjustment to new crops and cropping practices be fostered? Forecasts of climate and weather over a season or over a few years are improving, but the links to long-term change are still not well understood. What farmers are directly interested in, of course, is not climate itself but successful (agronomically and financially) agricultural practices. In the context of changing climates, successful practices may change. Climate will not be the only factor that changes over the next 100 years. There will be new crops, new technologies, changing local demands (as incomes rise), and changing input and output prices. The U.S. agricultural experiment station,

---

[8] Special problems are introduced when considering indigenous peoples who live in a subsistence manner. Preservation of their rights to their culture may conflict with an adaptative insurance strategy of economic development, diversification and integration. These very things may be antithetical to preserving the culture. Because regional conditions are likely to change, it is fairly certain that for some such peoples conditions will worsen while for others conditions may improve by some measures. The same questions arise as economic developments and population increase leads to pressure to intensify production on the lands of indigenous peoples. Climate change provides an added dimension to the problem because even complete preservation of land rights of indigenous peoples may not preserve the underlying productivity of their production systems. How or whether these people could be compensated for loss of culture is likely overwhelmed by the process of bargaining for compensation with unequal power among the bargaining participants and the difficulty of evaluating the extent of loss.

similar approaches in other countries, and the international network of crop research institutes[9] serve to integrate biologically the local climate information into the cultivar selection process by selecting those that do well at the site. This crop selection process can be an essentially autonomous adjustment process: the crop breeder and selector need not recognise that a choice was in part determined by the fact that climate had changed (or that CO_2 levels had increased). Countries with an effective network of such stations will be more likely to deliver to local farmers crop varieties which are better adapted to the changing climate. Thus supporting the existing network and expanding it appears to be a reasonable response.[10]

Seventh, climate change may introduce some additional considerations in the crop development system. Selection criteria among cultivars have generally been on the basis of yields under somewhat idealised conditions. Such an approach may be justifiable if one is comparing one cultivar against a set of unchanging alternatives and climatic conditions. Under a situation of climate change where the yields of all crops may be changing simultaneously, there is probably a need to compare farming strategies rather than the yield of a single crop against the existing standard. Such farming strategies may include switches to different crops, different cropping practices, new rotation schemes, etc. Comparing farming strategies with each other will further require a measure of success that is broader than yield (e.g. profitability). The need to make such comparisons is already recognised as part of the move to evaluate sustainable agricultural practices that involve comparing performance of a mixed farming enterprise with commercial agricultural enterprises that involve greater specialisation. Thus consideration of climate change encourages momentum in a direction that sustainable agriculture has already begun.

Eight, considerations of climate change may also suggest some alterations in the specific strategies for selecting experimental sites. In the absence of climate change, the experiment site tends to be located in the heart of a particular agronomic belt and crops are selected that are known to be

---

9 Examples include the International Maize and Wheat Improvement Center (CIMMYT); the International Rice Research Institute (IRRI); International Agricultural Research Centre (IARC); International Center for Agriculture Research in the Dry Areas (ICARDA). For one of the roles and future directions of such institutions, see FAO (1989).

10 For other approaches, see N.S. Jodha, `Potential strategies for adapting to greenhouse warming: Perspectives from the Developing World', in N.J. Rosenberg *et al*, *Greenhouse Warming: Abatement and Adaptation*, Resources for the Future, Washington, CD, 1989, pp 147-158; and Vernon W. Ruttan, David E. Bell and William C. Clark, `Climate change and food security: Agriculture, health and environmental security', *Global Environment Change*, Vol 4, No 1, 1994, pp 63-77.

relatively well-adapted to the 'known' climatic conditions. With the prospects of climate change, there may be an advantage in locating sites at the edge of agronomic zones or near the transition between zones. It may be an advantage to select a range of crops and practices that include some that may be maladapted to the 'known' climate. These may prove to be surprisingly successful if climate has, in fact, changed but the change has not been broadly detected or recognised.

5 Conclusions

Considerable progress has been made in evaluating the potential effects of climate change on global agriculture but significant uncertainties remain. Based on current understanding of climate change as represented by equilibrium GCM scenarios, the potential magnitude of impacts on world agriculture generally shows that the economic losses are manageable and in some cases the effects are positive. The reasonably modest negative – and, in some cases positive – impacts in these scenarios depend on the estimated positive direct effect of CO_2 on plant growth. If the direct fertilisation effect is far less positive than currently estimated, then the economic effects could be substantial. The analyses presented in this paper are based on analysis of equilibrium climate scenarios associated with a doubling of trace gas concentrations. The Intergovernmental Panel on Climate Change (IPCC) projected that the global temperature increase like that shown in the equilibrium scenario might not occur until nearly 2100.

To consider the implications of such changes on the agricultural policy, one must consider the nearer-term effects. Until specific transient climate runs can be evaluated, the nearer-term effects of climate change are speculative. The nature of climate change in transient scenarios remains a significant uncertainty in addition to many other uncertainties within the narrow scope of specific modelling studies of agricultural effects. Given these uncertainties, specific adaptive policies cannot be designed. We see several modest changes that existing national and international institutions might make to better prepare the agricultural system to respond to uncertain climate change. For the most part, these suggestions support existing efforts to redirect agricultural systems. Notably, reduced barriers on international trade in agricultural commodities as have been pursued under the GATT and farming system evaluation (as opposed to a narrow focus on yield) as necessitated by the need to evaluate sustainable agricultural systems would lead to an agricultural system better prepared to respond to climate change. Populations that are vulnerable to hunger are likely to benefit from greater integration into food markets combined with improved education and skill development that can provide these people with the means to purchase food, should changing climate degrade their ability to produce food themselves.

Acknowledgement

This paper is a reprinted version that first appeared in *Global Environmental Change* Volume 4, Reilly, J., Hohmann, N. and Kane, S., Climate change and agricultural trade: who benefits, who loses?, pp 24-36 (1994). Reprinted with permission from Elsevier Science Ltd, Butterworth-Heinemann Imprint, The Boulevard, Landford Lane, Kidlington OX5 1GB, U.K.

References

Adams, R. *et al.* (1990). Global climate change and US agriculture. *Nature*, **345**: 219-223.

Arthur, L. (1988). The greenhouse effect and the Canadian Prairies. In: G. Johnson, Freshwater, D. and Favero, P. (eds.) *Natural Resource and Environmental Policy Issues.* Westview, Boulder.

FAO (Food and Agriculture Organization) (1989*) Research Cooperation in Agriculture Between European and Developing Countries.* FAO, Rome.

Fischer, G., Frohberg, K., Parry, M.L., and Rosenzweig, C. (1994). Climate change and world food supply, demand and trade: who benefits, who loses? *Global Environmental Change*, **4** (1): 7-23.

Fischer, G., Frohberg, K., Parry, M.L. and Rosenzweig, C. (1995). Impacts of potential climate change on global and regional food production and vulneability. In: Downing, T.E. (ed.) *Climate Change and World Food Security.* Springer-Verlag, Heidelberg.

Houghton, J.T., Callander, B.A. and Varney, S.K. (1992). *Climate Change 1992: The Supplementary Report to the IPCC Scientific Assessment.* Cambridge University Press, Cambridge.

Kane, S., Reilly, J. and Tobey, J. (1992). An empirical study of the economic effects of climate change on world agriculture. *Climate Change*, **21:** 17-35.

Kates, R.W., Ausubel, J. and Berberian, M. (1985). *Climate Impact Assessments.* John Wiley & Sons, New York.

Kokoski, M. and Smith, V. (1987). A general equilibrium analysis of partial-equilibrium welfare measures: The case of climate change. *American Economic Review*, **7:** 331-341.

Mendelsohn, R. and Reilly, J. (1993). *A Research Agenda To Study Global Climate Change Impacts on Agriculture*, unpublished report to the Electric Power Research Institute, Pao Alto.

Mooney, S. and Arthur, L. (1990). The impacts of climate change on agriculture in Manitoba. *Canadian Journal of Agricultural Economics,* **38:** 685-694.

Parry, M., Carter, T. and Konijin, N. (eds) (1988). *The Impacts of Climatic Variations on agriculture,* Vol 1: *Assessments in Cool Temperatures and Cool Regions* and Vol 2: *Assessments in Semi-Arid Regions*, IIASA/UNEP, Kluwer, Dordrecht.

Reilly, J. and Hohmann, N. (1993). Climate change and agriculture: the role in international trade. *American Economic Review,* May: 306-312.

Roningen, V. Sullivan, J. and Dixit, P. (1991). *Documentation of the Static World Policy Simulation (SWOPSIM Modelling Framework.* Staff Report No. AGES 9151, US Department of Agriculture, Washington, DC.

Rosenzweig, C., Parry, M., Frohberg, K. and Fischer, G. (1993). *Climate Change and World Food Supply.* Environmental Change Unit Research Report No. 2. University of Oxford, Oxford.

Sonka, S. (1991). Methodological guidelines for assessing the socio-economic impacts of climate change on agriculture. In: *Climate Change: Evaluating the Socio-Economic Impacts.* OECD, Paris. pp. 21-45.

Tobey, J., Reilly, J. and Kane, S. (1992). Economic implications of global climate change for world agriculture. *Journal of Agricultural and Resource Economic*, **17:** 195-204.

PART III

VULNERABILITY AND MULTIPLE THREATS TO SUSTAINABLE AGRICULTURE

Climate Change and Food Insecurity: Toward a Sociology and Geography of Vulnerability

Thomas E. Downing
Environmental Change Unit
University of Oxford
Oxford, U.K.

Michael J. Watts
Department of Geography
University of California
Berkeley, California, USA

and

Hans G. Bohle
South Asian Institute
Im Neuenheimer Feld 330
69120 Heidelberg, Germany

1 Introduction

Global climate change presents a serious challenge to future livelihood strategies of third world peoples, especially for those social groups which are currently poor and vulnerable. Relatively modest adverse changes in community resource availability and/or economic output imply critical shifts in food security. Feasible strategies for coping with future climate change must be rooted in a full understanding of the complex structure and causes of present-day social vulnerability. In the case of global warming and its human implications, sophisticated models forecasting climate change and impacts, in our view, must begin with a compelling theory of contemporary vulnerability to hunger and famine.

The purpose of this paper is twofold. First, it seeks to explore, in theoretical terms, the characteristics and causal structure of vulnerability to hunger and food insecurity as a precondition for grasping the potential consequences of global climate change. We argue that the place and time-specific configuration of three key analytical variables–what we call human ecology, expanded entitlements and political economy–defines the dimensions (or social space) of vulnerability.

NATO ASI Series, Vol. I 37
Climate Change and World Food Security
Edited by Thomas E. Downing

Second, the paper addresses the conjuncture of social vulnerability to hunger and climate change. Global climate change interacts in synergistic and complex ways with other global processes. Accordingly, climate change may be refracted in unpredictable ways through markets, through new ecological stresses, and through multilateral agreements which have consequences for food production, trade, and relief. That these indirect consequences are unpredictable – and hence a source of worry and concern – should not detract from the problem at hand, namely promoting and enhancing entitlements for vulnerable communities.

2 Defining vulnerability

Vulnerability can be defined as an aggregate measure of human welfare that integrates environmental, social, economic and political exposure to a range of potentially harmful perturbations or threats (Downing, 1992). While it is mainly the poor who suffer from famine, hunger and malnutrition, not all poor people are equally vulnerable to hunger. Indeed it is not necessarily the poorest who face the greatest risk. In addition to income, there are a multiplicity of other factors that co-determine whether an individual will go hungry; in fact, this is the heart of Sen's notion of different commodity bundles, mixes of food sources derived from one's own production, labour or market exchanges, and donations or relief (Sen, 1981). The processes which endeavour to account for why some rather than others are more likely to experience hunger or starvation, define what is typically referred to in the literature as vulnerability (Downing, 1991a, 1991b). In as much as poor people are usually among the most vulnerable by definition, a nuanced understanding of vulnerability rests on a careful disaggregation of the structure of poverty itself; that is to say, the social relation of poverty and how particular patterns of poverty are reproduced over time. Defined in this way, vulnerability must include both a sensitivity to temporal (i.e. monthly, seasonal, inter-annual) and geographical (i.e. national, regional, village) variability, and also a recognition of how different groups in society experience particular sorts of 'risk' and mitigate hazards.

Unfortunately, vulnerability as a concept does not rest on a well developed theory; neither is it associated with widely accepted indicators or methods of measurement. One of the relatively early discussions of vulnerability, provided by Chambers (1989), starts from the properties of the system–in this case the food system–which give rise to vulnerability, rather than the specific empirical forms which they assume. He defines vulnerability as:

> the exposure to contingencies and stress, and difficulty coping with them. Vulnerability has thus two sides: an external side of risks, shocks and stress to which an individual or household is subject; and an internal side which is defencelessness, meaning a lack of means to cope without damaging loss (p. 1).

This definition suggests three basic co-ordinates of vulnerability or three sorts of risk vectors:

- The risk of *exposure* to crises, stress and shocks.
- The risk of inadequate *capacities* to cope with stress, crises and shocks (which implicitly subsumes timely and effective external interventions).
- The risk of severe *consequences* of, and the attendant risks of slow or limited recovery from, crises, risk and shocks.

From this vantage point, the most vulnerable individuals, groups, classes and regions are those most exposed to perturbations, who possess the most limited coping capacity and suffer the most from the impact of a crisis or environmental perturbations (such as climate change), and who are endowed with a tightly circumscribed potential for recovery. Vulnerability can be, in other words, defined in terms of exposure, capacity and potentiality (see also Anderson and Woodrow, 1989; Blaikie *et al.*, 1994).

The notion of exposure needs further clarification. We generally follow the conventional distinction in hazards literature between potential vulnerability and the risk of an extreme geophysical event (see UNDHA, 1992). Our focus is on societal vulnerability, not on the probabilities of specific extreme events. Exposure, therefore, refers to the construction of social, economic and political factors that place a specific population at-risk. For example, exposure to the adverse impacts of drought is quite distinct from the probability of rainfall deficits below a given threshold. In this case, exposure is more akin to 'climate sensitivity' in the climate impact literature (Carter *et. al.,* 1994) than to an all-embracing measure of hazardness.

3 The causal structure of vulnerability

What are the conditions and factors which govern vulnerability and which define the specific coordinates of risk exposure, coping capacity and recovery potential?

This is a complex question and has been addressed from a multitude of vantage points and at many scales of analysis (Blaikie *et al.*, 1994; Harriss *et al.*, 1990; Liverman, 1990; Offe, 1984; Palm, 1990). Vulnerability has been discussed in ecological terms, in relation to political economy and class structure, and as a reflection of social relations including ethnicity, caste, generation and gender. Alternatively, vulnerability can be expressed spatially–from the local to the regional to the transnational–and temporally both as a long-term structural baseline and as a short-term conjunctural condition.

Whatever the particularities of these different approaches, vulnerability is a complex and multi-dimensional social space defined by the determinate political, economic and institutional capabilities of people in specific places at specific times. In this sense a theory of vulnerability should be capable of mapping the historically and socially specific realms of choice and constraint–the degrees of freedom as it were–which determine risk exposure, coping capacity and recovery potential. In a narrow sense, this is about individual command over resources and basic necessities; in a wider sense it should identify the totality of individual rights and social entitlements (Curtis *et al.*, 1988). And in a still broader sense it should also speak to the structural properties of the political economy itself. These three attributes are rooted in distinctive approaches to vulnerability which, from different vantage points, shed light on the multi-dimensional space of vulnerability. They serve as the building blocks for our treatment of the causal structure of vulnerability.

In our view, the space of vulnerability (that is, the dimensions that determine risk exposure, coping capacity and recovery potential) is defined by three distinctive processes (Figure 1). Theoretically derived, we identify this tripartite causal structure of hunger–the three sides of our analytical triangle–as human ecology, expanded entitlements, and political economy. Each can only be grasped as congeries of social relations and hence each point of triangulation represents a network of ideas, a broad and complex literature which often carries important complementarities and areas of overlap with the other two coordinates (Bohle and Watts, 1993).[1] For the purposes of this paper, our causal structure of vulnerability is defined in the following way:

---

1 The large literatures pertaining to these three sides of our triangle are elaborated in much more detail, including case studies, in Bohle and Watts (1990). Terms such as capacity, vulnerability, and entitlements are employed in rather different ways by a panoply of authors drawn from a number of different disciplines. For the purposes of clarity of exposition we endeavour to define our key terms, albeit idiosyncratically, to ensure consistency.

Human Ecology: In the most general sense refers to the relations between nature and society. Our emphasis, however, underscores the ways in which the transformation of nature by human labour is rooted in the specific ecosystemic properties (climatic, pedological, ecological) of the environments of which society is part. In this way, human ecology is not so much the application of ecological concepts to the study of society as recognition of how social organisation and reproduction (encompassing, for example, population growth) have direct implications for sustainability and how the environment is experienced in terms of risks and threats (e.g. human-induced global warming, or short-term climatic variability such as droughts and floods). Human ecology is a way of both understanding the biophysical risks which vulnerable groups confront, and the 'quality' of their resource endowments (i.e. the extent to which land is degraded, the quality of the work environment).

Expanded Entitlements: According to Sen, food is obtained through production, exchange or donation: 'the entitlement of a person stands for the set of different commodity bundles that a person can acquire through the use of the various legal channels of acquirement open to someone in his position' (Sen, 1990, p. 36; see also Drèze and Sen, 1989; Sen, 1987). However, as Sen himself recognises, this narrow legal definition needs to be expanded not only to include non-legal, cultural, and intra-familial entitlements but ultimately to subsume what he calls the 'capabilities' and 'totality of rights' by which access to food (or some other aspect of human welfare) is conferred.[2] In our view, then, entitlements are employed in an expanded sense to include not only ownership bundles and endowments, but wider social entitlements and the necessary empowerment (enfranchisement) by which entitlements are secured, fought over and contested (Appadurai, 1984; Curtis *et al.*, 1988).

---

[2] Capabilities here refer to the abilities of people to undertake valuable 'doings and beings'; formally, this is a set of functioning entitlement bundles representing the various alternatives doings and beings that a person can achieve with his/her specific characteristics (Drèze and Sen, 1989, p. 12). Drèze and Sen (1989, p.11) note that 'extended entitlements is the concept of entitlements extended to include the results of more informal types of rights sanctioned by accepted notions of legitimacy.' In our view this can be extended still further to include what we call empowerment and enfranchisement which are implicit in Drèze and Sen's notion of legitimacy. As they note, entitlements refers only to the instrumental control over commodities rather than the more important issue of human capabilities.

Political Economy: Particular resource endowments, and patterns of entitlement are always embedded in a macro-structure provided by political economy, that is to say the national regime of accumulation which is obviously influenced by transnational processes. Each political economy is, as Wolff and Resnick (1987) argue, a product of 'fundamental class processes' (the processes by which members of society produce necessary and surplus labour) and 'subsumed class processes' (the distribution of surplus by the appropriators). Implicit in this definition of political economy is the notion that specific configurations of class processes (the regime of accumulation) confer crisis tendencies (for example market failures or crises of overproduction) which are important in grasping the specific risks and threats experienced by vulnerable groups (Watts, 1991).[3]

In Figure 1 we suggest that the intersection of these axes produces three parallel analytical concepts which are central to our explanation of food security and vulnerability to climatic change: endowments, class relations and empowerment and political ecology. Endowments is here understood, following Sen (1990), to be the original ownership bundle which determines a person's entitlement set. Swift (1989) elaborates this view to make it clear that ownership here embraces not only physical assets such as land but also human resources (the 'quality' of labour power for example), and social claims over resources based on informal rights and networks. In broad terms, endowments are distributional in nature since they implicitly refer to the link between population (and hence demographic growth) and total endowments in a community, region or nation state. But we also wish to include within endowments an ecological parameter which acknowledges environmental status (there is a large difference between a degraded peasant plot located in the drought prone Sahel and a well–maintained farm on the rich volcanic loams of the Kenyan highlands)–what Blaikie and Brookfield (1987, p. 6) refer to as the 'capability' of the land (or of an environmental resource). Endowments can therefore (see Figure 1) be located at the intersection of entitlements on the one hand and human ecology on the other. To this extent, endowments are directly related to risk exposure.

Class relations and empowerment, conversely, stand at the intersection of entitlements and political economy and provides the foundation stone for the second coordinate of

3 Albala-Bertrand (1993) provides a fresh view on the political economy of disasters, but one that focuses too narrowly on macroeconomic impacts to be equivalent to our concept here.

vulnerability, namely capacity. In so far as capacity identifies the ability to cope with stress and various shocks (climatic, economic, political), class relations root this ability in the social relations of production while empowerment looks to the (class-based) rights by which entitlement claims can be made. Property (the ownership of a particular asset) is thus both a rights question and an issue of class (what sorts of assets and how they enter into the production of necessary and surplus labour). We emphasise particularly how capacity is an empowerment question, namely how political rights determine how and whether entitlements can be claimed, contested, defended and lost. These rights reside in the sphere of politics, broadly put, which we see as inscribed in three domains: the domestic (gender and generational politics), work (production politics), and the public-civil sphere (state politics).

Finally, political ecology resides at the confluence of political economy and human ecology, and refers to the ways in which the relations of production shape the management of resources. Our notion of political ecology is, therefore, in line with that of Blaikie and Brookfield (1987, p.17) who combine 'the concerns of ecology with a broadly defined

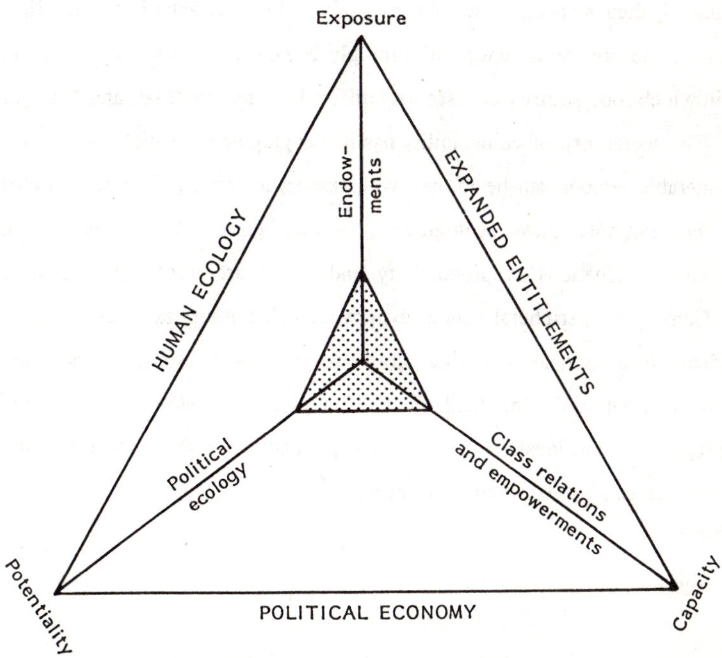

Figure 1 The causal structure of vulnerability. Source: Bohle *et al.* (1994).

political economy' as a means of grasping the dialectical and interactive effects between nature and society. In short, they link the sustainability of the environment, through 'chains of explanation', to the 'constantly shifting dialectic between society and land-based resources and also within classes and groups within society itself'. Political ecology, understood as a theory to locate environmental management in the relations of production, is, in our model (Figure 1), a way of grasping what we called potentiality (the third co-ordinate of vulnerability): namely, the resiliency of groups to shocks and perturbations.

From this vantage point, it is also possible to represent both vulnerable groups (social) and vulnerable regions (spatial) within a more embracing space of vulnerability. In the former, vulnerable individuals, groups and classes can be located according to the causal processes which present possibilities and constraints in the sphere of subsistence. For example, individuals and groups sensitive to climatic perturbations and unable to cope with food entitlement decline because they are endowed with environmentally degraded resources, are located with respect to the human ecological axis. In sketching these basic elements of the space of vulnerability, it needs to be reiterated that all three spaces of vulnerability exist simultaneously, although their respective 'weight' or analytical significance is an empirical question. Determining the precise weighting accordingly becomes an important device in assessing the way in which food security or insecurity differs between, say Mali and Niger or Kerala and China. The social map of vulnerability has its geographic or spatial counterpart; in other words, vulnerable regions can be located with respect to the tripartite structure of causal processes. For example, those ecologically marginal regions which regularly or sporadically experience fluctuations in productivity and prices are most liable to food entitlement crises. Conversely, peripheral regions experience vulnerability expressed through relations of dependency to a regional core which drains surpluses and resources away from the periphery (i.e. regions within the 'political space' of vulnerability). However, it is more useful to think of regions as the location of vulnerable populations, rather than inherently vulnerable due to resource and infrastructural constraints.

4 Who is vulnerable?

From the above perspective, we can now chart in detail the nature of vulnerable groups. The three sides of our causal structure of vulnerability (Figure 1) define specific sets of vulnerable peoples. While the precise boundaries of vulnerable groups is indeterminate, a common catalogue, based on livelihood, includes:

- *Rural smallholder agriculturalists*, with limited land and labour, may be resource-poor and especially exposed to external shocks if they are farming in marginal lands. Their empowerment varies considerably, often at short notice.
- *Pastoralists* utilise a range of ecologies, often sensitive to drought and pests, and subject to dramatic collapse if entitlements fail in a political context of marginality.
- *Rural wage labourers*, dependent on exchange entitlements with little or no direct agricultural production may be sensitive to market and political failures, particularly in remote agricultural lands.
- *Urban poor and informal sector workers* are in similar circumstances as rural wage labourers, but often form a more visible vulnerable group with some power and access to the political economy.
- *Refugees and displaced people* (in the broader sense of forced migrants) are often nearly destitute, but may command food through voluntary, national and international assistance, although they remain dependent on the political economy for their subsistence.
- *Destitute* groups may be distinguished from the poor and vulnerable by the degree of poverty–they are no longer capable of productive activity, are not likely to re-enter their former occupations, and are dependent on food aid from a variety of sources.

In addition to livelihood groups, individuals may be particularly vulnerable, due to their social status, labour availability and special nutritional needs. Such vulnerable groups, intersecting with livelihood, form populations at the highest risk:

- *Rural women* undertake arduous work, while being deprived of health, education and social services. Widowed, divorced or separated women, heading rural households, are disadvantaged by a shortage of labour. Pregnant and lactating women have higher nutritional needs.
- *Malnourished children*, particularly in the early years, are susceptible to minor fluctuations in food consumption; they are the first victims of a crisis.
- *Infirm and handicapped* people are likely to have fewer resources, while requiring assistance simply to survive.

- *Elderly* people may have a lower social status and receive less care than 'productive' members of the family, although cultural and community attitudes play a significant role.

Such vulnerable individuals, groups and classes can be located primarily according to our three coordinates. Individuals and groups vulnerable to food entitlement decline because they are resource and/or asset poor, such as many smallholder agriculturalists, may be located in the human ecological space of vulnerability. Conversely, if the likelihood of deprivation is rooted in problems of enfranchisement or political empowerment, which are inscribed in gender (patriarchal politics), work (production politics) and the public sphere (state politics), individuals and groups may be located in the political space of vulnerability determined by power and institutional relations. Pastoralists are often in this situation. And if deprivation arises from processes of surplus extraction and appropriation, individuals and groups are located in the structural-historical space of vulnerability given by specific configurations of class relations. For example, the urban poor in the informal sector may be defined in these terms.

5 Mapping vulnerable spaces

A world map of vulnerability to hunger and famine, incorporating the complex dynamics sketched above, has yet to be prepared. Indeed, it may be beyond our understanding of resource, economic, social and political systems. However, a relative ranking of countries based on three indicators provides one starting point for viewing the range of vulnerability. Figure 2 depicts national patterns of current vulnerability, based on:

- Food availability in kcal/day per capita (for 1986), reflecting variations in resource endowment or human ecology.
- Gross National Product (GNP) per capita (in 1987), a coarse measure of expanded (economic) entitlement.
- Childhood (under 5) mortality per 1000 (in 1987), a weak correlate with political economy for which more precise ratings are required.

Data for the three indicators were drawn from the World Bank (1989, 1990) for 172 countries. The nominal values of each indicator were normalised, and a composite food security index based on the three indicators was constructed. The index illustrates the diversity of countries vulnerable to hunger. The famine-prone countries of Africa (particularly countries, with a food security index of less than -1.00) comprise 37 countries with a total population of over 800 million in 1988. Conventional wisdom on vulnerability tends to focus on relatively narrow economic indicators of food poverty (FAO, 1993; Fischer *et al.*, 1995, this volume). For example, using the World Bank's country and regional rates of food poverty, almost 400 million people in the 37 famine-prone countries of Africa would be food poor, or vulnerable to recurrent food deprivation. At the world level, food poverty by such definitions affects 1,290 million people, 26 per cent of world population. However, our concept of social vulnerability is far broader than economic measures of food poverty. Considerable debate continues over who is hungry and where they live (in this volume, see Chen and Kates, 1995; Fischer *et al.*, 1995; UNDP, 1992).

Vulnerability maps also can be prepared at national or local scales. A burgeoning effort on vulnerability mapping is underway with methodological developments being reviewed by the Food and Agriculture Organisation, World Food Programme, International Decade for Natural Disaster Reduction, and International Federation of Red Cross and Red Crescent

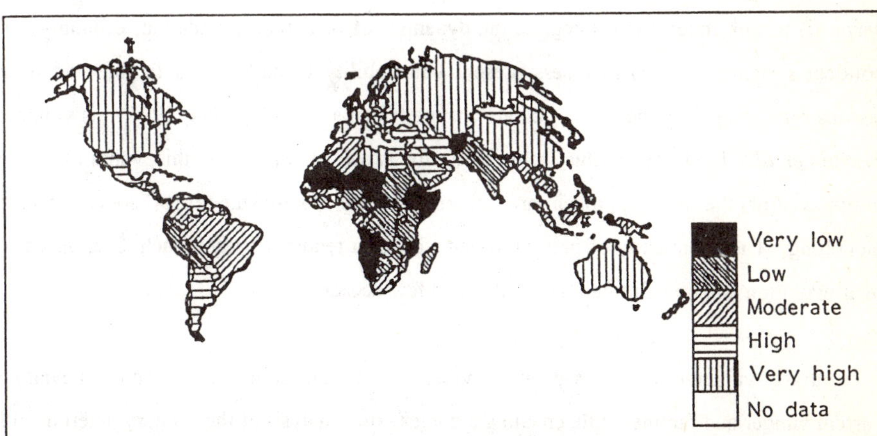

Figure 2 World food security. Source: Based on World Bank data (1989, 1990).

Societies, among others. Three lessons appear crucial at present: (i) vulnerability implies a clear conceptual definition as opposed to techniques of multiple-attribute mapping; (ii) vulnerability and vulnerability mapping should be based on social characteristics rather than Cartesian spatial units of analysis (such as administrative districts); and (iii) quantitative exercises should consider a wide range of analytical models, validate models against survey data and expert judgement, and be responsive to the needs of decision makers (see Downing, 1993).

6 Vulnerability and climate change

Climate change presents an uncertain but potentially serious threat to vulnerable populations. As thoroughly reviewed elsewhere, climate change will: increase the atmospheric concentration of carbon dioxide, with potentially positive effects on agriculture; increase temperatures and change rainfall patterns, with mixed effects depending on the local changes and cropping systems; affect multiple resources, including water supplies, vegetation and forests, biodiversity and health; and modify economic, social, and political systems as (real and potential) first-order effects stimulate responses to prevent climate change, mitigate its adverse impacts, or bear the consequences (see Houghton *et al.*, 1992, 1995; Schneider, 1995, this volume).

The implications of climate change for world food poverty may be assessed in at least two ways: (i) to link models that integrate the dynamics of resources, population, economy and political systems; and (ii) to assess present vulnerability to hunger and famine and infer lessons for coping with the future climate change (Chen and Kates, 1994, 1995, this volume; Fischer *et al.*, 1994, 1995, this volume; Reilly *et al.*, 1994, 1995, this volume). The limitations of the first approach are inherent in the state-of-the-art of socioeconomic modelling. The complexity of present systems belies a reductionist approach, even more so their evolution and transformation over the next few decades.

In contrast, vulnerability assessment provides a practical point of departure–alleviating current hunger and famine-while enabling a context-rich analysis of the primary determinants of vulnerability. That being said, the difficulties are still considerable, requiring: (i) an understanding of current vulnerability, as broadly portrayed above; (ii) assessment of trends in vulnerable groups, incorporating the causal dynamics, spatial extent and magnitude of vulnerability; (iii) portraying the risk of climatic variations, largely based on models of agricultural systems, but developed for vulnerable systems in the context of regional and global linkages; and (iv) review of potential responses and pathways that reduce vulnerability and enhance resilience.

The causal structure of present vulnerability enables us to begin to chart the evolution of vulnerable groups over the next few decades, on the same scale as global climate change. Climate change will directly affect human ecology and resource endowments. For example, the tropics and semi-arid areas may be subject to reduced crop yields, increased flood hazard and further land degradation. The indirect effects are most uncertain, but could reduce world food surpluses, which might limit aid to burgeoning refugee populations. At the same time, other driving forces of global change–such as population growth, environmentalism, economic growth, trade liberalisation, financial integration, and demographic political transitions–affect resource endowment, entitlement and the political economy. The outcome of the interaction of vulnerability and climate change depends on the articulation of driving forces and adaptive responses, and their effects on vulnerable populations. For example, increased economic wealth implies a shift in vulnerability from the human ecology of production to exchange entitlement. Or, commercial plant breeding, to take advantage of enriched CO_2 may benefit export-oriented economies in the OECD rather than smallholder farmers with little access to new technology and investment capital.

The potential evolution of vulnerability for specific vulnerable groups may be sketched in a broad sense:

- *Destitute* peoples, often displaced from their homes, have no resources, few entitlements and are typically politically disenfranchised. They are the most vulnerable and dependent group, requiring long-term support and social security. Some individuals and households will be able to re-enter productive employment by their own initiative, community support and sponsored training. However, as a class options for returning to agricultural production or entering the wage economy are limited.
- The *vulnerable with marginal resources* encompass rural producers, landless and urban poor who have low food security, on the edge of present poverty, but with sufficient resources to provide most of their requirements. These groups may be most sensitive to

future changes–a slight shift in endowment, empowerment or political ecology could dramatically affect their status. For example, reform of land tenure and liberalisation of the agricultural economy in China dramatically reduced food poverty in the early 1990s (although the improvements should not be considered stable at present). Since they have a considerable present capacity for productive activities, self-help and community development can be promoted to reduce vulnerability.

- The *moderately vulnerable* are able, at present, to sustain livelihoods above the level of poverty. Nevertheless, they are vulnerable to large-scale changes in the food system. The need is to protect their existing resources, entitlement and self-help capacity.
- *Groups with relatively low vulnerability* include those with a high command over food through markets and distribution systems and sufficient assets and wealth to ensure food security in the present and future. Often these groups occupy strategic positions in the food system, which ensures their ability to extract surplus production and dominate economic exchanges. It is conceivable that political reform associated with structural adjustment might reverse the fortunes of the ruling class, although the resulting political instability tends to increase social vulnerability in general (as shown in Rwanda, Burundi and Eastern Europe).

7 Vulnerability and climate change in Zimbabwe

Despite an upsurge in applied vulnerability assessments, research on vulnerability and climate change is at relatively early stage (Carter *et al.*, 1994; Downing,, 1992; Downing, 1991; Gleick, 1993; Liverman, 1990). Most assessments that go beyond literature reviews or theoretical frameworks describe present vulnerability. Or, in the context of climate change impact assessment, simply overlay the risk of future climate change. While this is a useful device, it fails to consider the evolution of vulnerability in response to the driving forces of global change in the world economy. This section outlines research in progress in Zimbabwe, elements toward reaching the aim of defining the present and future social space of vulnerability.

Christensen and Stack (1992) summarise regional and national surveys of food security in Zimbabwe, delineating four vulnerable livelihood groups--urban (unemployed, informal workers) and rural (communal farmers, landless, commercial farm workers, unemployed)

(Table 1). While the definition of vulnerable groups is robust and widely accepted, references points for enumerating the extent of vulnerability, food insecurity, poverty and malnutrition differ in time, region and measurement. However, a reasonable estimate of the nature and extent of vulnerability in Zimbabwe emerges.

Most of the vulnerable were in rural areas. Agricultural land use in Zimbabwe spans a range of natural farming regions from the productive specialised, intensive farming regions of central to eastern Zimbabwe (zones I and II), to the lowland, extensive farming zones of western and southern Zimbabwe (zones IV and V) (Figure 3a). Communal farmers are smallholder agriculturalists, concentrated in the semi-arid, extensive farming zones. Even in years of average production, food-insecure communal farmers number almost 7.5 per cent of the population (at least in 1991). Rural households with no land, dependent on agricultural labour (commercial farm workers) or unemployed, comprise over 15 per cent of the population. Female-headed households (i.e., divorces, separated or widowed), 3.5 per cent of the population, may be critically vulnerable due to less labour, restricted social relations, and high dependency ratios.

Some 5.5 per cent of the population are vulnerable to food insecurity in urban areas (in 1991), comprising the unemployed and households in the informal economy. Some 10 per cent of the urban households are headed by divorced, separated or widowed women--a particularly vulnerable group. The total vulnerable population, over 2.5 million, is over a quarter of the population of Zimbabwe.

Although food security is a priority for national policy, the causal structure of vulnerability is embedded in the human ecology, political economy and entitlement relations of post-independence Zimbabwe. Maize, the staple food, has increased in production, more than doubling among communal farmers between the 1970s and 1980s. Yet, only 10-20 per cent of communal farmers consistently produce a surplus (Mudimu *et al.*, 1990). The patterns of vulnerability outlined above, typified by this inequitable distribution of agricultural development, are due to: weak macroeconomic performance, inequitable land distribution, and misdirected social policy (Christensen and Thomas, 1993). External shocks of recurrent drought in the last ten years and structural adjustment have further stressed vulnerable groups (Magadza, 1994, 1995, this volume).

Table 1 Vulnerable livelihood groups in Zimbabwe

	1991		A: Conv. Wisdom 2050		D: Accelerated Devt 2050	
	Rate, %	Total	Rate, %	Total	Rate, %	Total
Driving Forces						
Population	2.3	10,100,000	2.128	34,983,735	1.518	24,567,531
Rural	72.00	7,272,000	54.00	18,891,217	36.00	8,844,311
Urban	28.00	2,828,000	46.00	16,092,518	64.00	15,723,220
GNP: $m	1.90	6,161	4.87	101,877	4.87	101,877
Per capita		610		2,912		4,147
AgGNP, %		20		15		10
AgGNP, per capita		169		809		1,152
Urban GNP, per capita		1,743		5,381		5,831
Food Insecurity						
Population	26.49	2,675,001	14.91	5,216,522	11.21	2,754,249
Rural Population	20.99	2,120,006	11.56	4,043,493	8.12	1,994,102
Unemployed	1.58	160,000	0.33	116,087	0.23	57,250
Landless	1.49	149,985	1.49	519,508	1.04	256,202
Communal						
I&II	0.89	89,991	0.89	311,705	0.63	153,721
III	0.99	99,990	0.99	346,339	0.70	170,802
IV&V	5.69	574,993	5.69	1,991,624	4.00	982,196
Commercial Farm workers	10.35	1,045,047	2.17	758,229	1.52	373,931
Urban Population	5.50	554,995	3.35	1,173,028	3.09	760,147
Unemployed	3.17	319,968	1.03	358,957	0.95	232,612
Informal workers	2.33	235,027	2.33	814,072	2.15	527,536

Source: Christensen and Stack (1992) for estimates of 1991 vulnerability; World Bank (1993), Pepper *et al.* (1992) and Achebe *et al.* (1990) for scenarios of driving forces.

The key question, is how will the 'driving forces' of global change affect the vulnerable in Zimbabwe, in the absence of climate change? To illustrate the issues, two reference scenarios have been constructed. 'A' corresponds to the conventional wisdom of a medium level of population growth and medium-high economic growth (Table 1). This scenario is based on Achebe *et al.* (1990), and reflects the Intergovernmental Panel on Climate Change's IS92a scenario, often called the 'business-as-usual' projection (Pepper *et al.*, 1992; IPCC, 1990). It entails a continuation of the current trends in social vulnerabilityv – relatively rapid urbanisation, declining share of agriculture in GDP, increased wealth and modest progress in the human ecology of production and exchange entitlements, but less so in the political economy. The projected population, some 35 million, is three times wealthier, dominated by the major urban centres. Zimbabwe in 2050, in this projection, would be in the upper-middle income group of countries, similar perhaps to Greece in the 1980s or Uruguay in the early 1990s.

It is intriguing to think how vulnerable groups might evolve in such a scenario. Vulnerability among communal farmers currently reflects social status and land quality–the prevalence of food insecurity would be difficult to reduce without large social welfare programmes. This implies that the rural sector becomes a major source for urban immigration, rather than increased impoverishment due to land scarcity. Households with greater exchange entitlements, the commercial farm workers and unemployed (both rural and urban) are likely to benefit from economic expansion. A reasonable assumption may be that the rate of food poverty declines commensurate with the increase in per capita income. The rural landless and informal workers may increase dramatically in numbers and vulnerability, as they become more destitute, with fewer skills and less access to the market and political economies. However, a more optimistic assumption, of no change in relative poverty, is assumed here–an expanded labour market should accommodate much of the low-skilled work force.

The resulting reference scenario for 2050 suggests that socioeconomic vulnerability continues to plague agricultural economies such as Zimbabwe. The overall rate of food insecurity is likely to fall, perhaps by almost half, reflecting progress in preventing hunger. However, the total number of the vulnerable almost doubles. Rural vulnerability continues to dominate, with major increases in the number of landless and communal farmers in semi-arid areas.

The second vision of the future is more normative, maintaining the economic growth rate of the first scenario, but with a low projection of population growth (loosely styled after the IS92d scenario). Per capita income is over a third larger than the first scenario, and much of this gain is in the agricultural sector. This scenario of accelerated development results in a country that may be more akin to Portugal than any of Africa's present economies.

The implications of the accelerated development scenario for the vulnerable groups is quite positive, assuming commensurate progress in the political economy of empowerment and exchange entitlement. The prevalence of food insecurity would decline in all groups. Indeed, the assumptions resulting in national vulnerability of some 10 per cent may be unduly pessimistic.

These two reference scenarios omit the surprises of environmental fluctuations, economic recession, resource conflicts and civil strife that plague food-insecure countries at present. Thus, a necessary question is, what potential surprises would affect the future of the vulnerable in Zimbabwe? Obviously, drought is currently a serious risk for the rural population. In a country where the moisture gradient between farming regions is modest, drought is recurrent. Among rural households, vulnerability doubles in times of drought, particularly if national institutions fail to provide timely support to the food system. In fact, the 1991/92 drought also affected many urban workers as water and power shortages resulted in factory closures and economic hardship (Benson and Clay, 1994).

Finally, does climate change present a serious risk to the vulnerable of Zimbabwe? We are not able to provide a definitive answer. Small changes in agroclimatic resources in Zimbabwe entail significant spatial shifts and reveal land use zones that are highly sensitive to change. A simple index of the atmospheric water balance, precipitation minus potential evaporation, gauges the extent to which agricultural land use may be subject to changes in water resources–often the limiting factor in semi-arid regions of Zimbabwe (Downing, 1992). The present balance shows the wetter highlands (with surplus water) and the extreme water deficits of southern Zimbabwe (Figure 3a). With a temperature increase of 2° C, the wet zones (with a water surplus) decrease by a third, from 9 percent of Zimbabwe to about 2.5 per cent (Figure 3b). The driest two zones double in area. A further increase in temperature, to +4° C, reduces the summer water-surplus zones to less than 2 per cent of Zimbabwe, approximately corresponding to the 1991/92 drought. In addition to a shrinkage of the agricultural area, crop yields in marginal zones would become more variable. Simulations for one semi-arid area indicate that with 2° C of warming, yields currently expected 70 per cent of the time would be exceeded only in less than 40 per cent of the years (Muchena, 1994).

Increased risk in food production directly affects vulnerable farmers. Surveys in Buhera, in the semi-extensive, semi-arid farming region near Chisumbanje show that household food security is affected by erratic rainfall, sandy and infertile soils and low levels of crop technology (Govereh and Mudimu, 1991). Maize yields are low–averaging 648 kg/ha. Farm sizes are relatively large, but with households of 10 people. Average gross margins are negative. With climate change, household food security would deteriorate, possibly with a 10-20 per cent decrease in food availability in vulnerable households.

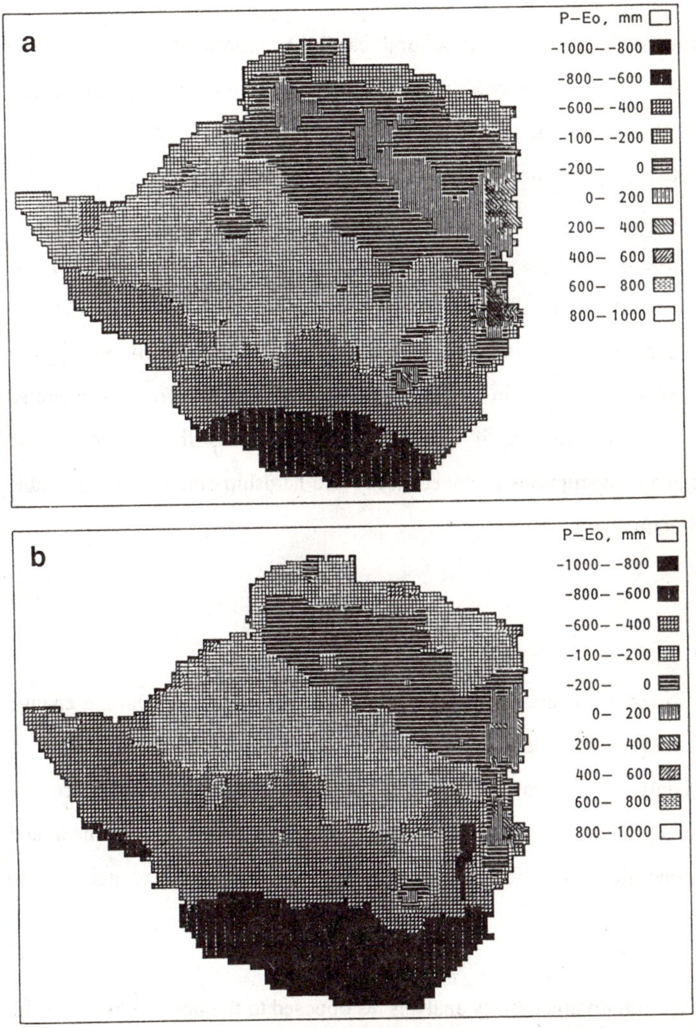

Figure 3 Seasonal atmospheric water balance in Zimbabwe: (a) current average conditions for November to March; (b) with +2°C warming. Source: Downing (1992).

The semi-extensive farming zone, on the margin of more intensive land uses, appears to be particularly sensitive to small changes in climate. Socioeconomic groups in this area, already vulnerable in terms of self-sufficiency and food security, would be further marginalised. Increased variations in rainfall and yields would alter the mix of appropriate response strategies. Successful farming systems would have to be responsive to good seasons, implying improved use of weather information, flexible markets for inputs and produce, and reliable drought responses.

More importantly, broad-scale shifts in agricultural capability would affect rural labour markets and the national economy. Climate change threatens each vulnerable group through the multiple effects that diminish resource endowment, and possibly increase resource conflicts and tension between agricultural and industrial/commercial sectors.

In conclusion, even gradual changes in mean conditions, against a backdrop of the conventional projections, could derail efforts to intensify agriculture in Zimbabwe. Equally, modest changes in potential yields, themselves a long way below actual production in Africa, are unlikely to have a large impact in the kind of world resulting from accelerated development. Somewhere in between, if climate change increases the frequency and persistence of drought, major disruptions to the economy and hardship among the vulnerable would be difficult to contain.

8 Conclusion

Feasible interventions to reduce vulnerability and ameliorate the impact of climate change (among the full range of global changes) revolve around supporting the capacity of vulnerable groups to maintain resources and productive activities, efforts to enhance recovery from food crises (spanning the range from the individual/household to regional famines), and promoting political economies that ensure access to resources and a safety net for the destitute.

On the research frontier, formal vulnerability analysis, as opposed to the pejorative use of the word, may provide a linkage between sectoral resource assessments, economic models based on undifferentiated populations, and the social relations of entitlement, empowerment, and enfranchisement embedded in the political economy.

The prescriptive and normative responses to vulnerability must reduce exposure to potentially harmful perturbations, increase ability to cope with crises and threats of various sorts, and strengthen processes of recovery, based on a sound understanding of the dynamics and relations that prescribe vulnerability for specific groups in the present and future. Our causal structure of vulnerability provides the theoretical basis for such an understanding. Four conclusions are warranted:

- The map of social vulnerability varies over time and space.
- The contours of the vulnerability map require careful specification in order to assess the number and character of those vulnerable, their capacities and exposure, and their resiliency.
- The impact of climate change can only be understood in relation to the complex ways in which global change will be experienced by the vulnerable–according to their resource endowment and entitlements related to gender, age, class, political empowerment and social status.
- Human ecology, entitlements and political economy provide a key analytical constituents for sorting, classifying and mapping differing sorts of vulnerability in relation to food security and, hence, for mitigating future unpredictable consequences of climate change.

Thus, we seek, in this paper, to promote efforts to reduce current vulnerability by focusing the debate on adapting to future climate change firmly upon an understanding of the present causal structure of hunger.

Acknowledgements

This paper is a substantially revised and extended version that first appeared in *Global Environmental Change* Volume 4, Bohle, H.G., Downing, T.E. and Watts, M.J., Climate change and social vulnerability: Toward a sociology and geography of food insecurity, pp 37-48 (1994). Reprinted with permission from Elsevier Science Ltd, Butterworth-Heinemann Imprint, The Boulevard, Landford Lane, Kidlington OX5 1GB, U.K.

References

Achebe, C., Okeyo, A.P., Hyden, G. and Magadza, C.H.D. (eds.) (1990). *Beyond Hunger in Africa*. Heinemann, Nairobi.

Albala-Bertrand, J.M. (1993). *The Political Economy of Large Natural Disasters: With Special Reference to Developing Countries*. Clarendon, Oxford.

Anderson, M.B. and Woodrow, P.J. (1989). *Rising from the Ashes: Development Strategies in Times of Disaster*. Westview, Boulder.

Appadurai, A. (1984). How moral is South Asia's economy? – a review article. *Journal of Asian Studies,* **43**(3): 481-497.

Benson, C. and Clay, E. (1994). *The Impact of Drought on Sub-Saharan African Economies: A Preliminary Examination*. Working Paper 77. Overseas Development Institute, London.

Blaikie, P., Cannon, T., Davis, I. and Wisner, B. (1994). *At Risk: Natural Hazards, People's Vulnerability, and Disasters*. Routledge, London.

Blaikie, P. and Brookfield, H. (eds.) (1987). *Society and Land Degradation.* Tavistock, London.

Bohle H.G. and Watts, M.J. (1993). The space of vulnerability: the causal structure of hunger and famine. *Progress in Human Geography,* **13**(1): 43-67.

Carter, T.R., Parry, M.L., Harasawa, H. and Nishioka, S. (1994). *IPCC Technical Guidelines for Assessing Climate Change Impacts and Adaptations*. University College London, London and Center for Global Environmental Research, Tsukuba.

Chambers, R. (1989). Vulnerability, coping and policy. *IDS Bulletin,* **20**(2):1-7.

Chen, R.S. and Kates, R.W.(1994). World food security: prospects and trends. *Food Policy,* **19**(2):192-208.

Chen, R.S. and Kates, R.W. (1995). Toward a food secure world: prospects and trends. In: Downing, T.E. (ed.) *Climate Change and World Food Security.* Springer-Verlag, Heidelberg.

Christensen, G. and Thomas, D. (1993). *Policy Analysis for Household Food Insecurity Case Study: Rural Zimbabwe*, Working Paper No. 7. Food Studies Group, Oxford.

Christensen, G. and Stack, J. (1992). *The Dimensions of Household Food Insecurity in Zimbabwe, 1980-1991.* Working Paper No. 5. Food Studies Group, Oxford.

Curtis, D., Hubbard, M. and Shepherd, S. (eds.) (1988). *Preventing Famine: Policies and Prospects for Africa*: Routledge, London.

Downing, T.E. (1991a). *Assessing Socio-Economic Vulnerability to Famine.* Research Report. Alan Shawn Feinstein Hunger Program, Brown University, Providence.

Downing, T.E. (1991b). Vulnerability to hunger and coping with climate change in Africa. *Global Environmental Change,* **1**(5):365-380.

Downing, T.E. (1992). *Climate Change and Vulnerable Places: Global Food Security and Country Studies in Zimbabwe, Kenya, Senegal and Chile.* Research Report No. 1. Environmental Change Unit, Oxford.

Downing, T.E. (1993). Concepts of vulnerability to hunger and applications for monitoring famine in Africa, In: Bohle, H.G., Downing, T.E., Field, J.O., and Ibrahim, F.N. (eds.) *Coping with Vulnerability and Criticality.* Verlag breitenbach, Saarbrucken.

Drèze, J. and A.K. Sen (1989). *Hunger and Public Action.* Clarendon, London.

Fischer, G., Frohberg, K, Parry, M.L. and Rosenzweig, C. (1994). Climate change and world food supply, demand and trade: who benefits, who loses? *Global Environmental Change,* **4**(1):7-23.

Fischer, G., Frohberg, K., Parry, M.L. and Rosenzweig, C. (1995). Impacts of potential climate change on global and regional food production and vulnerability. In: Downing, T.E. (ed.) *Climate Change and World Food Security.* Springer-Verlag, Heidelberg.

FAO (Food and Agriculture Organization) (1993). *Agriculture: Toward 2010.* FAO, Rome.

Gleick, P.H. (1993). *Water in Crisis: A Guide to the World's Fresh Water Resources.* Oxford University Press, Oxford.

Govereh, J. and Mudimu, G.D. (1991). Prospects for increasing household food security and income through increased crop productivity and diversification in low rainfall areas of Zimbabwe. In: Rukuni, M. and Wyckoff, J.B. (eds.) *Market Reforms, Research Policies and SADCC Food Security.* University of Zimbabwe, Harare.

Harriss, B., Gillespie S. and Pryor, J. (1990). Poverty and malnutrition at extremes of South Asian food systems. *Economic and Political Weekly:* 2783-2799.

Houghton, J.T., Callander, B.A. and Varney, S.K. (eds.) (1992). *Climate Change 1992: The Supplementary Report to the IPCC Scientific Assessment.* Cambridge University Press, Cambridge.

Houghton, J.T., Meira Filho, L.G., Bruce, J., Lee, Hoesung, Callander, B.A., Haites, E., Harris, N. and Maskell, K. (1995). *Climate Change 1994: Radiative Forcing of Climate Change and An Evaluation of the IPCC IS92 Emission Scenarios.* Cambridge University Press, Cambridge.

IPCC (Intergovernmental Panel on Climate Change) (1990). *Climate Change: The IPCC Response Strategies.* WMO, Geneva and UNEP, Nairobi.

Liverman, D. (1990). Vulnerability to global environmental change, In: R. Kasperson *et al.*, (eds.) *Understanding Global Environmental Change: The Contributions of Risk Analysis and Management*, Worcester, MA: Clark University, pp. 27-44.

Magadza, C.H.D. (1994). Climate change: some likely multiple impacts in southern Africa. *Food Policy* **19**(2):165-191.

Magadza, C.H.D. (1995). Climate change: some likely multiple impacts in southern Africa. In: Downing, T.E. (ed.) *Climate Change and World Food Security.* Springer-Verlag, Heidelberg.

Muchena, P. (1994). Implications of climate change for maize yields in Zimbabwe. In: Rosenzweig, C. and Iglesias, A. (eds.) *Implications of Climate Change for International Agriculture: Crop Modeling Study.* United States Environmental Protection Agency, Washington, DC.

Mudimu, G., Chopak, C., Chigume, S. ,Govereh, J. and Bernsten, R. (1990). Household income, food production and marketing in low-rainfall areas of Zimbabwe: status, constraints and opportunities. In: Rukuni, M., Mudimu, G., and Jayne, T.S. (eds.) *Food Security Policies in the SADCC Region.* University of Zimbabwe, Harare.

Offe, C. (1984). *The Contradictions of the Welfare State.* MIT Press, Cambridge.

Palm, R.I. (1990). *Natural Hazards.* Johns Hopkins, Baltimore.

Pepper, W., Leggett, J., Swart, R., Wasson, J., Edmonds, J. and Mintzer, I. (1992). *Emission Scenarios for the IPCC: An Update.* IPCC Working Group I, Bracknell.

Reilly, J., Hohmann, N. and Kane, S. (1994). Climate change and agricultural trade: who benefits, who loses? *Global Environmental Change* **4**(1): 24-36.

Reilly, J., Hohmann, N. and Kane, S. (1995). Climate change and agricultural trade: who benefits, who loses? In: Downing, T.E. (ed.) *Climate Change and World Food Security.* Springer-Verlag, Heidelberg.

Schneider, S. (1995). The future of climate: potential for interaction and surprises. In: Downing, T.E. (ed.) *Climate Change and World Food Security.* Springer-Verlag, Heidelberg.

Sen, A.K. (1981). *Poverty and Famines: An Essay on Entitlement and Deprivation.* Clarendon, Oxford.

Sen, A.K. (1987). Property and hunger. *Economics and Philosophy*, **4**:57-68.

Sen, A.K. (1990). Food, economics and entitlements. In: J. Drèze and A. Sen (eds.) *The Political Economy of Hunger*, Volume 1, Clarendon, Oxford. pp 34-50.

Swift, J. (1989). Why are rural people vulnerable to famine? *IDS Bulletin,* **20**(2):8-15.

UNDHA (United Nations Department of Humanitarian Affairs) (1992). *Internationally Agreed Glossary of Basic Terms Related to Disaster Management*. UNDHA, Geneva.

UNDP (U.N. Development Programme) (1992). *Human Development Report 1992*. Oxford University Press, New York.

Watts, M.J. (1991). Entitlements or empowerment?. *Review of African Political Economy,* **51**:9-26.

Wolff, R. and Resnick, S. (1987). *Economics: Marxian versus Neoclassical.* Johns Hopkins, Baltimore.

World Bank (1989). *Social Indicators of Development 1989.* World Bank, Washington, DC.

World Bank (1990). *World Development Report 1990.* World Bank, Washington, DC.

World Bank (1993). *World Development Report 1993*. World Bank, Washington, DC.

Changing Vulnerability to Food Insecurity and the International Response: The Experience of the World Food Programme

Bruce Crawshaw*
9 Muirfield Park
Gullane
East Lothian EH31 2DS
Scotland, UK

John Shaw*
112 Kenwood Drive
Beckenham
Kent BR3 2RB, UK

1 Introduction

Data produced by the Food and Agriculture Organisation of the United Nations (FAO) and the World Health Organisation (WHO) for the International Conference on Nutrition (ICN) in 1992 suggest that the prevalence of chronic undernutrition throughout the developing world is declining, albeit slowly. The number of chronically undernourished in the developing world is estimated to have fallen by one-sixth over the last two decades, from 941 million people to 786 million (see Table 1). As a proportion of their total population, the situation is estimated to have improved even further, declining from more than one-third in 1969-71 to one-fifth in 1988-90.

Malnutrition is closely linked to poverty. Although the situation is less clear, there appears to have been similar general progress in poverty reduction. World Bank figures show that consumption per capita in the developing world between 1965 and 1985 rose by almost 70 per cent (World Bank, 1990). Broader measures of well-being – such as life expectancy, child mortality and educational attainment – also seem to confirm this global improvement (UNDP, 1993). The World Bank suggests that the number of poor in the developing world could fall by more than 300 million by the end of the century.

* The views expressed in this paper do not necessarily reflect those of the World Food Programme.

NATO ASI Series, Vol. I 37
Climate Change and World Food Security
Edited by Thomas E. Downing

Table 1 Estimates of chronic undernutrition* in developing regions: 1969-71, 1979-81 and 1988-90

Region	Period	Total population (millions)	Proportion (per cent)	Undernourished (millions)
Africa	1969-71	288	35	101
	1979-81	384	33	128
	1988-90	505	33	168
Asia and the Pacific	1969-71	1880	40	751
	1979-81	2311	28	645
	1988-90	2731	19	528
Latin America and the Caribbean	1969-71	281	19	54
	1979-81	357	13	47
	1988-90	433	13	59
Near East	1969-71	160	22	35
	1979-81	210	12	24
	1988-90	269	12	31
All developing regions	1969-71	2609	36	941
	1979-81	3262	26	844
	1988-90	3938	20	786

* Chronic undernutrition is defined as 'chronic dietary energy deficiency'.

Source: FAO and WHO (1992).

However, this general improvement hides some worrying trends. Furthermore, the patterns of vulnerability to food insecurity have shifted quite dramatically over recent years. The ways in which governments and the international community have reacted to these shifts might have lessons for those looking at longer-term changes in global vulnerability, whether from climate change or other causes.

Based on the experience of the World Food Programme (WFP) in providing relief food aid to the victims of man-made and natural disasters, this paper concentrates on some of the more prominent recent changes in the incidence of vulnerability, namely:

- regional variations;

- the causes of vulnerability;
- the most vulnerable populations; and
- the ability to cope with both chronic and transitory food insecurity.

Finally, the paper looks at some of the implications of the changing relationship between transitory and chronic vulnerability.

2 Regional variations

Although global data mask significant inter- and intra-country differences, consideration of regional variations in vulnerability is instructive. According to FAO/WHO data, major improvements in world nutrition have occurred primarily in Asia and the Pacific region, where the number of chronically undernourished has declined by nearly one-third over the past two decades (Table 1). In Africa, on the other hand, the number of chronically undernourished has steadily increased, and was nearly two-thirds higher in 1988-90 than 20 years previously. (However, because of population increases, the proportion of people in Africa estimated to be malnourished has remained stable at about one-third of the total population.) The FAO study *Agriculture: Toward 2010* (FAO, 1993a) suggests that sub-Saharan Africa is the only region where increases in the production of food will not keep pace with population growth.

In the Latin American, Caribbean and Near East regions, reductions in chronic undernutrition achieved in the 1970s were followed by an increase in the 1980s (although, because of population increases, the proportion of the population affected remained unchanged in both regions).

FAO/WHO data for the ICN do not show the recent and dramatic decline in the situation of people in many of the countries of Eastern Europe, ex-Yugoslavia and the former Soviet Union. Latest available FAO data suggest that total per capita daily food supply in these countries, although still comfortably above the average for developing countries, has fallen significantly (FAO, 1993b). Furthermore, there have been major changes in diet, with consumption of animal proteins declining dramatically, to be replaced by cereals. In 1993, this region was the largest recipient of food aid, receiving 40 per cent of world food aid; the

Russian Federation alone received nearly half of all food aid provided to this region (or one-fifth of all food aid - see Table 2). Although less than 10 per cent of food aid provided to this region was categorised by donors as relief assistance in 1993 – mainly for refugees and people displaced by ethnic and civil conflict – much of the food aid was provided for free distribution to vulnerable and poverty-stricken groups.

Improvements in food availability in Asia have not only been dramatic; they have also been largely unpredicted. In the late 1960s and early 1970s, the main concern was with the spectre of famine in the Indian sub-continent and China. For example, Paul Ehrlich wrote in 1968: 'I have yet to meet anyone familiar with the situation who thinks India will be self-sufficient in food by 1971, if ever' (Ehrlich, 1968). In the book *Famine 1975* (Paddock and Paddock, 1967), India was classified as one of the countries that could not be saved. Bangladesh was referred to as a 'basket case' at the World Food Conference in 1974.

Despite dramatic improvements, the situation in Asia continues to give cause for concern. The region still contains the largest number of the developing world's chronic undernourished – 528 million according to FAO/WHO estimates, or two-thirds of the total. But when India was struck by widespread drought in 1987, affecting an estimated 285 million people, the country was able to cope mainly from its own resources, and did not need large-scale imports of relief food aid as in the past. This is not to say that many poor people did not suffer, but the widespread starvation predicted for India in the 1960s did not happen. The international community remains rightly concerned with problems of alleviating poverty in Asia, but there are now few fears concerning mass famine, at least for the foreseeable future.

The concern with Asia's food problems in the 1950s and 1960s was largely responsible for the major improvements achieved in the 1970s and 1980s. The allocation of significant resources for agricultural research, applied technology and irrigation development in the region, best exemplified in the Green Revolution, was partially stimulated by graphic depictions in the 1950s and 1960s of devastating mass starvation in Asia.

Just as improvements in food security in Asia were unforeseen 30 years ago, so too was the general decline in the food situation in sub-Saharan Africa. There were few, if any, warnings of potential problems; before the early 1970s, virtually no emergency food aid was provided

Table 2 Food aid deliveries by recipient region and category, 1988-93

	Year	Cereals (in grain equivalent) 1000 mt				Non-cereals 1000 mt			
		Total	Non-project	Relief	Project	Total	Non-project	Relief	Project
Sub-Saharan Africa	1988	4124.3	1658.6	1911.4	554.3	422.6	53.3	242.5	126.8
	1989	2581.8	1068.2	1025.4	488.1	295.7	53.9	137.2	104.5
	1990	2907.5	752.2	1716.8	438.4	346.2	28.2	227.3	90.7
	1991	3592.9	963.4	2127.4	502.1	397.2	24.4	284.4	88.4
	1992	5472.4	1833.8	3151.2	487.4	502.6	29.6	389.0	84.0
	1993	4359.0	1153.7	2609.6	595.7	547.7	13.8	417.6	116.4
Asia and Pacific	1988	4363.0	1528.6	900.5	1933.9	691.9	418.0	75.9	197.9
	1989	3031.1	1174.8	854.2	1002.1	336.2	193.6	34.3	108.3
	1990	2540.4	811.6	557.3	1171.4	348.5	241.2	33.9	73.5
	1991	2866.1	796.4	638.3	1431.4	188.5	71.8	25.0	91.7
	1992	2433.2	816.9	623.8	992.5	283.6	154.0	46.0	83.6
	1993	1666.9	601.8	303.3	761.7	318.1	204.8	46.6	66.7
Latin America and Caribbean	1988	2250.1	1648.9	34.7	566.5	324.6	154.4	19.3	150.9
	1989	1869.9	1218.4	64.2	587.3	173.5	67.3	5.3	100.9
	1990	2092.8	1520.3	46.1	526.5	207.5	86.9	8.9	111.6
	1991	1807.0	1328.7	23.2	455.1	212.3	75.5	15.1	121.8
	1992	1548.0	976.5	21.5	549.9	269.4	111.5	1.8	156.0
	1993	1835.0	1223.8	30.8	580.4	203.6	56.8	4.2	142.6
North Africa and Middle East	1988	2414.9	1890.2	72.8	451.9	256.7	158.2	24.0	74.6
	1989	2635.2	1992.1	230.7	412.3	184.0	78.5	36.9	68.5
	1990	2819.0	2300.6	140.2	378.1	135.3	48.1	40.3	46.9
	1991	2946.4	2367.8	319.4	259.2	148.4	59.5	58.7	30.2
	1992	1486.2	1006.5	250.2	229.5	149.6	74.7	42.6	32.3
	1993	1034.4	523.6	342.5	168.3	106.4	32.7	43.2	30.5
Eastern Europe and former USSR	1988	-	-	-	-	-	-	-	-
	1989	609.4	609.4	-	-	17.0	17.0	-	-
	1990	2102.9	2102.8	0.1	-	146.2	146.0	0.2	-
	1991	770.0	765.2	4.8	-	229.7	196.2	33.5	-
	1992	2465.3	2239.2	226.1	-	592.3	352.2	240.1	-
	1993	6384.8	5823.2	561.7	-	668.4	510.2	158.2	-
World total	1988	13152.3	6726.3	2919.4	3506.6	1695.9	783.9	361.7	550.2
	1989	10727.3	6063.0	2174.4	2489.9	1006.3	410.4	213.7	382.2
	1990	12462.6	7487.6	2460.6	2514.5	1183.7	550.4	310.5	322.8
	1991	11982.4	6221.6	3113.1	2647.8	1176.1	427.3	416.7	332.0
	1992	13405.1	6873.1	4372.8	2259.3	1797.4	722.0	719.5	355.9
	1993	15280.0	9326.2	3847.8	2106.1	1844.3	818.3	669.7	356.2

Table 3 Cereal production, imports, exports, total supplies per capita and food aid, 1984/85 to 1993/94

	Unit	1984 / 85	1985 / 86	1986 /87	1987 /88	1988 /89	1988 /89	1989 /90	1990 /91	1991 /92	1992 /93	1993 /94*
Developing countries												
Production[1]	million t	771.0	773.7	791.7	786.2	823.5	823.5	839.6	869.5	876.6	901.9	904.2
Per-capita production[1]	kg	215.9	212.0	213.0	207.5	212.3	212.3	212.0	215.1	212.9	214.7	211.4
Imports[2]	million t	107.7	97.6	108.2	116.5	116.4	116.4	122.8	113.1	129.0	127.0	122.9
Exports[2]	million t	37.1	39.0	31.9	27.5	32.6	32.6	27.5	31.4	46.0	42.7	41.7
Net imports[2]	million t	70.6	58.6	76.3	89.0	83.8	83.8	95.3	81.7	83.0	84.3	81.2
Food aid	million t	12.5	10.9	12.5	13.6	11.3	11.3	11.1	13.0	12.6	15.2	14.3
Food aid/ imports	%	11.6	11.2	11.6	11.7	9.7	9.7	7.8	9.0	9.8	12.0	11.6
Total supply per capita[3]	**kg**	**235.6**	**228.1**	**233.5**	**231.0**	**233.9**	**233.9**	**236.1**	**235.3**	**233.0**	**234.7**	**230.4**
Sub-Saharan Africa												
Production[1]	million t	36.4	50.3	53.2	48.8	57.8	57.8	55.0	50.9	55.6	51.6	57.3
Per-capita production[1]	kg	88.2	117.5	122.1	109.1	125.6	125.6	115.7	103.9	110.1	99.3	107.2
Imports[2]	million t	12.6	10.2	9.7	9.1	8.2	8.2	8.0	10.3	10.4	14.0	11.7
Exports[2]	million t	0.3	1.1	1.2	1.8	1.4	1.4	1.6	1.4	1.2	0.8	0.8
Net imports[2]	million t	12.3	9.1	8.5	7.3	6.8	6.8	6.4	8.9	9.2	13.2	10.9
Food aid	million t	4.9	3.9	3.3	3.8	2.9	2.9	2.6	3.4	4.3	5.0	3.3
Food aid/ imports	%	38.9	38.2	34.0	41.8	35.4	35.4	32.5	33.0	41.3	35.7	28.2
Total supply per capita[3]	**kg**	**118.0**	**138.8**	**141.6**	**125.5**	**140.4**	**140.4**	**129.2**	**122.1**	**128.3**	**124.8**	**127.6**

[1] Data refer to the calendar year of the first year shown. Rice is in terms of milled rice.

[2] For total grain, season beginning 1 July of the first year shown; for rice, the calendar year of the second year shown.

[3] All per capita supplies are based on production and net imports, and do not consider stock changes.

* Provisional.
Source: WFP/INTERFAIS database.

to Africa. Although famines occurred, they were mostly localised. But over the past two decades such incidents have become increasingly widespread. Drought in the early 1970s principally affected the Sahelian countries of west Africa, but many more countries were afflicted in the mid-1980s. Sub-Saharan Africa became a net food importer in 1962 (Singer, 1989) and per capita food production there has shown an almost steady decline since 1974 (although total food production has increased over the period – see Table 3). However, current international concern with the food security situation in sub-Saharan Africa may be traced from the 1983-85 food crisis.

Since then, the multiple and pressing problems of Africa have increasingly come to dominate the attention of the international community (World Food Programme and African Development Bank, 1987; World Bank, 1989a, 1989b; World Bank and World Food Programme, 1991; IFPRI, 1991). The seemingly endless crises faced by Africa, increasingly man-made, and the lack of apparent success in resolving them, have resulted in 'donor fatigue'. Food aid deliveries to sub-Saharan Africa declined from 1992 to 1993, but this is explained by the particularly high deliveries to southern Africa in 1992 to forestall the effects of widespread drought in that sub-region; 1993 food aid deliveries to sub-Saharan Africa were the second highest on record (see Table 2). However, these data do not account for additional aid that might otherwise have been made available to help resolve Africa's problems. There is evidence that resources provided to the countries of Eastern Europe and the former Soviet Union have come, at least in part, at the expense of aid to Africa.

Recently, attention has shifted to Eastern Europe, ex-Yugoslavia and the former Soviet Union as a result of the problems posed by the change from centrally planned to market economies and the break-up of the Union of Soviet Socialist Republics, often accompanied by outbreaks of ethnic violence. The nature of food insecurity in the countries of the Commonwealth of Independent States has changed fundamentally in the past few years (FAO, 1993c). Current problems are mainly related to low incomes and inadequate purchasing power following price deregulation and sharp increases in retail prices of staple foods; the widespread supply shortages that characterised 1991/92 have eased. Although food availability has improved, high prices restrict access of poor and disadvantaged households to the food they need. The proportion of the vulnerable population varies from country to country, and their access to food is likely to be made more difficult as unemployment increases and further planned price rises occur. Considerable pockets of malnutrition have emerged among less favoured regions and population groups. A study prepared by leading economists in early 1994 on the social and economic effects of reform over the past two years is reported to suggest that one in ten of the population of the former Soviet Union now earns below 'starvation wages' and that one-third is below the poverty line (*The Guardian*, 1994). It reports that in 1990 food bills represented 30 per cent of average incomes whereas, in 1994, 60-70 per cent of family income was needed for food. Those on fixed incomes, such as pensions, have been worst affected. Problems are exacerbated in the republics of the Caucasus and central Asia, where insecurity is widespread and transport infrastructure is inadequate.

The increase in the number of people affected by chronic malnutrition in the Latin American, Caribbean and Near-East regions during the 1980s is also a cause for concern. Particularly worrying is that, by and large, national governments and the international community have not reacted to these changes with increased assistance or new types of aid programmes. Exceptions are the concern for street children and attention given to the idea of self-supported people's canteens, both initiatives aimed at helping to alleviate symptoms of urban poverty in Latin America, and the high level of relief assistance provided to displaced people and vulnerable groups in the aftermath of the Gulf War. The fairly common perception holds that the nutrition problems of both these regions are primarily the consequences of misguided national and international economic policies (McAffee, 1991; Walton, 1989). This has implications for other regions undergoing, or likely to undergo, far-reaching economic restructuring (e.g. sub-Saharan Africa, Eastern Europe etc.).

3 The causes of vulnerability

The main cause of the African food crisis in the mid-1980s was drought (although the effects of drought were often worsened by warfare and civil strife). As a result, considerable effort has been put into developing drought prevention and preparedness schemes and early-warning systems in sub-Saharan Africa. If the response to the drought in southern Africa in 1992 is any measure, that effort would seem to have paid off (WFP, 1994). In Africa at least, the international community has shown that it is able to respond quickly and effectively to food shortages that are drought-induced. Response to other natural causes of food shortages throughout the developing world (e.g. floods, hurricanes, earthquakes, crop diseases, locust plagues) has often been less publicised, but also largely effective.

There is still room for improvements in response to natural disasters, however, especially in the reaction time of donors to information obtained from early-warning systems (Buchanan-Smith *et al.,* 1992). However, it is unclear whether further improvements in the design of early-warning systems (rather than in response time) would be worth the additional effort and expense required. Those drought-affected countries that had stable and democratic governments, generally peaceful conditions and a free national press, coped well with the food crises of the mid-1980s. For those countries, improvements in early-warning systems might not have resulted in more rapid response. The sub-Saharan African countries whose populations suffered most were those where access to information was restricted and where the flow of relief assistance was severely disrupted by warfare or civil unrest, which compounded the impact of drought.

Since the mid-1980s, armed conflict and civil unrest have come to dominate the food insecurity problems of sub-Saharan Africa. Countries such as Angola, Burundi, Ethiopia, Liberia, Mozambique, Rwanda, Somalia and Sudan have been torn apart by internal dissension. Elsewhere in the world, current major food emergency situations are also primarily conflict-driven, e.g. Afghanistan, Cambodia, Iraq, and countries of the former Soviet Union and ex-Yugoslavia. An estimated 90 per cent of relief food aid provided by WFP is currently for victims of armed conflict: less than 10 per cent is for those solely affected by natural disasters.

The international community is still searching for ways of resolving conflict, and of assuring the speedy and safe passage of humanitarian assistance to civilians caught up in conflict zones, despite some early successes, as with Operation Lifeline Sudan (Minear, 1990) and similar early operations in Angola and Mozambique. The use of international or United Nations peace-making forces (rather than peace-keeping forces) to try to impose some sort of order to enable the delivery of humanitarian relief (as in Liberia, Iraq, Somalia and ex-Yugoslavia) so far seems to have resulted at best in no more than temporary arrangements that have little impact in the longer term. Internal conflict and civil unrest pose major challenges to improving food security in many countries.

In other countries, emergency situations have arisen as a result of major economic disruptions (which may, in turn, either result from, or aggravate, civil turmoil and conflict). Poverty, still the main underlying cause of hunger and malnutrition, results, in large part, from adverse economic conditions. But in some regions (especially in Latin America and the Caribbean, but also many African and Eastern European countries), a rapid worsening in the economic condition of the most vulnerable, often resulting from sudden changes in government policies and inappropriately designed and misapplied structural adjustment measures (Cornia *et al.*, 1987), has resulted in increased vulnerability of better-off population groups not normally considered at risk. The urban poor and lower-middle class, especially of societies in transition from subsistence to cash economies, have been particularly affected.

In many cases, deteriorating economic conditions affect the poorest people first, but when there are severe jolts to the economy (such as rapid inflation, large fluctuations in exchange rates, collapse of banking systems, etc.) the emerging middle class may be worst hit, both because it is more reliant on the formal economic system (in terms of savings, reliance on fixed salaries) and because it has fewer informal coping mechanisms. In a number of developing countries (for example, Bolivia, Ghana, Lebanon, Peru) and in the countries of Eastern Europe, such economic crashes have come with little warning. In these situations, people can appear to be wealthy in terms of owning consumer goods (such as colour televisions, cars or refrigerators) – for which there is little resale value – but lack the purchasing power to buy food. Such economically-induced emergency situations may become more common in the coming years, particularly as the planned economies of formerly socialist countries undergo further transformation.

4 Population groups at risk

Poor rural populations continue to remain the most numerically affected by food insecurity. The landless, women-headed households, small-scale, single-crop farmers, small-scale pastoralists and subsistence farmers are among the socio-economic groups most at risk. They are the groups on whom relief agencies currently concentrate.

Half of the world's population now lives in urban areas; the proportion of total population living in urban areas in developing countries ranges from 27 per cent in southern Asia to 70 per cent in Latin America (Von Braun *et al.*, 1993). This rapid urbanisation has been accompanied by increasing urban poverty, and associated food and nutrition problems. The urban poor have often been overlooked by relief agencies. The situation is particularly acute in peri-urban slums and shanty towns, areas that usually house recent immigrants from rural areas looking for work. In conflict situations, urban populations are often swollen by people displaced by fighting in the countryside seeking shelter and assistance. Because urban areas are more dependent on functioning markets and stable economic conditions, they become particularly vulnerable when these are disrupted, whether by changed government policies, warfare, civil unrest or changing international economic circumstances.

Urban populations typically have become more used to, and usually expect, more expensive commodities. In sub-Saharan Africa, a strong tendency has been observed in urban areas for the so-called 'preferred' cereals of wheat, rice and maize, to displace traditional African staples, such as sorghum and millet. More highly milled grain (a whiter and finer product with a reduced nutritional content) also tends to be preferred. In Latin America, consumption of wheat, sugar, oil and animal protein has increased, whereas that of the more traditional staple foods, based on maize, beans or other legumes, and tubers, has decreased (Von Braun *et al.*, 1993). Thus, the food and nutrition problems of the urban poor are substantively different from those of the rural poor (Hussain, 1990).

Being more reliant on processed foods, urban people are less likely to have any means of processing raw foods, such as mills or querns to grind grain to flour (although there are regional differences – small hand-milling equipment is still fairly common in many African cities). The food aid provided has to consist of the more expensive processed foods (such as flour rather than grain and canned food items).

Targeting assistance on urban populations as a whole is not a major problem. Many urban-assistance programmes have adopted a macro-approach, often trying to improve the conditions of the urban poor through policy and economic changes. However, it is less easy to target particular segments of the urban population (such as the poorest 10 per cent, or particular ethnic, age or gender groups), except through micro-projects benefiting very small numbers of people (such as projects by non-governmental organisations for small numbers of street children). Untargeted food subsidy programmes have been shown to be an expensive way of improving food consumption and nutrition among the poor, although the cost of highly targeted subsidy schemes may be low (Pinstrup-Andersen, 1988); food stamp programmes have been subject to fraud, counterfeiting and bribery, and in addition require stable government and efficient administrative structures (Lorge-Rogers, 1988). More should be done to design self-targeting mechanisms for assistance, and especially food-aided, programmes in urban areas, through, for example, the provision of food commodities only, or largely, consumed by the poor.

Some assistance has been channelled to the urban poor through labour-intensive, public works programmes. For example, WFP has used food aid to help pay workers employed in urban renewal programmes, including improving sanitation facilities, housing, roads and footpaths. In Zambia, a WFP-supported urban project has created employment for more than 3000 women, who are paid in daily food rations. Food for their families is more attractive to these women than cash, because of high food costs. Food is also preferable to cash because, in the unanimous opinion of project authorities, many women would not be allowed by their husbands to keep cash earnings.

Although urban renewal through food-for-work is an improvement over free food distribution, and the improved health and sanitation conditions that result are a useful contribution, these programmes do not by themselves result in a sustained improvement in the economic situation of the urban poor. The design of urban assistance programmes that do more than alleviate immediate symptoms of poverty and hunger remains a challenge. Given the expectation that the urban population in developing countries will continue to increase, and that problems of urban poverty are likely to worsen (UNDP, 1991), more effective and efficient means of assisting the urban poor to make permanent improvements in their lives and livelihoods should become a higher priority (Shaw *et al.*, 1994).

Within rural and urban households, the people who remain especially vulnerable to food problems are young children, women, the elderly and the handicapped. Designing programmes and projects that direct assistance to these groups, without unduly interfering with traditional markets, social structures and the independence of the household, remains a major challenge.

5 Changes in coping ability

The poor have to be skilled at devising coping strategies that make the best use of their resources and opportunities to reduce their vulnerability to hunger. If they were not, many would not survive. Elements of the anti-hunger strategies of poor households and communities are to be found in various combinations at times when, for whatever reason, access to food is limited. When times become especially difficult, as during the seasonal hungry period, or during years of acute food shortage and famine caused by continuous drought or war and civil unrest, the result is an extension and modification of the coping strategies employed to overcome the effects of poverty on access to food in more normal times, as well as the adoption of new strategies to help meet new requirements.

For the sake of analysis, the anti-hunger strategies of poor households and communities may be divided into two broad and interrelated categories: (a) accumulation strategies and (b) diversification strategies (WFP, 1989).

Accumulation strategies aim to increase a household's resource base. They include land and other asset acquisition and food stocks, as well as less obvious elements, such as increasing the size of the household, expanding links with extended family members or other social groups and improving education and health. Accumulation strategies often have an anti-hunger component (i.e. the reserve accumulated may be drawn upon in times of hunger) but are not necessarily confined to anti-hunger mechanisms.

Strategies devised for diversification, which is a form of insurance, seek to promote a variety of income sources, so as to avoid the risks associated with relying on a single source. Diversification strategies in rural areas include crop diversification, seasonal hunting and gathering, and off-farm employment. An important aspect is that the different sources of income should not be exposed to the same vulnerability of risk.

Anti-hunger strategies may involve several interrelated activities that together are designed to ensure that food consumption is stable and sufficient. Coping strategies may have taken generations to develop. They have been adapted over time to changing circumstances. But they are often fragile, and may be overwhelmed by catastrophes caused by nature or by man.

In many situations this has resulted in the reduction or disappearance of the anti-hunger options previously available to poor households.

As traditional coping mechanisms have come under increasing pressure, many societies have found themselves unable to contend with food shortages, resulting in an increase in both the number of people dependent on national and international assistance and in the period for which they rely on that assistance. Coping mechanisms appear to have become less effective both because social and economic transitions make them less applicable (for example, population increases; breakdown in traditional family and ethnic support mechanisms) and because many societies have been under pressure for so long that they have exhausted whatever assets and coping mechanisms they were able to bring to bear.

The example of the nomadic pastoralists in Kenya illustrates both these aspects (Sachers, 1992). One of the ways in which nomadic pastoralists in Kenya were able to cope with the severe 1984-85 drought was by migrating to better rangeland conditions and water supplies in neighbouring countries. This option was no longer feasible in 1992, when Kenyan pastoralists were again severely afflicted by drought. Political instability in nearly all of the countries neighbouring Kenya meant that not only was migration to those countries precluded, but also that large numbers of refugees from those countries, including pastoralists and their herds, flooded into the border areas of Kenya, thereby creating extra demand on already scarce food, water, grazing and fuelwood resources. The easy access to weapons, supplied in large quantities to the various warring factions in the neighbouring countries, combined with traditional enmities, led to a rapid decline in the local security situation, reducing still further the availability of already scarce areas where pastoralists could find peaceful grazing for their herds. As a result, many pastoralists were compelled to take their animals into tsetse fly zones, increasing the risk to their weakened and emaciated herds.

Increasing insecurity, including attacks on traders' trucks, meant that commercial supplies of food and other essential commodities became more scarce in pastoral areas and prices rose. At the same time, there were no market outlets available to enable pastoralists to sell their animals. Nor were there alternative employment opportunities for pastoralists. A WFP/FAO drought assessment mission in February 1993 estimated that, as a consequence, about half of pastoralists' herds had died since 1990, and approximately one million people from pastoralist households would require emergency assistance for the coming year.

One effect of the breakdown in traditional coping strategies is to increase the flow of rural poor to urban areas. The increase in urban poverty has resulted in a population with few, if any, traditional mechanisms to cope with sudden food shortages. Seeking a better lifestyle, or displaced by fighting or severe drought, poor rural migrants flock into squatter settlements, often on the periphery of urban areas, where sanitation facilities are poor or non-existent and community infrastructure and employment opportunities are limited. Local governments often neglect these areas, particularly when their populations may differ in ethnic or religious composition from the settled urban population and there may be major cultural differences between the two population groups.

Whereas most rural people – farmers and townsfolk – maintain some sort of food in storage for lean periods, urban populations normally rely on the availability of regular incomes, functioning markets and frequent shopping for their food supplies. Even if urban households wish to store food supplies against possible future shortages, most urban habitations have little or no space to maintain such food stocks.

Recent migrants to urban areas, however, often maintain links with rural kinsfolk. These links with rural family members often provide small food parcels, useful supplements to incomes. However, these links can be easily broken, especially in times of conflict.

Because they are so reliant on the sound functioning of economic systems and markets, with few coping systems to fall back on, the urban poor can go from relative plenty to almost total reliance on relief food aid in a remarkably short period of time. Crises afflicting the urban poor often come with little or no warning indicators – at least of the sort relief agencies have come to rely on. As a result, relief agencies may have little time to arrange for the import of relief food aid and the establishment of effective food distribution mechanisms. Food imports may be important because local markets have been disrupted; any attempt to purchase food locally may only worsen the situation, by further raising prices. Although established market mechanisms may provide an effective means of reaching the bulk of the urban population, the food may still be too expensive for many of the urban poor, particularly in crisis situations. In a number of cases, assistance agencies have tried to channel relief food aid to the urban poor through schools and health centres. These facilities may provide useful distribution mechanisms, but such formal institutions tend to cater for the relatively better-off.

Further work is needed to develop both more appropriate early indicators of urban food problems and to design more effective distribution systems for relief food aid in urban areas. Relatively little work has been done on either of these aspects. Particularly in less democratic systems, where the poor have few ways of making their plight known, special methods may be required both to identify those in need, and to provide assistance packages.

6 Chronic and transitory vulnerability[1]

The past five years have seen a massive increase in the need for, and provision of, international relief food aid. Globally, relief food aid now accounts for some four million tons annually, an increase of some 50 per cent over the previous five-year annual average (Table 2). Relief food aid accounts for more than one-quarter of all food aid.

With the massive increase in total relief food aid needs, there has been concern that resources for development, and for overcoming chronic hunger and malnutrition, might decline. The latest available data do not show a significant sustained decline in either programme or project food aid.[2] Programme food aid, although fluctuating over the past five years, significantly increased in 1993, as a result of large shipments to the countries of the former Soviet Union. Project food aid appears to have levelled off after a decline from peak levels in the late 1980s

---

[1] Food insecurity is either chronic or transitory. Chronic food insecurity involves a continuously inadequate diet. Transitory food insecurity is a temporary decline in a household's access to enough food.

[2] *Programme food aid* is provided bilaterally as a grant or on soft repayment terms for balance of payments and budgetary support to help fill the gap between demand at existing income levels and the supply of food from domestic production and commercial imports. *Project food aid* is aimed at transferring income directly to the poor or satisfying their minimum nutritional needs in normal years.

(Table 2). On the basis of food aid flows,[3] the indications are that (at least so far) the relief needs of the 1990s have been largely met by additional resources.[4]

Nevertheless, given the general expectation that aid budgets will decline in real terms, and the expected increase in the needs of some of the countries of Eastern Europe, ex-Yugoslavia and the former Soviet Union over a protracted period, there is widespread concern that the pressing political and humanitarian need to meet immediate relief requirements will jeopardise the continued flow of resources to overcome chronic vulnerability through development programmes.

What does seem to be true (although it has not so far been possible to quantify the shift) is that relief needs are assuming greater importance within aid agencies, attracting additional staff and attention. Because most aid agencies have had staff levels frozen (or even reduced), this must inevitably result in decreased attention to development assistance to overcome chronic vulnerability. Even if resource flows to development activities remain unaffected, therefore, the impact of those resources may be reduced as less staff time is available to ensure that they are used in the most effective and creative manner to overcome chronic vulnerability. The resources of national governments may be similarly diverted by increasing pressure to meet relief needs.

This shift in attention from chronic to transitory food insecurity is occurring despite much recent work pointing to the complex interrelationships between the different causes of chronic undernutrition, and hence the more sophisticated approaches needed to overcome it (see, for example, Drèze and Sen, 1989; Bohle *et al.,* 1995, this volume). Thus, at the very time when more is being understood about the causes of vulnerability and chronic food insecurity, it

---

[3] It is unclear whether food aid flows are a satisfactory indicator of overall aid flows or of donor intentions, but unfortunately, because of the time lapse in publishing data on aid flows, reliable figures on total aid are not recent enough to show latest developments.

[4] Benson *et al.* (1994) suggest that current high levels of programme food aid to the countries of the former-Soviet Union reflect the practice of donors turning to food aid as a knee-jerk reaction to problems, providing food aid as an initial, politically determined response. Thus, as donors aid programmes for the countries of Eastern Europe mature, the flow of food aid to these countries may substantially decline.

appears likely that fewer resources, manpower and attention will be made available to help fight the problem.

7 Conclusions

Food insecurity and vulnerability are not static, yet the international community has not shown itself to be able to cope well with the recent shifts in vulnerability that have been outlined in this paper. Too little attention in the past has been paid to the dynamics of vulnerability.

New approaches are needed to identify the changing incidence of poverty and vulnerability as national and international situations evolve, because of climate change or other factors. New indicators of vulnerability, taking into account changing conditions, are also required. Mapping population groups and areas that are particularly vulnerable may be a useful first step in helping to target assistance, but special efforts will be required to keep such data up to date and relevant.

It has been claimed that 'generals always fight the next war with the tactics of the last'. As a generalisation, that sentiment also seems to have been true of development and relief agencies. The challenge will be to change this situation in the future.

References

Benson, C., Clay, E. and Dhiri, S. (1994). *Additionality and Attribution: Food Aid and Export Credits to Eastern Europe and the Former Soviet Republics.* Overseas Development Institute, London.

Buchanan-Smith, M., Davies, S. and Petty, C. (1992). *Famine Early Warning Systems and Response: The Missing Link? Summary of Findings and Conclusions.* Save the Children Fund (UK) and Institute of Development Studies, University of Sussex, Brighton, UK.

Cornia, G.A., Jolly, R. and Stewart, F. (1987). *Adjustment with a Human Face: Protecting the Vulnerable and Promoting Growth.* Clarendon Press, Oxford.

Downing, T.E., Watts, M.J. and Bohle, H.G. (1995). Climate Change and Food Security: Towards a Sociology and Geography of Vulnerability In: Downing, T.E. (ed.). *Climate Change and World Food Security.* Springer-Verlag, Heidelberg.

Drèze, J. and Sen, A. (1989). *Hunger and Public Action.* Clarendon Press, Oxford.

Ehrlich, P. (1968). *The Population Bomb.* Ballentine, New York.

FAO (Food and Agriculture Organization) (1993a). *Agriculture: Toward 2010.* FAO Conference Document C93/24, Rome.

FAO (Food and Agriculture Organization) (1993b). *The State of Food and Agriculture.* FAO, Rome.

FAO (Food and Agriculture Organization) (1993c). *Assessment of the Current World Food Security Situation and Recent Policy Development.* Document CFS: 93/2, January. FAO, Rome.

FAO (Food and Agriculture Organization) and WHO (World Health Organisation). 1992. *Nutrition and Development – A Global Assessment.* FAO, Rome. Table 2, p.7.

Hussain, M.A. (1990). Nutrition policy and the urban poor in developing countries. *Food Policy,* **15** (3): 186 - 192.

IFPRI (International Food Policy Research Institute) (1991). *A Policy Agenda for Famine Prevention in Africa.* Food Policy Report. IFPRI, Washington, DC.

Lorge-Rogers, B. (1988). Design and implementation considerations for consumer-oriented subsidies. In: Pinstrup-Andersen, P. (ed.) *Food Subsidies in Developing Countries: Costs, Benefits, and Policy Options.* Johns Hopkins, Baltimore. pp. 127-146.

McAffee, K. (1991). *Storm Signals, Structural Adjustment and Development Alternatives in the Caribbean.* Zed Books, London.

Minear, L. (1990). *Humanitarianism under Siege – A Critical Review of Operation Lifeline Sudan.* The Red Sea Press and Bread for the World Institute on Hunger and Development, Washington, DC.

Paddock, W. and Paddock, P. (1967). *Famine, 1975! America's Decision: Who Will Survive*? Little Brown, Boston.

Pinstrup-Andersen, P. (ed.) (1988). *Food Subsidies in Developing Countries, Costs, Benefits and Policy Options.* Johns Hopkins University Press, Baltimore.

Sachers, W. (1992). *Assessment of Food Aid Requirements of Kenya.* World Food Programme, Rome.

Shaw, J., Crawshaw, B. and Fortier, F. (1994). The use of food aid to help alleviate urban poverty: Issues and challenges. *GeoJournal, Special Issue: Cities of Hunger: Urban Food Systems in Developing Countries,* **31** (11): 31-45.

Singer, H. (1989). The African food crisis and the role of food aid. *Food Policy,* **14** (3): 196-206.

The Guardian (1994). Shock therapy meant mass starvation, Yeltsin told. 3 February.

UNDP (United Nations Development Programme) (1991). *Cities, People and Poverty, Urban Development Cooperation for the 1990's.* Strategy Paper. UNDP, New York.

UNDP (United Nations Development Programme) (1993). *Human Development Report 1993.* Oxford University Press, Oxford.

Von Braun, J., McComb, J., Mensah, B. and Pandya-Lorch, R. (1993). *Urban Food Insecurity and Malnutrition in Developing Countries, Trends, Policies and Research Implications.* International Food Policy Research Institute, Washington, DC.

Walton, J. (1989). Debt, protest and the state in Latin America. In: Eckstein, S. (ed.) *Power and Popular Protest: Latin American Social Movements.* University of California Press, Berkeley.

World Bank (1989a). *The Challenge of Hunger in Africa. A Call to Action.* World Bank, Washington, DC.

World Bank (1989b). *Sub-Saharan Africa: From Crisis to Sustainable Growth.* World Bank, Washington, DC.

World Bank (1990). *World Development Report, 1990: Poverty.* Oxford University Press, New York.

World Bank and World Food Programme (1991). *Food Aid in Africa: An Agenda for the 1990's.* World Bank, Washington, DC, and WFP, Rome.

WFP (World Food Programme) (1989). *Anti-hunger Strategies of Poor Households and Communities: Roles of Food Aid.* WFP/CFA:27/P/INF/1, Add.1. WFP, Rome.

WFP (World Food Programme) (1994). *Terminal Evaluation Summary Report on Umbrella Operation Southern Africa Drought.* Terminal Evaluation Summary Report on Umbrella Operation Southern Africa Drought (Emergency Operations Nos. 5052-60) – assistance to drought victims, CFA:37/SCP:12, Document 12/6-B. FAO, Rome.

World Food Programme and African Development Bank (1987). *Food Aid for Development in Sub-Saharan Africa.* Report on a seminar held at African Development Bank Headquarters, Abidjan, Cote d'Ivoire, 8-11 September 1986. WFP, Rome and ADB, Abidjan.

The Conjunction of Threats to Regional Food Production: How Serious Are Environment, Economy, Population and Climate?

David Norse
Department of Geography
University College London
26 Bedford Way
London WC1H 0AP

1 Introduction

Environmental degradation, economic growth, the population 'explosion' and climate change have been put forward as major ongoing or potential threats to developing region or global food security (Brown and Kane, 1994; Doos, 1994; Fischer *et al.*, 1994). Much of the evidence for such views comes from the extrapolation of local or sub-regional measurements and analyses, and commonly involves the application of computer models that, although numerically complex, tend to be functionally simplistic in terms of their ability to represent natural processes and economic systems. Can we, therefore, be any more confident in these prognoses than those made 200 or 20 years ago? They have proved to be too pessimistic, because of their absolute or relative neglect of key factors like technology and external trade in food (Malthus, 1798; Carson, 1962; Meadows *et al.*, 1972; IFPRI, 1987; U.S. Council on Environmental Quality, 1982).

How can we assess the evolution of these threats over time, given that all except climate change are problems of today in some areas of all of the developing regions, and given large uncertainties about future economic and population growth rates? For example, it may be at least ten years before computer models will have the data and structure to give reliable advanced warning of climate change and 50 years before the negative aspects of climate change on some aspects of food production are clearly apparent.

Is it possible on the other hand, that projections have underestimated the risks which these threats might pose to food production? In general the above prognoses stem from studies that looked at each threat relatively independently and did not consider their possible conjunction, which could seriously compound their impact on food security. These negative interactions include, for example, the complex links between poverty and environmental degradation, which tend to force

NATO ASI Series, Vol. I 37
Climate Change and World Food Security
Edited by Thomas E. Downing

poor people to adopt unsustainable food production techniques, and those between economic growth and food security given the importance of individual and national economic access to food, i.e. purchasing power, rather than simply the resources or ability to produce more food.

But then the conjunction of these threats does not necessarily lead to negative consequences. Population pressure may bring about the correction of environmental degradation and the adoption of technologies which raise food production potential (Boserup, 1965, 1993; Tiffen *et al.*, 1994). Economic growth may raise incomes, thereby increasing the ability of food deficit countries to import what they need and improve education levels, particularly those of women, leading to a more rapid decrease in fertility rates and hence slower population growth.

The food situation in the developed regions differs greatly from that of the developing ones. First, their aggregate food production potential may not be seriously affected by climate change, and could even be a net beneficiary given appropriate responses by them to shift production spatially and change technologies. Second, most of their environmental problems are ones of social choice, in that they are within their economic and technical means to resolve, given appropriate changes in lifestyle or consumption preferences, and in their regulatory environment. Third, they have sufficient economic strength to buy or invest their way out of any domestic food supply constraints. Fourth, their populations are nearly stable or growing slowly. There is, however, one possibility that runs counter to these four positive features. It has been argued that most developing countries will find it impossible to produce much of the food needed to meet their projected long-term demand (Carruthers, 1993). If this were the case then the main suppliers would be the major developed country cereal producers, since they have the potential capacity, but the required increase in production would be so high that it would inevitably place serious pressure on their environments (see Crosson and Anderson, 1995, this volume).

Consequently, this chapter first considers the developing region food problem in more detail; then looks at some of the methodological difficulties in undertaking a balanced assessment of the four threats; moves on to examine the nature of each of the threats as they apply to the developing countries, emphasising the environment; and finally presents some conclusions.

2 Regional food production

There is little doubt that many developing countries will need to double their food production or raise their ability to pay for food imports if they are to feed the additional 1.5 billion mouths expected over the next 25 years (McCalla, 1994). Nor is there much disagreement that poverty and the lack of appropriate agricultural technologies are major factors behind the 10 million hectares deforested every year through shifting cultivation, and that this leads to a range of environmental problems. It is also evident that environmental degradation has forced people to migrate to other rural or urban areas to find new land or employment to provide either the physical means or the purchasing power to grow or buy adequate food (El-Hinnawi, 1985; von Braun and Pandya-Lorch, 1991). Such generalisations, however, do not automatically support the above views that the four threats may seriously undermine regional or global food security.

Some of the regional evidence is irrefutable:

- Food security in Asia is heavily dependent on irrigation, yet between 10 and 20 per cent of the irrigated land is threatened by salination. India has already lost around 7 million hectares owing to salination. Newly developed irrigation barely compensates for that lost through waterlogging and salination. If this situation continues, food production will be severely affected.
- Over the past 10-20 years per capita incomes, and therefore food purchasing power, have declined widely in sub-Saharan Africa (and to a lesser extent in Latin America), and short-term economic prospects are not good.
- All of the developing regions have substantial areas where rural population densities are now higher than can be fed by the available land and technologies.
- Climate variation in the main regions, particularly in the form of episodic droughts and floods, continues to provoke famines and compound longer-term food shortages stemming from institutional distortions and other factors.
- There are strong grounds for the view that global warming would intensify climate variation, and add to food supply problems.

On the other hand there are too many unknowns in both the natural and social sciences to be sure of even the magnitude of some of the positive and negative feedback relationships that exist between food production, climate change, the environment, the economy and population growth.

Moreover, past or present performance is not necessarily a sound indicator of future possibilities. In sub-Saharan Africa, for example, although population growth has exceeded food production since the 1960s, most of the agricultural problems and the consequent food insecurity have stemmed from institutional factors rather than any of the four threats (FAO, 1986). If these institutional factors can be overcome, the region has sufficient land and water resources to feed its projected population provided that:

- natural resource management improves considerably;
- agricultural research is expanded appreciably to generate sustainable, yield enhancing technologies appropriate for a wide range of locations with rather specific needs; and
- barriers to intra- and extra-regional trade are removed to permit food deficit countries physical and economic access to the potentially large food surpluses of countries like Angola and Zambia which could become major exporters to the rest of Africa once their current technical and infrastructural difficulties are removed.

This leads to four other limitations of narrow regional assessments of threats to food production and food security. First, in the coming decades, as in the past, many countries are likely to choose to increase their dependence on the world food market and on other regions and countries with greater comparative advantage in food production. If this is so, the key issue will be global rather than regional food production prospects.

Second, it is wrong just to consider food production, or to emphasise the potential impact of climate change on cereal production as most studies have done (Rosenzweig and Parry, 1994). For the main regional decision-makers, namely farmers and countries, the key issue will be whether non-food crop production, e.g. cotton, tea or coffee, is adversely affected, since they provide the income or the foreign exchange to buy or import food. Such changes are critical to make up deficits, use relative comparative advantage, and obtain a wider variety of foods.

Third, and related to the latter issue, food security is not simply about food and agricultural production. It also concerns the wider question of food purchasing power, and therefore the potential impact of climate change on incomes and employment both inside and outside agriculture. This is of equal or even greater importance than the impact on supply only.

Fourth, such assessments do not give adequate consideration to the possible conjunction of the four threats. Although it is still not clear how and where climate change may affect regional food production (Houghton *et al.*, 1995), conjunction of the other threats is already apparent, and seems likely to intensify. The links between poverty and environment are complex and involve both negative and positive aspects, but poverty undoubtedly drives some poor people to degrade the environment. Similarly, there is evidence for the negative impact of environmental degradation on food production, though again the relationship is complex and non-linear. Population growth compounds the problems of slow economic growth by reducing the income share going to each person, and increasing the proportion of public income that has to go on physical infrastructure rather than on direct welfare gains. Thus sub-Saharan Africa, which has a higher proportion of problem soils than any other developing region, is projected to have the fastest population growth, the lowest income growth, and the greatest risk from climate change. It is therefore important that regional assessments are carried out in a more dynamic and multisectoral framework.

3 Improving the assessment of the multiple threats

Although part of the gap between the pessimists and the optimists can be ascribed to differing conceptual models of the world and how it is changing, a large proportion clearly stems from the major methodological difficulties inherent in making balanced assessments. First, there is the lack of scientific understanding of key processes, as exemplified by the following. Although knowledge of the relationship between regional climate and sea surface temperature has improved and is now incorporated into global circulation models (GCMs), we have only a limited understanding of the role of the deep oceans in this relationship and its dynamics. Thus, there is still no consensus as to whether sub-Saharan Africa will get wetter or drier, though the general feeling tends towards the latter. Similarly, although there are many anecdotal statements about the magnitude of soil erosion and other forms of land degradation, there are few well conducted assessments of the impact of such environmental changes on crop yields, factor productivity, grassland output or on livestock productivity.

Second, there are difficulties in developing computer models that are able to capture the dynamic properties of real world economic, social and biological systems. A number of the computer models used to assess potential climate change impacts on food security project substantial cereal price increases. The more extreme results indicate price increases of 150-600 per cent, compared with the static or declining real prices of the past 50 years or more. Such results tend to be based on extreme assumptions, for example that neither farmers nor governments respond to climate change impacts on agriculture, or ignore the scientifically proven CO_2 fertilisation effect whereby higher atmospheric concentrations improve the efficiency of plant growth processes and may increase crop yields by 10-40 per cent (Kimball *et al.*, 1995). Even with more balanced assumptions, other models project cereal price increases of up to 100 per cent (Fischer *et al.*, 1994, 1995, this volume), yet it can be argued that farmers, grain markets and the agricultural research community would respond strongly to counteract production losses or price changes as they have in the past. In 1970, for example, when U.S. maize production was greatly reduced by a fungal disease, with 50 per cent losses in some states, resistant varieties of seed were quickly produced and distributed with the result that output was back to normal within about two years. Similarly, in 1972/73 when the USSR quietly cornered the world, and in essence the U.S. grain markets, causing a doubling of world maize prices, the use of maize for livestock fell rapidly, the area under maize increased substantially and other countries expanded their exports. Prices soon stabilised and then fell back to their long-term trend. Behavioural equations and market clearing mechanisms to reproduce such responses have yet to be introduced into climate impact assessment models.

A related methodological problem is that of projecting food crop yields in the absence of robust causal models of technological change. Assessment of the way each of the four main threats may affect food production is very sensitive to the assumptions made about the nature and rate of change in technological growth in the future, and how it influences the interactions between threats. World Bank and Food and Agricultural Organisation (FAO) studies suggest that the bulk of future incremental food production will come from intensification of production on existing land, and not from the development of new land (Crosson and Anderson, 1992, 1995, this volume; Alexandratos, 1995). But if this intensification is to raise food security without increasing environmental degradation, the new technologies will have to be more sustainable than those of today. If technological growth does not accelerate, economic development will be held back, with additional negative consequences for the environment and food security. Finally, producers and consumers will not have exhausted all of the possibilities for substituting other commodities or inputs for those that may be affected by climate change. Thus, if rice prices increase relative to wheat or maize prices, they could slowly change their consumption patterns. If feedgrain prices increase, livestock producers could switch to cassava and protein feeds, or use bovine somatotrophin (BST) and similar growth stimulants to lower feedgrain needs. If sugar prices rise, consumers or the food processing industry could switch to artificial sweeteners with not insignificant human health gains. Such limitations in the current models point to the need for more comprehensive impact analysis models that exhibit more closely the dynamics of the real world.

Projected regional food balances are consequently very sensitive to small changes in the underlying assumptions. Thus when regional climate changes projected by three of the most advanced GCMs were introduced into a global food production and trade model (Fischer *et al.*, 1994), cereal production in developing countries in circa 2060 was estimated to be 9-11 per cent less than in the situations with no climate change. Given that these projections were for a 70-year period, such a small supply gap could be closed by minor changes in the model's assumptions. Replacing the 0.7 per cent per annum cereal yield growth rate assumption of the model, which is about half the historic rate, with a 0.9 per cent growth rate would close the projected cereal supply gap. This illustrates the relatively small difference that exists between a static comparison (the reference and climate change scenarios) and a dynamic, goal-seeking assessment.

4 Multiple threats to regional food production

4.1 Environment

The environmental threats take two main forms. First, and by far the most important given our present understanding of the processes involved, are the threats to regional food production stemming from the impact of agriculture on the environment. Second, the potential impact of climate change on the environment, e.g. on rainfall intensity and erosion, is not well understood. This will be considered later when discussing the climate change threat.

Land degradation

A recent assessment of agricultural land degradation provides for the first time a global/regional picture that has been built up using a common methodology, global assessment of soil degradation (GLASOD) rather than from the pooled results of non-standardised estimations (Oldeman *et al.*, 1990). It provides estimates of the current extent and intensity of degradation in the main regions, with soil erosion being the most widespread cause of degradation and water the principle agent of erosion (Table 1). Asia has the greatest area of degraded land, but Africa has the highest proportion.

Table 1 Human-induced soil degradation (classified as moderately to excessively affected), million hectares.

	Erosion		Degradation		Total
	Water	**Wind**	**Chemical**	**Physical**	
Regions					
Africa	170	98	36	17	321
Asia	315	90	41	6	452
South America	77	16	44	1	138
North & Central America	90	37	7	5	139
Europe	93	39	18	8	158
Australasia	3	–	1	2	6
Total	748	280	147	39	1214
Major causes (in per cent)					
Deforestation	43	8	26	2	384
Overgrazing	29	60	6	16	398
Mismanagement of arable land	24	16	58	80	339
Other	4	16	12	2	93
Total	100	100	100	100	1214

Source: Adapted from Oldeman *et al.*, 1990.

The important issue for regional food production and security, however, is the intensity of degradation rather than the extent thereof. Two aspects are of particular importance. How has the intensity of degradation changed over time? What will be its impact on future crop and livestock productivity? Many of the data required to answer these two questions do not exist, and our understanding of the processes involved in intense erosion is limited. The information gaps are both physical and economic. For example, there are few data on the spatial patterns of deposition of eroded soil, yet if deposition occurs in the same region from which is was eroded, the net effect on food production might be minimal.

When regional *soil losses* from erosion are expressed in terms of billions of tonnes per year, the image created is one of great risk to food production and food security. However, much of the soil loss may be relatively localised and concentrated in steeply sloping or other marginal areas, which commonly play a secondary role in aggregate food production. And in per hectare terms the losses might be quite small and correctable by a combination of improved approaches to land management and soil conservation and productivity improvements.

The empirical evidence for heavy and extensive soil erosion comes from two main sources. Small catchment and plot experiments indicate losses from agricultural land that are greatly in excess of 50 tonnes of soil per hectare, and generally greater than those from natural ecosystems. These losses may be five or more times the natural rate of soil formation. However, there is little evidence that such point measurements are a valid basis for calculating average national or regional losses, which are likely to be considerably less (Biot *et al.*, 1992). The other source of evidence is suspended sediment levels in rivers. As with plot losses, large numbers can be misleading. First, when suspended sediment levels are translated into area loss terms, the implied erosion rates per hectare may be quite small. Second, there is increasing evidence that some of the extreme sediment levels which have been blamed on human activity are largely a consequence of natural tectonic movements. In China, for example, severe erosion of the loess plateau has been attributed largely to agriculture and deforestation, but there is now evidence that some 60 per cent of the erosion is probably due to movements in the earth's crust.

In the United States for example, it has been calculated that the continuation of soil erosion at the rates prevailing in the 1970s for a further 100 years would reduce crop yields by between three and ten per cent (Crosson, 1986; Alt *et al.*, 1989). Such yield reductions could be balanced by a fraction of a percentage point improvement in crop productivity over the projection period, which is quite feasible given recent rates of technical progress. There is a risk that the technological improvements might increase other environmental threats if the productivity gains arose from the greater use of conventional fertilisers and pesticides. However, this seems unlikely given the increasing regulatory and economic pressures against the environmentally damaging conventional approaches, which are a powerful force behind the current shifts to integrated plant nutrition systems and integrated pest management.

Technological compensation, however, may not be feasible in other regions where erosion rates are higher, and corrective measures face a wide range of constraints. This is particularly the case in most parts of sub-Saharan Africa, where productivity growth has been poor for several decades, and is likely to remain slow in the short- to medium-term. The decline in agricultural research funding at both the national and international level, and the wide range of institutional and infrastructural constraints, seem likely to constrain improvements in crop production techniques.

The uncertainties about soil erosion rates and their impact on food crop yields are compounded by uncertainties about the costs of rehabilitation. The GLASOD study estimated that nearly one billion hectares of agricultural land in developing countries are so degraded that productivity is being moderately or severely affected, and that some 9 million hectares world-wide, of which 5 million are in Africa, have reached a point where rehabilitation is probably uneconomic. However, this latter aspect is difficult to assess in that some farmers and rural communities in all developing regions have demonstrated that land which GLASOD classifies as unrestorable can in fact be rehabilitated. Farmers in the Machakos district of Kenya and in China, for example, have rehabilitated formerly abandoned or heavily degraded land at modest cost given appropriate incentives and technologies, though the conventional approaches assumed by GLASOD would probably have been uneconomic (Pretty, 1993; Tiffen *et al.*, 1994; Reijk,1992; Altieri, 1992). Thus, although there is no justification for complacency over the problem of land degradation, it would be wrong to make linear projections into the future and to suggest that it will inevitably compound food production problems arising from climate change. National policies and farm practices are being modified to correct at least part of the soil and water degradation problem, though the full benefits of such changes will not be felt in the short term.

Moreover, the impact of erosion on crop yields or production has not been well established in physical or economic terms even though there have been many attempts to do so. The relationship between erosion and productivity loss is more complex than previously thought, as is that between human-induced and natural erosion. Similarly, the experimental techniques once used to quantify these losses are less effective than previously thought.

The relationship is not linear – that is, it is not directly proportional to the thickness of the layer lost, nor to the type of particles lost. Moreover, the impact on soil structure, particularly the impact on the size of the air spaces, can be more important than that on the chemical composition. It is also possible that yield loss in one area may be compensated for by gains further down the slope, valley or plain where the soil is eventually deposited. But here again the threat to food production is not straightforward. The deposition of eroded soil may have negative external consequences,

Table 2 Siltation of selected Indian reservoirs

	Acre-feet per annum		
	Assumed rate of reservoir siltation	**Observed rate of reservoir siltation**	**Expected life as percentage of designed life**
Bhakara	23,000	33,4757	68
Maithon	684	5,980	11
Mayurakshi	538	2,080	27
Nizam Sagar	530	8,72	56
Panchet	1,982	9,533	21
Ramganga	1,089	4,366	25
Tungabhadra	9,796	41,058	24

Source: Data from Ministry of Agriculture, Government of India (1990).

such as the siltation of reservoirs and irrigation canals and reduction of their life span by 75 - 90 per cent (Table 2).

Notwithstanding the difficulties in estimating rates of soil erosion and its impact on productivity and food crop yields, certain regional generalisations are possible. In aggregate, the direct threat to food production could be quite small. In Asia, which accounts for over one half of the world's population, food security is heavily dependent on wheat and rice. These two staples commonly provide about 50 per cent of per capita calorie intake, and are generally either grown under controlled irrigation on naturally flat or levelled land, or in bunded paddies where the risk of soil erosion is very low (Parikh and Ghosh, 1991). These methods of cultivation are not strongly prone to erosion. Upland rice, which is commonly grown on steep slopes, is prone to erosion but accounts for less than three per cent of total production. The main threat to these crops from soil erosion is an indirect impact that results from the siltation and reduced life of reservoirs and irrigation canals as noted above. This could be very serious given the critical role that irrigation plays in Asian cereal production and food security (Norse, 1992a).

The situation in sub-Saharan Africa and Latin America is different. First, irrigated food production is much less important than in Asia. Second, some of the most food-insecure rural people have to farm steep slopes either in areas of intense and highly erosive rainfall (de Janvry and Garcia,

1988), or in dryland areas where the physical and chemical impoverishment of inherently marginal soils exposes them to declining and highly variable crop yields (van Keulen and Breman, 1990).

Uncertainty about the consequences of land degradation is not restricted to crop production, but also exists in the livestock sector where the methods for assessing the impact of overgrazing on land degradation and rangeland productivity are very weak. These uncertainties, together with poor understanding of traditional management systems, have commonly led to gross over-estimates of the impact of soil erosion on rangeland productivity and sustainability in certain situations (Behnke *et al.*, 1993). The relationship between erosion and productivity appears to be less direct than previous estimates. Simulations for Botswana suggest that semi-arid rangelands may be able to support present livestock densities for several hundred years in contrast to earlier projections of imminent collapse. For example, a stocking rate of one live unit per 5 hectares on a slope of 2.5 per cent yields soil losses of 0.16 mm/yr, whereas the same stocking rate on a slope of 15 per cent yields ten times the equivalent soil loss. The main erosion problem is thus on steep slopes (Table 3).

Soil nutrient depletion is another growing threat to food production (Smaling, 1991). Although commonly associated with soil erosion, soil nutrient depletion is generally a consequence of the shortening of fallows and prolonged crop harvesting without adequate technological responses to replace the soil nutrients removed by the crops. Unless these nutrients are replaced by organic or mineral fertiliser inputs, leguminous crops, nitrogen-fixing algae and so on, the nutrient status of soils is lowered, leading to declining crop yields, and unsustainable agricultural production.

The problem appears to be most acute in sub-Saharan Africa, but it is also serious in other regions. It concerns both major nutrients like nitrogen and phosphate, and micro-nutrients like boron and manganese. In the mid-1980s all countries in the region were estimated to be suffering from nutrient mining to a greater or lesser degree. The most serious problems occurred in semi-arid areas where livestock manures were in short supply and mineral fertilisers very expensive because of poor infrastructure and inefficient marketing bodies. But, as with soil erosion, the magnitude of the threat is not clear in that nutrient depletion does not appear to be causing major decreases in crop yields. Part of the impact may be masked by fertiliser applications, but it is also possible that natural rates of soil nutrient release from the parent material may be greater than currently assumed. Similarly, nutrient inputs from the atmosphere or from soil organisms may be

Table 3 Simulations of sustainability of semi-arid rangelands in Botswana

	Slope (%)			
	Slight slope (2.5 %)		Steep slope (15 %)	
Stocking rate (LU/ha)	0	1:5	0	1:5
Net soil loss (mm/yr)	0.14	0.16	1.46	1.65
System life (years)	4992	3616	468	340

Source: Adapted from Biot (1993).

greater than currently estimated. It is clear, however, that in the long-term, nutrient inputs must balance removals if food production is to be sustained, and the current scientific uncertainties on nutrient cycling must be resolved.

Nutrient depletion can threaten regional food production in a number of ways. First, it can lower subsistence food production. Second, it leads to lower farm incomes and food purchasing power through lowered yields and/or increased costs of production. Third, it reduces the water holding capacity of the soil and increases susceptibility to drought or rainfall shortages by lowering soil organic matter levels.

Salination is a growing environmental threat in almost all regions. It is primarily a problem of irrigated areas, but also occurs in hot dry zones where strong evaporation brings salts to the surface. In irrigated areas it is usually the consequence of poor drainage arising from bad design, inadequate investment, poor maintenance, inefficient management and energy or water subsidies leading to excessive application rates and seepage from water courses. The end result is waterlogging, salination and eventually, if corrective action is not taken, loss of land for agriculture through abandonment and desertification (Szabolcs, 1992). Some 10-25 per cent of the world's irrigated area is already affected. The threat to food security takes three main forms. The cropped area is reduced, and land is permanently lost (which according to some estimates may be 0.5 -1.5 million hectares a year) with some 30 million hectares at risk (United Nations, 1992). Depressed crop yields are in the range of 10-20 per cent although they may be higher for some sensitive crops (Euroconsult, 1989). Economic penalties result from the accelerated loss of major investments in irrigation and from the reduced yields, which raise the unit costs of production and lower farm incomes.

Salination is all the more serious when considered in association with population and food production projections, as well as possible climate change impacts. The former indicates growing dependence on irrigation in the absence of climate change. The latter underlines the possibility of greater aridity and higher evapotranspiration rates in regions already subject to, or at risk from, salination.

Desertification is the remaining threat to be considered. It can be broadly defined as land degradation in dryland areas. Much of the attention has been in Sudano-Sahelian Africa where deserts were once reported to be advancing by five to six kilometres a year, though such estimates are now challenged (see below). It is not, however, just an African problem. All the major regions face desertification (Table 1) with both cropland and rangeland under threat (Dregne *et al.*, 1991). China alone is thought to have approximately four million hectares of cropland and approximately five million hectares of rangeland under threat.

Two aspects of the desertification threat need to be stressed. First, there has been a major shift in scientific thinking on this topic in recent years, with a growing consensus for the view that the area affected by desertification has been greatly over-estimated (Nelson, 1988; Warren and Agnew, 1988; Bie, 1990), largely because of methodological weaknesses. It is now recognised that drylands are more resilient to drought and to human abuse than previously thought. Second, the contribution of drylands and particularly desert margins to regional food production is small. In sub-Saharan Africa, for example, low rainfall areas accounted for only about 12 per cent of cereal production in the early 1980s (Table 4).

Although the role of human activities in desertification is still not well established, there is little doubt that soil nutrient depletion and the over-cultivation of fragile soils does lead to dryland degradation and desertification. Land and fertiliser use projections indicate increasing pressures of this type (Alexandratos, 1995). Amelioration of these pressures will be dependent on improvements in farm practices, such as soil moisture conservation, and on the development of leguminous live-mulches and other techniques. These improvements would have the advantage of tackling current problems arising from recurrent drought as well as reducing vulnerability to any reduction in rainfall or soil moisture that might arise in the future from climate change. However, the overall impact on food security will not be great. First, deserts and drylands, though of vital importance to relatively small populations of vulnerable peoples, are projected to continue to make only a small contribution to regional food production and consumption (Alexandratos, 1995). Second, it seems likely that these regions will continue to experience marked out-migration as their populations seek higher and more secure incomes outside agriculture, particularly in urban areas (von Braun and Pandya-Lorch, 1991).

Contaminated water

The future water supply is threatened both by quantitative constraints arising from competition for water between agriculture, industry and the urban sector, and qualitative problems arising within agriculture which are a threat to crop yields and to human, livestock and wildlife health. There are four principle threats of agricultural origin:

- rising salt concentrations in irrigated areas;
- fertiliser and pesticide contamination of surface;
- fertiliser and pesticide contamination of groundwater; and
- discharges of organic effluents from intensive livestock units and fish-farms.

All four sources are projected to increase, because of the long period of time required to achieve appropriate corrective actions.

The intensification of irrigation will increase the degree of water re-use and hence the accumulation of salt concentrations. Crop yields and the sustainability of irrigation will be risked if corrective actions are not taken. Fortunately, the primary irrigated cereals (rice and wheat) are relatively tolerant to low or moderate salinity, but are subject to yield losses of approximately 10 per cent at high levels (that is, above 5-7 micromoles/cm).

The greater applications of organic and mineral fertilisers are essential to prevent soil nutrient mining and to raise crop yields. In many developing countries application rates will remain below those likely to cause major pollution problems, but in some areas with either very high application rates (e.g. the Punjab) or thin sandy rock strata above aquifers (as in parts of Sri Lanka), the risks of pollution could be significant if corrective measures are not taken. Application of mineral fertiliser per hectare is projected to exceed 100 kg of nitrogen in some areas by 2010. Heavy rates will thus have been applied for at least 20-30 years. This is the same time span over which the developed countries started to face severe groundwater nitrate problems, underlining the importance of devising and introducing integrated plant nutrition systems over the next decade or so if agricultural progress is to be sustained without major environmental penalties.

The threats are not only from the crop sector. Intensive dairying, pig and poultry units are projected to account for an increasing proportion of total output, and so their effluents will also grow.

Poor pest management

Three changes associated with the unavoidable intensification of food production could create pest problems, and if these were badly managed, they would pose serious threats to the environment and food security. First, further reduction in the length of fallow periods stemming from the need to crop land more frequently could endanger soil fertility as noted above, and, in the absence of suitable corrective actions, lead to more serious and more frequent weed, insect and disease attacks, because of their survival in greater numbers from one cropping season to the next. Moreover, this problem could be compounded by global warming if higher temperatures extend the geographic range of pests and diseases, as well as increasing their growth rates. Second, the increase in the area of land carrying two or even three crops a year, could have similar effects to fallow reduction. Finally, the rise in the demand for vegetables, and to a lesser extent fruit, could lead to greater pollution and health risks from excessive use of insecticides. These crops tend to receive heavy applications of insecticides, often close to harvest, either as insurance against the risk of losing a high value crop, or in order to improve their cosmetic appearance and hence their price. Such excessive application rates can pose numerous risks – to the people applying them, to natural predators of vegetable pests, to those consuming the treated crops, and to drinking water supplies. Though this is an issue of growing seriousness in all regions, the situation is worst in Asia.

It is reasonable to assume, however, that this threat will diminish in the future. A combination of rising pesticide prices, higher crop production costs, greater emphasis on integrated pest management (IPM) and concerns about public health risks and ecosystem protection leading to regulatory action, will continue the present trend of reducing pesticide use. The rates and frequency of application will continue to be reduced and environmental persistence and mammalian toxicity of insecticides and pesticides in general continue to decline. These positive trends are evident. Egypt, for example, has banned the use of a number of toxic pesticides, and IPM has been successfully introduced against the brown rice hopper in Indonesia, bringing widespread environmental and economic benefits.

5 Climate change

Climate change is the most uncertain element in the possible conjunction of threats. It is important to underline the point that climate change is a threat to food production in some developing regions and countries, and a possible stimulus in others. Increases in ambient temperature in the high altitude tropics, for example, may widen the range of crops that can be grown or increase the length of the growing season so as to permit the use of higher yielding cultivars (Downing, 1992).

Given the methodological problems discussed earlier, it is impossible to be categorical about potential regional impacts of climate change and on food security. Cereals are the most critical crops for people, countries and regions vulnerable to food security problems, and hence have been analysed more closely than other foods. The best estimates to date suggest that developed region cereal production could be adjusted to compensate for the climate change projected for the mid-21st century, but that some developing countries could face major reductions. Production in Africa, Central and South America and Asia is estimated to decline by about 33, 22 and 10 per cent respectively by 2060. However, this decline could be largely offset by higher production in the developed countries, together with relatively small improvements in the growth rate of technology in the developing countries given that the projections are based on conservative yield assumptions.

It is clear that sub-Saharan Africa is the region most at risk in terms of food security. It is more vulnerable to reduced rainfall, to rainfall variability and to greater evapotranspiration than any other region. About half of its arable land is already arid or semi-arid, only 23 per cent of its cropland is irrigated, and the high cost of irrigation development limits its use for low cost basic staple foods (FAO, 1986). Many of its soils have low water holding capacity and this could be reduced further through higher soil temperatures leading to greater rates of soil organic matter breakdown. On the economic side, the anticipated low GDP growth rates imply that people and countries will generally be unable to overcome domestic food production problems by raising food imports.

In Asia, problems could arise from the present and commonly growing dependence of food security on irrigation. Climate change raises the possibility of greater aridity and higher evapotranspiration rates in areas already subject to, or at risk from, salination, so more irrigated land could be lost or degraded. Conversely, more intense rainfall through climate change could increase soil erosion leading to lower crop yields and faster rates of siltation of irrigation dams and canals. The significance of the former, however, is difficult to estimate. In Britain, for example, it has been estimated that climate change could increase annual soil erosion rates by about 20 per cent, that is, from about 4.7 to 5.3 cubic metres per hectare (Boardman and Favis-Mortlock, 1993). Such low impacts would have little influence on crop yields over a 50 - 100 year timeframe, and could be balanced by a very small increase in the rate of technological growth. The impact could be more significant in the higher rainfall areas of the tropics where soil erosion rates are already up to ten times higher than those prevailing in Britain. But, again, the lack of information about the current impact of soil erosion on food security, and uncertainties regarding the rates and sustainability of future crop yield increases, make it difficult to assess the potential consequences for food security.

Strategic planning should therefore balance the uncertain costs and benefits of some climate change response options against the unambiguous social and environmental gains from alternative investments in current and clearly defined development problems. Regional population and economic growth come together as a driving force for climate change, because of their role in greenhouse gas emissions. Asia, for example, is the dominant producer and consumer of rice. Irrigated and flood rice cultivation systems are a major source of methane, a powerful greenhouse gas, and are already heavily concentrated in the Asia region. Modest or slow population growth in Asia will therefore be one of the main driving forces for methane emissions, but this will be tempered by income growth, which tends to bring about a shift in food consumption patterns away from rice to wheat. Conversely, high population growth in sub-Saharan Africa, combined with low population densities and weak income growth, seem likely to continue as major factors behind the dominance of extractive production systems which fail to maintain soil fertility. This would lower soil organic matter levels and hence reduce the role of soil as a sink for carbon dioxide, and increase the vulnerability of food production to drought or low rainfall. Further treatment of the threat of climate change is deferred to other papers in this volume (for example, Appendini and Liverman; Fischer *et al.*; Futang and Zhao; Magadza; Riley *et al.*; Schneider; Schulze *et al.*) and the forthcoming assessment of the Intergovernmental Panel on Climate Change.

6 Economy

The greatest threat to regional food production and food security in the developing world for the foreseeable future is the lack of economic growth, and particularly low income growth amongst the poor. In the main, it is the billion or so people in households with average per capita incomes of less than one dollar a day who are food-insecure. Low economic growth is a more immediate and serious threat to food production and food security than the pattern of economic growth per se, and the environmental consequences thereof. Early global models tended to project the collapse of food production in certain regions, because of a combination of high economic and population growth rates leading to serious pollution and the exhaustion of key raw materials (Meadows *et al.*, 1982). Later models and desk analyses, however, have highlighted the dynamic properties of the global system, the role of technological responses to natural resource constraints, and the importance of institutional constraints (HMSO, 1977; Parikh *et al.*, 1988; Fischer *et al.*, 1990; Waggoner, 1994).

In most developing countries, food insecurity is largely a problem of inadequate food purchasing power and slow or no income growth in both agricultural and non-agricultural populations, and not the result of the lack of food production potential or environmental threats to it (Alexandratos, 1995). Moreover, the lack of economic growth may compound the perceived threats from the environment and population growth. First, many countries are constrained in their response to current food production problems arising from environmental degradation and climate variation by shortages of human and financial capital for investment in research and development for sustainable agriculture, and for the correction or alleviation of the degradation.

Farmers, for example, fail to react to some forms of land degradation, because the available corrective measures carry heavy short-term economic penalties, and this commonly relates to the lack of profitable markets for their crops. Second, decisions on family size are strongly influenced by income and education levels, rather than simply by access to contraception.

Although even short-term projections of regional or national economic growth are hedged with great uncertainty, and those for the longer-term are little better than educated guesses, there is general agreement on the likely pattern of evolution. Those used by the World Bank suggest high to very high growth rates for China and the rest of Asia, and slow or modest growth in Africa and

Latin America (Table 4). Over the long-term, economic growth tends to slow down as a lagged response to reduced population growth. This is already apparent in the developed countries, and is likely to be generally the case in the developing countries by the mid-21st century. Nonetheless, the overall picture is of substantial economic growth, which carries a number of important implications for regional food production and food security.

First, the structure of economies will change in line with the historic trend (Table 5), with countries becoming less dependent on agriculture for wealth and welfare, while agriculture will continue to be an important engine of growth in the short to medium term. This implies that countries will be better able to import rather than produce those foods for which they have no comparative advantage, that policies of self-reliance rather than self-sufficiency in food will be followed, and that people will become more dependent on bought rather than home-produced foods.

Table 4 Projected per capita income growth in developing countries

	(per capita GDP in 1990 US$)	
	1990	**2030**
Sub-Saharan Africa	340	500
Asia and Pacific	490	2000
Latin America and Caribbean	2180	5500
Middle East and North Africa	1800	4000

Source: World Bank, 1992

Table 5 Agricultural contribution to gross domestic product (%)

	Percentage of GDP	
	1965	**1990**
India	44	31
Malawi	50	33
Thailand	32	12
Italy	11	3
UK	3	2

Source: World Bank, 1992.

Second, and following on from the first, a significant proportion of today's small farmers and landless labourers will move wholly or partially out of agriculture, and many of the new entrants into the labour force will go directly into the manufacturing and service sectors. This is likely to have four important consequences:

- Some of the marginal lands that are vulnerable to environmental degradation and climate change are likely to go out of crop production and revert to more sustainable pastures, rangelands and forests, following the pattern apparent in many developed countries.
- Farm consolidation will take place in some areas, improving economies of scale, and possibly raising investment in measures to restrict vulnerability to any adverse climate change impacts.
- Rising wage rates will limit the use of very labour intensive crop production systems such as complex relay planting systems, and those involving heavy applications of locally collected organic fertilisers, which could lead to a number of environmental problems and constrain food output.
- Farm families will have more diversified agricultural and non-agricultural incomes and be less dependent on farm incomes, and consequently less vulnerable to drought-induced or other seasonal and annual fluctuations in food purchasing power, which currently are a major cause of food insecurity.

Third, income growth will lower considerably the share of food in total personal consumption expenditure. Whereas many poor families or people in developing countries currently spend over 70 per cent of their income on food, by the mid-21st century most of them will have followed the pattern of the developed countries and the rich in their own countries and will spend only 15-20 per cent of their income on food. Thus, people will be increasingly able to absorb food price increases, and prices per se will be less of a threat to food security. Sub-Saharan Africa is the only region which seems unlikely to follow this pattern of economic growth (Table 5), and its agriculture will continue to be severely constrained by the lack of effective demand (FAO, 1993).

Such economic and income growth, however, carries a number of threats to food production and the environment. It will commonly increase the demand for livestock commodities, most of which are likely to come from intensive production systems that can be a major direct and indirect cause of environmental degradation, and an important factor for global warming through the release of methane. Rapid urbanisation and industrialisation will probably cause air and water pollution that has negative impacts on food production, for example through ozone emissions.

However, developing countries are able to benefit from the past mistakes and environmental research of the developed countries, and may adopt cleaner technologies at an early stage. Finally, economic growth will intensify the competition for limited water resources and may reduce the amount of water available for irrigation. Food production in North Africa, the Near East and Asia is heavily dependent on irrigation. In India, for example, irrigation provides over 50 per cent of staple food production. This threat is thus a very serious one. Fortunately, however, there are many opportunities for increasing water use efficiency in agriculture, though the high capital cost of some methods may limit their use for low-value staple food crops.

It is clear that one must not assume that climate is the only factor that is changing. Future economic changes will in the main make people less food-insecure. Such changes will increase the ability of countries and the international community to invest in research, infrastructure and institutions to overcome environmental degradation, and will lower the risk of adverse climate change impacts on regional or global food production.

7 Population

There are those who believe that population growth, environmental degradation, climate change and food production are on a collision course, with over-population being the principle issue and poverty the dominant driving force. We have gone from 18th century warnings of population collapse (Malthus, 1798) to 20th century statements on the population bomb (Ehrlich, 1968), the population explosion (Ehrlich and Ehrlich, 1990), and predictions of widespread famine in 1975 (Paddock and Paddock, 1967) which proved to be incorrect, notwithstanding the regional famine caused by the Sahelian drought of the early 1970s. Technology based on indigenous knowledge and on hard science has postponed collapse, but it has led others to argue that the pattern of technological growth is combining with population growth to magnify the nature of the threat to food production and could lead to irreversible environmental degradation (U.S. National Academy of Sciences and the U.K. Royal Society, 1992). Where does the balance lie?

There is no doubt that growths in population, income and food production are all driving forces for climate change (Norse, 1991). Moreover, population growth can reduce the resilience of food production systems through its impact on environmental degradation, thereby lowering food security. But it can also be a positive force for enhancing food production and safeguarding the environment.

Population and income growth are the main driving forces for fossil and non-fossil fuel energy demand, for food production and for timber extraction. They are therefore the main cause of greenhouse gas emissions, and hence contribute to the risk of agriculturally unfavourable shifts in regional climates. Food production, however, is directly responsible for less than 10 per cent of emissions, although it contributes indirectly through the role of shifting cultivation and more permanent agricultural land development in deforestation.

Population growth introduces or strengthens a number of positive feedback loops that reduce regional food security, notably by lowering the resilience of food production systems to external shocks and to longer term structural changes. However, the strength of these loops is largely unknown (Norse, 1992b).

The dominant positive feedback loop is population pressure against finite land resources. This is strongest in Asia and weakest in sub-Saharan Africa and Latin America, although there are considerable intra-regional differences (Alexandratos, 1995). This limits the area of undeveloped land that can be cleared for agriculture, which in turn commonly forces farmers to shorten or eliminate fallows that are essential to maintain soil fertility in traditional low-external input farming systems. If farmers compensate for the lack of fallow by changes in their production systems, including the use of mineral and/or organic fertilisers, then there need not be a threat to food production. However, suitable technologies do not exist for all situations, they may not be profitable to farmers, or may be unavailable because of weaknesses in farmer support services, such as credit and marketing (Merril-Sands, 1986; Carr, 1989). This can have a number of consequences for food production. First, it can reduce the incentive for farmers to produce more than their own subsistence needs. Second, it can lower food production and depress crop yields through soil erosion. Third, it can push food production on to increasingly marginal lands with unfavourable chemical or physical soil characteristics, or on to steeper slopes such that erosion and other forms of land degradation are intensified. Fourth, it generally lowers soil organic matter levels, and hence rainfall/moisture retention capacities fall, leading to greater vulnerability of food production to drought. However, as mentioned earlier, our knowledge of the dynamics of these processes is inadequate to establish how serious this threat is to long-term food production (Biot *et al.*, 1992).

Fortunately, there are other population-related responses that enhance rather than reduce food production. Thus there are negative feedback mechanisms, such as migration and the intensification of land use, which can raise local, regional and national food security. A study of Machakos, Kenya (Tiffen *et al.,* 1994) illustrates the important role that these negative feedback mechanisms can play:

- In the 1930s it was thought that environmental degradation in the Machakos District arising from human and livestock population growth would lead to ecological disaster. In the 1990s the same land is either directly or indirectly meeting the food, fuel, shelter and income needs of a population nearly six times greater.
- Much of the district is semi-arid and has inherently poor soils which would be regarded as only marginally suited to food crop production with a low population supporting capacity under the FAO's agroecological zone classification.
- Population pressure in the district and income growth in local or overseas markets have combined to promote the wider adoption of effective soil conservation and more sustainable production systems.
- A large part of the initiative and finance for agricultural improvements came from the farmers and communities themselves.

This study supports the hypothesis first raised by Boserup (1965) that population pressure stimulates positive technological improvements that raise food output and population supporting capacity. It also lends support to the idea that environmental degradation may accelerate in the early stages of development, but eventually investments are made to safeguard the environment and the incomes derived from natural resources.

The Machakos situation is not fully replicable in all marginal areas in that it is relatively close to a major city, Nairobi, which provides a ready market for its agricultural produce, and has good domestic and international communications. Kenya also has a progressive and dynamic economy compared with most African countries. Machakos, however, is not a unique example, but it is perhaps the most thorough longitudinal analysis currently available of such an environmental transformation. In the Middle Ages, population pressure combined with slow technological responses in parts of Germany caused similar environmental degradation to that of Machakos in the 1930s (Russel and Russel, 1968). Today there is little sign of that degradation, although the past problems of rapid soil erosion have been replaced in part by those of soil compaction and nitrate accumulation in groundwater. In this century farmers in China, India and many other regions of the world have demonstrated that population pressure and associated environmental threats can be overcome.

This is not to argue that rapid population growth or high population densities are unimportant but to suggest that at the regional level they are not the most pressing threat to food production or food security. The main danger lies in the way population growth dilutes the welfare and development benefits of economic growth. It slows down increases in per capita incomes and hence reduces the ability of people and governments to invest in the improvements in education and infrastructure that are vital for motivating people to have fewer children.

The regional food production/population dimension comes in at various levels. Asia, for example, is the dominant producer and consumer of rice. Flood rice cultivation systems are a major source of methane, a powerful greenhouse gas, and are heavily concentrated in the Asia region. Population growth in Asia will therefore be one of the main driving forces for methane emissions. High population growth in sub-Saharan Africa (Table 6), combined with low population densities and weak income growth, are major factors behind the dominance of extensive production systems which lower soil organic matter levels and hence reduce the role of soil as a sink for carbon dioxide, and increase the vulnerability of food production to drought or low rainfall. Conversely, a further slowing down of population growth, first in Asia, then in Latin America and finally in Africa, combined with higher population densities and/or greater

Table 6 Regional population projections

	(UN medium series – millions)				
	1990	**2000**	**2050**	**2100**	**2150**
Africa	642	867	2265	2391	3090
China	1139	1299	1521	1405	1389
India	853	1042	1699	1870	1949
Other Asia	1121	1372	2379	2703	2807
Latin America	448	538	922	1075	1117
Developed countries	1089	1143	1233	1202	1191

Source: UN, 1993.

rural to urban migration, will create regional differences in the ability to invest in more sustainable food production systems.

8 Conclusions

Environmental degradation, economic and population growth and climate change are all threats to regional food security, but not in a straightforward way. The greatest threats are sub-regional, and commonly linked to poverty, so it is the conjunction of threats which pose the greatest risks and the greatest challenge.

Sub-Saharan Africa is the region most under threat. It has the most fragile environment, the poorest economic growth prospects, the fastest population growth and the greatest vulnerability to climate change. The first three threats are problems of today and can only be resolved in the long term. They will be intensified by climate variation whether or not the threat of climate change becomes reality. But in essence the central problem in Africa is income-related food insecurity, that is poverty and the lack of food purchasing power. Similar threats in other regions include those Asian countries where current food surpluses simply mask the unmet needs of low-income groups, which lack either the land or the money to ensure food security.

This paper attempts to present a balanced view of the threats to regional food production arising from the environment, economy, population growth, and the way they may interact with each other and with climate change. There seems to be little substance to arguments that mankind is on the edge of a precipice. Global warming and the loss of biodiversity, for example, are clearly destabilising features that are linked in a number of ways to environment, economic and population changes. However, our understanding of the global system is insufficient to substantiate claims that the future of the planet is in balance. In some areas, humankind is pressing ecosystem resilience to its limits, but in the main it seems possible to move back from these limits, and develop sustainable food production systems that could feed a world population stabilising at around 12 billion, and which could accommodate slow changes in climate of the magnitude currently projected.

It is not simply a question of technological change, although more rapid and sustainable technological progress is vitally important for food security in all regions. On the contrary, the most important changes are of an institutional or behavioural nature. In addition, there are serious gaps in our scientific knowledge, particularly regarding feedback loops between environmental degradation and food production. Thus, there is no justification for complacency or for moving away from the precautionary principle embodied in the Rio Declaration adopted at United Nations Commission for Economic Development (UNCED). Our models of natural systems are still partial, and so positive and negative surprises cannot be ruled out. Our understanding of the interplay between poverty and environmental degradation is incomplete, as is that on the long-term impact of such degradation on food security, given that it can drive people to migrate to other areas where they can meet their food needs, notwithstanding the social and cultural disruption that are consequences of migration.

This lack of scientific understanding or objectivity has fed scientific and socioeconomic uncertainty, which in turn has contributed to the emergence of excessively polarised views regarding the potential impact of climate change, environmental degradation, economic development and population growth on present or long-term regional food security. At one extreme there are those who claim that these four factors constitute an unprecedented challenge to world food security, and commonly base their arguments on excessively linear extrapolations of current social and technological responses. At the other extreme it is argued that population and income growth will drive technological change so as to remove food production constraints and reach an economically and environmentally sustainable growth path. Both extremes can be challenged.

Acknowledgement

This paper is a substantially revised and extended version that first appeared in *Food Policy* Volume 19, Norse, D., Multiple threats to regional food production: environment, economy, population?, pp 133-148 (1994). Reprinted with permission from Elsevier Science Ltd, Butterworth-Heinemann Imprint, The Boulevard, Landford Lane, Kidlington OX5 1GB, U.K.

References

Alexandratos, N. (ed). (1995). *World Agriculture: Towards 2010.* John Wiley, Chichester.

Alt, K., Osborn, C. and Colaccio, D. (1989). Soil erosion: what effect on agricultural productivity? *Agricultural Information Bulletin. No. 556.* U.S. Department of Agriculture, Economic Research Service, Washington DC.

Altieri, M.A. (1992). Where the rhetoric of sustainability ends, agro-ecology begins. *Ceres,* **24** (2): 33-39.

Behnke, R.H., Spooner, I. and Kercen, C. (1993). *Range Ecology at Disequilibrium: New Models of Natural Variability and Pastoral Adaptation in African Savannahs*. Overseas Development Institute, London.

Bie, S. W. (1990). *Dryland degradation Measurement Techniques,* World Bank Environment Department Working Paper 26, World Bank, Washington DC.

Biot, Y. (1993). How long can high stocking densities last? In: Behnke, R.H., Spooner, I. and Kerven, C. (eds). *Range ecology at disequilibrium: new models of natural variability and pastoral adaptation in African savannahs.* Overseas Development Institute, International Institute for Environment and Development and the Commonwealth Secretariat, London.

Biot, Y., Lambert, R. and Perkin, S. (1992*). What's the Problem? An Essay on Land Degradation, Science and Development in Sub-Saharan Africa.* School of Development Studies Discussion Paper No. 222. University of East Anglia, Norwich.

Boardman, J. and Favis-Mortlock, D.T. (1993). Climate change and soil erosion in Britain. *The Geographical Journal,* **159** (2): 179-183.

Boserup, E. (1965). *The Conditions of Agricultural Growth: The Economics of Agrarian Change under Population Pressure*. University of Chicago Press, Chicago.

Boserup, E. (1993). *The Conditions of Agricultural Growth: The Economics of Agrarian Change Under Population Pressure*. Earthscan, London.

Brown, L.R. and Kane, H. (1994). *Full House: Reassessing the Earth's Population Supporting Capacity.* W.W. Norton and Co., New York.

Carr, S.J. (1989). *Technology for Small-scale farmers in sub-Saharan Africa.* World Bank, Washington, DC.

Carruthers, I. (1993). Going, going, gone! Tropical agriculture as we knew it. *Tropical Agriculture Association Newsletter*, **13** (3).

Carson, R. (1962). *Silent Spring*. Houghton Mifflin, Boston.

Crosson, P. (1986). Soil erosion and policy issues. In: Phipps, T., Crosson, P. and Price, K. (eds.). *Agriculture and the Environment,* Resources for the Future, Washington D.C.

Crosson. P. and Anderson, J.R. (1992). *Resources and Global Food Prospects: Supply and Demand for Cereals to 2030*. World Bank Technical Paper No. 184, Washington DC.

Crosson, P. and Anderson, J.R. (1995). Demand and supply: trends in global agriculture. In: Downing, T.E. (ed.) *Climate Change and World Food Security.* Springer-Verlag, Heidelberg.

de Janvry, A. and Garcia, R. (1988). *Rural Poverty and Environmental Degradation in Latin America: Causes, Effects and Alternative Solutions.* Paper for the IFAD sponsored international consultation on environment, sustainable development and the role of small farmers, Rome, 11-13 October 1988.

Doos, B.R. (1994). Environmental degradation, global food production and risk for large-scale migrations. *Ambio*, **23** (2).

Downing, T.E. (1992). *Climate Change and Vulnerable Places: Global Food Security and Country Studies in Zimbabwe, Kenya, Senegal and Chile*. Environmental Change Unit Research Report No. 1. Environmental Change Unit, Oxford.

Dregne, H., Kassa, M. and Rozanov, B. (1991). A new assessment of the world status of desertification. *Desertification Control Bulletin* **20**. UNEP, Nairobi.

Ehrlich, P. (1968). *The Population Bomb.* Ballantine, New York.

Ehrlich, P. and Ehrlich, A. (1990). *The Population Explosion*. Simon & Schuster, New York.

El-Hinnawi (1985). *Environmental Refugees.* UNEP, Nairobi.

Euroconsult (1989). *Agricultural Compendium for Rural Development.* Elsevier, Amsterdam.

FAO (Food and Agriculture Organization) (1986). *African Agriculture: The Next 25 years.* FAO, Rome.

FAO (Food and Agriculture Organization) (1990). *Committee for Food Security*. FAO, Rome.

Fischer, G., Froberg, K.,Keyzer, M.A., Parikh, K.S and Tims, W. (1990). *Hunger - beyond the reach of the invisible hand.* IIASA, Laxenberg.

Fischer, G., Froberg, K., Rosensweig, C. and Parry, M.L. (1994). Climate change and world food supply, demand and trade. *Global Environmental Change,* **4** (1): 7-23.

Fischer, G., Froberg, K., Rosensweig, C. and Parry, M.L. (1995). Impacts of potential climate change on global and regional food production and vulnerability. In: Downing, T.E. (ed.) (1995). *Climate Change and World Food Security.* Springer-Verlag, Heidelberg.

HMSO (1977). SARUM 76 Global modelling project. Department of the Environment Research Report 19. HMSO, London.

Houghton, J.T., Meira Filho, L.G., Bruce, J., Hoesung Lee, B.A., Callander, B.A., Haites, E., Harris, N. and Maskell, K. (eds.) (1995). *Climate change 1994. Radiative Forcing of Climate Change and an Evaluation of the IPCC IS92 Emission Scenarios.* Cambridge University Press, Cambridge.

IFPRI (International Food Policy Research Institute) (1987). *Food Needs of Developing Countries: Projections of Production and Consumption to 1990.* International Food Policy Research Institute. Research Report No. 3, Washington DC.

Kimball, B.A., Pinter, P.J., Wall, G.G., Hunsake, D.J., Garcia, R.L. and LaMorte, R.L. (1995). Progress Report on Free-Air CO_2 Enrichment (FACE) wheat experiments. *Global Change Newsletter* 21B.

Malthus (1798). *First Essay on Population.* Macmillan, London.

McCalla, A.F. (1994). *Agriculture and Food Needs to 2025: Why We Should be Concerned.* Sir John Crawford Memorial Lecture Consultative Group on International Agricultural Research, Washington, DC.

Meadows, D.H., Meadows, D.L., Randers, J. and Behrens, W.W. (1972). *Limits to Growth.* Universe Books, New York.

Meadows, D.H., Richardson, J. and Bruchman, G. (1982). *Groping in the dark.* John Wiley, Chichester.

Merril-Sands, D. (1986). *The Technology Applications Gap: Overcoming Constraints to Small-farm Development.* FAO, Rome.

Ministry of Agriculture, Government of India (1990). *National Watershed Development Project for Rainfed Areas - Guidelines.* Department of Agriculture and Cooperation, New Dehli.

Nelson, R. (1988). *Dryland Management. The Desertification Problem,* World Bank Environment Department Working Paper 8, World Bank, Washington, DC.

Norse, D. (1991). Population and global climate change. In: Jaeger, J. and Ferguson, H.L. (eds). *Climate Change: Science Impacts and Policy - Proceedings of the Second World Climate Conference.* Cambridge University Press, Cambridge.

Norse, D. (1992a). *Policies for Sustainable Agriculture: Getting the Balance Right.* Staff Working Paper 3. IFAD, Rome.

Norse, D. (1992b). A new strategy for feeding a crowded planet. *Environment,* **34:** 5.

Oldeman, L.R., Hakkeling, R.T.A. and Sombroek, W.G. (1990). *Global Assessment of Soil Degradation (GLASOD).* ISRIC/UNEP, Wageningen and Nairobi.

Paddock, W. and Paddock, P. (1967). *Famine - 1975.* Little, Brown, Boston.

Parikh, K.S. and Ghosh, U. (1991). *Natural Resources Accounting for Soils: Toward an Empirical Estimate of Costs of Soil Degradation for India.* Indira Gandhi Institute of Development Research Discussion Paper No. 48, Dehli.

Parikh, K.S., Fischer, G., Froberg, K. and Gulbrandsen, G. (1988). *Toward Free Trade in Agriculture.* Martinus Nijhoff, The Hague.

Pretty, J.N. (1993). *Regenerating Agriculture: Policies and Practices for Sustainability and Self-reliance.* Earthscan, London.

Reijk, C. (1992). *Soil and Water Conservation in sub-Saharan Africa: Towards Sustainable Production by the Rural Poor,* IFAD, Rome.

Rosensweig, C. and Parry, M.L. (1994). Potential impact of climate change on world food supply. *Nature*, **367:** 133-138.

Russel, C. and Russel, W.M.S. (1968). *Violence, Monkeys and Man.* Macmillan, London.

Smaling, E.M.A. (1991). Two scenarios for the sub-Sahara: one leads to disaster. *Ceres: the FAO Review,* **22** (2).

Szabolcs, I. 1992. Salination of soils and water and its relation to desertification. *Desertification Bulletin* **221**.

Tiffen, M., Mortimore, M. and Gichuki, F. (1994). *More People, Less Erosion – Environmental Recovery in Kenya.* Wiley, Chichester.

U.S. Council on Environmental Quality. (1982*). The Global 2000 Report to the President: Entering the 21st Century*. Washington DC.

U.S. National Academy of Sciences and the U.K. Royal Society of London (1992). *Population Growth, Resource Consumption and a Sustainable World.* Joint statement released on 27 February 1992.

United Nations (1992). *International Conference on Water and the Environment Development Issues for the 21st Century.* United Nations Environment Programme, Nairobi.

United Nations (1993). *World Population Prospects*. UN, New York.

van Keulen, H. and Breman, H. (1990). Agricultural development in the west African Sahelian region: a cure against land hunger? *Agriculture, Ecosystems and Environment,* **32:** 117-197.

von Braun, J. and Pandya-Lorch, R. (eds). (1991). *Income Sources of Malnourished People in Rural Areas: Microlevel Information and Policy Implications.* Working papers on commercialisation of agriculture and nutrition, No. 5. International Food Policy Research Institute, Washington DC.

Waggoner, P.E. (1994). *How Much Land can Ten Billion People Spare for Nature*? Task force 121 U.S. Council for Agriculture, Science and Technology, Washington DC.

Warren, A. and Agnew, C.T. (1988). *An Assessment of Desertification and Land Degradation in Arid and Semi-arid Areas.* International Institute for Environment and Development. Paper 2, London.

World Bank (1992). *World Development Report 1992.* Oxford University Press, Oxford.

Climate Change and Involuntary Migration: Implications for Food Security

JoAnn McGregor
St Antony's College
University of Oxford
Oxford, U.K.

1 Introduction

The prospect of mass involuntary population displacement resulting from climate change has fuelled a multitude of studies over the past 10 years (Myers, 1993; Houghton *et al.*, 1990, 1992; Tickell, 1989, 1989a, 1990; El-Hinnawi, 1985; Westing, 1992). One of these accounts has evoked the spectre of '150 million destitutes abandoning their homelands' (Myers, 1993: p. 17). Some authors have hypothesised that involuntary migration may be the most severe of the potential effects of global warming due to its negative impact on food security and because of its destabilising effect on global politics.

This paper does not attempt to quantify the scale of potential migrations accompanying climate change. Rather, it aims to question some of the assumptions on which existing alarmist projections are based. First, the way in which these studies conceptualise the link between environmental change and migration is considered. Writing on environmental refugees portrays climate change, environmental degradation and natural disasters as external push factors, or 'generators' for refugees to which migration is the mechanical response. This denies any role for human agency by obscuring both the political nature of many environmental problems and the differential way in which people living in hazardous environments incorporate risk into their livelihood and is so doing shape the contours of the risk itself. The second part of the paper turns to the premise underlying the link drawn between migration and food insecurity.

Much of the literature considers these relationships at a very general level: the frame of reference has tended to be continents, sub-continents or at least whole countries. The arguments made have tended to be implicitly Malthusian. At the same time, they have drawn on little or none of the growing body of localised empirical research on migrants' coping strategies and the economic and ecological changes in areas receiving displaced populations.

NATO ASI Series, Vol. I 37
Climate Change and World Food Security
Edited by Thomas E. Downing

The final section of the paper considers ongoing changes to the legal and institutional structures of humanitarian relief and the way intervention to assist different categories of displaced people is legitimised.

2 Environmental change and forced migration

Studies of environmental refugees often assume a simple and direct cause-and-effect link between climate change and migration. This premise goes against much recent work in the social sciences. For example, Sen's understanding of the relation between drought and famine is mediated by shifts in exchange entitlements (Sen, 1981; Drèze and Sen, 1989). Other writers have analysed the relation between environmental change and society by focusing on the role of social institutions in resource access (Berry, 1989), or by using concepts of labour (Mandala, 1983) or culture (Croll and Parkin, 1992) as mediators: in so doing they avoid the pitfalls of environmental determinism on the one hand and 'social relations determinism' on the other.

The geographical literature on natural hazards has repeatedly stressed the role of human agency either in causing the disaster itself, or in causing population to be more vulnerable to disasters (Wijkman and Timberlake, 1984; Hewitt, 1983). Groups may become vulnerable when their coping strategies have been undermined directly or indirectly by the state, or their recovery prevented by failure to provide insurance and relief, as in the Dust Bowl disasters in the United States in the 19th century (Warrick, 1983) and in much of the developing world today. War itself also commonly interferes with people's strategies for coping with environmental variability. In the case of drought, for example, famine most commonly occurs in those countries affected by war (Duffield, 1991). The increasing use of the term 'environmental persecution' to describe the relation between states and minority indigenous peoples, acknowledges the political nature of some environmental changes and the way they can be entwined with the abuse of human rights.

The relationship between a hazard and those who live with it is also obscured by the way nature and culture are cast as discrete entities having an impact on one another rather than being instantiated in each other (Croll and Parkin, 1992). As Rosalind Shaw has argued in relation to floods in Bangladesh: 'Floods not only have varying consequences for rich and poor, men and women, and rural and urban dwellers, but their very nature as hazards is constituted by these and other forms of human social difference' (Shaw, 1992: p 200).

It is not analytically useful to separate environmental from political, economic or ethnic causes of migration. A recent text on civil war and internal strife as causes of displacement (Cohen, 1991) cites examples of countries where military conflict has led to war-displacement. The same countries are cited in another recent text as examples of situations where environmental disturbance led to flows of environmental refugees (El-Hinnawi, 1985). The countries include Sudan, Ethiopia and Mozambique amongst others. Such broad categorisations are inevitably reductive, and fail to reflect the interaction between conflict, environmental change and migration.

The role of particular state politics, new technologies, and conflict over resources is usually considered in projections of migration following climate change. In Bangladesh, for example, investment in hi-tech embankment construction may have been counter-productive (Shaw, 1992). On the cause of the 1987-88 Ethiopian famine, Clay *et al.* (1988) note the correlation between famine areas and specific government policies: in Tigray and Eritrea, famine prevailed in areas outside government control and under military attack; in Tigray and Wollo, famine occurred in areas of forced resettlement; in northern Bale, Hararghe and Shoa, famine occurred as a result of the government villagisation programme; and in Wolega, Illubabor and other administrative regions the forcibly resettled were themselves unsettled and local production disrupted.

Studies of migrants' actual decisions to flee show that they are commonly much more complex than a simple environmental push as implied in studies on the affects of climate change. Migration is usually only one of a variety of survival strategies pursued by families either simultaneously or consecutively with other strategies such as selling assets, wage-labour, eating bush foods or undertaking short distance migration. Migration theory has incorporated these insights by shifting away from 'push-pull' analyses, and focusing instead on social networks to understand population migration (Portes and Bach, 1985; Portes and Barocz, 1989).

Pankhurst's (1992) study of livelihood changes in the 1987-88 drought in Ethiopia shows how coping strategies other than migration were undermined by state restrictions on travel, declining opportunities for both rural and urban wage-labour and increasingly unfavourable terms of trade as grain prices rocketed. He reveals how aid itself placed pressure on peasants to migrate. The existence of family or social networks outside the region of habitual residence can facilitate the decision to flee (see Colson, 1989, for a general text). Such ties can arise from a history of migrant labour or trade, shared political alliances, ethnic identity or myths of origin, and structures of patri- or matri-locality. For Vietnamese leaving the agricultural district of Yen Hung, salination of the soil leading to low rice yields was one factor among many in their decision to leave: other considerations included the unavailability of local employment, misinformation and rumours spread by confidence tricksters and boat organisers, and a fundamental lack of confidence in the future of the Vietnamese economy. In describing the on-going exodus from this agricultural district, Hitchcox (1991) notes that:

> The main anchor to the community is the family group which can be very large, consisting of several households. When this group begins to divide, some relatives left behind in the original settlement, some abroad and in Hong Kong, the decision to move becomes a less weighty and difficult decision, especially if living conditions are poor.

A study of communities affected by the 1988 Bangladesh flood describes how some families stayed on the roof of their homes whilst others migrated to a relief camp. The decision to move involved balancing different risks: the risk of having possessions stolen or squatters in the home and the reduced privacy and purdah for women in a camp, against food insecurity, dirty water, the presence of pirates and the threat of snake bites which life on the rooftop entailed. Disease was a problem in both environments, albeit for different reasons. Shaw notes that it was usually severe illness or a near fatal accident of a child (such as falling into the flood water) which precipitated their move (Shaw, 1992).

3 Migration and food security

This part of the paper looks at the relationship between migration and food security in areas which have received sizeable displaced populations. As the most common assumption is that refugees have a negative impact on the host economy, studies of situations where the reception of displaced populations has been an asset to the host economy in the mid or long term, despite being initially disruptive, is reviewed. This is followed by a discussion of the waves in which assistance affects food security. This section reveals some of the problems in generalising about the food security implications of displaced populations.

A key concept in justifying international intervention to assist refugees and other displaced people is the notion that they are a burden on the host community, and that this burden should be shared by other states. The media use the image of destitute and needy populations to evoke sympathy and funds. The nature of this burden is not generally made explicit. In the 1960s, refugees' negative impact was usually defined in terms of a strain on economic development and infrastructure. Since the energy crisis of 1973 and the concern expressed thereafter about the 'other energy crisis' (fuelwood shortages in the developing world), environmental impact has increasingly become an explicit part of the notion of host burdens. In the West, in contrast, the main concerns have been the strain on welfare services and contribution to unemployment.

In some situations the capital, skill and business links refugees brought with them have been beneficial to hosts. For example, the expatriation of Cuban capital and the business class to the United States with the first wave of refugees after 1959 was significant in developing the Miami area (Portes and Bach, 1985). Similarly, the departure of the colonial élite from former Portuguese territories benefited both the metropole and the other receiving countries such as South Africa and Swaziland. In Nepal, the carpet industry initiated by Tibetan refugees has become the country's largest exchange earner (Hagen, 1980).

In other contexts, public policies initiated by the host state have stimulated economic change. Zetter (1987: pp 173-184) has shown how Cyprus was able to transform the severe economic disequilibrium which accompanied the Turkish invasion and arrival of 180,000 refugees into a development asset. This was brought about by a large state-directed refugee housing programme which was launched immediately and played a significant role in the process of reconstruction. It was linked to longer-term economic policies which shifted the economy from an agricultural to an industrial base, and achieved full employment.

More recently, the involuntary migrations accompanying the Gulf Crisis have been analysed by Van Hear (1992, 1993). He describes how the return of 360,000 Jordanians (most of whom were Palestinian) to Jordan in the wake of the Iraqi invasion of Kuwait led to short-term disruptions as it strained state-subsidised foodstuffs and exacerbated water shortages. However, the influx brought benefits in the medium to long term. Van Hear calculates that the inflow of financial resources which accompanied the influx was almost equivalent to Jordan's foreign debt, whilst estimates of compensation due amount to US$3 billion. The surprise economic boom of 1992 may be partly attributable to their arrival, although the effects of the coincidental structural adjustment programme need also to be taken into account.

Other studies have detailed the beneficial role of migrants in agricultural development (Mollett, 1991). Kok (1989) and Kuhlman (1990) document how cheap Ethiopian labour fuelled the expansion of commercial agriculture in Eastern Sudan; whereas Funnell (1991) describes how the development of the sugar and paper industries in Swaziland were predicted on cheap Mozambique labour. In peasant farming systems displaced populations can remove constraints on production by providing much needed labour at critical times in the agricultural cycle.

Rural migrants can also contribute to receiving economies through trade and their own production. In Western Equatorial region of southern Sudan in the early 1980s, Ugandan refugee settlements became the main locus of rural marketing, and stimulated the growth of larger urban markets, whilst Ugandan peasant tobacco producers contributed to the doubling of tobacco production between 1982 and 1984 (McGregor, 1985; Harrell-Bond, 1986).

The Tanzanian government used agricultural settlement schemes for refugees to develop selected parts of the country (Daley, 1993). Daley describes how local politicians in one such area were keen to have refugees to boost their clientele, to provide cheap labour for local farmers and to increase tobacco production. Refugees were also settled in an area the World Bank characterised as low-density, high-potential which thereafter became one of the 'big four' food productive regions of Tanzania – a shift directly attributable to the presence of refugees (Daley, 1993).

Hosting societies are not homogeneous, and different socio-economic groups are not equally affected by the changes accompanying a mass influx. Even if the aggregate economy benefits from cheap labour, a fall in local labour prices may adversely affect those dependent on labouring. Similarly, traders may benefit from food scarcities and price rises whilst other socio-economic categories suffer.

When refugees arrive in rural areas, their impact on food security may depend on the seasonality of their arrival. De Waal (1989) describes a situation in western Sudan in 1984-85 where the arrival of refugees coincided with the local harvest and their effect on food availability and prices was temporarily masked. Yet a famine was precipitated several month after their arrival.

The assistance programme itself can influence food security in the hosting areas. For example, interventions may not address legal constraints on displaced people's livelihoods. In most refugee situations, refugees face constraints on movement, employment and trading. This directly undermines their attempts to secure food. It has been argued that attention to these legal problems would enhance food security and reduce the need for costly and logistically difficult food relief operations (Wilson, 1992; Keen, 1992).

Food aid tends to be the main disbursement made to refugees and other forcibly displaced groups, despite growing interest in monetisation. Refugees often use these rations to meet many and varied non-food needs in addition to providing meals. For example, the sale of food aid may constitute their main source of income for soap, firewood, shelter, clothing, protection, transport, etc. (Wilson, 1992; Keen, 1992). For most African refugee populations, rations comprise a dry cereal and erratic supplies of other dried commodities (beans, fish, salt, skimmed milk, oil). These are usually insufficient in quantity or quality by the time they arrive, thereby contributing to the nutritional crisis experienced by many refugees living on international relief. Furthermore, the administration of food aid is also often made contingent on residence in a camp. Even when this is not the case, travelling and queuing for a ration competes with other food acquisition strategies for refugees' time (Hansch, 1992).

In addition to its contribution to refugee livelihoods, food aid can have an impact on local host economies. Such impacts have been analysed in non-refugee contexts in terms of changes to food prices and 'disincentive effects' on local production (Clay and Stokke, 1991; Stewart 1986). The same processes may be equally relevant in areas with large relief-dependent refugee populations because food aid is commonly sold or exchanged, and even in closed camps the degree of 'leakage' outside is known to be very high (as documented by Hansch (1992) for Latin American camps, and Reynell (1989) for camps on the Thai/Cambodian border).

The process of encampment is the norm in assistance to refugees and disaster victims, despite the fact that camps are known to constitute a health hazard and undermine the potential for self-support (according to the United Nations High Commission for Refugees' (UNHCR) guidelines they are to be created only as a last resort). Comparative studies of nutritional status among refugees within and without camps have often shown a higher prevalence of malnutrition in camp populations. For example, a study of Mauritanian refugees in Senegal showed that malnutrition rates were highest amongst refugee populations in camps than amongst the self-settled (Ritmeijer, 1991). Similarly during the pellagra epidemic in Malawi in 1988/9, vitamin deficiency was more common amongst Mozambican refugees in camps than amongst those who were self-settled (Moren, 1991). Longitudinal studies have shown that a relatively healthy population on arrival sometimes experiences deteriorating nutritional status after living on international support in camps (Nieburg *et al.*, 1992; Toole and Bhatia, 1992).

Population concentrations in camps commonly overwhelm local markets, minimising opportunities for engaging in piecework or other wage labour. The localised depression in wage rates can cause extreme bad feeling from local labourers (Kok, 1989). In comparison to more dispersed settlements, camps may exacerbate pressure on natural resources such as firewood, timber and water, once again leading to conflict with locals. Refugee populations and the relief goods and health and education services they receive are highly visible and can engender hostility if goods are not also distributed to local populations. The ongoing debate over the pros and cons of refugee encampment also touches on issues of state security and agency accountability (see Kibreab, 1991, for a review). The debate cannot be resolved in absence of specific contexts.

Whilst some planned agricultural settlements have been economically viable, many have not (Kibreab, 1991) and 'income generating settlements' for refugees have usually failed to overcome the legal constraints on refugee income-generating activities noted above. In some instances, target dates for self-sufficiency have led to food being phased out before settlement populations are food secure.

The maintenance of a good relationship with local hosts for displaced populations' access to food, other resources, employment and security has been described as 'the cardinal principle in refugee assistance' (Wilson, 1992). Interventions and structures which fail to facilitate this relationship can jeopardise refugees' livelihoods. For example, the location of refugee camps is often determined on the basis of a national level decision and without soliciting the goodwill of locals: refugees forcibly resettled in the remote East Awin area of West Papua were faced with acute antagonism from the local population who staged road blockages and denied refugees access to food and labour markets (Refugee Participation Network, 1991). In Sierra Leone, interventions to secure refugees' land rights similarly threatened to jeopardise refugees' livelihood by disrupting crucial relationships between them and their hosts.

The importance of funding projects for 'refugee-affected areas' which address developmental needs of whole regions and which do not discriminate between refugees and hosts has been recognised since the early 1980s. However, budgets created for such projects by the European Commission (under Lome III article 204 and IV article 255) and by the United Nations Development Programme (UNDP) have sometimes been channelled into projects which are segregationist in nature.

The food security of areas where there are involuntary migrants is context specific. Its understanding requires an analysis of the local economy and ecology, the relationships between migrants and their hosts, the nature of the assistance programme and the structure of broader political relations and conflicts. Failing such understanding, the question can be wrongly posed in Malthusian terms as a battle between an increased population and resources, rather than a negotiated (and often conflictual) partitioning of resources between different groups.

4 Trends in humanitarian relief for different categories of involuntarily displaced

Only a fraction of the world's involuntary migrants are currently targeted as legitimate beneficiaries of international assistance or protection. This section looks at the legal and institutional issues in assistance for displaced people. It discusses the terminology used to describe different categories of involuntary migrants in the light of the ongoing debate over the legal definition of a refugee and reviews recent shifts in the way interventions are justified. The implications of these debates for people potentially forced to move by climate change are considered.

Involuntary migrants displaced by climate change, natural disasters, degradation and pollution, a collapse in food security or economic hardship are currently excluded from the categories protected by instruments of international refugee law, both in terms of the text and intent of the drafters, and in terms of much current practices, particularly by Western states (see also Refugee Policy Group and IOM, 1992).

Originally intended to deal with refugees from communism following World War II, the current refugee definition (detailed in the 1951 Geneva Convention and 1967 Protocol Relating to the Status of Refugees) can be used to limit refugee status to those outside their country of origin with a well-founded fear of persecution, the latter being defined in narrow political terms. Such a definition is inappropriate for the root causes of flight in many developing nations, as a narrow political interpretation of 'persecution' can exclude those suffering economic and social persecution or the effects of war, as well as victims of natural disasters in countries where the state offers no protection. Migrations attributed to climate change would be similarly excluded. Hathaway has expressed the nature of the current *status quo* in the following way (Hathaway, 1991: p 114):

> Refugee law as codified in the 1951 Convention and 1967 Protocol relating to the Status of Refugees not only continues the original rejection of the notion of comprehensive assistance for all involuntarily displaced persons, but it allies international law with a series of strategic limitations determined by Western political objectives.

Some legal theorists are arguing for the definition of a refugee to be rooted in human rights. This argument can be grounded in different ways. Firstly, persecution can be defined in terms of human rights violations (as suggested by Coles, 1988; Hathaway, 1991) which, under the Convention on Economic, Social and Cultural Rights, includes the right to an adequate standard of living and the right to food (Lawyers Committee, 1991). The inclusion of some form of right to a healthy or safe environment is currently being debated (Commonwealth Institution and Law Department, SOAS, 1993). If this human rights approach is to be more widely adopted, those forcibly displaced across international boundaries for 'environmental' reasons could be eligible for international assistance and protection according to whether suffering amounted to a first order violation of human rights.

Alternatively, in can be argued that, as disasters and environmental change themselves or an individual's vulnerability to them are commonly the result of human actions, rather than 'acts of God', the state has a duty to protect its citizens from them. If the state is negligent or indifferent to meeting its obligations to protect its citizens' basic needs, this breach of the contract with the state could be grounds for international assistance (Shacknove, 1985)

In practice, UNHCR has long assisted a broader group than those included by the narrow Convention definition in its mandate. Since the late 1950s, UN General Assembly resolutions have repeatedly been passed to authorise the High Commissioner for Refugees to lend his/her 'good offices' to assist persons outside the HCR's competence (Kibreab, 1991). With respect to Sudan in 1972, the Economic and Social Council (ECOSOC), and the General Assembly (GA) passed resolutions to enable the inclusion of internally displaced people (Goodwin-Gill, 1988). Later in 1976, ECOSOC and GA resolutions recognised UNHCR's activities in the 'context of man-made disasters (UNHCR had been key in assistance in India/Bangladesh and Cyprus in the 1970s). By 1992, UNHCR recognised groups protected under its mandate to include internally displaced, war displaced and even 'other needy groups' (EXCOM, 1992).

However, it is important to note that the class of beneficiaries has been expanded *without* any corresponding broadening of states' legal obligations (Goodwin-Gill, 1988). Which groups are and which are not included is highly political.

Regional agreements also broaden the category included as refugees. For example, within Africa, the Accord of the Organisation of African Unity (OAU, 1969) recognises as grounds for asylum claims; 'external aggression, occupation, foreign domination or events seriously disturbing public order'. It differs from the Geneva definition in that it refers to objective conditions in the country of origin, and allows refugee status to be recognised for groups rather than individuals (Rwelamira, 1983). The Accord was drawn up in the context of liberation struggles against colonial powers and 'disturbances' were assumed by the drafters to be of a political nature (Greenfield, 1986). Disasters and climate change were not specifically intended in the drafting. Although the scope of this expanded definition is habitually applied to those for whom economic protection by the state is not forthcoming, a restrictive interpretation excluding victims of climate change, ecological and man-made disasters, is by no means precluded (Rwelamira, 1983).

The notion of 'environmental refugees' has gained currency in describing groups displaced by environmental change (Myers, 1983; El-Hinnawi, 1985; Jacobson, 1988; Tickell, 1989; 1989a; 1990; Filho, 1993; Trolladen, *et al.*, 1992; Lassailly-Jacobs and Zmolek, 1992). The definitions of this term are broad enough to include migrations attributed to climate change. One definition of the term is:

> People who have been forced to leave their traditional habitat, temporarily or permanently because of a marked environmental disruption (natural and/or triggered by people) that jeopardised their existence and/or seriously affected the quality of their life (El-Hinnawi, 1985: p 4).

Those who use the term are for the main part environmental agencies and lobbyists. Many migration and refugee specialists and agencies have now explicitly rejected the term (McGregor,

1993; Refugee Policy Group and IMO, 1992; Surkhe and Visentim, 1991; Kibreab, personal communication, 1993). They have done so because of the conceptual, legal, institutional and political problems which accompany the term's use. Conceptual problems arise from the problem of separating 'environmental' causes from political and economic causes of flight, as documented in the outset of this article.

Legal and institutional problems arise because refugees currently receive protection which goes beyond the assistance given to disaster victims. Legal obligations on the part of refugee-hosting states are well defined. Guards against refugees being sent back to the persecuting state are at the core of refugee protection. In contrast, states' responsibilities with respect to those in humanitarian need are much less well defined. MacAlistair-Smith (1985) noted the inadequacies of international humanitarian law relating to relief operations in so-called natural disasters, and the need to transform 'functional humanitarianism responsibilities into legal obligations':

> There is still no international convention setting out obligations of States concerning the donation or acceptance of humanitarian assistance or regulating the coordination of relief in peace time.

International provision for refugees' protection and assistance is made through UNHCR. In contrast, disaster planning and relief is coordinated by the United Nations Disaster Relief Organisation (UNDRO) as the 'focal point in the UN system for disaster relief matters', often through local UNDP offices (Siegel and William, 1991; MacAlistair-Smith, 1985). UNHCR's protection services go beyond the relief, assistance and disaster preparedness undertaken by the host of UN institutions (United Nations Environment Program (UNEP), the Food and Agriculture Organisation (FAO), World Health Organisation (WHO), UNDP) and other regional organisations (such as the Intergovernmental Authority for Drought and Desertification (IGADD), or non-governmental organisations (NGOs), which are coordinated by UNDRO. The term 'environmental refugee' thus confuses different types of service and institutional responsibility.

Criticism of poor coordination within and without the United Nations in emergencies mounted following the Gulf emergency (Cuenod and ECOSOC, 1991; Childers and Urquhart, 1991). This led to the creation of the Department of Humanitarian Affairs (DHA) to coordinate and strengthen emergency response. The debates leading up to the DHA's creation acknowledged and did not rule out responding to 'the possibility of massive food losses through plagues of insects resulting from climate change, environmental degradation or chemically-induced mutations' (Childers and

Urquhart, 1991: p 48). The DHA became functional in 1992 and is headed by an Under Secretary-General.

Whilst it is important to highlight environmental problems and their association with migration pressures, in so far as the term 'environmental refugee' conflates the idea of disaster victim and refugee, its use brings with it the danger that key features of refugee protection could be undermined and the lowest common denominator adopted. Because 'environmental' can imply outside politics, states may use the term in much the same way as 'economic migrant' to reduce their obligation to protect and assist.

By reducing the complexity of real situations, the term 'environmental refugee' can also reinforce images of Malthusian squeeze. Based on generally inaccurate conceptions of the relation between population change and environmental resource use, Malthusianism commonly underlies current literature on population and the environment in the Third World (as argued by Williams, 1992, with reference to World Bank reports). Such ideas can contribute to the paranoia of 'fortress Europe' and hence reinforce the trend towards restrictive asylum practices.

Recently, UNHCR has developed a 'new strategy' in its protection and relief operations. This change has occurred as a result of a convergence of events and the end of the Cold War, the United Nations' action against the Iraqi regime in the Gulf, the UN action in the former Yugoslavia, and the appointment of a new High Commissioner for Refugees, Mrs. Sadako Ogata, whose experience is in human rights. The new strategy has been conceptualised by an internal UNHCR Working Group on International Protection (EXCOM, 1992).

The new approach reinforces the trend towards extension of UNHCR's mandate, but at the same time, the new concept of 'temporary protection' weakens the principle of durable asylum which formerly was the mainstay of refugee protection. Jaeger (1993) argues that this change is a compromise in the face of intransigent, restrictive Western states, a move which he describes as a 'perceptible and disquieting compromise between principles and *Realpolitik*'. The two elements of this approach are: general support for the early warning peace-making and peace-keeping activities on the UN coupled with in-country protection. In conjunction, repatriation has been promoted as a solution to refugee situations, and has taken prominence over other solutions such as local integration and resettlement in the West (Harrell-Bond, 1989). The Declaration of Principle of

International Law on Compensation to Refugees by the International Law Association in 1992 also addresses the problem of population displacement from the perspective of the responsibility of the country of origin (Lee, 1993).

Desirable as it is to prevent conflicts, Jaeger rightly questions whether UNHCR will be in a position to bring about the desired changes, as it operates in a country only with government consent or under the authority of the Secretary-General of the Security Council. This brings with it the risk of compromise. Furthermore, preventive protection will entail addressing economic and environmental problems, issues well outside UNHCR's experience to date.

A common interpretation of these trends is that what we are witnessing is a crumbling of the system of nation states and the concept of sovereignty. Alternatively, it has also been suggested that this is merely a new form of *Realpolitik*, as (differently justified) intervention was also a feature of the Cold War period (Dacyl, 1993). High profile interventions to assist internally displaced and national minorities are made only in some places. At the same time, there are more people externally displaced than ever before for whom no intervention is forthcoming, including refugees, stateless persons, and expellees. Restrictionism in the West takes the form of carrier sanction, the imposition of visa requirements, the blocking of access to asylum procedures and *refoulement* and has been matched by the *ad hoc* granting of temporary asylum. States are bolstering attempts to control their borders and maintain control over which foreign citizens are to be allowed to enter their sovereign territories.

5 Conclusion

The recognition that displacement has been associated with environmental change and may be associated with climate change had made an important contribution to recent debates over who should legitimately benefit from international relief. However, many of the existing studies on migration and the environment have misrepresent the nature of the relationships between environmental change and migration on the one hand, and between migration and food security on the other.

Theoretical insights and empirical studies on the interrelations between politic and environmental change and on the nature of migrants' actual decisions to flee indicate the inadequacy of the idea of environmental push factors and the notion of carrying capacity being exceeded. Use of the term 'environmental refugee' to describe those displaced by climate change has been argued against on the grounds that the term is conceptually, legally and institutionally unfounded. The growing body of literature on refugee livelihoods and on the economic and ecological transformations in refugee receiving areas point to diverse experiences in different places. The nature of assistance itself influences changes in hosting countries, often in unintended ways. The challenge in responding to future displacements is to remove the constraints on displacees' own attempts to improve their situation and ultimately to work towards the prevention of forced displacement.

The legal and institutional structures which mediate assistance to different categories of forcibly displaced are subject to ongoing review: and current debates over the provision of assistance to displaced populations are moving towards the inclusion of groups previously on the fringes of the refugee definition. Importantly, however, these changes in the provision of assistance have not been accompanied by binding legal obligations on states. At the same time that attention is drawn to formerly forgotten groups in humanitarian needs, some of the bulwarks of refugee protection are being undermined.

Acknowledgements

This paper is a reprinted version that first appeared in *Food Policy,* Volume 19, McGregor, J., Climate change and involuntary migration: implications for food security, pp 120-132 (1994). Reprinted with permission from Elsevier Science Ltd, Butterworth-Heinemann Imprint, The Boulevard, Landford Lane, Kidlington OX5 1GB, U.K.

References

Berry, S. (1989). Social institutions and access to resources. *Africa,* **59**(1): pp 41-55.

Childers, E. and Urquhart, B. (1991). *Strengthening International Response to Humanitarian Emergencies.* Dag Hammerskjold Foundation/Ford Foundation, New York.

Clay, J., Steingraber, S., and Niggli, P. (1988). *The Spoils of Famine: Ethiopian Policy and Peasant Agriculture.* Cultural Survival Incorporated, Cambridge, MA.

Clay, J. and Stokke, O. (1991) *Food Aid Reconsidered: Assessing the Impact on Third World Countries.* Frank Cass, London.

Cohen, R. (1991). *Human Rights Protection for Internally Displaced Persons.* Refugee Policy Group, Washington, DC.

Coles, G. (1988) The human rights approach to the solution of the refugee problem: a theoretical and practical inquiry. In: Nash, A. (ed.) *Human Rights and the Protection of Refugees Under International Law.* Institute for Research on Public Policy, Halifax, Nova Scotia.

Colson, E. (1989). In good years and in bad. *Anthropological Research,* **35**.

Commonwealth Institute and Law Department, SOAS (School of Oriental anad African Studies) (1993). *Human Rights Approaches to Environmental Protection in the Commonwealth and Beyond.* Conference Report. Commonwealth Institute, London, 27-28 May.

Croll, E. and Parkin, D. (eds.) (1992). *Bush Base: Forest Farm. Culture, Environmental and Development.* Routledge, London.

Cuenod, J. and ECOSOC (United Nations Economic and Social Council) (1991). *Report on Refugees, Displaced Persons and Refugees.* ECOSOC, Geneva.

Dacyl, J. (1993). *Sovereign and Human Rights: Dilemmas in the Post War Responses to People of Humanitarian Concern.* Draft paper, Refugee Studies Programme, University of Oxford, Oxford.

Daley, P. (1993). From the Kibande to the Kibali: the incorporation of refugees and labour migrants in Western Tanzania 1900-1987. In: Black, R. and Robinson, V. (eds.) *Geography and Refugees: Patterns and Processes of Change*. Belhaven, UK, pp 17-33.

De Waal, A. (1989) *Famine that Kills: Darfur, Sudan, 1984-1985.* Oxford University Press, Oxford.

Drèze, J. and Sen, A. (1989). *Hunger and Public Action.* Clarendon Press, Oxford.

Duffield, M. (1991) The internationalization of public welfare: conflict and the reform of the donor/NGO safety. Paper presented to the *Workshop on the Prospect for Peace, Recovery and Development in the Horn of Africa.* Institute of Social Studies, The Hague (February).

El-Hinnawi, E. (1985) *Environmental Refugees.* UNEP, Nairobi.

EXCOM (1992) *Note on International Protection.* UNHCR Executive Committee, 43rd Session, (October 1992) Document A/AC. 96/799. EXCOM, Geneva.

Filho, W.L.(1993) An overview of the environmental degradation processes contributing to the increase in the problem of environmental refugees in NE Africa. Paper presented at the conference, *Human Rights Approaches to Environmental Protection in the Commonwealth and Beyond*, Commonwealth Institute and School of Oriental and African Studies, London, May 27-28 1993.

Funnell, D. C. (1991). *Under the Shadow of Apartheid: Agrarian Transformation in Swaziland.* Avebury, Aldershot.

Goodwin-Gill, G. (1988). *Refugees: The Expanding Mandate of the Office of the United Nations High Commission for Refugees*, Unpublished paper.

Greenfield, R. (1986). *The OAU and African Refugees.* Reprinted extracts from Yassin El-Ayouty and Zartman, W (eds) (1984) *The OAU After Twenty Years of Praeger*, Johns Hopkins, New York.

Hagen, T. (1980) *Integration of the Tibetan Refugees (Wege und Irrwege der Entwicklungshilfe)* (forthcoming in English translation).

Hansch, S. (1992). Patterns of diet and ration consumption in Central American camps. *Refugee Studies,* **5** (3/4): pp 300-313.

Harrell-Bond, B. (1986). *Imposing Aid: Emergency Assistance to Refugees.* Oxford University Press, Oxford.

Harrell-Bond, B. (1989) Repatriation: under what conditions is it the most desirable solution for refugees? *African Studies Review,* **32** (1): pp 41-69.

Hathaway, J. (1991). Reconceiving refugee law as human rights protection, *Refugee Studies,* **4**(2): 113-131.

Hewitt, K. (1983). *Interpretations of Calamity.* Allen and Unwin, London.

Hitchcox, L. (1991). *Returnees Within their Communities: A Pilot Study of Local Attitudes and Economy Related to Development Policy.* Unpublished report.

Houghton, J.T., Callander, B.A. and Varney, S.K. (eds.) (1992). *Climate Change 1992: The Supplementary Report to the IPCC Scientific Assessment.* Cambridge University Press, Cambridge.

Houghton, J.T., Jenkins, G.C. and Ephraums, J.J. (eds.) (1990). *Climate Change: The IPCC Scientific Assessment.* Cambridge University Press, Cambridge.

Jacobson, J. (1988). *Environmental Refugees: A Yardstick of Habitability.* World Watch Institute, Washington DC.

Jaeger, G. (1993). The recent concept and policy of preventive protection. *Refugee Participation Network,* **14**: 20-22.

Keen, D. (1992). *Refugees: Retaining the Right to Life. The Crisis in Emergency Relief.* Zed Books, London.

Kibreab, G. (1991). *The State of the Art Review of Refugee Studies in Africa.* Uppsala papers in economic history, Research Report No. 26. University of Uppsala, Uppsala.

Kok, W. (1989). Self-settled refugees and the socio-economic impact of their presence in Kassala, Eastern Sudan. *Journal of Refugee Studies,* **2**(4): pp 419-441.

Kuhlman, T. (1990). *Burden or Boom? Study of Eritrean Refugees in the Sudan.* Vrije Universiteit Press, Amsterdam.

Lassailly-Jacob, V. and Zmolek, M. (1992). Environmental refugees. *Refuge* **12** (1) (Special Issue on Environmental Refugees).

Lawyers Committee (1991). *The Human Rights of Refugees and Displaced Persons: Protections Afforded Refugees, Asylum Seekers and Displaced Persons under International Human Rights, Humanitarian and Refugee law.* Briefing paper (May 1991).

Lee, L. (1993). The declaration of principles of international law on compensation to refugees: its significance and implications. *Journal of Refugee Studies* **6**(1): pp 65-71.

Low, J., Hume, M., Sulikova, I., Xaji, M., Issaq, X. and Maxamood, M. (1985). *Results of a Baseline Survey in Sablaale District, Lower Shabeelle Region, Somalia.* Euro Action Accord, London.

Mandala, E. (1983). *Work and Control in a Peasant Economy: A History of the Lower Tchiri Valley in Malawi 1859-1960.* University of Wisconsin Press, Wisconsin.

MacAlistair-Smith, P. (1985). *International Humanitarian Assistance. Disaster Relief Actions in International Law and Organization.* Martinus Nijhoff, The Netherlands.

McGregor, J. (1985). Marketing in Yei District. In: *The Lutaya Expedition.* Occasional Paper No 1, Refugee Studies Programme, Oxford.

McGregor, J. (1993). Refugees and the environment. In: Black, R and Robinson, V. *Geography and Refugees: Patterns and Processes of Change.* Belhaven, Chichester.

Mollett, J. A. (ed.) (1991). *Migrants and Agricultural Development. A Study of Intra-Rural Migration.* Macmillan, London.

Moren, A. (1991). Epidemiological Surveillance among Mozambican Refugees in Malawi. Paper presented at the *International Symposium Responding to the Nutritional Crisis Among Refugees: The Need for New Approaches.* Refugee Studies Programme, Oxford (October 17-20).

Myers, N. (1993). Environmental refugees in a globally warmed world: estimating the scope of what could well become a prominent international phenonmenon. *Bioscience,* **43**(11): 752-761.

Nieburg, P., Person-Karell, B. and Toole, M. (1992). Malnutrition/morality relationships among refugees. *Journal of Refugee Studies,* **5**(3/4): pp 247-257.

OAU (Organisation of African Unity) (1969). *Convention Governing the Specific Aspects of Refugee Problems in Africa* 10.9.69, 1001, UNTS, 14691, entered into force 20.6.74. OAU, Addis Adaba.

Pankhurst, A. (1992). *Resettlement and Famine in Ethiopia: The Villagers' Experience.* Manchester University Press, Manchester.

Portes, A. and Bach, R. (1985). *Latin Journey. Cuban and Mexican Immigrants in the United States*. University of California Press, Berkley.

Portes, A. and Barocz, C. (1989). International migration: theoretical perspectives on its determinant and modes of incorporation. *International Migration Review,* **23**(2): 606-630.

Refugee Participation Network (1991) Voluntary repatriation? Refugees from West Papua. *Refugee Participation Network,* **11.**

Refugee Policy Group (RPG) and the International Organization for Migration (IOM) (1992) *Migration and the Environment.* (June 1992), RPG/IOM, Washington DC.

Reynell, J. (1989). *Political Pawns: Refugees on the Thai-Kampuchea Border*. Refugee Studies Programme, Oxford.

Ritmeijer, K. (1991). Refugees' food acquisition strategies in times of inadequate ration: the case of Mauritanian refugees in Senegal. Paper presented at the international symposium *Responding to the Nutritional Crisis Among Refugees: The Need for New Approaches.* Refugee Studies Programme (17-20 October 1991), Oxford.

Rwelamira, M. (1983). Some reflections on the OAU convention on refugees: some pending issues. *Comparative International Law Journal of South Africa* **16**: 153-178.

Sen, A. (1981). *Poverty and Famine: An Essay on Entitlement and Deprivation.* Clarendon Press, Oxford.

Shacknove, A. (1985). Who is a refugee? *Ethics,* **95**: 275-284.

Shaw, R. (1992). Nature, culture and disasters; floods and gender in Bangladesh. In: Croll, E. and Parkin, D. (eds.) *Bush Base, Forest Farm. Culture, Environment and Development.* Routledge, London.

Siegel, R. and William P. (1991). UNDP coordination of disaster and development planning. Managing Natural Disasters and the Environment. International Bank for Reconstruction and Development/World Bank, Washington DC.

Stewart, F. (1986). Food aid: pitfalls and potential. *Food Policy* **11** (4).

Suhrke, A. and Visentim, A. (1991). The environmental refugee: a new approach. *Ecodecision* (September 1991).

Tickell, C. (1989). Climate change could cause world refugee crisis. *British Oversees Development* 7 (16).

Tickell, C. (1989a). *Environmental Refugees: The Human Impact of Global Climate Change.* Lecture delivered at the National Environment Research Council Annual Lecture, Royal Society, London.

Tickell, C (1990) *Environmental Refugees: The Human Impact of Global Climate Change.* National Environment Research Council, Swindon.

Toole, M. and Bhatia, R. (1992). Somali refugees in Hartisheik A Camp, Eastern Ethiopia. *Journal of Refugee Studies* **5** (3/4): 313-327.

Trolladen, J.M., Birkeland, N.M., Bryen, J. and Scott, P.T. (1992). *Environmental Refugees: A Discussion Paper*. World Foundation for Environment and Development/Norwegian Refugee Council, Oslo.

USCR (1991) *World Refugee Survey.* US Committee for Refugees, Washington, DC.

Van Hear, N. (1992). *Consequences of the Forced Mass Repatriation of Migrant Communities: Recent Cases from West Africa and the Middle East.* UNRISD Discussion Paper DP 38 (November 1992). UNRISD, Geneva.

Van Hear, N. (1993). Seminar paper, Refugee Studies Programme (Trinity term 1993), Oxford.

Warrick, R. A. (1983). Drought in the U.S. Great Plains, shifting social consequences. In: Hewitt, K. (ed.) *Interpretations of Calamity.* Allen and Unwin, London.

Westing, A. H. (1992). Environmental refugees: a growing category of displaced persons. *Environmental Conservation* **19** (3).

Wijkman, A. and Timberlake, L. (1984). *Natural Disasters: Acts of God or Acts of Man?* Earthscan, London.

Williams, G. (1992). *Modernizing Malthus: the World Bank, population control and the African environment.* Paper presented at the International Meeting on Population Movements, Food Crisis and Community Responses, New Delhi, India, 11-13 January 1992.

Wilson, K. (1992). Enhancing refugees' own food acquisition strategies. *Journal of Refugees Studies,* **5** (3/4): 226-247.

Zetter, R. (1987). *Rehousing the Greek-Cypriot Refugees from 1974.* A study of institutional access and labellings. D. Phil (unpublished), University of Sussex, Brighton.

Scenarios of Sustainability: The Challenges of Describing Desirable Futures

Dale S. Rothman
Sustainable Development Research Institute
University of British Columbia
B5-2202 Main Mall
Vancouver, BC V6T 1Z4
Canada

Robert Coppock
German-American Academic Council
1055 Thomas Jefferson Street, Suite 2020
Washington, DC 20007
USA

1 Introduction

The 2050 Project was a multi-year effort launched by the World Resources Institute, the Brookings Institution and the Santa Fe Institute.[1] It was undertaken to explore the question: how can we achieve a sustainable existence on this planet by the middle of the next century? To answer this, however, one must first ask: what are the essential components of a sustainable existence? Food security is a fundamental aspect of any desirable sustainable future, and the implications for food security of potential climate change and other global changes must be considered in any scenario of the future.

This paper explores the challenges of developing scenarios of sustainability. The value of developing scenarios is succinctly summarised by Benjamin *et al.* (1994, p. 11): 'Creating multiple visions – confronting a diversity of goals, values, and structures – allows for ongoing clarification, communication, contestation, and critique of alternative futures. It also provides a forum for examining possible ramifications of current decisions, envisioned actions, and proposed new pathways.'

1 For a complete list of participating institutions and individuals, contact Allen Hammond, chair of the 2050 Project's steering committee at World Resources Institute.

NATO ASI Series, Vol. I 37
Climate Change and World Food Security
Edited by Thomas E. Downing

The next section addresses our philosophy and approach to developing scenarios in the context of the 2050 Project. After that we discuss the steps taken in developing the scenarios and the problems we had along the way. We conclude with our expectations for further work and general thoughts on scenario development.

2 Forecasting versus backcasting - blindly follow or boldly go?

Two general types of scenario development can be characterised: forecasting and backcasting (Robinson, 1988, 1990; Sunter, 1987). Forecasting is generally portrayed as an objective exercise. Starting with current conditions and trends, the goal of forecasting is to determine the most likely future. Key variables of interest are projected and, in most cases, multiple projections or ranges of uncertainty are also included in the forecasted scenario. In contrast, backcasting begins with an end-state, or picture of the future, and attempts to identify plausible paths connecting this state to current conditions and trends. The goal is to envision a range of alternative futures and to analyse these for their desirability and feasibility, and implications for policy. It is an explicitly normative approach.

Recent scenario development efforts have begun to turn away from forecasting and towards backcasting for a number of reasons (Robinson, 1989; Sunter, 1987; Huntley *et al.*, 1989; Schwartz, 1991; Petersen, 1993; Chen and Kates, 1994a). Firstly, there are justifiable concerns about the objectivity and reliability of forecasts. Much of this is due to the inherent inability of forecasts to handle surprise and discontinuous changes (Brooks, 1987; Robinson, 1988, 1990; Ayres, 1994). Secondly, in the context of planning for the future, what is most likely may not necessarily be the most desirable (Robinson 1988, 1990; Sunter, 1987). Finally, forecasts tend to be 'passive' and backcasts 'active' in that the latter are more amenable to the explicit recognition that the future is largely what we make it (Sunter, 1987; Chen and Kates, 1994b, 1995, this volume).

Long-term trends and well-understood cycles can be useful in backcasting, however, by defining what Ayres (1994) terms the 'boundaries of the possible' and Sunter (1987) the 'universe of possibilities' or 'cone of uncertainty'. It is within this space that the end-states used in backcasting must be found. By maintaining a broad perspective, a wide array of possible futures can be discovered within this realm.

3 Envisioning alternative futures

Care must be taken in envisioning a range of alternative futures. The process must be diversified, with both regional and disciplinary diversity being emphasised. Wide participation in the development of the scenarios will enhance both of these aspects. Too often in the discussion of sustainability there has been an imbalance as social aspects have been ignored at the expense of ecological and economic issues (Benjamin *et al.*, 1994; Norse, 1994, 1995, this volume). Similarly, the views of persons from a particular region are often conspicuously absent, or hidden, in scenarios that depict the future of that region.

One example of how these concerns can be addressed is the Visions Exercise of the 2050 Project. This exercise 'solicited ideas from respected thinkers on the meaning and relevance of a sustainable future to the people of their respective regions, addressing long-term concerns and desires, valued culture-specific institutions and relationships, and how these values are reflected in production, investment and consumption decisions. The essays convey a sense of the way people feel about their future, not what they *expect* of the future, but what they *desire*, and consider to be within the realm of possibility as an alternative to the current trajectory of global developments' (Nagpal and Foltz, 1995; WRI, 1994a, p. 3). As part of the Visions Exercise, interviews were also collected from a diverse group of persons.

These essays and interviews inform the elaboration of alternative futures in the 2050 Project. The details of any scenario, particularly quantitative data, must be fleshed out in much greater detail than can be expected from any single essayist or interviewee. Nor is it expected that any of these will address all of the issues that we feel will be required to describe a complete scenario. The trick is to use the preferences and images expressed by the participants in the Visions Exercise to depict the broad outlines of plausible futures, then to fill in the gaps and flesh out details. Thus, the essays must be augmented by reviews or previous studies and other material in the literature, and by input from regional and sectoral experts.

A second key step, obtaining the input of regional and sectoral experts, would begin at a series of regional workshops. The participants would be presented with postulated futures, which they would be asked to comment on and revise. This type of policy exercise brings interested and knowledgeable parties into the process and has been used in a number of recent studies (Toth, 1988; Robinson, 1989; Jäger *et al.*, 1990; Svedin and Aniansson, 1987; Achebe *et al.*, 1990). Together with Environnement et Développement du Tiers Mond (ENDA-Zimbabwe) and the Rural Development Authority (ARDA, Zimbabwe), the 2050 Project assembled a group of individuals from Nigeria, Ghana, Senegal, Côte d'Ivoire, Tanzania, Kenya, Burundi, and Zimbabwe in October 1994 to review the project's preliminary scenarios for the region and to develop policy strategies addressing problems hindering sustainability in Africa. This methodology is also being used by the United Nations Development Programme's African Futures Project in collaboration with the United Nations University's Millenium Project Feasibility Study. As any single workshop will have limited participation, early versions of the scenarios would be shared with other individuals, who provide further commentary. The participation of these people, both those attending the workshop and others, extends beyond exploring the content of the scenarios to include their basic form and structure.

Another consideration in developing regional sustainability scenarios is the need to incorporate mechanisms capable of altering social, economic and political institutions that can exhibit considerable resistance or 'cultural momentum'. Scenarios prepared as input to business planning, in contrast, typically are limited to conceivable sequences of events that could affect the firm in significant ways, with special attention given to possibilities that are 'surprising' in that they fall outside the firm's conventional thinking about itself and its environment (Schwartz, 1991; Rainbow, 1995). When exploring regional sustainability, however, there is no 'firm'. Instead there usually are several nation-states, each with its own government, industries and various kinds of public and private organisations. Rather than trying to identify the conventional wisdom for this array in each country, and then to challenge those visions, we used a different approach.

We developed a set of alternative scenarios for a region, each based on a fundamentally different mechanism capable of substantially altering the overall human systems in that region. Each scenario describes a possible course of events in which a single mechanism comes to dominate the social and economic affairs of the region. We found four 'generic' mechanisms useful: one based on the combination of technological and economic forces, a second relying on conflict and combative solutions, a third involving the emergence of a charismatic leader capturing the imagination and motivation of people in ways that dramatically alter relevant behaviours, and a final mechanism using cultural and social changes that stimulate new institutional arrangements and incentives. None of the scenarios, however, are expected to characterise actual future developments in the region. Rather each scenario is intended to highlight a potentially important mechanism and explore its implications. All four mechanisms are likely to affect actual events as they unfold, and all four should be considered in choosing policies regarding the region. This approach is being used in the 2050 Project for sub-Saharan Africa, the region of Islamic cultural influence ranging from North Africa to South-Central Asia, and the East Asia region including China. It may be applied to additional regions in a later phase.

It is obviously necessary to adopt a broad perspective when developing scenarios of sustainability. Sustainability is a system property, demanding a holistic view of society, economy and environment for its definition, as does a description of the transition to such a state (Cochlin, 1994; Gell-Mann, 1994). In describing the workshop on which this volume is based, Chen and Kates (1994b, p. 4) note that it 'highlighted the complex interactions associated with evolving environmental, agricultural, and socio-economic systems that are heterogeneous in space and time and multidimensional in nature.'

Components of a system may appear to be sustainable in isolation, but their behaviours may be in conflict. If the interlinkages between sectors are ignored, threats to sustainability may be missed. The same holds true for regions. Some challenges to sustainability may be missed if all regions and the linkages among these are not considered. Also, the notion of sustainability is inherently temporal. Activities that maintain the current integrity of a system at the expense of future integrity are by definition unsustainable.

To draw conclusions about food security, for example, it is necessary to know the facts about production, for example, the amount and quality of agricultural land, supply of water, energy intensity and availability, and abundance of seeds and fertiliser. One needs to know about storage and distribution, and how much food is lost to spoilage. It is also necessary to understand patterns of consumption, the number of people in various areas, their dietary patterns and whether they have enough money to buy food. But even high agricultural yields and access to food do not necessarily keep a population well nourished. For example, a significant percentage of total nutrition consumed in some countries is lost through dietary ailments, primarily diarrhoea. The provision of clean water, where it is not available, is the most effective way to reduce the incidence of diarrhoea. Yet population growth is currently outstripping infrastructure development, so the proportion of the world's population with access to clean water is shrinking (World Bank, 1994).

Thus, a vision of a sustainable food and agriculture system must be qualified, and cannot be divorced from similar visions for other physical and social systems. A review of recent studies of the prospects for agriculture and food security shows the need to pay attention to such diverse areas as water quality and quantity, land quality and quantity, fertiliser use, pest control, energy, population, health of land-based and marine ecosystems, short- and long-term impacts of air pollutants (climate change, stratospheric ozone depletion, troposphere ozone impacts, etc.) on the biophysical side, and entitlements, economic incentives, nutrition, diets, labour supply, technological advancements (particularly in biotechnology) and management practices on the socio-political axis (FAO, 1993; Crosson and Anderson, 1994, 1995, this volume; Kendall and Pimental, 1994; WRI, 1994c; Norse, 1994, 1995, this volume; Chen and Kates, 1994a, 1995, this volume; Ruttan *et al.*, 1994, 1995, this volume).

At the same time, interregional linkages in the food system must be considered in developing any scenario. Global trade in agricultural commodities is a major component of many economies and a major source of food for many nations. Food security in a sustainable world will be much more a function of self-reliance than self-sufficiency (Norse, 1994, this volume; Bender, 1993, 1995, this volume; Fischer *et al.*, 1994, 1995, this volume; Reilly *et al.*, 1994, 1995, this volume).

It is important to emphasise the need to move beyond scenarios consisting solely of biophysical parameters (Norse, 1994, 1995, this volume). These are necessary for establishing feasibility at one level, but they ignore the social aspects of sustainability. To paint the fuller picture that is a 'complete' or 'integral' scenario, they need to be evaluated from a broader, qualitative or socio-political perspective. Otherwise, 'they do not ask about intrinsic cultural and individual qualities, about human experiences or about that which is valued. As such, they provide little or no guidance to the central directions by which people can create dignity, liberation, or fulfilment or how these may be possible in different nature-society settings' (Benjamin *et al.*, 1994, p.2).

4 Conventional development scenarios

An initial step was the preparation of a set of regional scenarios of conventional development to the year 2050. The scenarios reflect prevailing thought as provided by several recent international studies and assume that there will be no significant efforts to redirect the current paths of development. This exercise served two purposes. The first was to examine these previous scenarios from a methodological perspective: what are their stated and unstated assumptions, what is addressed and what is left out, what were the goals of the scenario developers, and how were the scenarios developed? From a more substantive perspective, the scenarios were investigated for internal inconsistencies and major uncertainties that may make the expressed outcomes highly unlikely. This is related to the former purpose in that many of the concerns about the feasibility of the scenarios is connected to those elements deemed necessary for sustainability that were not addressed. The issue of desirability was deliberately excluded from this exercise.

Quantitative estimates of a small set of demographic, economic, energy, water, land use, agriculture and human capital data derived from current data and two previously published scenarios of conventional development were prepared for 11 regions. Current data and projections were taken from the United Nations Population Division (UNPD, 1993) and the World Resources Institute (1994c, and associated database). The first scenario, the 1992A scenario developed for the Intergovernmental Panel on Climate Change, Working Group 1, has as its primary focus the estimation of emissions of greenhouse gases (Pepper *et al.*, 1992; Pepper, 1993). The second was formulated by the International Institute for Applied Systems Analysis to provide a single, internally consistent scenario of global changes in population, agriculture and energy over the next century for use as a 'conventional wisdom' reference case (Toth *et al.*, 1989). (A detailed description of the data and methods used in the 2050 Project is available from the authors.)

These data formed the basis of a two-day working session in January 1994 at the Tellus Institute in Boston, Massachusetts, USA, involving seven researchers within the 2050 Project. Participants in the workshop were asked to review these data and to reflect on them in a broader context. Single-page summaries were prepared for each region discussing the data, the implicit assumptions behind the projections including qualitative factors, internal contradictions, and potential pitfalls for the depicted scenarios. The group also attempted to describe the general assumptions, explicit and implicit, that make up the conventional development paradigm. As indicated earlier, the initial critique of these scenarios focused on their believability. Thus, although the group looked for internal inconsistencies and major uncertainties within the scenarios that may make the expressed outcomes highly unlikely, the desirability of the scenarios was not addressed. These data and the associated stories were presented to a broader group at the project meeting in February 1994 and were reviewed by a larger group of participants within and outside the 2050 Project.

5 Alternative futures

Nearly 50 essays, representing almost 40 countries, were submitted as part of the Vision Exercise. These were supplemented by close to 20 interviews. A volume has been prepared covering not only the content of these essays and interviews, but also the value of the exercise itself (Nagpal and Foltz, 1995). The essays and interviews from Africans were used in the development of scenarios for that region, as described below.

As was expected, the essays and interviews showed some degree of consensus, but with variations within and among regions. Themes such as increased access to education, greater equity and an end to absolute poverty, and a more harmonious relationship between mankind and nature recurred throughout many of the essays. The scope of analysis and the degree of emphasis and detail accorded to specific issues varied among the essays and interviews, reflecting regional and personal differences in visions of the future. Of importance for the development of scenarios is that, although technological development and biophysical issues were widely mentioned, the participants in the Visions Exercise appeared far more concerned with the human dimensions of change in the future (WRI, 1994b).

A paper describing the typology of scenarios used in the 2050 Project is being prepared. It examines different types of mechanisms that affect possible futures, including how these are likely to differ along a set of dimensions including population, economy, technology, systems of governance, degree of cultural homogenisation, degree of modification of natural systems, etc. Will regional and global assemblages of nations take firm root or will increasingly balkanised political entities dominate? Will current inequities within and between nations grow, further separating rich and poor? Will population growth and resource use approach Malthusian limits, requiring fundamental changes in behaviour, or will rapid advances in technology allow people in wealthy nations to continue their current rates of consumption and others to increase theirs? Will greenhouse warming and other environmental changes have impacts that are negligible to moderate, or dramatic?

These and other dimensions must be considered individually and in combination. The purpose was to classify scenarios on a broad scale, thus providing images, and forcing the scenario-developers to consider a broad range of possible futures.

Following the above strategy, a set of positive futures was developed for sub-Saharan Africa and continues to be developed for other regions. First-stage scenarios used the Visions essays and interviews, data collected by the project's base studies and others, an examination of previously published and current work on possible future developments. In general, for example, an Africa is depicted that is more stable, self-reliant and integrated than currently exists. The African scenarios differ in the degree to which the continent is tied into the global economic and political systems, the rate of overall advancement, and the relative progress of different groups within the continent, as well as in other facets. In a similar way, relevant characteristics are being developed for other regional scenario sets.

The preliminary scenarios are evaluated by regional and sectoral experts. The preliminary African scenarios provided input for a regional policy workshop in Harare, Zimbabwe. The African scenarios, and the other regional sustainability scenarios being developed in the 2050 Project, may provide a model for other exercises.

6 Conclusions

By adopting a holistic perspective and listening to a multitude of voices, the 2050 Project has attempted to develop scenarios of the future that are realistic and informative. We are not producing scenarios because we think we know what the future *will* be. We do so because they help us to think about what the future *could* be. Perhaps more importantly, the process of preparing scenarios can help us to recognise what the future *cannot* be. Once we learn to discard the latter, we can picture futures that we should do all we can to prevent from occurring or futures that we should strive to make happen. In either case, having clearly in mind a destination to aspire to, or to avoid, can guide decision-makers in developing the policies and actions of today. Scenarios can also help us to be alert to, and perhaps to take advantage of, the inevitable surprises that will befall us along the way.

References

Achebe, C., Okeyo, A.P., Hyden, G. and Magadza, C. (1990). *Beyond Hunger in Africa: Africa 2057 An African Vision*. Heinemann, Nairobi.

Ayres, R U. (1994). *Socio-Economic Modelling and Forecasting*. INSEAD, Fontainebleau.

Bender, W.H. (1994). An end use analysis of global food requirements. *Food Policy* **19** (4): 1-15.

Bender, W.H. (1995). An end use analysis of global food requirements. In: Downing, T.E. (ed.) *Climate Change and World Food Security*. Springer-Verlag, Heidelberg.

Benjamin, P., Emel, J.L., Aspersion, J.X., Kasperson, R.E., and Rocheleau, D. (1994). Social visions of future sustainable societies. In: Kasperson, J.X. and Kasperson, R.E. (eds.) *Global Environmental Risk*. United Nations University Press, Tokyo.

Brooks, H. (1987). The Typology of Surprises in Technology, Institutions, and Development. In: Clark, W.C. and Munn, R.E. (eds.) *Sustainable Development of the Biosphere*. Cambridge University Press, Cambridge. pp. 325-348.

Brown, L. (1994). Facing food insecurity. In: Brown, L. *et. al.*, *State of the World 1994*. W. W. Norton & Company, New York. pp. 177-197.

Chen, R.S. and Kates, R.W. (1994a). World food security: prospects and trends. *Food Policy*, **19**(2): 192-208.

Chen, R.S. and Kates, R.W. (1994b). Climate change and world food security. *Global Environmental Change*, **4**(1): 3-6.

Chen, R.S. and Kates, R.W. (1995). Towards a food secure world: prospects and trends. In: Downing, T.E. (ed.) *Climate Change and World Food Security*. Springer-Verlag, Heidelberg.

Conklin, C.R. (1993). What does cumulative effects analysis have to do with sustainable development? *Canadian Journal of Regional Science*, **16**(3): 453-480.

Crosson, P. and Anderson, J.R. (1994). Demand and supply: trends in global agriculture. *Food Policy*, **19**(2): 105-119.

Crosson, P. and Anderson, J.R. (1995). Demand and supply: trends in global agriculture. In: Downing, T.E. (ed.) *Climate Change and World Food Security*. Springer-Verlag, Heidelberg.

Food and Agricultural Organisation (1993). *Agriculture: Towards 2010*. FAO, Rome.

Fischer, G., Frohberg, K., Parry, M.L. and Rosenzweig, C. (1994). Climate change and world food supply, demand and trade - who benefits, who loses?. *Global Environmental Change*, **4**(1): 7-23.

Fischer, G., Frohberg, K., Parry, M. L. and Rosenzweig, C. (1995). Impacts of potential climate change on global and regional food production and vulnerability. In: Downing, T.E. (ed.) *Climate Change and World Food Security*. Springer-Verlag, Heidelberg.

Gell-Mann, M. (1994). *The Quark and the Jaguar: Adventures in the Simple and Complex*. W. H. Freeman and Company, New York.

Huntley, B., Sigfried, R. and Sunter, C. (1989). *South African Environments into the 21st Century*. Human & Rousseau, Tafelberg.

Jäger, J., Sontag, N., Bernard, D., and Kurz, W. (1990). *The Challenge of Sustainable Development in a Greenhouse World: Some Visions of the Future*. Report of a Policy Exercise held in Bad Bleidberg, Austria September 2-7, 1990. Stockholm Environmental Institute, Stockholm.

Kendall, H.W. and Pimental, D. (1994). Constraints on the expansion of the global food supply. *Ambio*, **23**(3): 198-205.

Nagpal, T. and Foltz, C. (eds.) (1995). *Choosing Our Future: Essays and Interviews on Sustainability*. WRI: Washington, D.C.

Norse, D. (1994). Multiple threats to regional food production: environment, economy, population? *Food Policy*, **19**(2): 133-148.

Norse, D. (1995). The conjunction of threats to regional food production: how serious are environment, economy, population and climate? In: Downing, T.E. (ed.) *Climate Change and World Food Security*. Springer-Verlag, Heidelberg.

Pepper, W. (1993). Personal communication.

Pepper, W., Leggett, J., Swart, R., Wasson, J., Edmonds, J., and Mintzer, I. (1992). *Emission Scenarios for the IPCC - An Update - Assumptions, Methodology, and Results*. Prepared for The Intergovernmental Panel on Climate Change, Working Group 1.

Petersen, J.L. (1993). *The Road to 2012: Looking Toward the Next Two Decades*. The Arlington Institute, Arlington.

Rainbow, R. (1995). Personal communication.

Reilly, J., Hohmann, N. and Kane, S. (1994). Climate change and agricultural change: who benefits, who loses? *Global Environmental Change*, **4**(1): 24-36.

Reilly, J., Hohmann, N. and Kane, S. (1995). Climate change and agricultural change: who benefits, who loses? In: Downing, T.E. (ed.). *Climate Change and World Food Security*. Springer-Verlag, Heidelberg.

Robinson, J. B. (1990). Futures under glass: a recipe for people who hate to predict. *Futures*, October: 820-841.

Robinson, J. B. (1989). *Exploring a Sustainable Future for Canada: The Next Step in the Conserver Society Discussion* . Paper presented at a meeting of the Leaders of National 21st Century Studies, July 16-21, co-sponsored by the Institute for 21st Century Studies and the World Futures Society, Washington, D.C.

Robinson, J. B. (1988). Unlearning and backcasting: rethinking some of the questions we ask about the future. *Technological Forecasting and Social Change*, **33**: 325-338.

Ruttan, V.W., Bell, D.E., and Clark, W. C. (1994). Climate change and food security: agriculture, health and environmental research. *Global Environmental Change*, **4**(1): 63-77.

Ruttan, V.W., Bell, D.E., and Clark, W. C. (1995). Climate change and food security: agriculture, health and environmental research. In: Downing, T.E. (ed.). *Climate Change and World Food Security*. Springer-Verlag, Heidelberg.

Schwartz, P. (1991). *The Art of the Long View*. Doubleday, New York.

Sunter, C. (1987). *The World and South Africa in the 1990s*. Human & Rousseau, Tafelberg.

Svedin, U. and Aniansson, B. (eds.) (1987). *Surprising Futures*. Notes from an International Workshop on Long-Term World Development. Swedish Council for Planning and Coordination of Research, Stockholm.

Toth, F., Hizsnyik, E. and Clark, W.C. (eds.) (1989). *Scenarios of Socioeconomic Development for Studies of Global Environmental Change: A Critical Review*. RR-89-4. International Institute for Applied Systems Analysis, Laxenburg.

Toth, F. (1988). Policy exercises: objectives and design elements. *Simulation and Games,* **19**: 235-255.

UNPD (United Nations Population Division) (1993). *Interpolated National Populations, The 1992 Revision*. UNPD, New York.

World Bank (1994). *World Development Report 1994: Infrastructure for Development*. Oxford University Press, Oxford.

WRI (World Resources Institute) (1994a). *Newsletter of The 2050 Project*, **1**(1), Winter.

WRI (World Resources Institute) (1994b). *Newsletter of The 2050 Project*, **1**(2), Spring.

WRI (World Resources Institute) (1994c). *World Resources 1994-95: People and the Environment*. Oxford University Press, Oxford.

PART IV

LOCAL IMPACTS AND RESPONSES TO GLOBAL CHANGE

Climate Change and the Agro-ecosystems in China

Futang Wang and Zong-ci Zhao
Chinese Academy of Meteorological Sciences
Baishiqiao Road, No. 46
Western Suburb
Beijing 100081
People's Republic of China

1 Introduction

Climate change can be divided into two parts: natural change and change induced by human activity. Working Group I of the Intergovernmental Panel on Climate Change (Houghton *et al.*, 1990; Houghton *et al.*, 1992) has found that global warming of about 0.3-0.6°C has occurred over the last 100 years. Warming of about 0.6-0.8°C in East Asia over the last 100 years has been calculated by Hulme *et al.* (1992). Global climate change scenarios due to the greenhouse effect have indicated that global temperature and precipitation might increase (Houghton *et al.*, 1990; Houghton *et al.*, 1992). Regional climate change scenarios have also demonstrated that the temperature and precipitation in East Asia and China might increase due to the greenhouse effect (Hulme *et al.*, 1992).

Agricultural production serves as an essential precondition for the continuous and stable development of the national economy in China. At present, development and progress in China are facing the challenges of population increase, shortage of resources and degradation of the environment. With the rapid development of science and technology in the world, the anthropogenic impact on the environment and on climate is becoming increasingly serious. Changes in the global climate arising from the increased emission of CO_2 and other greenhouse gases to the atmosphere can potentially change the current environment-climate conditions for agricultural production and cause difficulties for sustainable development in China. Studies, as presented here, of the possible impact of climate change on cropping patterns and on agricultural production in China are therefore of profound significance.

NATO ASI Series, Vol. I 37
Climate Change and World Food Security
Edited by Thomas E. Downing

2 Methods and data

In order to estimate regional climate change scenarios for China, two global projections were used, following methods described in Hulme *et al.* (1992). A simple global socioeconomic climate-impact model (GSECIM) was used to predict global-mean warming for the future. Seven scenarios derived from seven general circulation models (GCMs: GFDL, GISS, LLNL, MPI, OSU, UKMO-L, UKMO-H) and a composite scenario were chosen to provide regional patterns of changes (Hulme *et al.*, 1992).

To investigate the possible impacts of climate change on Chinese cropping patterns, special models for cropping patterns, based on a set of climatic parameters (see Table 1), were used. The distribution and changes in cropping patterns under current (1951-80) and future projected climates (2050) were mapped using three baseline datasets (Climatic Research Unit [CRU], International Institute of Applied Systems Analysis [IIASA] and State Meteorological Administration [SMA]). The East Asia and China region extends from 15-60°N and 70-140°E, with a 0.5° x 0.5° gridded resolution for the climate impact assessment.

Table 1 Active accumulated temperature above 0°C

	Chinese cropping patterns	**AAT***	
		Northern boundary	**Southern boundary**
I	Single cropping area	-	4000
	Spring wheat, maize		
II	Double cropping area	4000	5800
	Wheat-maize	4000	4800
	Wheat-cotton,-rice	4800	5500
	Rape-rice, rice-maize,-rice	5500	5800
III	Triple cropping area	5800	8000
	Wheat-rice-maize	5800	6100
	Rice-rice-wheat-rape	6100	7000
	Rice-rice-sugar cane-subtropical fruit	7000	8000
IIId	Hot triple cropping	8000	-
	Rice-rice-rice-tropical crop		

*AAT is the active accumulated temperature – the sum of daily temperatures above a given threshold temperature (e.g. 0 °C).

3 Climate change in the past

In recent years, climatologists in China have made great efforts to study and analyze characteristics of historical climate change in China. They have suggested that, in the past 500 years, the climate in China has experienced three cool periods (1470-1550, 1620-1720 and 1840-1890) and two warm periods (1550-1600 and 1720-1830) (Zhu, 1975). The temperature changes range from 0.5°C to 1.0°C between cool and warm periods.

Some periods of lower rainfall (less than 600 mm) and some periods of higher rainfall (more than 700 mm) have been observed since the 17th century. In Beijing, for example, the amount of precipitation has varied greatly, particularly since the 1950s. In fact, northern China has had periods of drought since the 1950s and rainfall has decreased during the same period.

According to the instrumental annual global-mean temperature record since 1854 (Figure 1), a warming trend occurred up to about 1940 (1950 for East Asia), followed by three decades of little change. After the mid-1970s, the temperature warmed up again rather rapidly. The six warmest years in the entire series have all occurred since 1980, with the single warmest year being 1990. Overall, the world has warmed by about 0.45 ± 0.15°C since the late 19th century (Folland *et al.*, 1992). There has been a similar pattern of change in annual mean temperature over East Asia and China, except that the warming since the 19th century has been somewhat greater than the global mean.

The observed warming over East Asia and China, however, shows quite complex seasonal behaviour (Figure 2). Summer temperatures have remained constant over this period, whereas winter temperatures have warmed considerably – by over 0.8°C. Spring and autumn temperatures have risen slightly, by between 0.2°C and 0.3°C. Investigation of the spatial pattern of these temperature changes over China in the last 40 years demonstrates that annual warming has been greatest (+0.3°C) over northeastern and northwestern China, but that a large proportion of central and eastern China has cooled slightly.

Precipitation is more spatially and temporally variable than temperature. Based on the datasets described above, global and East Asian time series of precipitation were constructed and

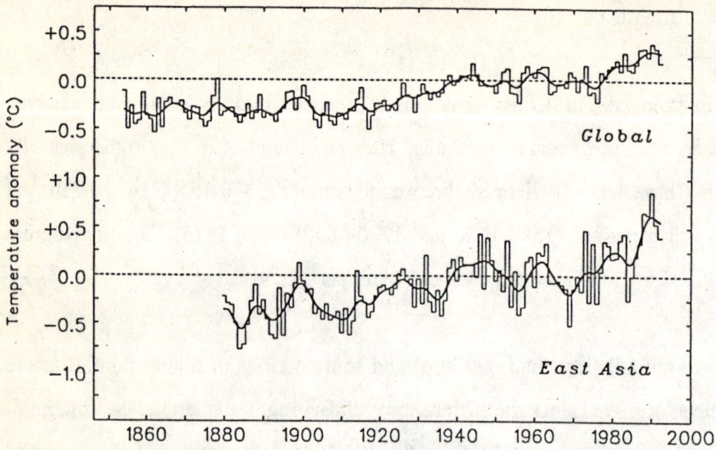

Figure 1 Annual mean temperature for global land and marine areas (1854-1992, Jones *et al*., 1991) and for East Asia (1880-1992, calculated from Jones, 1988; updated). The individual yearly values are anomalies from the 1951-80 reference period and the smooth curves result from applying a filter that suppresses variations on time-scales of less than ten years.

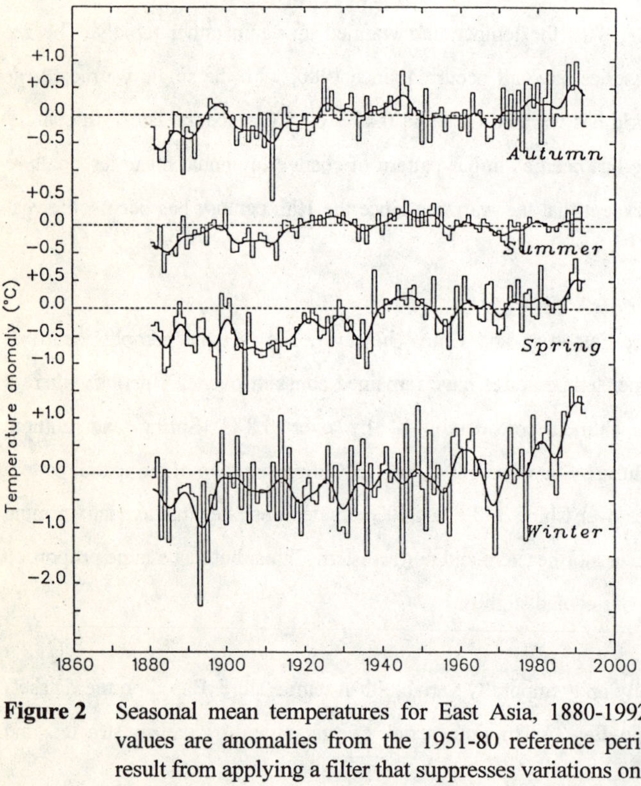

Figure 2 Seasonal mean temperatures for East Asia, 1880-1992. The individual seasonal values are anomalies from the 1951-80 reference period and the smooth curves result from applying a filter that suppresses variations on time-scales of less than ten years.

expressed as percentages of the common reference 1951-80 mean. Figure 3 in both cases indicates a weak tendency for increasing annual precipitation over the last 100 years.

The four seasonal time curves for East Asia (Figure 4) indicate the much larger relative variability of winter precipitation compared with that in summer (mean interannual variability of 25.7% and 6.3% respectively), although in absolute terms summer precipitation variability is larger (10.2 mm and 20.1 mm respectively). Over the last 40 years, much of southern and eastern China has experienced reduced annual precipitation, whereas northwestern China has seen a slight increase in precipitation.

4 Climate change scenarios for China

Regional climate change in the future for East Asia and China has been calculated using the global warming projection simulated by the GSECIM model combined with seven GCMs (Table 2). The assessment shows that a global mean warming of 1.2°C is likely to occur by 2050. According to this scenario (Figure 5), the annual mean temperature in east and southeast China would increase by between 1.0°C and 1.5°C; in northeast, north and west China by 1.5-1.8°C; and in Taiwan and Hainan by slightly less than 1.0°C. Precipitation in northwest China would increase by between 5 and 8 per cent relative to present annual precipitation, and in other parts of China would increase by less than 5 per cent. As shown in Table 3, temperature in East Asia and China might increase by about 0.88°C in 2030, 1.40°C in 2050 and 2.95°C in 2100. Precipitation might also increase, by about 2.6 per cent in 2030, 4.2% in 2050 and 8.9 per cent in 2100. Changes in temperature due to human activity would be more obvious than the changes in precipitation in East Asia and China.

Table 2 Climate change scenarios for East Asia and China for 2050

	Temperature (°C)					**Precipitation (%)**				
GCM	**Annual**	**DJF**	**MAM**	**JJA**	**SON**	**Annual**	**DJF**	**MAM**	**JJA**	**SON**
UKMO-1	1.33	1.33	1.43	1.17	1.39	4.0	3.6	7.1	4.8	2.8
UKMO-H	1.82	1.85	1.66	1.84	1.92	5.9	8.2	7.6	5.4	3.7
GFDL	1.34	1.46	1.47	1.46	1.34	2.6	3.9	1.8	3.0	1.8
GISS	1.14	1.32	1.06	1.01	1.16	3.3	5.7	4.0	2.1	3.1
OSU	1.25	1.34	1.21	1.22	1.21	5.1	6.8	5.1	3.4	6.8
LLNL	1.47	1.42	1.31	1.74	1.40	4.4	4.8	5.8	3.1	4.9
MPI	1.35	1.66	1.47	0.97	1.26	4.0	5.7	4.0	9.8	1.6
Composite GCM	1.40	1.48	1.37	1.34	1.38	4.2	5.5	5.1	4.5	3.5

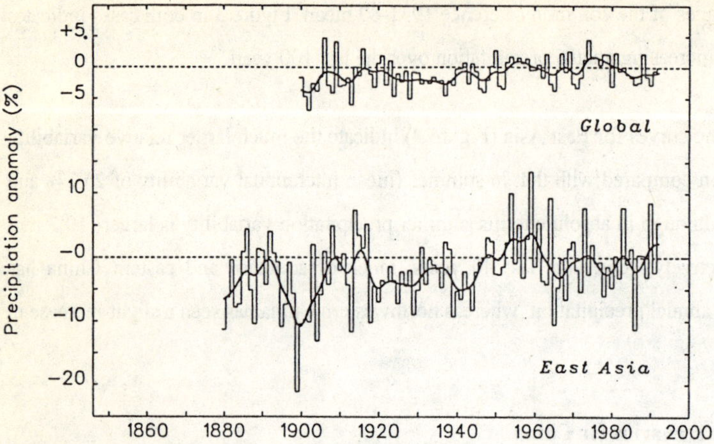

Figure 3 Annual mean precipitation for global land areas (1900-92) and for East Asia (1880-1992, calculated from the dataset of Hulme, 1991). The individual yearly values are per cent anomalies from the 1951-80 reference period and the smooth curves result from applying a filter that suppresses variations on time-scales of less than ten years.

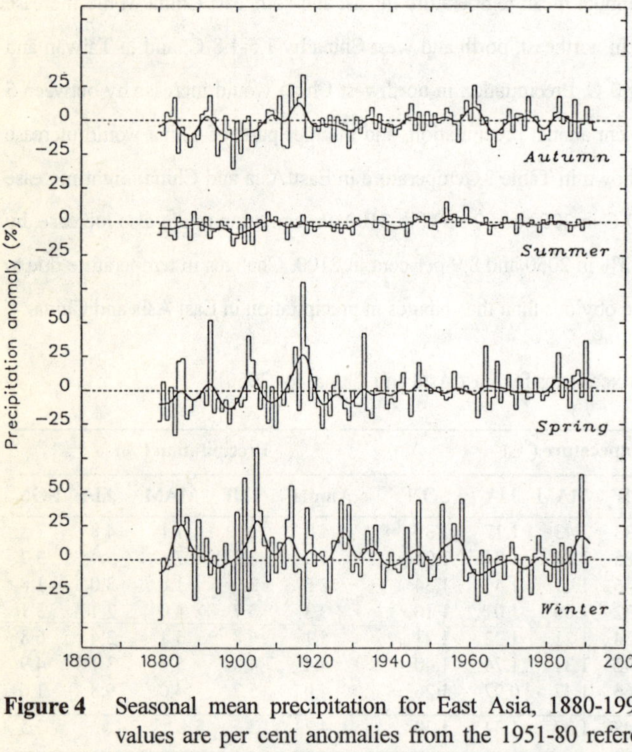

Figure 4 Seasonal mean precipitation for East Asia, 1880-1992. The individual seasonal values are per cent anomalies from the 1951-80 reference period and the smooth curves result from applying a filter that suppresses variations on time-scales of less than ten years.

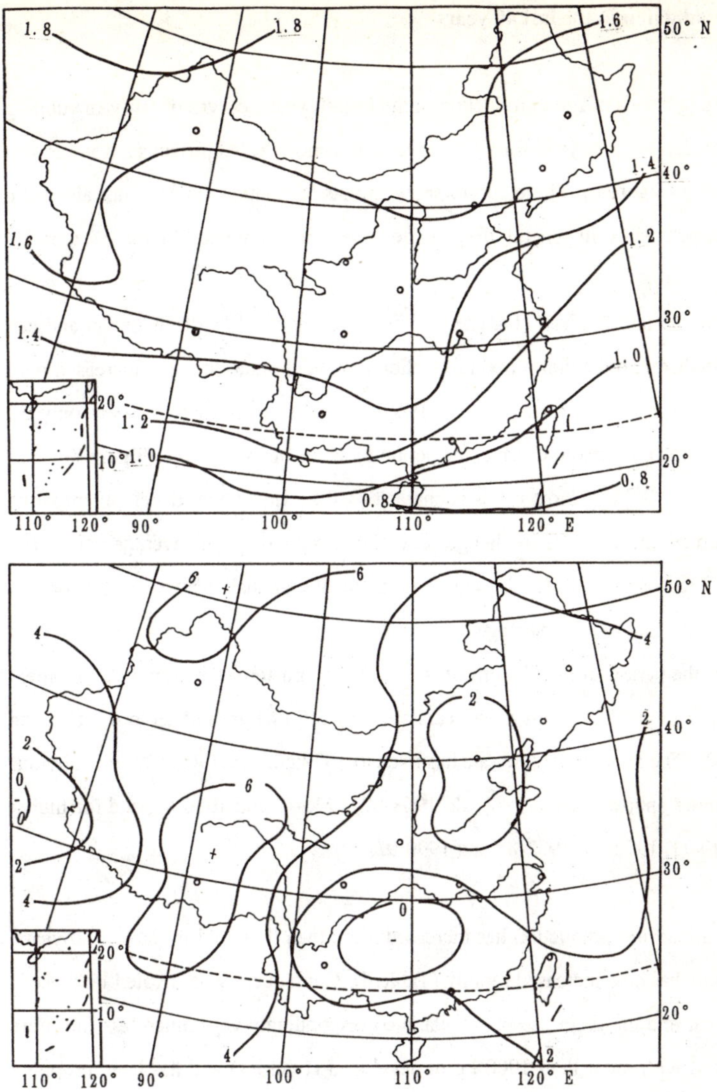

Figure 5 Estimated mean temperature (top, °C) and precipitation (bottom, %) changes in East Asia by the year 2050 with respect to the present climate (1951-80).

Table 3 Climate change scenarios for East Asia and China in the future (1991-2100)

Years	2000	2010	2020	2030	2040	2050	2060	2070	2080	2090	2100
Temperature change (°C)	0.20	0.35	0.65	0.88	1.06	1.40	1.64	2.01	2.30	2.66	2.95
Precipitation change (%)	0.6	1.1	1.9	2.6	3.2	4.2	4.5	5.5	6.3	7.2	8.9

5 Agricultural production in the last 40 years

To analyse the changing trends of food production in the last 40 years, curves of ten-year running mean of increases and decreases in yield for all of China and some selected provinces were drawn, following Wang (1982) (Figure 6). These curves show similar fluctuation patterns, but also have different regional characteristics. In general, they can be divided into three development stages:

I The period before the late 1960s is characterised by a prolonged, slow development and can be defined as low-level production. Food production remained rather low and increased only slightly. Over about 20 years (from 1950 to 1969), the annual increase rate was only 32 kg/ha/yr (i.e. 2 kg/mu). In absolute terms, the per unit yield is below 1.8 t/ha (Figure 7).

II The period from the 1970s to early 1980s is characterised by fluctuating development while yields were at a medium level. Yield changes also fluctuated around the average value. The per unit yield increased to 2.8 t/ha for this period; i.e. the annual increase was about 86 kg/ha/yr greater than that in the first stage.

III The third stage is the period from 1980 onwards, which is characterised by rapid development with high yields. The yield levels in most years are over 3.5 t/ha/yr, and up to 4.0 t/ha and more in the 1990s. The annual increase rate has risen to 107 kg/ha/yr (Figure 7). It is obvious that the lower values appear to be in 1960-62, 1968-69, 1976-77 and 1980-81, and the higher yields are in 1970-71, 1978-79, 1982-84 and 1990-92.

However, although agricultural production has increased more than threefold in the last 40 years, the population of China has also increased rapidly (Table 4). Consequently, cultivated land use is decreasing year by year and the grain use per capita is correspondingly very much less (no more than 400 kg, compared with more than 1000 kg in developed countries) and has been dropping even lower since 1983-84 (Figure 8). The development of Chinese agriculture and the national economy is facing two serious challenges: population increase and shortages of natural resources, such as arable land and water. Given these threats, climate change has important implications for Chinese agriculture and the national economy.

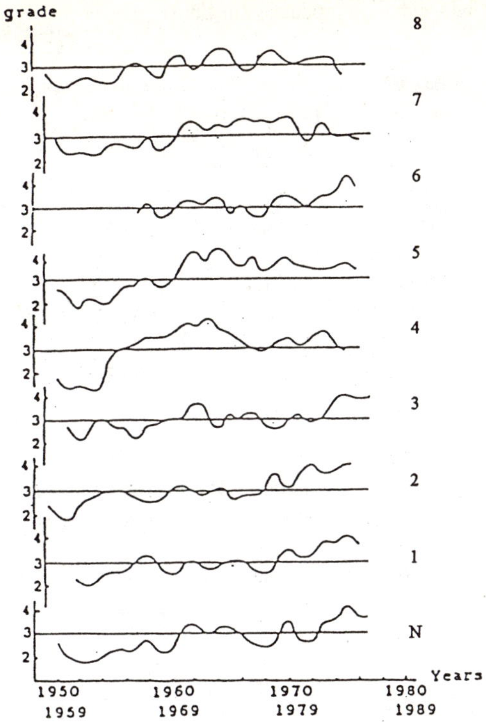

Figure 6 Curves of the ten-year running mean of increases and decreases in yield per unit area in the last 40 years. N indicates the yield per unit area of the whole country. 1: Jilin province; 2 and 3: Jianxi and Hubei provinces; 4 and 5: Henan and Shanxi provinces.; 6,7 and 8: Sichuan, Guangxi and Zhejiang provinces respectively.

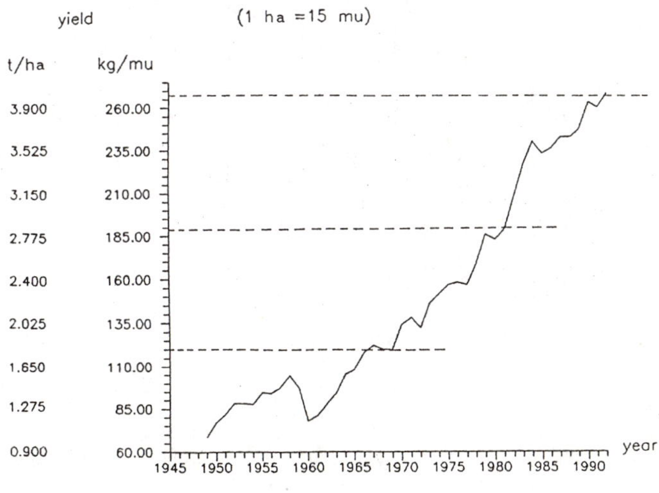

Figure 7 Evolution of the yield per unit area for China in the last 40 years.

Table 4 Changes in population, cultivated land and grain production in China

Year	Population (thousands)	Cultivated land (ha)	Area/ person (ha)	Grain production (10^6t)	Grain production / person (kg)
1949	541,670,000	97881333	0.18	112.18	207.10
1952	646,530,000	107918667	0.22	163.92	253.54
1957	646,530,000	111830000	0.21	195.05	301.68
1962	723,000,000	102903333	0.18	160.00	221.30
1965	725,380,000	103594000	0.17	194.53	268.17
1970	824,000,000	101134667	0.14	239.96	291.21
1975	919,700,000	99708000	0.13	284.52	309.36
1977	940,000,000	99247333	0.12	282.73	300.77
1979	970,920,000	99389333	0.10	332.12	342.06
1980	982,550,000			320.56	326.25
1982	1,008,175,288		0.10	354.50	351.62
1987	1,069,164,090			404.73	378.54
1988	1,086,540,493			395.90	364.36
1990	1,133,682,501		0.09	435.00	383.70

Source: Markham and Yihui (1992).

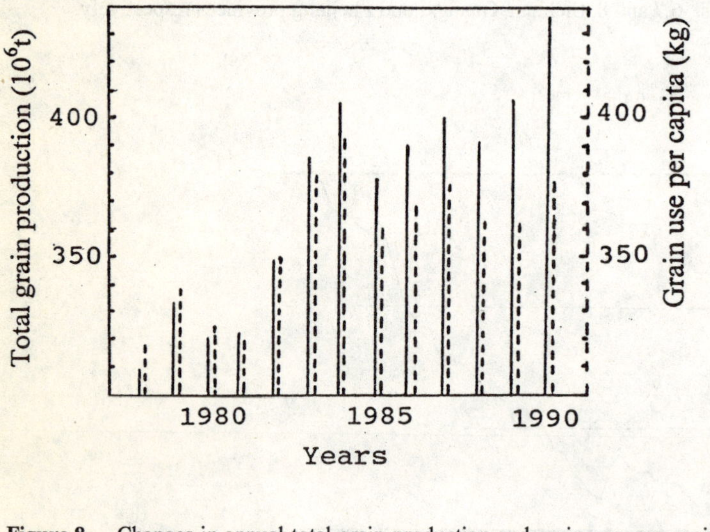

Figure 8 Changes in annual total grain production and grain use per capita in China in recent years.

6 Distribution, change and implications for agriculture

In China, the development of agriculture over several thousand years has meant that large areas of forest and grassland have been converted into arable lands. Agriculture is now a very important part of vegetation in China. To illustrate the present-day distribution and the possible change in cropping patterns in the future, a cropping system was modelled rather than individual crop varieties. The main patterns of the cropping system are defined by the number of different crops that can be grown successively and successfully during a single growing season. AAT data (above 0°C) have been used to determine the major Chinese cropping patterns (see Table 1).

The eastern part of China is the most important area for Chinese agricultural production. From Figure 9, it can be seen that, from north to south, the single cropping pattern changes to multiple cropping and is more diverse with increases in temperature and precipitation – the two major climatic factors influencing crop growth and agricultural production. Figure 9 also shows that AAT values of 4000 and 5800 are significant in distinguishing between the single-double and

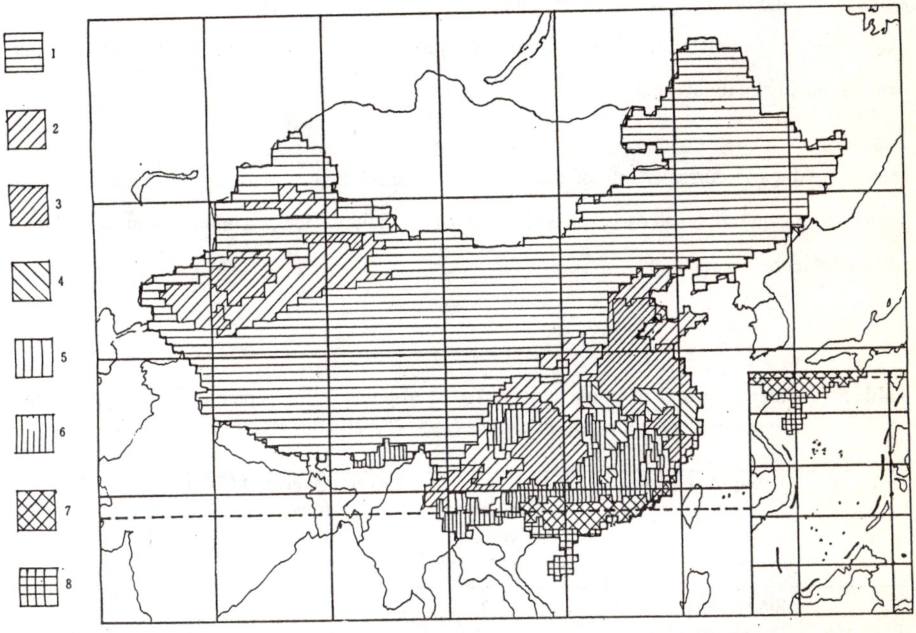

Figure 9 Distribution of the cropping system in China under the current climate (1951-80). 1: single cropping; 2-4: different combinations of double cropping; 5-8: different combinations of triple cropping (see Table 1).

cropping area to be replaced by different triple cropping patterns, and the current double cropping area would shift northwards towards the central part of the present single cropping area. More explicitly, the northern boundary of the triple cropping area would shift from its current border at the Changjiang River towards the Huaihe River – a shift of more than 5° latitude. As a result, the current area of single cropping would be reduced by about 23.1% and the triple cropping area could be extended by roughly 22.4%. The double cropping area would remain almost the same, as likely under the current climate (see Table 5).

To a certain extent, the changes described above seem to indicate that climate warming would be favourable to Chinese agriculture due to a diversification of cropping patterns. In fact, however, the area suitable for cultivating rice and wheat (the major two crops in China) would increase but mean yields may decrease due to reduced water availability. That is, by the year 2050, the net balance between precipitation and evapotranspiration would be negative and moisture stress would be more severe than today, although precipitation by 2050 would increase somewhat (Table 4). Apart from this, the calculation of the LGS for the future climate (by 2050), projected by the composite GCM scenario of climate change, also shows that it would be reduced by 5-25% in South China and in large parts of North China because of no relative precipitation increase. So the likely impacts of climate warming on agricultural production in some eastern regions of China would probably be unfavourable.

Northeast China is currently one of the oldest agricultural areas where low temperatures often prevent crop maturation and early frosts result in harvest failure. Here, a climate warming by 2050 will surely increase crop yield.

Table 5 Percentage area of China under different cropping patterns predicted by the cropping system model for the current climate (1951-80) and the year 2050 and its changes

	Cropping pattern	**Percentage area of China**		
		Current climate (1951-80)	**Future climate (2050)**	**Change (%)**
I	Single cropping	62.3	39.2	-23.1
II	Double cropping	24.2	24.9	+0.7
III	Triple cropping	12.4	28.6	+16.2
IIId	Hot-triple cropping	1.1	7.3	+6.2

double-triple cropping patterns, respectively. Within double and triple cropping areas, several combinations of specific crops can be easily distinguished (see Table 1 and Figure 9). However, the vast western areas are dominated by a single cropping pattern due to lower precipitation and relatively higher temperatures in the northwestern region (Xingjian area) and lower temperatures in the southwest (Qing-Zang Plateau). Generally, this simulated map fits well with the actual distribution of cropping patterns in China (Academy of Meteorological Sciences, 1981; Ye, 1992).

Another important indicator for agricultural production is the length of the growing season (LGS). According to the FAO's definition of LGS (FAO, 1980), calculations for the current climate show that LGS values of 120-150 days and 240-270 days are the boundaries between the single-double and double-triple cropping patterns, respectively.

Potential changes in cropping patterns projected by the composite GCM scenario in the year 2050 are demonstrated in Figure 10. Large changes would occur almost everywhere in China except for the very high and cold Qinghai-Xizang Plateau in southwestern China and the northern part of Northeast China. The warmer climate causes a large part of the present double

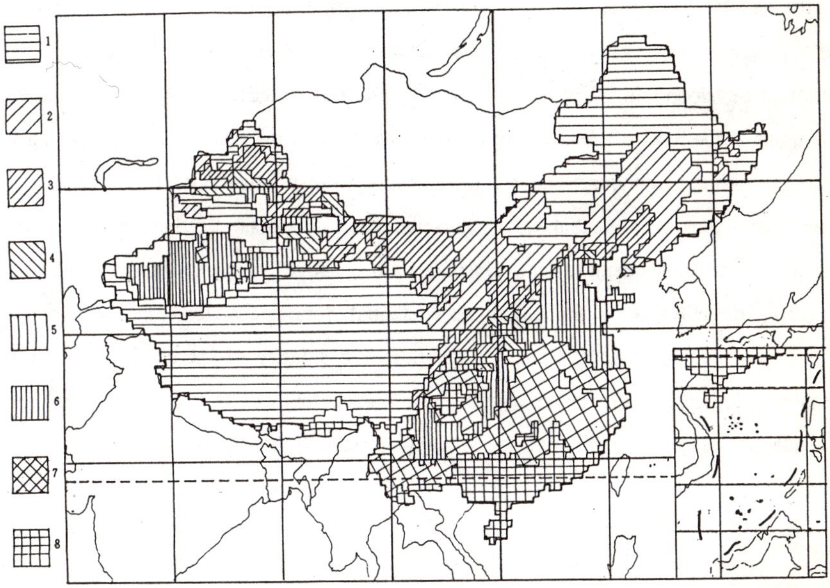

Figure 10 Cropping system in China under climate change projected by the composite GCM scenario for 2050. 1: single cropping; 2-4: different combinations of double cropping; 5-8: different combinations of triple cropping (see Table 1).

7 Discussion

The uncertainties in the investigations of climate change have three roots: the uncertainty of future human activity (such as population growth, emissions and land use); the uncertainty of the natural climate change; and the uncertainty of the scientific estimation methods (such as limitations of the GCMs, global averaged and impact models). To investigate the changes in the agro-ecosystem induced by human activity and climate change, regional climate changes should be analyzed. This depends on the local climatic characteristics and is difficult to study using GCMs. Further investigations are needed to portray adequately the range of climates that might occur in China in the next few decades.

In summary, climate warming induced by an increase in greenhouse gases in the atmosphere would have a substantial impact on agro-ecosystems almost everywhere in China. However, it is difficult to estimate and assess, either quantitatively or qualitatively, the impacts of climate warming, especially on agricultural production in China. There are several reasons for this. First, Chinese farming and crop patterns are very diverse. The different crops and cropping systems could be interchanged in agricultural practice. It seems that Chinese agriculture would adapt well. Second, the impacts depend not only on the various climatic factors affected by the increase of greenhouse gases in the atmosphere, but also on the social and economic conditions of agricultural development. Advances in technology, crop varieties and farm management would mitigate the effects of climate warming on the agricultural system. Third, the major difficulty is to predict greenhouse-related climate change with sufficient confidence, because of the many uncertainties mentioned above. Such a prediction is a basic precondition of targeted adaptive strategies. Consequently, the analysis described above gives only some idea of the aspect, scale, direction and implication of potential future changes for the Chinese agro-ecosystem. For the present, China should pursue a strategy of increasing agricultural development and research with the aim of being able to adapt to a variety of climatic conditions. Effective adaptive responses could increase per capita production and improve the prospects of the poor.

Acknowledgements

We would like to thank Dr. M. Hulme (Climatic Research Unit, University of East Anglia) and Dr. R. Leemans (National Institute for Public Health and Environmental Protection) for their great help and cooperation in this research.

References

Academy of Meteorological Sciences (1981). *Agricultural and Climate Resources and Planting Systems in China.* Agricultural Publishing House. (In Chinese.)

FAO (Food and Agriculture Organization) (1980). *Report on the Agro-ecological Zones Project.* World Soil Resources Report No.48, Vol.IV, Methodology and results for Southeast Asia. FAO, Rome.

Folland, C.K., Karl, T.R., Nicholls, N., Nyenzi, B.S., Parker, D.E. and Vinnikov, K. Ya (1992). Observed climate change and variability. In: Houghton, J.T., Callander, B.A. and Varney, S.K. (eds.), *Climate Change 1992: The Supplementary Report to the IPCC Scientific Assessment.* Cambridge University Press, Cambridge. pp. 136-170.

Houghton, J.T., Callander, B.A. and Varney, S.K. (eds.) (1992). *Climate Change 1992: The Supplementary Report to the IPCC Scientific Assessment.* Cambridge University Press, Cambridge.

Houghton, J.T., Jenkins, G.J. and Ephraums, J.J. (eds.) (1990). *Climate Change: The IPCC Scientific Assessment.* Cambridge University Press, Cambridge.

Hulme, M. (1991). *Analysis of Worldwide Precipitation Records and Comparison with Model Predictions.* Final report for the UK Department of the Environment, Contract PECD/7/10/198. Climate Research Unit, Norwich.

Hulme, M., Leemans, R., Zongci Zhao, Futang Wang, Markham, A., Wigley, T., Yihui Ding and Tao Jiang (1992). *Climate Change due to the Greenhouse Effect and its Implications for China.* CRU/WWF/SMA, Banson Production, London.

Jones, P.D. (1988). Hemispheric surface air temperature variations: recent trends and an update to 1987. *Journal of Climate,* **1**: 654-660.

Jones, P.D., Wigley, T.M.L. and Farmer, G. (1991). Marine and land temperature data sets: a comparison and look at recent trends. In: Schlesinger, M.E. (ed.), *Greenhouse-gas-induced Climatic Change: A Critical Appraisal of Simulations and Observations.* Elsevier, Amsterdam. pp.153-172

Leemans, R. and Cramer, W. (1991). *The IIASA Database for Mean Monthly Values of Temperature, Precipitation and Cloudiness on a Global Terrestrial Grid.* RR-91-18, IIASA, Laxenburg.

Markham, A. and Yihui, D. (eds.) (1992). *Proceedings of the SMA/WWF Conference on Environment and Climate Change: The Challenge for China.* WWF International, Gland, Switzerland. (Chinese version published in 1994 by China Meteorological Press.)

Wang Futang (1982). On the variation of accumulated temperature during the last 100 years and crop yield in China. *Acta Geographical Sinica,* **37** (3): 272-280. (In Chinese).

Ye Duzheng (1992). The research of global change in China. *Bulletin of CNC-IGBP,* **2** (1): 6-23.

Zhu Kezhen (1975). A preliminary study on the climate variation during the last 5000 years in China. *China Science,* February: 168-189. (In Chinese).

Agricultural Vulnerability and Poverty Alleviation in Bangladesh

Z. Karim
Bangladesh Agricultural Research Council
New Airport Road
Farmgate
Dhaka 1215
Bangladesh

1 Introduction

Bangladesh has the lowest per capita arable land in the world (less than 0.1 hectare) and a high population density. Over 50 per cent of rural households are functionally landless, owning less than 0.20 hectares of land. Its limited resource base, large population and vulnerability to frequent natural calamities have all contributed to widespread poverty. It is one of the poorest countries of the world, with a per capita income of US$190 per annum. Recurrent floods, droughts and cyclones, especially in recent years, have weakened the economy of rural people. In 1988, Bangladesh had the largest flood of the century immediately after serious flooding in 1987. Then, the country suffered devastating cyclones in November 1988 and April 1991, drought in March and April 1989, and damaging tornadoes in April 1989 and 1992.

It seems that Bangladesh is increasingly prone to frequent natural calamities. Poverty, unemployment, malnutrition, illiteracy, rapid population growth and frequent natural disasters remain endemic problems. Poverty is not only a state of deprivation, but also a condition of vulnerability. Mainstream development thinking tends to treat poverty as a problem of insufficient income only and seriously ignores structural weaknesses and individual exposure to hazards.

The prospect for gradually providing adequately remunerative employment and alleviating the poverty of the large and growing population depends largely on sustained, rapid economic growth. However, gross domestic product (GDP) growth averaged only about 4 per cent per annum during the 1980s. This indicates an average growth of only 1.6 per cent per capita per annum. What is more important is that this growth stemmed mainly from the expansion of the service and construction sectors. The contribution of the productive sectors, such as agriculture and manufacturing, has been limited. During the 1980s, agriculture grew at less than 3 per cent per annum and manufacturing at 3.1 per cent (World Bank, 1992).

NATO ASI Series, Vol. I 37
Climate Change and World Food Security
Edited by Thomas E. Downing

It is important to note that agriculture is central to the economy, contributing nearly 37 per cent of GDP. Its growth critically affects other sectors through links with input supply, income and employment generation and creating effective demand. The manufacturing sector, although small (only 9 per cent of GDP), is equally important in creating remunerative employment opportunities for the growing labour force and in diversifying exports. The sluggish performance of agriculture and manufacturing represents a continuation of a longer-term stagnation (World Bank, 1992). It raises fundamental questions about the effectiveness and appropriateness of agricultural policies adopted by the government during the last decade.

This paper reviews the vulnerability of the poor in Bangladesh and their exposure to natural hazards and climate change. The paper provides the foundation for the national synthesis of vulnerability provided by Huq *et al.* (1995, this volume). Additional insight is provided by (Bangladesh Unnayan Parishad, 1994; BCAS, 1994).

2 Vulnerability of the poor

The recent Household Expenditure Survey conducted by the Bangladesh Bureau of Statistics (BBS) shows that 47 per cent of the population (in particular, 48 per cent of rural and 44 per cent of the urban population) cannot afford a basic diet considered to be the minimum necessary by international standards. Nearly 30 million people (about 27 per cent of the population) can afford only 1805 kcal per capita per day, which is about 85 per cent of the recommended intake of 2122 kcal per capita per day (Hossain, 1992). Although a recent study by the Bangladesh Institute of Development Studies (BIDS) indicates a slight improvement in overall poverty in 1989/90 compared with 1987/88, even then about 60 per cent of the rural population live below the poverty line (Hossain, 1992). Moreover, the improvement occurred mostly for the moderate poor whose proportion fell from 35 to 27.3 per cent (UNDP/FAO, 1988). The proportion of hard-core poor, on the other hand, has increased from 24.3 to 28.1 per cent. The deterioration of hard-core poverty is highly pronounced for the landless and marginal landowners, agricultural and non-agricultural wage labourers. This change in the poverty situation could be attributed to the abnormal economic situation following the 1987 flood.

The BIDS study indicates that, apart from the problem of insufficient income, the rural poor also face an equally important problem of downward mobility arising from vulnerability to crisis (Rahman, 1992). Environmental factors are particularly important, but the study also identified three broad categories of factors responsible for such downward mobility pressure. They are: (i) life-cycle events such as death or loss of earners or an increase in the number of dependants; (ii) crises such as natural calamities, illness and insecurity; and (iii) structural factors such as lack of economic growth, inflation and bad initial conditions (Table 1). Apart from these unanticipated threats to the welfare of the households and individuals, about 68 per cent of all households face some seasonal deficit and the minimum duration appears to be two months. Deficit-proneness appears to be highest among the landless, marginal and small landowners. In fact, half of the landless and functionally-landless households face a minimum of four deficit months a year; a similar percentage of marginal and small landowners face a minimum of three deficit months. Overall, there appears to be one distinct lean season on the annual calendar covering mid-September to mid-November and two subsidiary lean months covering mid-March to mid-April and mid-June to mid-July. A study by BIDS indicates that this pattern of seasonal deficit has not registered any major change over the recent past, with 83 per cent of all households surveyed reporting no change.

Table 1 Factors explaining downward mobility

Explanation	**percentage of responses**				
	Deficit household	**Occasional deficit**	**Break-even household**	**Surplus household**	**All households**
Life-cycle factors	**56**	**40**	**29**	**15**	**46**
Structural factors	**17**	**14**	**12**	-	**17**
Low opportunities	12	5	8	-	9
Bad initial conditions	2	2	-	-	2
Inflation	3	7	4	-	6
Crisis factors	**24**	**46**	**59**	**85**	**37**
Natural disaster	9	30	37	57	20
Illness expenditure	3	3	2	14	3
Property loss	7	8	8	14	8
Ceremonial expenses	3	2	6	-	3
Litigation	2	3	6	-	3
Others	**3**	-	-	-	-
	100	100	100	100	100

Source : Rahman (1992).

These anticipated seasonal deficits are part of a normal cycle and hence allow households to undertake, although inadequately, some compensatory preparation. However, the unanticipated threats to household welfare, such as natural calamities, permit little compensatory preparation and involve a tangible loss of one kind or another. More importantly, these unanticipated losses either directly threaten survival or impair a household's capacity to graduate from poverty.

Table 2 indicates that the highest incidence of crisis is in the category of natural disasters (66 per cent of households). These include crop failure, damage to the land and other assets by flood, cyclone or drought and loss of land due to river erosion. Naturally, farmers are most vulnerable to such crises. All the natural calamities occur either during or just before the months when most of the rural poor face seasonal deficits.

3 Agroecological resources

Four levels of environmental information (physiography, soils, land levels in relation to flooding and agroclimatology) were combined to define 30 agroecological regions with 88 sub-regions (Figure 1) in the country (UNDP/FAO, 1988). Environmental parameters vary widely. There are 34 physiographic units and sub-units. About 500 distinct soil series are grouped into 21 general soil types with broadly similar appearance and characteristics. The third level of information on agroecological resources indicates the distribution of land types in relation to seasonal

Table 2 Major types of crisis faced by rural households by occupational status

Occupational category	percentage of households in each category subject to:		
	Natural disasters	Health crisis	Insecurity crisis
Farmer	83	49	31
Agricultural labourer	50	46	16
Non-agricultural labourer	55	46	21
Non-farm self-employed	54	46	20
Service	55	50	25
All groups	66	48	25

Source : Rahman (1992).

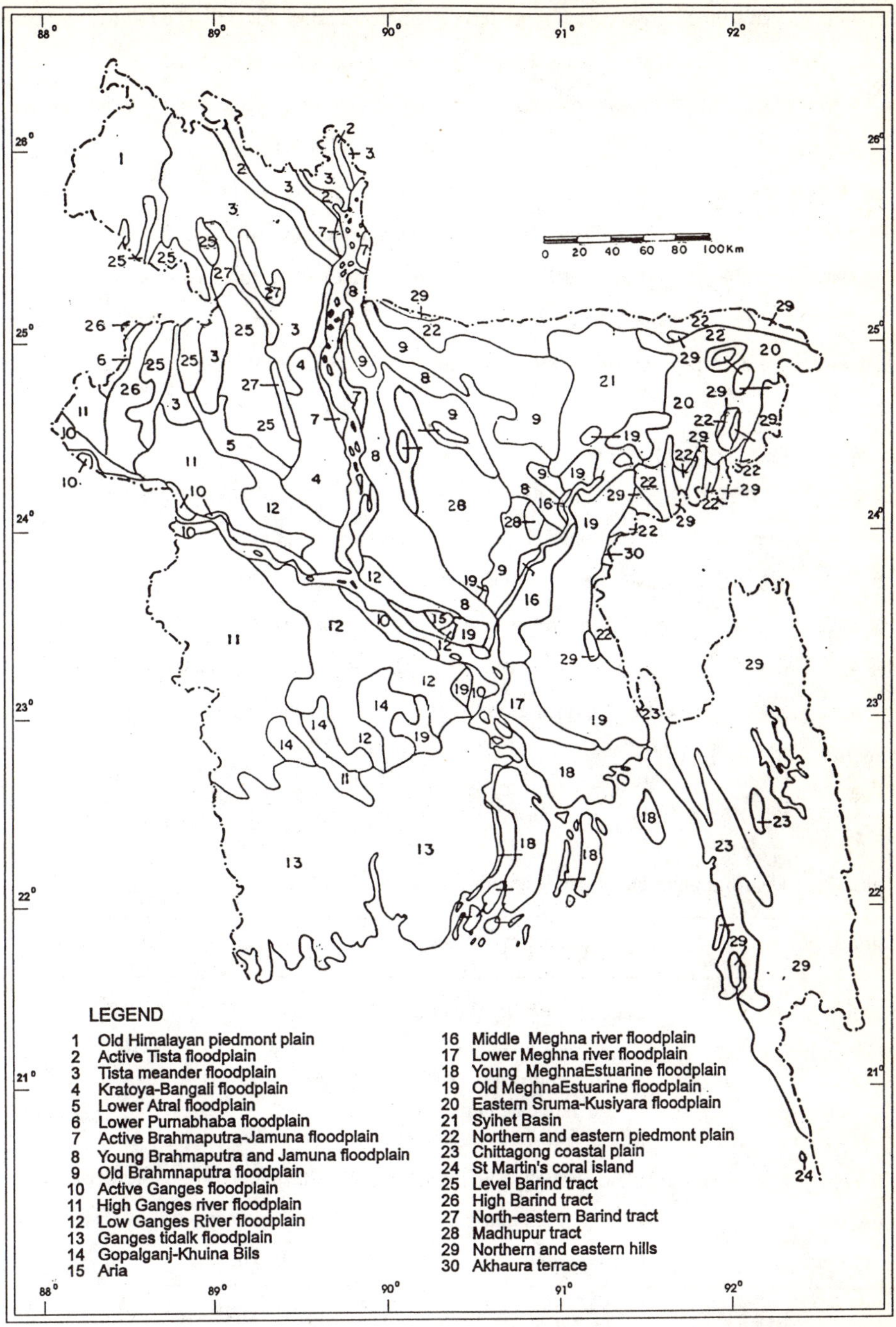

Source: SOIL RESOURCES OF DEVELOPMENT INSTITUTE (SRDI)

Figure 1 Agroecological regions of Bangladesh

flooding, and five broad general land types have been identified (Table 3). The following major distinct features of the climate are considered in delineating agroecological zones:

- average length of the pre-*kharif*[1] period when the rain-fed soil moisture supply is intermittent and uncertain;
- average lengths of the rain-fed *kharif* (wet) and *rabi*[2] (dry) growing periods;
- average number of days in a year with minimum temperatures below 20°C;
- average number of days in a year with maximum summer temperatures above 40°C.

At present, about 74 per cent of the effective land area of the country is under crop production. The total cropped area is about 13.16 million hectares and the cropping intensity is recorded to be 160.0 per cent. About 78 per cent of the agricultural value addition is created through crop production. Livestock and poultry account for 7.6 per cent of agricultural incomes; fisheries and forestry account for 7.5 per cent and 6.6 per cent respectively. Currently, the agricultural sector (crop, livestock, forestry and fisheries) employs about 65 per cent of the labour force. More than 100 different kinds of crops are presently grown in Bangladesh. Rice is the staple food and by far the most important crop in Bangladesh (grown in all three growing seasons of the year and covering about 78 per cent of the total cropped area), although wheat is becoming increasingly important.

Table 3 Distribution of land type by inundation

Land types	**Characteristics**	**Area (ha)**
Highland (H)	Land above normal inundation	4,199,952
Medium highland (MH)	Land normally inundated up to 90 cm deep. For some purposes, this is divided into: MH-1: Inundated up to 30 cm deep; MH-2: Inundated 30-90 cm deep	5,039,724
Medium lowland (ML)	Land normally inundated up to 90-180 cm deep	1,771,102
Lowland (L)	Land normally inundated up to 180-300 cm deep	1,101,560
Very lowland (VL)	Land normally inundated deeper than 300 cm	193,243

[1] *Kharif* - wet season, when moisture from rainfall and soil storage is sufficient to support non-irrigated crops.

[2] *Rabi* - dry growing season, when plants are dependent on stored soil moisture after the monsoon.

Cereal grains account for some 85 per cent of the calories and most of the protein in the average diet. Other important crops are jute, potato, oilseeds, pulses, tobacco, sugarcane and vegetables. It is understandable that food security is generally equated with a quest for self-sufficiency in cereal grain production.

Data on the area and production of these major crops for the last 20 years published by the BBS have been used for simple productivity analysis. During the last 25 years (1965-90), agricultural production in Bangladesh grew at an annual rate of 1.8 per cent, with some higher rates (2.6 per cent) in cereal production. Foodgrain production has increased from little over 9 million tonnes in 1960 to 18.5 million tonnes in 1989-90. This trend in growth had a narrow base. Progress has been entirely due to irrigated modern *boro*[3] rice and wheat. In recent years, the yield of modern rice has started to decline in all the growing seasons. The decline was pronounced for the *aus*[4] season, with a current yield level of 1.82 t/ha from 2.70 t/ha in 1969-70 (Figure 2). The yield of wheat stagnated at about 1.94 t/ha. The growth in cereal production has been achieved partly at the expense of oilseeds, sugarcane, pulses and spices. The area under non-cereal crops has declined in absolute terms and their production has increased at a rate of about 0.9 per cent per annum. Growth rates, however, were negative for pulses and spices. Grain yield of pulses decreased slightly and stagnated at about 0.68 t/ha. The yield of sugarcane showed a gradual decline: it was 43.7 t/ha during 1969-74, and 42.2 t/ha during 1984-89. Similar stagnation is marked in the livestock, fisheries and forestry sub-sectors.

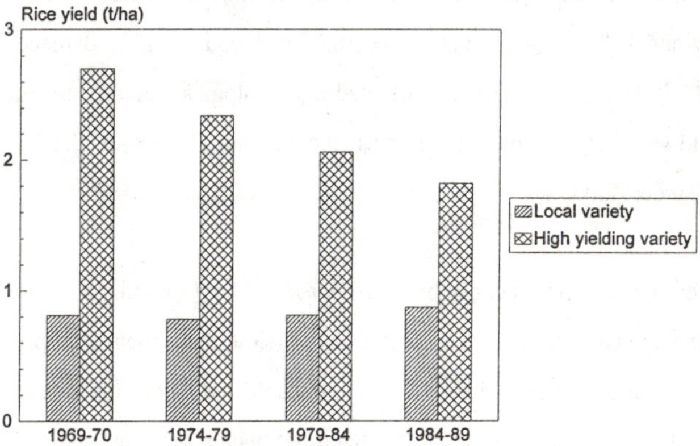

Figure 2 Yield trend for *aus* rice

[3] *Boro* - paddy cultivated in the dry season.
[4] *Aus* - paddy planted before the monsoon and harvested during the monsoon season.

4 Environmental vulnerability

Bangladesh is the world's largest delta of the major river systems of the Ganges, the Brahmaputra and the Meghna. The catchment area is 1.66 million km^2, nearly 92.5 per cent of which is located outside Bangladesh. Natural hazards such as floods, droughts, cyclones, norwester and tidal surges are frequent and all cause considerable damage to crops. The country's agricultural production system is largely influenced by these natural events. During recent years, floods and cyclones have become more frequent and the loss of lives, crops and property has been heavy (Mahtab and Karim, 1992). Major environmental hazards including floods, droughts, tidal and coastal salinity, and loss of soil productivity, are posing a serious threat to the sustainability of the life-support systems in the country.

Floods of different types occur in Bangladesh: flash floods, riverine floods and rainwater floods, or combinations of these. Since independence in 1971, Bangladesh has experienced major floods in 1974, 1984, 1987 and 1988, and droughts in 1979, 1981, 1982, 1989 and 1992. Some 1.32 and 5.05 million hectares are in severely and moderately flood-prone areas respectively. Areas under flash floods vary considerably in different years. Flash floods are characteristic of gentle sloping piedmont plains and basins of the hills, and mainly damage *boro* paddy. River floods affect floodplains and damage *aus*, deep-water *aman*[5], transplanted *aman* rice and jute. Rainwater floods result from heavy rainfall during the pre-monsoon and early monsoon seasons; they damage *boro* and *aus*, deep-water *aman* rice, and summer vegetables. Frequent devastating floods also occur in Bangladesh; the floods of 1987 and 1988 are the worst on record. The floods of 1988 damaged 2.11 million tonnes of rice (BBS, 1989). The severely affected agroecological regions are the Brahmaputra floodplains, the lower Ganges River floodplains, the Jamuna floodplain, the Old Meghna Estuarine and Tista Meander floodplains.

In Bangladesh, droughts of different intensities occur in both the *kharif* and *rabi* seasons (Karim *et al.*, 1990a). During *kharif*-season drought, transplanted *aman* rice, which is the principal crop of the country, is mainly affected. About 2.32 million hectares of the transplanted *aman* rice area are very severely or severely affected in one in five years, when often more than 3.5 million tonnes of foodgrain production are lost. During the *rabi* and pre-*kharif* seasons, wheat, potato, mustard and B. *Aus* are mainly affected. For supplementary irrigation for transplanted *aman* rice and for *rabi*

[5] *Aman* - paddy grown during the monsoon and harvested after the monsoon season.

crops, an appropriate irrigation schedule needs to be developed, matching crop water requirements and other management practices for the specific agroecological regions of the country.

In Bangladesh, over 30 per cent of the net cultivatable area (2.85 million hectares) is on the coast, but 0.83 million hectares are affected by different degrees of seasonal soil salinity. Agricultural land use in the area is very poor; local transplanted *aman* rice is the principal crop cultivated with an average yield of 1.27 t/ha (Karim *et al.*, 1990b). Cyclones and tidal surges are also common in the area. The worst-ever cyclone and tidal surges on 29 April 1991 destroyed 1.8 million hectares of standing crops and caused colossal damage to human life, livestock, aquatic resources and infrastructures of the area (BCAS, 1994).

The dry season flow of the major rivers, such as the Ganges and the Brahmaputra, is reduced in Bangladesh because of upstream water diversion. This has disturbed the delicate balance of the hydrological cycle and the ecosystems that rely upon it. The effect of the Farakka barrage in India on the Ganges River in India has aggravated this problem. Recent evidence shows that the groundwater level in many of the northern parts of Bangladesh has dropped, affecting plant growth and the availability of drinking water. The vulnerability of Bangladesh, being the downstream and deltaic region of a huge watershed, is increased by the low terrain of the delta. Salt water ingress from the Bay of Bengal is reaching farther and farther inland owing to low flows in the river systems during the dry season. The intrusion of salinity has adversely affected the south-western regions and the Sundarban forest areas of the country. The problem of salinity will be further aggravated with the rise in sea level due to global warming.

Soil erosion and depletion of soil fertility are two main causes of low productivity in Bangladesh. Land under different degrees of wetness and in active floodplains is subject to moderate to heavy erosion. Loss of topsoil on hilly slopes due to faulty cultivation practices and lack of appropriate conservation measures causes severe erosion. In the active floodplains, severe loss of land resources occurs due to changes in river courses.

Because of increasing intensity and cultivation of modern varieties of crops, the net removal of nutrients has largely increased, exceeding the nutrient supply. As a result, there are widespread nutrient-deficiency problems. At present, there is a sulphur-deficiency problem in about 4.0 million hectares, and a zinc-deficiency problem in 1.75 million hectares of arable lands.

In addition there is an acute crisis in Bangladesh of biomass fuel, which constitutes 75 per cent of total energy consumed. The per capita supply of biomass fuel is declining. Increased use of crop residue and dung as fuel diverts valuable nutrients and organic matter away from the soil, accelerating the decline of organic matter in cultivated soils. The issues of sustaining increased crop production are intricately linked with the biomass crisis in the country.

To assess the effect of natural calamities (mainly floods and droughts) on foodgrain production, a semi-logarithmic trend line has been estimated on a time series for the period 1973-88 and the deviation of actual production from the trend for the affected years has been measured. The analysis has been conducted separately for the monsoon crops (*aman* and *aus* rice) and dry season crops. The *aus* production during 1973-88 is too erratic to establish a trend. This may be because the area under *aus* crop to a great extent depends on the area under jute cultivation, which fluctuates inversely with the international price of jute and jute goods. Moreover, *aus* is a risky crop and could be affected by either drought or floods or both. For the present study, the trend in *aus* production was derived by deducting production trends of monsoon (*aman* rice) and dry season crops (wheat and *boro* rice) from the total cereal production trend (Table 4).

Table 4 Deviation of foodgrain production from trend by season (1973-88)

Year	**Absolute deviation (1000 tons)**				**Deviation as percentage of trend**			
	Early monsoon (*aus*)	**Monsoon season (*aman* (*boro*+ wheat)**	**Dry season (*aus*+ *aman*+ *boro*+ wheat)**	**All seasons (*aus*)**	**Early monsoon (*boro*+ wheat)**	**Monsoon season (*aman*)**	**Dry season (boro+ wheat)**	**All season (total *aman*+ boro+ wheat)**
1973-74	-424	437	424	437	-13.14	6.98	22.26	3.84
1974-75	-388	-413	289	-512	-11.95	-6.44	13.92	-4.36
1975-76	-31	478	241	688	-0.95	7.28	10.66	5.69
1976-77	-254	181	-557	-630	-7.78	2.69	-22.62	-5.06
1977-78	-154	535	-99	282	-4.73	7.77	-3.69	2.20
1978-79	49	377	-405	21	1.51	5.35	-13.87	0.16
1979-80	-340	81	57	-259	-10.80	1.12	1.79	-1.90
1980-81	79	441	200	720	2.50	5.96	5.78	5.14
1981-82	125	-479	283	-71	4.04	-6.32	7.5	-0.49
1982-83	9	-273	461	147	0.29	-3.52	11.22	1.32
1983-84	268	-132	16	152	9.23	-1.66	0.36	0.99
1984-85	-36	-329	417	52	-1.30	-4.01	8.56	0.33
1985-86	209	210	-591	-172	7.98	2.52	-11.14	-1.06
1986-87	694	-263	-676	-245	28.50	-3.08	-11.70	-1.46
1987-88	775	-1047	-512	-784	34.94	-11.98	-8.14	-4.55

Seasonal production is subject to more variation from year to year than is yearly production, as indicated by the higher value of the coefficient of variation. This implies that the shortfall due to natural calamities is to some extent compensated for by above-average production in the following season and so the effect on total annual production for the year is less severe. It is important to note that, although floods generally affect monsoon crops, they affect the yearly monsoon crop as well. Drought can affect all three crops over two consecutive years.

In the two severe flood years of 1974 and 1987, the shortfalls in production from the trend were about 0.8 and 1.05 million tonnes respectively. During 1984, flood affected both *aus* and *aman* crops, and the shortfall was 0.365 million tonnes. In the severe drought year of 1979, the shortfall from the trend was about 0.745 million tonnes (*boro* of 1978-79 and *aus* of 1979-80). During 1981 and 1982, drought affected the production of the monsoon crop (*aman*) only, and the shortfalls from the trend were 0.479 and 0.273 million tonnes respectively. For the dry season, there were substantial shortfalls in 1977, 1985 and 1987 that cannot be attributed to floods and droughts. These shortfalls might have been due to economic reasons. Experience suggests that economic factors can have a severe effect on rice-based foodgrain production. In fact, during the late 1980s, there was hardly any increase in total cereal production, although part of the reason for such stagnation in cereal production may be attributed to floods. Shortfalls in dry season crops are probably due to a drastic reduction in the subsidy on fertilizer and minor irrigation.

5 Drought impacts

Further assessment of the current drought hazard is warranted. In Bangladesh, annual rainfall ranges from 1400 mm in the dry north-west region to over 5000 mm in the wet north-east region. About 90 per cent of the precipitation generally occurs during four months (June to September) but the distribution within these four months varies greatly. Another complexity is the occurrence of different soil and land types in the country. There is a large variety of soils, varying considerably in moisture-retention capacities, infiltration and other moisture-related properties. These soils have been broadly grouped into floodplain, terrace and hill soils, which are distributed over five major land types. Crop suitability is largely dependent on the different land types, as it relates to the annual, normal flooding conditions of the country. Highland and medium highlands are most suitable for transplanted modern *aman* rice cultivation in the country. But, because of uncertain rain and its uneven distribution, drought of different intensities occurs during the growing period of transplanted *aman* (Karim *et al.*, 1990a). Drought in the transplanted *aman* season has a great impact on the agricultural production of the country for the following reasons:

- Early drought delays transplanting of *aman* rice and shifts the harvesting time, which affects the growing of *rabi* crops after transplanted *aman*. Consequently, a large area remains fallow after transplanted *aman*.
- Early drought retards tillering and vegetative growth, which in turn reduces the achievable yield of the crop.
- The maximum susceptibility factor of transplanted *aman* is during the heading, panicle initiation and milking periods.

Thus, drought in the later growth period of transplanted *aman* may cause very large reductions in yield. In all areas, a drought severity of one in five years affects the transplanted *aman* yield as this crop has maximum stress between panicle initiation and maturity. The maximum values were found in the vulnerable area (VA) zones of NW and SW of very severe drought-prone areas (Figure 3). These areas have potentially high yields due to favourable temperature and sunshine during the growing period of transplanted *aman*, but a higher intensity of drought greatly reduces the yield potential. Yield losses may exceed 50 per cent in the very severe drought areas (Table 5). In the severe drought areas, yield losses vary from 35 to 45 per cent, depending on rainfall, soil types and planting time of the crop. In moderate drought-prone areas, yield losses range from 20 to 35 per cent. The minimum yield reduction (8 per cent) is likely to occur in the wettest district of the country.The losses may be more than half of the achievable yields in some regions of the country.

Table 5 Intensity of drought, yield reduction and supplementary irrigation needs of modern *aman* rice cultivars

Drought classes	Area (ha)	Present average yield* (t/ha)	Reduction over no-stress yield (%)	Stages vulnerable to moisture stress	Supplementary irrigation need (cm)	Yield with no stress (t/ha)
Very severe	573534	1.7-2.5	>45	Pre-planting,+ tillering, PI to milking	30-40	4.5-5.5
Severe	1747597	2.0-2.8	35-45	Pre-planting+ PI to milking	20-30	4.0-5.0
Moderate	2178263	2.5-3.5	20-35	PI to milking	15-20	4.5-5.5
Slight	2595636	3.0-4.0	>20	PI to milking	8-10	3.5-5.0

* With limited irrigation practised by the farmers.

\+ For early planting in June/July.

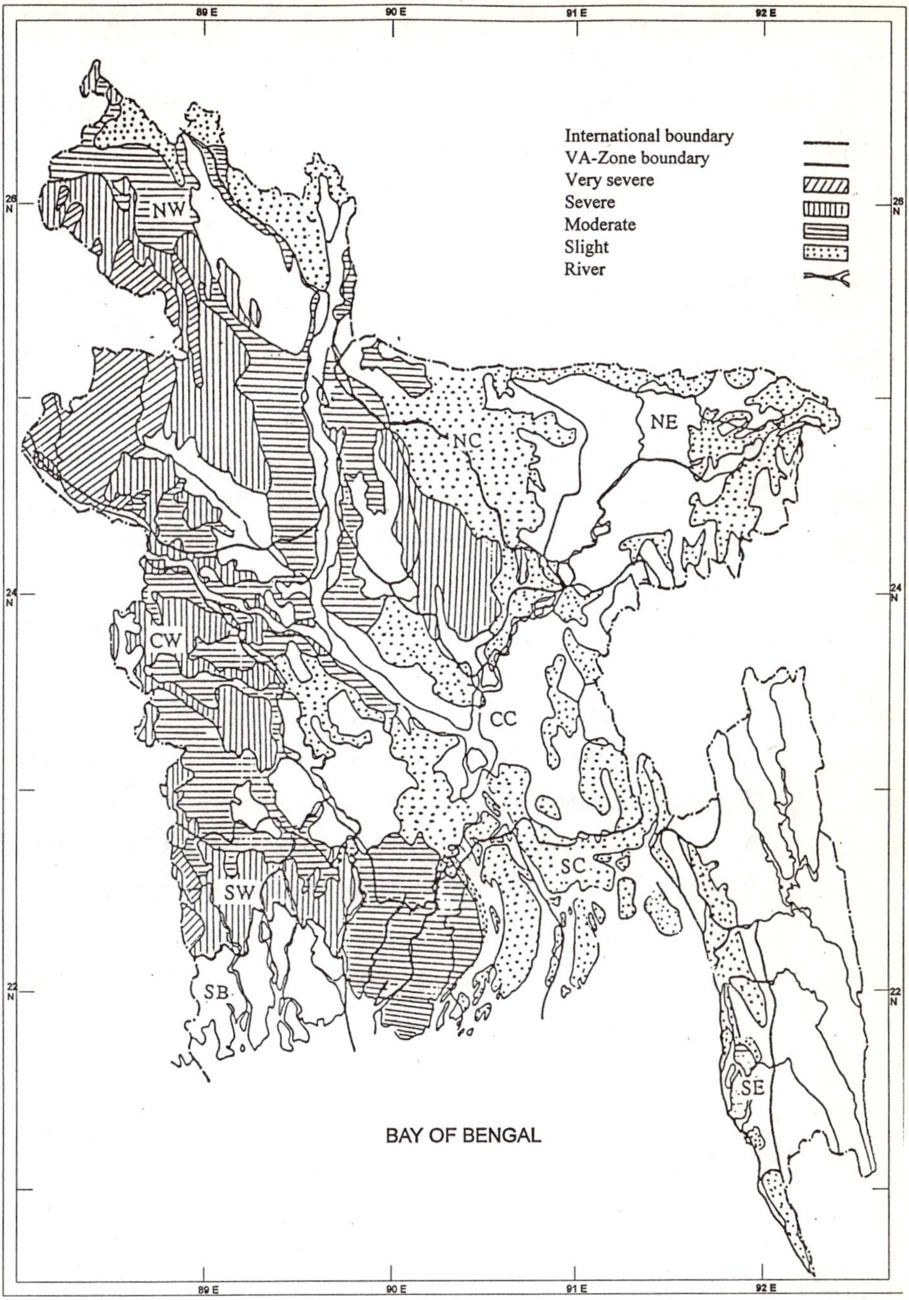

Figure 3 Drought intensity in Bangladesh for the *kharif* season under current conditions

During the *rabi* season, major crops grown are wheat, potato, mustard, *boro* rice, vegetables and pulses. During the *rabi* growing period (mid-October to late February), precipitation is minimal, whereas during the pre-*kharif* transition period (24 March to 18 May) rainfall is irregular and unreliable. Owing to this unfavourable agroclimate, the *rabi* crop in general suffers from moisture stress.

Where there is sufficient rainfall in the later part of the *kharif* period, *rabi* crop establishment using residual soil moisture is possible if sowing/planting can be done just after the transplanted *aman* rice is harvested. *Rabi* crops grown then will most frequently suffer from moisture stresses during the middle and later stages of growth. High-yielding *boro* is primarily an irrigated crop. Thus, yield reduction is not because of drought but due to a breakdown of the water supply, resulting in moisture stress. Soil moisture stress during the pre-*kharif* transition delays sowing of *aus* and jute. *Aus* and sugarcane suffer from drought at their early stages of growth, and non-irrigated crops suffer from drought during the later part of the dry season.

The intensity of *rabi* and pre-*kharif* droughts and the yield reduction of some important crops are given in Table 6. Yield reduction in the very severe class is in the range of 50-70 per cent. In the severe drought areas (Figure 4), the number of dry days is more than in the very severe class but, because of the favourable soil moisture situation, the yield reduction varies from 30 to 70 per cent depending on the crop and the time of planting. Moderate drought-prone classes also include high and low hills. Depending on the topography, type of crop grown and soil characteristics, the yield reduction varies from 20 to 60 per cent. The areas in the less moderate class have on average few days above 40°C, but, because of the low moisture-holding capacity of the soils of hills and charlands, suffer from water stress affecting yield by 10-50 per cent. The northern and eastern

Table 6 Intensity of rabi and pre-kharif droughts and yield reduction of crops

Drought classes	Area (ha)	Yield reduction of crops (%)			
		Wheat	Potato	Mustard	B.*Aus*
Very severe	263,855	60-70	>70	<35	>40
Severe	864,063	50-60	60-70	40-50	30-40
Moderate	3,284,698	40-50	50-60	30-40	20-30
Less moderate	145,740	30-40	40-50	20-30	10-20
Slight	4,352,410	<30	30-40	<20	<10

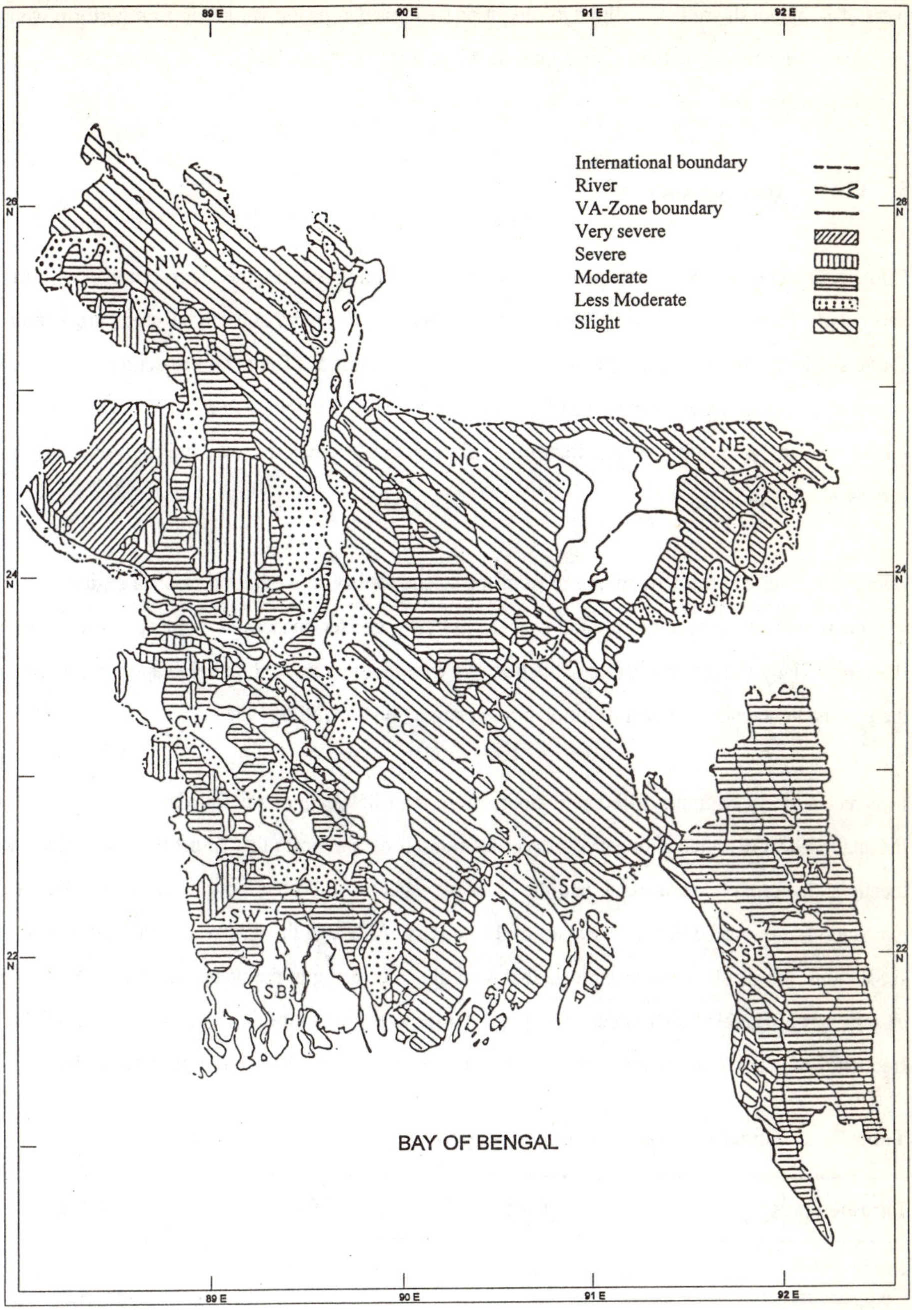

Figure 4 Drought intensity in Bangladesh for the *rabi* season under current conditions

parts of Bangladesh are in the slightly drought-prone class. Crops grown during *rabi* perform well but, due to adverse agroclimatic behaviour, the yield may vary from less than 10 per cent to 40 per cent in extreme cases.

6 Climate change scenarios

Climate change and sea-level rise due to global warming within the next century will be particularly hard-felt in deltaic regions such as Bangladesh where substantial areas are barely above sea level and are vulnerable to frequent natural calamities (Table 7). Projections of global sea level rise range from 20 cm by 2130 and 65 cm by 2100 (Houghton *et al.*, 1990, 1992, 1995; see Schneider, 1995, this volume, for further discussion of climate change and the potential for surprises).

There is a general agreement among most of the scientists that, in the humid tropical region, there will be an increase in summer rainfall by 5-15 per cent, greater chances of tropical storm surges, flooding of larger areas and risk of salination. As warming increases potential evapotranspiration, there could be a tendency towards more drought stress (WHO/UNEP, 1988).

In a recent study with six different stations, more than 85 years of rainfall data were analyzed (Mahtab, 1989). Results of linear regression of rainfall data indicate that four out of six selected locations depict an upward trend, and the other two stations indicate a downward trend. The last two stations are located in the driest part of the country. Although the data were analyzed for only six stations, the results seem to indicate that rainfall is increasing in the wetter areas mostly in the eastern part of Bangladesh and decreasing in the drier, western part of the country. Analysis of the frequency of heavy rainfall indicates that the increase or decrease in rainfall is mainly due to

Table 7 Climate change scenarios for Bangladesh

Parameters	**Unit**	**CCS-1**	**CCS-2**
Sea level rise	cm	30	100
Temperature	°C	2	4
Precipitation (monsoon)	%	8	15
Precipitation (winter)	%	10	20
Evaporation (monsoon)	%	18	33
Evaporation (winter)	%	12	22

an increase or decrease in the rainfall intensity. Analysis of rainfall data by season seems to indicate a change in the spatial and temporal distribution of precipitation. The study also noted that the coefficient of variation of rainfall is higher in the drier area and in the pre- and post-monsoon periods.

Two scenarios of climate change in Bangladesh were evaluated (see Huq *et al.*, 1995, this volume):

- Moderate changes (CCS-1): sea level rise of 30 cm; increased precipitation of 18 per cent (monsoon) and 12 per cent (winter); increased evaporation by 8 per cent (monsoon) and 10 per cent (winter); peak river discharges increased 6 per cent and low flows decreased 12 per cent; and a 10 per cent increase in cyclone intensity.
- Severe changes (CCS-2): sea level rise of 100 cm; increased precipitation of 33 per cent (monsoon) and 22 per cent (winter); increased evaporation by 15 per cent (monsoon) and 20 per cent (winter); peak river discharges increased 13 per cent and low flows decreased 22 per cent; and a 25 per cent increase in cyclone intensity.

7 Drought impacts due to climate change

Future drought impacts were studied for the two climate scenarios, as given in Table 8. The intensity of drought will increase in both scenarios (Figures 5-8). About 12,219 km^2 will be transformed to the very severe class under CCS-2, compared with 3638 km^2 during the *rabi* season. During *kharif,* four times more area will be changed to the very severe class from the severe and moderate areas (Tables 8 and 9).

Table 8 Change in area affected by drought due to climate change in *rabi*/pre-*kharif* season (km^2)

Drought classes	Current	Climate change scenarios	
		CCS-1	CCS-2
Very severe	3,638.69	8,636.12	12,219.64
Severe	8,580.95	10,873.53	15,302.72
Moderate	18,161.03	15,694.34	10,779.59
Less moderate	14,571.20	9,747.17	19,814.20
Slight	43,367.28	43,367.28	30,203.00

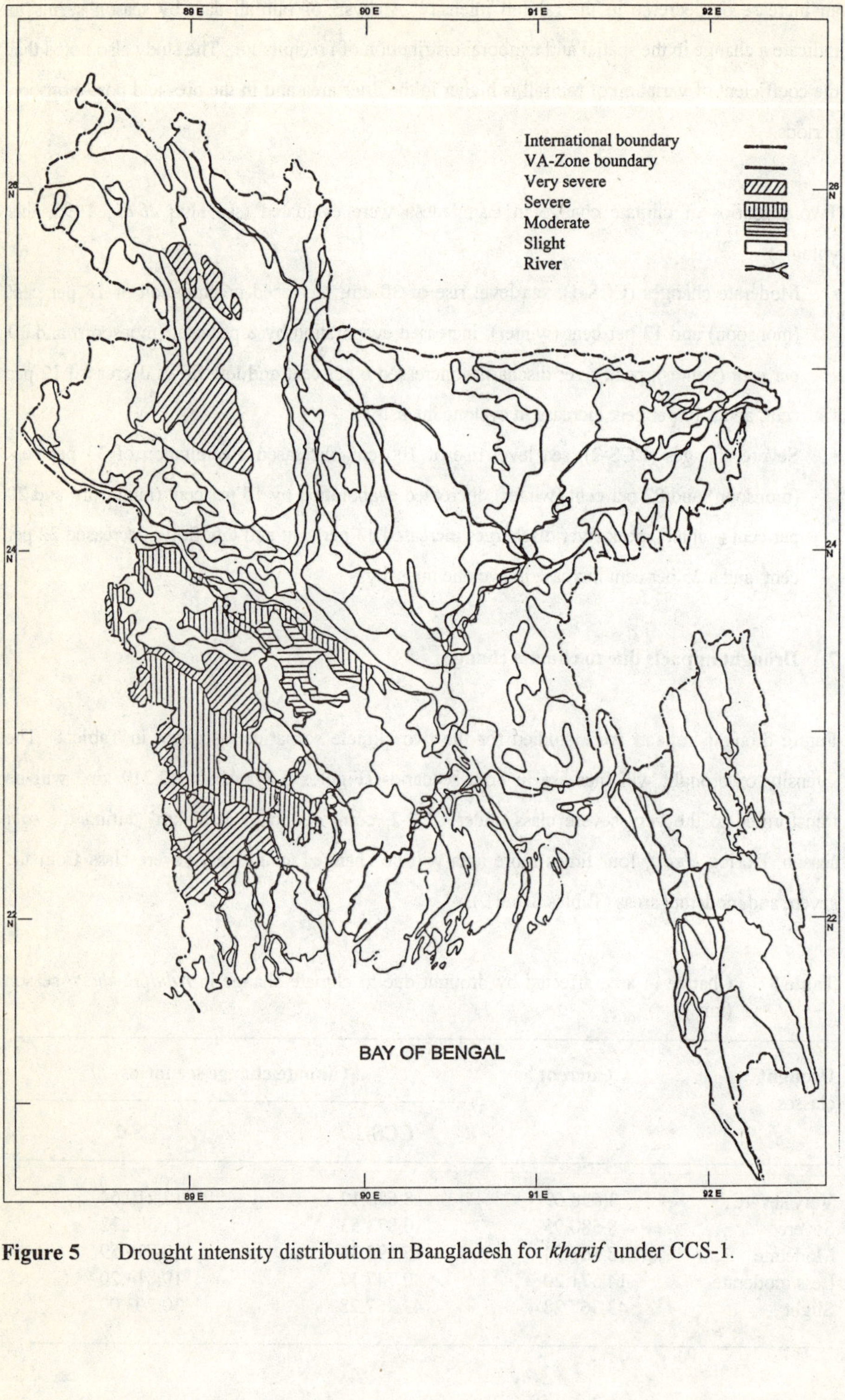

Figure 5 Drought intensity distribution in Bangladesh for *kharif* under CCS-1.

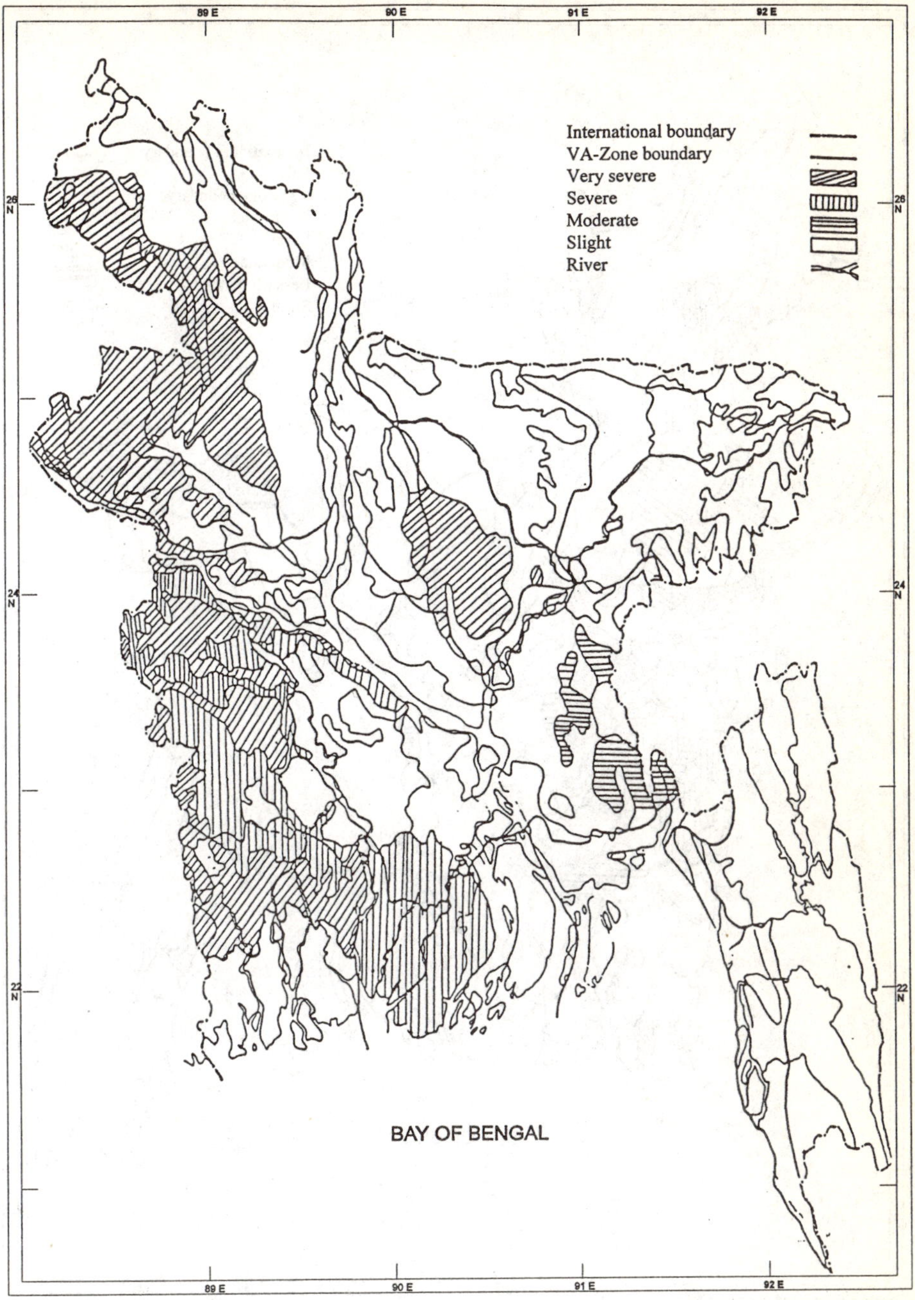

Figure 6 Drought intensity distribution in Bangladesh for *kharif* under CCS-2.

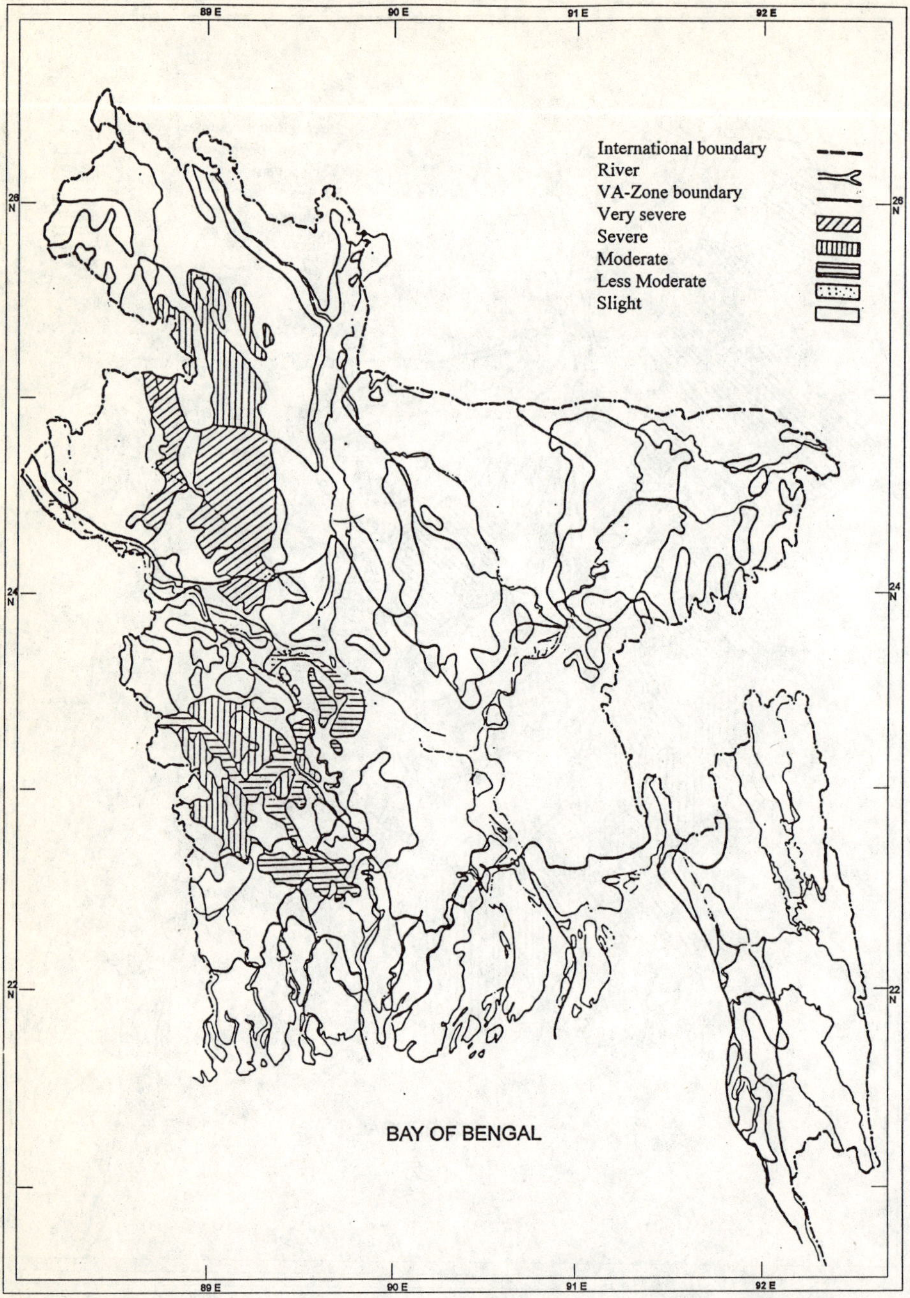

Figure 7 Drought intensity distribution in Bangladesh for *rabi* under CCS-1.

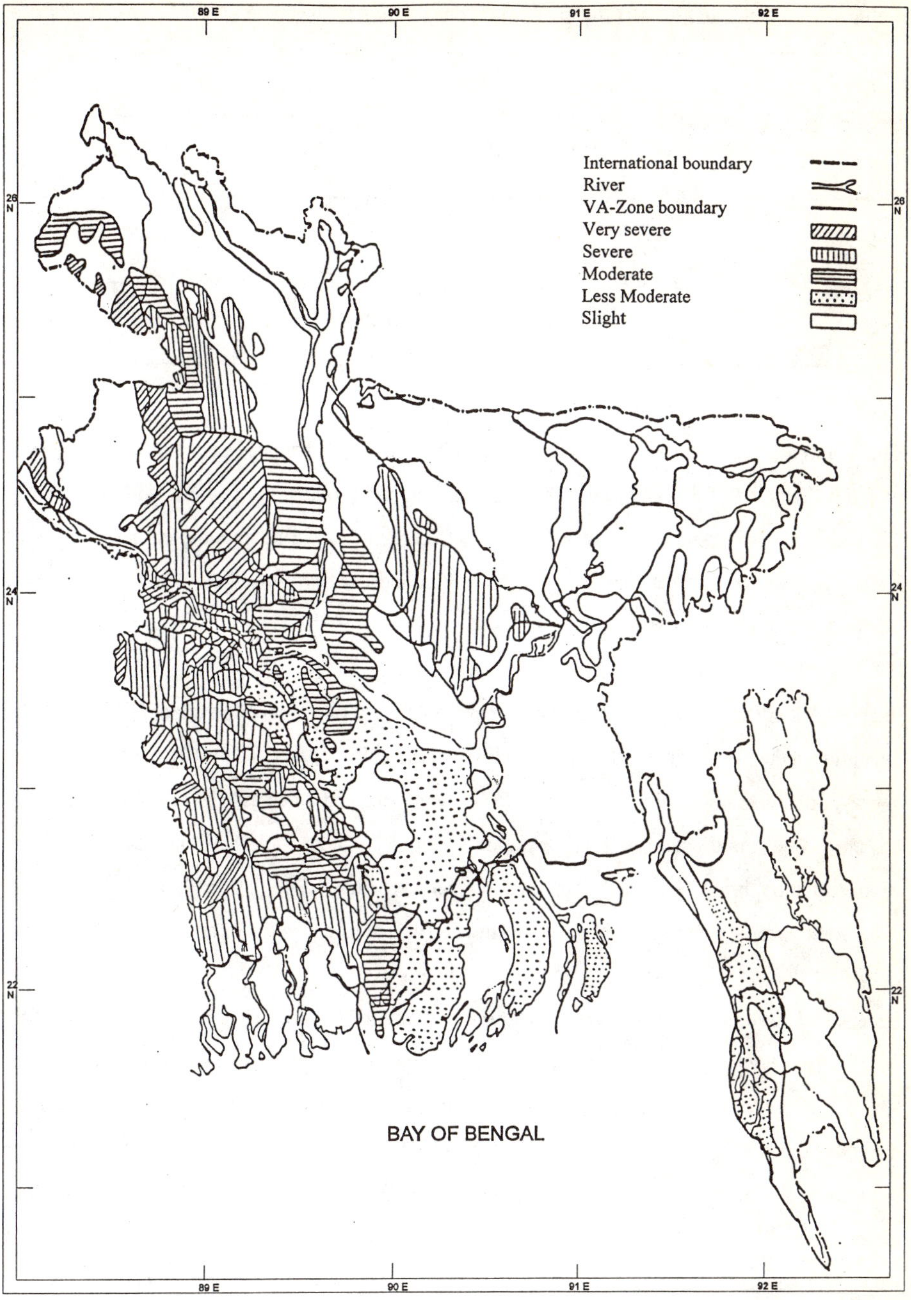

Figure 8 Drought intensity distribution in Bangladesh for *rabi* under CCS-2.

Table 9 Change in area affected by drought due to climate change in *kharif* season (km^2)

Drought classes	Current	Climate change scenarios	
		CCS-1	CCS-2
Very severe	5,735.55	12,065.91	22,664.40
Severe	17,476.20	16,897.90	10,539.83
Moderate	21,782.73	16,864.61	13,372.71
Slight	26,207.81	25,170.25	24,625.35

The worst affected VA zones are NW, CW and NC, in which irrigated and rain-fed crop yields would be reduced (Tables 10 and 11). Both the area and production of wheat will be seriously affected by a rise in temperature of 4°C. No wheat cultivation will be possible in the SC and SW VA zones because of insufficient thermal growing periods (temperatures below 20°C). Potato will also be affected in these zones.

In general, there could be more than 30 per cent reduction in rice production under CCS-2 with the maximum reduction in the modern high-yielding variety (HYV) *aus* (Table 12). This reduction is due to both a decrease in the area suitable for growing rice and a decrease in per hectare yields of this crop. Wheat and potato will have very little potential. There could be more than 50 per cent yield reduction in CCS-1 and more than 60 per cent in CCS-2 (Tables 10-12). In general, reduction of production was pronounced in irrigated agriculture (Tables 13 and 14).

Because of the increase in drought intensity, there would be more demand for irrigation (Table 15). The per hectare requirements may increase to up to 310 cm from the existing 200 cm for HYV *boro* in very severe areas. The irrigation requirement of HYV *aus* will also increase by more than 50 per cent from the present level. The prospect of growing more irrigated crops will be reduced by the limited availability of irrigation water under CCS-1 and CCS-2. This will reduce the cropping intensity in many parts of the country, particularly in the most severe vulnerable zones. Further effects of the two scenarios and for sea level rise are shown in Huq *et al.* (1995, this volume).

Table 10 Area and production of crops under non-irrigated conditions (area in 1000 ha; production in 1000 t)

VA Zone	Crops	Current climate			Climate change scenario					
					CCS-1			CCS-2		
		Area	Yield	Prod	Area	Yield	Prod	Area	Yield	Prod
SE	HYV *aus*	1.78	2.58	4.59	1.60	2.58	4.13	1.34	2.58	3.46
	HYV T. *Aman*	54.01	3.07	165.81	54.01	3.07	165.81	45.01	2.92	157.71
	Wheat	0.00	0.00	0.00	0.00	0.00	0.00	0.00	0.00	0.00
	Potato	0.62	8.73	5.41	0.00	0.00	0.00	0.00	0.00	0.00
SC	HYV *aus*	21.25	2.58	54.83	21.25	2.58	54.83	19.13	2.58	49.34
	HYV T. *Aman*	65.28	3.07	200.41	65.28	3.01	196.40	65.28	2.92	190.39
	Wheat	4.43	1.62	7.18	0.00	0.00	0.00	0.00	0.00	0.00
	Potato	7.00	8.73	8.73	0.00	0.00	0.00	0.00	0.00	0.00
SW	HYV *aus*	5.19	2.31	11.96	4.15	2.19	9.09	3.37	1.73	5.83
	HYV T. *Aman*	74.33	2.74	203.29	72.84	2.60	189.27	66.90	1.91	128.07
	Wheat	5.85	1.32	7.72	0.00	0.00	0.00	0.00	0.00	0.00
	Potato	7.14	8.12	57.99	0.00	0.00	0.00	0.00	0.00	0.00
CC	HYV *aus*	20.42	2.58	52.68	17.36	2.32	40.30	15.32	1.94	29.63
	HYV T. *Aman*	123.54	2.83	349.62	123.54	2.83	349.62	123.54	2.55	314.66
	Wheat	105.09	1.53	160.79	52.55	1.15	60.30	42.04	0.92	38.59
	Potato	14.07	7.50	105.53	7.04	6.00	42.21	5.63	5.63	31.66

(continued on next page)

Table 10 (continued) Total area and production of crops under non-irrigated conditions (area in 1000 ha; production in 1000 t)

VA Zone	Crops	Current climate			Climate change scenario					
					CCS-1			CCS-2		
		Area	Yield	Prod	Area	Yield	Prod	Area	Yield	Prod
CW	HYV *aus*	0.98	2.03	1.99	0.78	1.73	1.35	0.74	1.52	1.12
	HYV T. *Aman*	54.64	2.64	144.25	51.91	2.11	109.63	49.18	1.58	77.89
	Wheat	21.28	1.32	28.09	12.77	0.92	11.80	7.45	0.66	4.92
	Potato	6.81	8.33	56.73	3.75	7.08	26.52	3.41	6.66	22.69
NE	HYV *aus*	19.84	2.54	50.39	19.84	2.54	50.39	18.85	2.54	47.87
	HYV T. *Aman*	66.66	2.87	188.65	66.66	2.85	188.65	66.66	2.83	188.65
	Wheat	5.54	1.53	8.48	4.16	1.30	5.40	3.60	1.15	4.13
	Potato	8.38	7.50	62.85	5.45	6.75	36.77	4.61	6.00	27.65
NC	HYV *aus*	26.27	2.54	66.73	23.64	2.54	60.05	21.02	2.03	42.70
	HYV T. *Aman*	126.96	2.83	359.30	126.96	2.75	348.52	120.61	2.41	290.13
	Wheat	37.93	1.53	58.03	30.34	1.38	41.78	26.55	1.15	30.47
	Potato	4.85	7.50	36.38	3.15	6.75	21.28	2.67	6.00	16.01
NW	HYV *aus*	19.14	2.00	38.28	13.40	1.70	22.78	11.48	1.40	16.08
	HYV T. *Aman*	114.58	3.14	673.78	203.85	2.51	512.07	171.66	1.73	296.46
	Wheat	87.75	1.33	116.71	70.20	1.20	84.03	65.81	1.06	70.02
	Potato	20.61	6.25	128.81	15.46	5.94	91.78	13.40	5.31	71.17

Table 11 Area and production of crops under non-irrigated conditions (area in 1000 ha; production in 1000 t)

VA Zone	Crops	Current climate			Climate change scenario					
					CCS-1			CCS-2		
		Area	Yield	Prod	Area	Yield	Prod	Area	Yield	Prod
CW	HYV *aus*	64.02	3.42	218.95	57.62	2.91	167.50	51.22	2.57	131.37
	HYV T. *Aman*	139.72	4.08	570.06	125.75	3.26	410.44	111.78	2.45	273.63
	HYV *boro*	169.01	4.63	782.52	160.57	4.63	743.39	152.11	4.63	704.26
	Wheat	69.46	2.32	161.15	34.73	1.62	56.40	20.84	0.93	19.34
	Potato	3.17	12.00	161.15	3.75	7.08	26.52	3.41	6.66	22.69
NE	HYV *aus*	2.66	3.61	9.60	2.66	3.61	9.60	2.53	3.61	9.12
	HYV T. *Aman*	23.49	3.89	91.38	23.49	3.89	91.38	23.49	3.89	91.38
	HYV *boro*	181.56	3.85	699.01	181.56	3.85	699.01	176.11	3.85	678.04
	Wheat	4.83	2.44	11.79	3.86	2.32	8.96	3.14	1.83	5.75
	Potato	1.70	12.00	20.40	1.11	10.80	11.93	0.93	9.60	8.98
NC	HYV *aus*	20.76	3.61	74.94	18.68	3.61	67.45	16.61	3.25	53.96
	HYV T. *Aman*	108.82	3.89	423.31	103.38	3.31	341.82	97.94	2.92	285.73
	HYV *boro*	264.46	3.85	1018.17	251.24	3.85	967.26	238.01	3.85	916.35
	Wheat	19.33	2.44	47.17	15.46	2.20	33.96	12.56	1.95	24.53
	Potato	4.89	12.00	58.68	3.18	10.80	34.33	2.69	9.60	25.82
NW	HYV *aus*	115.23	3.50	402.73	92.18	3.15	289.96	80.66	2.62	211.43
	HYV T. *Aman*	369.35	3.94	1455.24	332.42	3.35	1113.26	177.01	2.36	654.86
	HYV *boro*	709.26	4.09	2900.87	567.41	3.89	2204.66	496.48	3.27	1624.49
	Wheat	119.03	2.43	290.43	71.42	1.95	139.41	59.52	1.95	116.17
	Potato	31.15	8.5	264.78	18.69	7.23	135.04	15.58	6.80	105.91

(continued on next page)

Table 11 (continued) Total area and production of crops under non-irrigated conditions (area in 1000 ha; production in 1000 t)

VA Zone	Crops	Current climate			Climate change scenario					
					CCS-1			CCS-2		
		Area	Yield	Prod	Area	Yield	Prod	Area	Yield	Prod
SE	HYV *aus*	3.63	3.23	11.72	3.45	3.32	11.14	3.27	3.23	10.55
	HYV T. *Aman*	46.86	4.17	195.41	46.86	3.96	185.64	46.86	3.75	75.87
	HYV boro	69.79	4.00	279.16	69.79	4.00	279.16	66.30	4.00	65.20
	Wheat	2.53	2.60	6.58	0.00	0.00	0.00	0.00	0.00	0.00
	Potato	0.50	12.79	6.40	0.00	0.00	0.00	0.00	0.00	0.00
SC	HYV *aus*	0.70	3.23	2.26	0.70	3.23	2.26	0.70	3.23	2.26
	HYV T. *Aman*	13.47	4.17	56.17	13.47	3.96	53.36	13.47	3.75	50.55
	HYV *boro*	23.54	4.00	94.16	23.54	4.00	94.16	22.83	4.00	91.34
	Wheat	0.57	2.60	1.48	0.00	0.00	0.00	0.00	0.00	0.00
	Potato	0.31	12.79	3.96	0.00	0.00	0.00	0.00	0.00	0.00
SW	HYV *aus*	9.55	3.42	32.66	8.12	3.25	26.37	7.16	2.74	19.60
	HYV T. *Aman*	43.26	4.08	176.50	41.10	3.67	150.91	34.61	2.86	98.84
	HYV *boro*	62.74	4.63	290.49	59.60	4.63	275.96	56.47	4.63	61.44
	Wheat	7.89	2.32	18.30	0.00	0.00	0.00	0.00	0.00	0.00
	Potato	0.92	12.00	11.04	0.00	0.00	0.00	0.00	0.00	0.00
CC	HYV *aus*	50.53	3.23	163.21	45.48	2.91	132.20	40.42	2.42	97.93
	HYV T. *Aman*	122.07	3.89	474.85	122.07	3.70	451.11	122.07	3.11	379.88
	HYV *boro*	393.57	3.85	1515.24	393.57	3.85	1515.24	373.89	3.85	439.48
	Wheat	63.70	2.44	155.43	25.48	1.95	49.74	19.11	1.34	25.65
	Potato	11.32	12.00	135.84	5.09	9.00	45.85	3.40	9.00	30.56

Table 12 Reduction in area and production due to climate change (area in 1000 ha; production in 1000 t)

Crops	Current climate		Climate change scenario			
			CCS-1		CCS-2	
	Area	Production	Area	Production	Area	Production
HYV *aus*	381.95	1197.52	51.04 (13.4%)	248.12 (20.7%)	88.13 (23.0%)	465.27 (38.9%)
HYV T.*Aman*	1647.04	5728.03	73.45 (4.5%)	870.14 (15.2%)	201.97 (17.7%)	2073.33 (6.2%)
HYV *boro*	1873.93	7579.62	166.65 (8.9%)	800.78 (10.6%)	291.73 (15.6%)	1599.02 (21.2%)
Wheat	555.21	1070.33	234.24 (42.0%)	587.55 (54.4%)	294.59 (53.1%)	739.76 (68.5%)
Potato	123.44	1001.57	58.93 (47.7%)	540.64 (53.9%)	70.12 (56.8%)	649.71 (64.8%)

Table 13 Non-irrigated area and production as affected by climate change scenario (area in 1000 ha; production in 1000 t)

Crops	Current climate		Climate change scenario			
			CCS-1		CCS-2	
	Area	Production	Area	Production	Area	Production
HYV *aus*	114.87	281.45	102.02	242.92	91.25	196.03
HYV T. *Aman*	780.00	2285.11	765.05	2059.97	717.84	1643.96
Wheat	267.87	387.00	170.02	203.31	145.45	148.13
Potato	69.48	462.43	34.85	218.56	29.45	169.18

Table 14 Irrigated area and production affected by climate change (area in 1000 ha; production in 1000 t)

Crops	Current		Climate change scenario CCS-1		CCS-2	
	Area	Production	Area	Production	Area	Production
HYV *aus*	267.08	916.07	228.89	706.48	202.57	536.22
HYV T. *Aman*	867.04	3442.92	08.54	2797.92	727.23	2010.74
HYV *boro*	1873.93	7579.62	1707.28	6778.84	1582.20	5980.60
Wheat	287.34	692.33	50.95	288.47	115.17	191.44
Potato	53.96	539.14	29.66	242.37	23.87	182.68

Table 15 Stress-free yield and irrigation requirements: changes due to climate change

Drought classes	Crops	Stress-free yield (t/ha)	Yields (t/ha)			Irrigation requirements (cm)		
			Current	CCS-1	CCS-2	Current	CCS-1	CCS-2
Very severe	HYV *aus*	5.0	3.5	3.1	2.6	150	180	225
	HYV T. *Aman*	5.5	3.3	3.4	2.4	35	40	65
	HYV *boro*	6.0	4.1	3.7	3.3	200	230	310
	Wheat	4.5	2.4	2.0	1.9	20	25	35
	Potato	18.0	8.5	8.0	6.5	30	35	70
Severe	HYV *aus*	4.5	3.4	2.9	2.6	150	175	210
	HYV T. *Aman*	4.0	4.0	3.3	2.5	30	35	60
	HYV *boro*	5.5	4.6	4.6	4.6	200	225	20
	Wheat	4.5	2.3	1.6	0.9	15	18	32
	Potato	20.0	12.0	9.6	9.0	25	30	65
Moderate	HYV *aus*	4.5	3.6	3.6	3.3	125	150	205
	HYV T. *Aman*	5.0	3.9	3.3	2.9	20	30	60
	HYV *boro*	5.5	4.0	3.9	3.8	150	185	250
	Wheat	4.5	2.4	2.2	1.9	12	18	30
	Potato	25.0	12.0	10.8	9.6	20	30	65
Slight	HYV *aus*	4.0	3.2	2.9	2.4	125	140	200
	HYV T. *Aman*	4.5	3.9	3.7	3.1	8	20	50
	HYV *boro*	5.0	4.0	3.8	3.8	150	170	190
	Wheat	3.5	2.3	1.9	1.3	6	15	25
	Potato	30.0	12.0	9.0	9.0	12	25	60

8 Implications of climate change

An important factor that has contributed to the long-term stagnation of the economy is the high rate of population growth. The slow growth in production and income over the years failed to contribute to the well-being of the people, especially the large masses in the rural areas, due to a rapid increase in their numbers. Recent studies indicate that the country has a crude birth rate (CBR) of 35.2 and the total fertility rate (TFR) is now 4.5, compared with 12.9 and about 6.0 respectively just over a decade ago. Although these achievements are appreciable, they are far short of what is desirable. As a result, due to past high fertility, the population of Bangladesh increased from 50.76 million in 1961 to 115 million in 1990.

Apart from the size of the present population, there is a tremendous growth potential built into its age structure as a consequence of past high fertility. Population below the age of 15 years is about 46 per cent of the total population and females of reproductive age (15-49 years) represent 48.4 per cent of the total female population. Even in 1990, the proportion of the population aged 0-4 years to the total population has not changed much compared with the 1981 level.

As the climatic parameters such as temperature, evaporation and day length are likely to change with changing scenarios, there would be a large reduction in production from the present level. Therefore, the potential of using existing technologies for maximising production may be considerably reduced by the intensity of climate change. A reduction of 30 per cent foodgrain from the current production level would bring a deficiency of about 6 million tonnes of foodgrain. This means that not only the potential of increased production will be impaired but also there will be a substantial reduction from the present production level, resulting in serious sociopolitical and environmental crises.

9 Population growth and poverty alleviation: mitigation strategies

The trend in environmental vulnerability is increasing and would likely be aggravated by climate change due to global warming. The future projections of population can be based on alternative assumptions regarding the trends in fertility and mortality and other relevant parameters. However, to reveal the magnitude of the problems, available projections are instructive. According to the Planning Commission, by the years 2000 and 2010 AD, the population will increase to 137 million and 154.5 million respectively. However, the World Bank's estimate, based on the assumption that the growth rate will decrease from the present 1.9 to 1.0 by 2025, indicates a population of 145 million in 2000, 177 million in 2010 and 232 million in 2030.

Using existing agricultural technology and present potential biophysical resources, the production of cereals, pulses and oilseeds can be greatly increased in the period 2000 to 2030 (Figures 9-11 and Table 16). However, the demand for the food items will also increase owing to the growth of the population, and cereal production would be deficient after 2010. However, pulses and oilseeds have better prospects for increased production.

Population growth has reduced land resources to 0.11 hectare per capita in 1990 from 0.13 hectare per capita in 1981. With the increase in population, it will decline further. This will lead to a further increase in the landless population, who comprised 52 per cent of the total population in 1988. The rapidly growing population will also result in ecological imbalances, leading to environmental hazards and greater resources spent on infrastructure instead of production.

The above facts clearly indicate that the agricultural resources in Bangladesh are already under severe environmental strain. It is absolutely necessary to reverse this disastrous trend, rebuild and, where possible, augment the productive capability of our scarce and essential agricultural resource base.

To meet the demand for food beyond 2010, there is an urgent need to develop rice cultivars with high photosynthetic efficiency, which are capable of yielding more than 8.0 t/ha and suitable for 4.5 million hectares of highland and medium highland of phase 1 (Table 3). For the 2.6 million hectares of land that is inundated with more than 180 cm, deep-water modern rice cultivars and short-duration heat-tolerant wheat cultivars with a yield of more than 5.0 t/ha at the farm level need to be developed. Wider germplasm bases and modern management practices should be made available to 2.85 million hectares of coastal land and drought-prone areas of the country. Management packages should be adapted by the farmers to maximise production in about 3.4 million flood-free lands to harvest more than 20.0 t/ha of cereal annually. Bangladesh must have wider choices of both germplasm and management practices for oilseeds, pulses, vegetables and fruits. Genetic engineering and other basic research efforts should be increased.

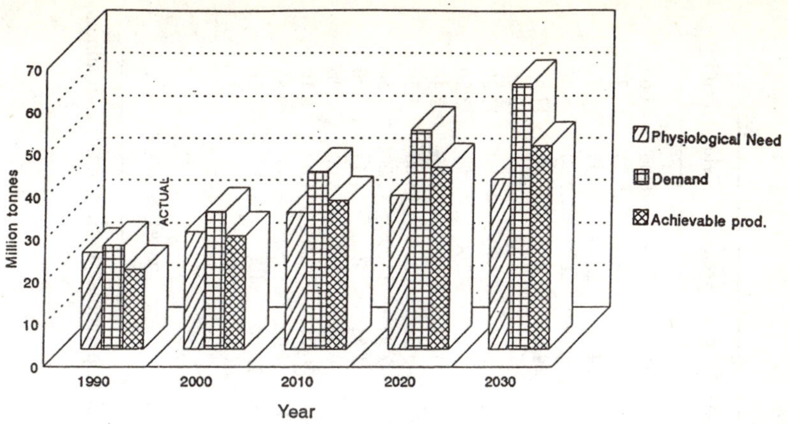

Figure 9 Production necessary to meet physiological requirements and demand for cereals: actual and maximum achievable production.

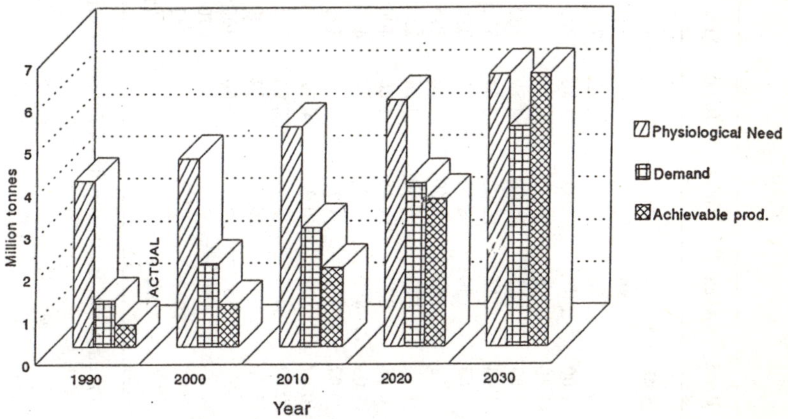

Figure 10 Production necessary to meet physiological requirements and demand for pulses: actual and maximum achievable production.

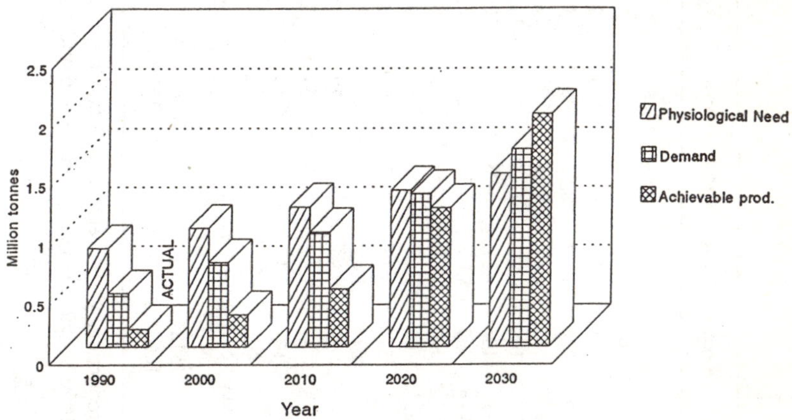

Figure 11 Production necessary to meet physiological requirements and demand for edible oils: actual and maximum achievable production.

Table 16 Maximum possible production of major crops

Crop	1990 Area	1990 Yield	1990 Prod	2000 Area	2000 Yield	2000 Prod	2010 Area	2010 Yield	2010 Prod	2020 Area	2020 Yield	2020 Prod	2030 Area	2030 Yield	2030 Prod
Rice	10440	1.71	17.85	10400	2.45	25.48	10290	3.25	33.44	9300	4.41	41.01	7600	6.00	45.69
Wheat	599	1.68	1.00	536	2.48	1.33	486	3.30	1.60	441	4.47	1.97	400	6.00	11.70
Maize etc.	11.2	0.72	0.68	170	2.50	0.43	350	3.85	1.35	710	6.00	4.26	1300	9.00	11.70
Pulses	728	0 .72	0.52	980	1.02	1.00	1400	1.35	1.86	1985	1.76	3.49	2800	2.30	6.44
Sugarcane	187	40.20	7.68	200	57.00	11.40	215	76.00	16.34	230	108.00	24.84	250	150.00	37.50
Potato	169	9.98	1.78	195	14.50	2.83	225	25.00	4.50	260	28.50	7.41	300	40.00	12.00
Winter Veg.	105	7.21	0.76	150	10.90	1.64	215	16.80	3.61	321	26.00	8.06	550	40.00	22.00
Summer Veg.	65	5.05	0.33	100	6.50	0.65	150	8.75	1.31	230	12.00	2.76	200	16.00	3.20
Fruit Per.	100	5.21	0.52	120	6.50	0.78	145	8.00	1.16	175	9.81	1.72	200	12.00	2.40
Fruit Annu.	69	13.47	0.93	110	16.00	1.76	175	18.50	3.24	250	21.50	5.38	300	25.00	7.50
Oilseed	570	0.79	0.45	660	1.00	0.68	870	1.45	1.26	1400	2.09	2.93	1500	3.00	4.50
Jute	584	1.65	0.96	490	2.10	1.03	460	2.70	1.24	430	3.50	1.51	400	4.40	1.76
Cotton	19	0.86	0.02	70	1.00	0.07	170	1.75	0.30	325	2.72	0.88	550	4.00	2.20
Tobacco	38	9.89	0.03	55	0.88	0.07	67	0.92	0.22	82	0.96	0.08	100	1.00	0.10
Spices	147	2.17	-	212	2.94	0.62	445	3.92	0.06	775	5.24	4.06	1350	7.00	9.45
Forage	N/A	-	-	100	15.00	1.50	280	20.00	1.74	500	26.00	13.00	1000	35.00	35.00
Green manure	N/A			250			350		5.60	460			600		
Others	N/A			550			550			670			810		
Total Cr.	15184			16208			17913			19843			22110		
Crop In.	166.1%			177.3%			196%			217%			241%		

Prod = production, Veg = vegetable, Total Cr = total cropped area, Crop In. = cropping intensity; Fruit Per. = fruit perennial; Fruit Annu. = fruit annual
Area in 1000 ha; yield in t/ha; production in million t.

Alleviating poverty requires not only raising agricultural production and the income of the poor but also countering downward mobility. This calls for a two-pronged strategy: on the one hand, providing technologies to intensify agricultural production after any environmental crisis and, on the other hand, creating more off- and on-farm activities for coping with crisis with additional incomes. Immediately after natural calamities, the agricultural production system needs reactivating. The type of agricultural programme depends on the nature of the hazards, the agroecological conditions of the regions and the availability of technological options to the farmers. The short-term measures to mitigate vulnerabilities are rapidly revitalising agricultural production with existing technologies developed by the national agricultural research system or innovation by farmers. Environmental vulnerabilities and their effects could be mitigated by the following strategic actions:

- rapidly revitalising agricultural production with available agronomic innovations;
- promoting cultivation of non-conventional food crops;
- enhancing rural agro-processing activities;
- medium-term measures through restructuring cropping patterns.

9.1 Revitalising the agricultural production system with available agronomic innovations

The national agricultural research system has developed some technologies to mitigate flood, drought salinity and storm hazards in Bangladesh (Table 17). Some of these technologies have been adapted by farmers, but many need to be tested.

For floods of different kinds, risk-minimising technologies are cultivation of short-duration modern *boro* rice in the areas affected by early flash floods and late transplanting of *aman* rice varieties in the regions affected by late flash floods and river water floods. Splitting the tillers from the mother plant of transplanted *aman* rice and using aged seedlings in areas of late transplanting due to flood are some of the management practices used to mitigate flood hazards. Other means of mitigating flood effects are the cultivation of quick-growing homestead vegetables, pulses, modern short-duration mustard, maize field peas inter crop, potato under zero tillage, and rice-fish culture. These cultivation practices would intensify crop production immediately after the flood and should generate cash incomes and provide food for the affected population.

Table 17 Technological options for the management of agricultural vulnerability

Types of vulnerability	**Major crop damage**	**Risk-minimising technologies**
Early flash floods April- May	HYV *boro, aus* rice (*kharif* vegetables)	Cultivation of short-duration HYV *boro* rice
Late flash floods August-September	T.*Aman* (HYV and local)	Late T.*Aman* varieties, *rabi* crops intensification
River water floods June-September	Deep water and T.*Aman, kharif* vegetables, sugarcane	Improved management practices
Late river water floods Late August-early September	T.*Aman*	Late variety T.*Aman* cultivation. Zero tillage cultivation of potato, maize, mustard, wheat, maize and grass pea
Tidal floods August-September	T.*Aman*	Cultivation of local tall T.*Aman*. Adjustment of planting time. Improved cultural practice for intensification of *rabi* crops
Drought (*kharif*-I) April - May	B.*Aus*, summer vegetables, jute, wheat,mustard, lentil, chickpea, potato and sugar cane	Improved method and early planting (dibbling etc.). November planting of wheat with vitaves and Furadan. October planting of mustard. November planting of lentil and chickpea. Tillage manipulation. Drilling of wheat and mustard. *Rabi* crops relaying with T. *Aman*
Drought (*kharif*-II) September - October	T.*Aman* (HYV and local)	Supplementary irrigation. Water harvesting in mini-ponds, raised and firm ails of plots. Chickpea-lentil inter-cropping after early HYV T.*Aman*
Drought (*rabi* season) February - March	*Rabi* crops	Improved management: tillage, mulching, inter-cropping and partial irrigation

For drought-affected areas, improved methods and early planting by dibbling in the pre-monsoon season and supplementary irrigation in transplanted *aman* rice are effective. In drought-prone areas, special types of cultivation techniques would increase the soil moisture-holding capacity. Minimum or no-tillage techniques such as dibbling crop seeds directly into the stubble of a previous crop field, dry transplanting *aus* or *aman*, might enable a dryland *rabi* crop to be sown after harvesting the rice crop. Such practices deserve priority adaptation.

9.2 Promoting cultivation of non-conventional food crops

A large variety of non-conventional food crops is grown in the different agroecological regions of Bangladesh. The commonly grown plants are yams, aroids, beans, aquatic food crops, buckwheat and *bajra*. Yams and aroids are rich in carbohydrates. Among the non-conventional beans, sword beans, winged beans and velvet beans are promising sources of vegetable protein. Some aquatic plants (water lily, lotus etc.) are also used as food by poor people during famine. The non-conventional food crops can be grown well in agroecological vulnerable zones and in x. To cope with the current trend of vulnerability, there is an urgent need to improve these non-conventional food crops, particularly those endemic to growing in water. These flood-resistant crops will supplement the conventional food crops in the vulnerable regions.

9.3 Enhancing rural agro-processing activities

Modern water-seed-fertilizer technology to increase production has recently been expanded, but this covers only one-quarter of the farms. Even under the most favourable scenario, the available land cannot absorb all surplus labour. One-third of the labour force is estimated to be unemployed (or equivalent). High population pressure on limited agricultural land, low levels of investment in infrastructure and off-farm activities, as well as seasonal underemployment, exacerbates the problem. Small-scale and cottage industries employ between two and three million workers. Food manufacturing, textiles, wood products and post-harvest paddy processing are among the more important agro-processing activities in the country. But the potential for rural industrialization anchored to agriculture remains largely untapped. Bangladesh produces a wide variety of agricultural commodities (many of them perishable), of which fruits and vegetables are important, amounting to 936 and 1467 thousand tons of winter and summer vegetables and fruits respectively. Of the total production of vegetables and fruit, however, 25-50 per cent is lost, sometimes at the farm gate. The country has a vast potential for agro-based industries with the improvement of present technologies. Processing and storage technologies of cereal, jute, oilseeds, fruits, vegetables and fish could be improved and made remunerative to the rural people.

9.4 Medium-term measures through restructuring cropping patterns

Farmers must keep their cropping plans and patterns flexible because of the occasional droughts and floods over the agroecological regions of Bangladesh. For this reason, different cropping plans matched with environmental changes need to be available to farmers. In flood-prone areas, the major emphasis is on intensifying cropping during dry periods, either in soils with adequate residual moisture or with irrigation. Some of the potential cropping patterns suitable in different agricultural zones with varying flood hazards are given in Figures 12 and 13. In irrigated flood-prone areas, HYV *boro* rice should be the dominant crop in the dry season. Adjustments should be made according to the nature of flood hazards and land types. Newly developed short-duration rice variety, BR 23, should be grown from the end of September after the monsoon flood. The suggested pattern for one rice crop would be BR 14 during the *boro* season which is relatively tall and quick maturing after mustard/potato. Early planting of potato/mustard would be desirable. Soil moisture conditions in this type of soil should permit early planting of *rabi* crops. In the medium highlands, and in certain cases in the lowlands, lands remain dry for six to seven months during November to April. In that situation, transplanting of modern varieties of *boro* should be made during early January to ensure a safe harvest. Irrigation investments should be made in those land types to ensure flood-risk-free *boro* cropping. In the medium highlands, BR 3 should be the dominant variety; in the medium lowlands, two HYV rice crops should be grown in the *boro* and *aman* seasons under irrigation instead of the existing local *aus* and local transplanted *aman*.

Drought-prone soils are dominated by light texture in the top layers and heavy texture in the subsurface horizon with less available moisture-holding capacity. Moisture availability varies from 100 to 200 mm within the effective profile depth, which does not meet the evapotranspiration demand of most of the *rabi* crops (such as wheat, mustard, and potato). In the high Ganges River flood plain, *rabi* maize and sorghum/millet could be extensively grown, but in the level Barind Tract, pulses, potato and wheat could be grown. For wheat and potato, partial irrigation would be indispensable from the late vegetative stage of the crop. In the high Ganges River flood plain, especially in the highland and medium highland, HYV *aus* could be a successful crop. To avoid failure of germination, planting by dibbling is to be practised. According to soil moisture and other major agroecological variables, a variety or *rabi* crops could

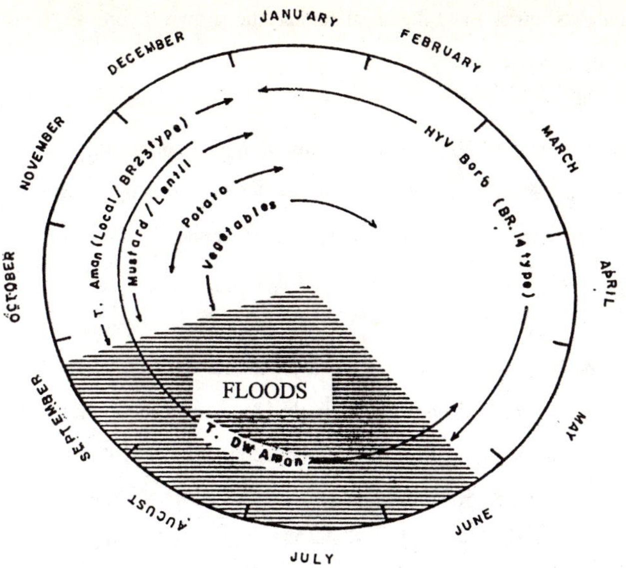

Figure 12 Restructuring of cropping patterns for moderately flooded lands with no flash-flood hazard

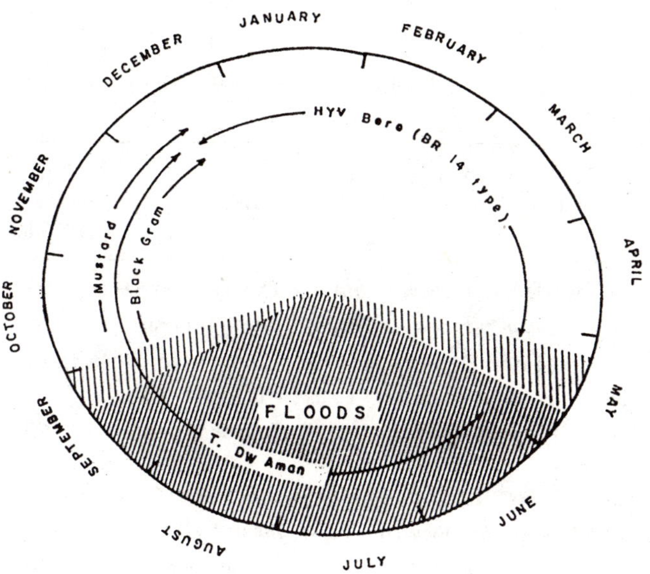

Figure 13 Restructuring of cropping patterns for moderately flooded lands with flash-flood hazard

be grown. In the medium highlands, quick-maturing mustard may be grown before the *boro* season, but BR 14 should be the desired rice variety.

For the coastal saline areas, productivity could be increased by intensifying cropping with relatively salt- and tidal-tolerant varieties, an improved irrigation and draining system, and with appropriate cultural practices such as land levelling, tillage, methods of planting and pre-plant irrigation.

Some improved cropping patterns have been developed by the national agricultural research system for the varied agroecological regions of the country (Table 18). A substantial amount of additional income could be obtained by practising these patterns. An additional US $740 and $350 per ha per yr could be earned by adopting a rotation of maize-*aman*-wheat in the low Ganges River floodplain and mung bean-transplanted *aman*-wheat in the Tista Meander floodplain. These patterns are suitable under irrigation, replacing the existing flood- and drought-prone patterns. These developed patterns would provide opportunities for crop intensification in the risk-free periods in the vulnerable regions and ensure greater employment of farm labourers with more income.

Table 18 Cropping patterns proposed by the national agriculture research institutes for vulnerable regions

AEZ region	Cropping patterns		Additional income (US $/ha)
	Current	Improved	
Level Barind Tract	*Boro* rice- fallow-T.*Aman*	*Boro* (HYV) -GM-T.*Aman*	221
	Boro - T.*Aman** - fallow	*Boro* - T.*Aman** - mustard*	85
Low Ganges floodplain	B.*Aman* - T.*Aman** - wheat	Maize - T.*Aman** - potato*	741
Active Ganges floodplain	Fallow - T.*Aman** - wheat	T.*Aus** - T.*Aman** - wheat	189
Old Bramaputra floodplain	*Boro** - T.*Aman** - fallow	*Boro** - T.*Aman** - mustard	77
Tista Meander floodplain	B.*Aus* - T.*Aus** - fallow*	Mung bean - T.*Aman** - wheat	349
Tista floodplain	T.*Aus** - T.*Aman** - tobacco	Jute - T.*Aman* *- wheat*	230
North and Eastern Hills and Chittagong coastal plain	*Boro* - T.*Aman* - fallow	*Boro* - T.*Aman* - mustard	1188

* High yielding varieties

10 Conclusions

The level of poverty and vulnerability of the poor must be addressed by the management of environmental vulnerabilities, generation and adaptation of wide choices of technologies, and promoting agro-processing activities. The status of economic growth and the level of agricultural production should be improved with special emphasis on the adoption of proven technologies.

The rural poor derive two-thirds of their income from agriculture. Therefore, intensifying agricultural production suited to the poor will alleviate community poverty. At present, few technological options are available to the poor. A wide choice of alternative technologies should be made available. The types of technologies needed should be labour-intensive and generate cash rapidly. Cultivation of quick-growing vegetables, kitchen gardening, spices production and agro-processing, poultry raising under scavenging conditions with improved breeds, backyard pond culture with common carps, intercropping with various crop combinations suited for smallholdings and creating conditions to promote agribusiness entrepreneurs are promising tools for the rural poor. Farmers need to make an inventory of their innovative risk-mitigating practices for vulnerable regions and these innovative practices should be improved for wider adaptation. Details of these management practices for averting risk should be made available to the group of cultivators and necessary training and technology support packages should be provided.

To cope with the increasing threats of environmental vulnerability and rapidly increasing population, there is an urgent need to develop technologies to manage environmental degradation and to sustain increased agricultural production from the finite natural resources of the country. Crop and management practices should be developed that are resilient to climatic fluctuations. Suitable crops and cultivars should be developed to reduce the environmental stresses of drought, flood and temperature. Combining high yields with drought-resistant reduction of duration of weakly photosensitive cultivars of rice for flood-prone areas would increase productivity. Developing suitable rice cultivars for about 2.6 million hectares of land that are inundated (more than 180 cm) and a greater thrust to develop germplasm and management practices for oilseeds, pulses, vegetables and fruits are desirable. The scope for multiple cropping in about 1.8 million hectares of flood-free, good-quality land needs to be evaluated, while combinations of relay, mixed and multiple cropping need to be developed for vulnerable regions.

Acknowledgements

The author wishes to express his heart-felt thanks to Dr. Mafizur Rahman of Bangladesh Sugar and Food Industries Corporation, and Mr. M. Anwar Iqbal, Mr. Abid Hossain Chowdhury and Mr. Fazlul Quader Chowdhury of BARC for their assistance and cooperation at all stages of preparing this report.

References

Bangladesh Bureau of Statistics (1989). pp. 131-132.

BCAS (Bangladesh Centre for Advanced Studies and Nature Conservation Movement) (1994) *Wetlands of Bangladesh.* Khan, M.S., Haq, E,m Huq, S., Rahman, A.A., Rashid, S.M.A. and Ahmed, H. (eds.). Bangladesh Centre for Advanced Studies and Nature Conservation Movement, Dhaka.

BUP/CEARS/UEA (Bangladesh Unnayan Parishad/Centre for Environmental and Resource Studies/University of East Anglia Climate Research Unit) (1994). *Bangladesh: Greenhouse Effect and Climate Change*, Briefing Documents, No. 1-7. BUP, CEARS, University of Waikato, Hamilton and Climate Research Unit, UEA, Norwich.

Hossain, M. (1992). *Structure and Distribution of Household Income and Income Dimension of Poverty. Re-thinking Rural Poverty.* Bangladesh Institute for Development Studies, Dhaka.

Houghton, J.T., Callander, B.A. and Varney, S.K. (eds.) (1992). *Climate Change 1992: The supplementary report to the IPCC Scientific Assessment.* Cambridge University Press, Cambridge.

Houghton, J.T., Jenkins, G.J. and Ephraums, J.J. (eds.) (1990). *Climate Change: The IPCC Scientific Assessment.* Cambridge University Press, Cambridge.

Houghton, J.T., Meiro Filho, L.G., Bruce, J., Hoesung Lee, Callander, B.A., Haites, E., Harris, N. and Maskell, K. (1995). *Climate Change 1994: Radiative Forcing of Climate Change and An Evaluation of the IPCC IS92 Emission Scenarios.* Cambridge University Press, Cambridge.

Karim, Z., Ibrahim, A.M., Iqbal, A. and Ahmed, M. (1990a). *Drought in Bangladesh Agriculture and Irrigation Schedules for Major Crops.* BARC Soils Publication 34, pp. 1-15. Bangladesh Agricultural Research Council, Dhaka.

Karim, Z., Hossain, G.G. and Ahmed, M. (1990b). *Salinity Problems and Crop Intensification in the Coastal Region of Bangladesh.* BARC Soils Publication 33, pp. 1-63. Bangladesh Agricultural Research Council, Dhaka.

Mahtab, F.U. (1989). *Effect of Climate Change and Sea-level Rise on Bangladesh.* For Expert Group on Climate Change and Sea-level Rise. Commonwealth Secretariat, London.

Mahtab, F.U. and Karim, Z. (1992). Population and agricultural land use: towards a sustainable food production system in Bangladesh. *Ambio*, **21** (1):50-55.

Rahman, H.Z. (1992). *Crisis and Insecurity: The 'Other' Face of Poverty. Re-thinking Rural Poverty.* Bangladesh Institute for Development Studies, Dhaka.

UNDP/FAO (United Nations Development Programme/Food and Agriculture Organisation) (1988). *Land Resources Appraisal of Bangladesh for Agricultural Development.* UNDP/FAO, Rome.

WHO/UNEP (World Health Organisation/ United Nations Environment Programme) (1988). *Developing Policies for Responding to Climatic Change.* WCIP-1, WHO/TD-No. 225, pp. 14-15. WHO/UNEP, Geneva.

World Bank (1992). *Bangladesh: Selected Issues in External Competitiveness and Economic Efficiency.* Report No. 10265-BD. World Bank, Dhaka.

Vulnerability of Bangladesh to Climate Change and Sea Level Rise

Saleem-ul Huq and Ahsan U. Ahmed
Bangladesh Centre for Advanced Studies (BCAS)
House 620, Road 10/A
Dhanmondi Dhaka 1209
Bangladesh

Rob Koudstaal
Resource Analysis
Zuiderstraat 110
2611 SJ Delft
The Netherlands

1 Purpose of the study

The Government of Bangladesh, a signatory to the Framework Convention on Climate Change (FCCC), sought to support the international community in understanding vulnerability to climate change and sea level rise and, at the same time, to formulate ways to mitigate the hazards and find modalities in developing institutions to deal with coastal zones. A pilot study of vulnerability was commissioned under the auspices of the Intergovernmental Panel on Climate Change (IPCC) Coastal Zone Management Subgroup (CZMS). One of the objectives of the pilot study was to develop tools and techniques for use by the Government of Bangladesh for future coastal resource planning and management. Another objective was to identify both the policies and the technical capacity that will be needed within Bangladesh in order to deal with climate change related issues and problems on an ongoing basis. In particular, the study addressed the following questions.

- What are the likely primary physical effects of climate change and sea level rise on the coastal areas of Bangladesh?
- What will be the likely consequences of these primary physical effects on the population, physical infrastructure, agriculture and natural ecosystems of Bangladesh?
- What institutional arrangements and capabilities exist within Bangladesh to deal with and respond to the challenge of climate change and sea level rise?
- What are the interactions and sensitivities of different development options within and outside Bangladesh to the possible threats of climate change and sea level rise?
- What needs to be done in the future in order to reduce the vulnerability of Bangladesh to effects of climate change and sea level rise?

NATO ASI Series, Vol. I 37
Climate Change and World Food Security
Edited by Thomas E. Downing

The study places climate change in the context of other long- and short-term developments, both within Bangladesh and in upstream countries. The methodology developed has general application in analysing long-range impacts in support of short-term management decisions. It provides an operational tool to account for such impacts.

Although the technical study attempts to quantify many of the possible effects of climate change and sea level rise, such as changes in inundated areas, flood depths and agricultural production, the conditional nature of the analysis should be stressed. The study examines the likely consequences for Bangladesh if certain climatic changes and sea level rises should occur. It concentrates on several major agents of climate change using a set of assumptions that are admittedly highly uncertain. The results of these analyses should therefore not be regarded as predictions of situations that will occur; they merely indicate the magnitude of changes that are possible. (See BCAS, RA, CZM Centre and Approtech (1994) for the full report of the study.)

2 The problem: scenarios of climate change and sea level rise

With its low lying deltaic landmass, high population density and low level of development, Bangladesh already has to cope with a number of physical and man-made problems. These include natural hazards such as cyclones, floods and droughts, and socioeconomic problems such a low levels of literacy, poor health delivery systems, high unemployment and poverty. In the future, Bangladesh will also have to face adverse impacts of development across its borders – which, among other things, are expected to reduce the availability of water during low river flows – and has to cope with the possibility of climate change and sea level rise.

Changes in climate and sea levels are expressed in terms of variables such as temperature, evaporation and river discharges. In the Bangladesh pilot study, the following agents of change are considered because of their assumed effect on the natural systems of Bangladesh:

- sea level rise;
- temperature;
- precipitation;
- evaporation;
- river discharges;
- cyclone intensity.

Estimates of these agents of change based on IPCC's 'business-as-usual' scenario (Houghton *et al.*, 1992; IPCC, WMO and UNEP, 1992). Further information has been obtained from a study of climate change and its implication for China, which provides estimates of regional climate change in East Asia (WWF, 1992). Values for moderate and severe estimates for the year 2100 are given in Table 1.

Precipitation and evaporation are expected to increase in both the winter and monsoon period. They are closely related to changes in temperature. Seasonal changes increase the deficit in the water balance during the dry period. In absolute terms, a 12 - 22 per cent increase in precipitation is much less than a 10 - 12 per cent increase of evaporation in the dry period as the amount of rainfall is low. This results in a further decrease of low flows, estimated at 12 and 22 per cent for the two climate scenarios, respectively.

With global warming, the sea surface temperature will rise, which may affect the formation of cyclones. This is a complex process in which a threshold temperature of 27°C of surface water is considered to be important. However, there is little evidence that increased sea surface temperatures will increase the formation and frequency of cyclones. On the other hand, once formed, the intensity is expected to increase and with sea level rise the area exposed to inundation will change.

Table 1 Agents of climate change and their variations considered in the pilot study

		Climate change scenario (CCS)	
Agents of changes		**1: Moderate**	**2: Severe**
Sea level rise	(cm)	+ 30	+100
Precipitation	(%)		
Monsoon		+ 18	+ 33
Winter		+ 12	+ 22
Evaporation	(%)		
Monsoon		+ 8	+ 15
Winter		+ 10	+ 20
River discharges	(%)		
Peak flows		+ 6	+ 13
Low flows		- 12	- 22
Cyclone intensity	(%)	+ 10	+ 25

Note: [1] Only as result from climate change, no upstream developments.

Source: BCAS, RA, CZM Centre and Approtech (1994).

3 Methodology and framework for analysis

3.1 Vulnerability profiles and zones

The *primary physical effects* of climate change are experienced in Bangladesh in the form of increased inundation, reduced low river water flows, increased salt water intrusion and a higher intensity of cyclones. These primary effects affect the country's ecological and human systems. Important *ecosystems* in Bangladesh which are affected include the Sundarbans in the southwest, the Haor wetlands in the northeast, and the wetlands along the coasts. Impacts on *human systems* focus on population, infrastructure and foodgrain production.

Vulnerability depends on the dimension of the impacts on the natural and human systems, as well as on the capability of the society as a whole to cope with these impacts in a social, economic, financial, technical and administrative sense and to develop and implement response strategies. The combination of these impacts and the feasibility to implement response strategies composes the *vulnerability profile*.

The country was divided into nine vulnerability zones (VA-zones) to be able to investigate regional vulnerabilities (Figure 1). The zones were selected based on agro-ecological zones with only the Chittagong Hill Tracts excluded:

- *The Sundarbans zone* (SB) is a mangrove-type coast. Possible impacts: flooding due to sea level rise (SLR), storm surges and severe salinity intrusion.
- *The South West zone* (SW) consists of the Ganges tidal plains used for agriculture. Possible impacts: flooding due to SLR and storm surges, salination of surface and groundwater, and moderate to severe drought.
- *The South Central zone* (SC) includes the offshore islands and the Meghna estuarine floodplain. The area, built up by alluvial deposits, is subject to impacts of severe tropical storms and cyclones. Cyclonic storm surge usually occurs in the transition between the dry and monsoon seasons, generally in the period April/May and October/November. Morphologically this zone is extremely active, caused by the changing flow and silt and

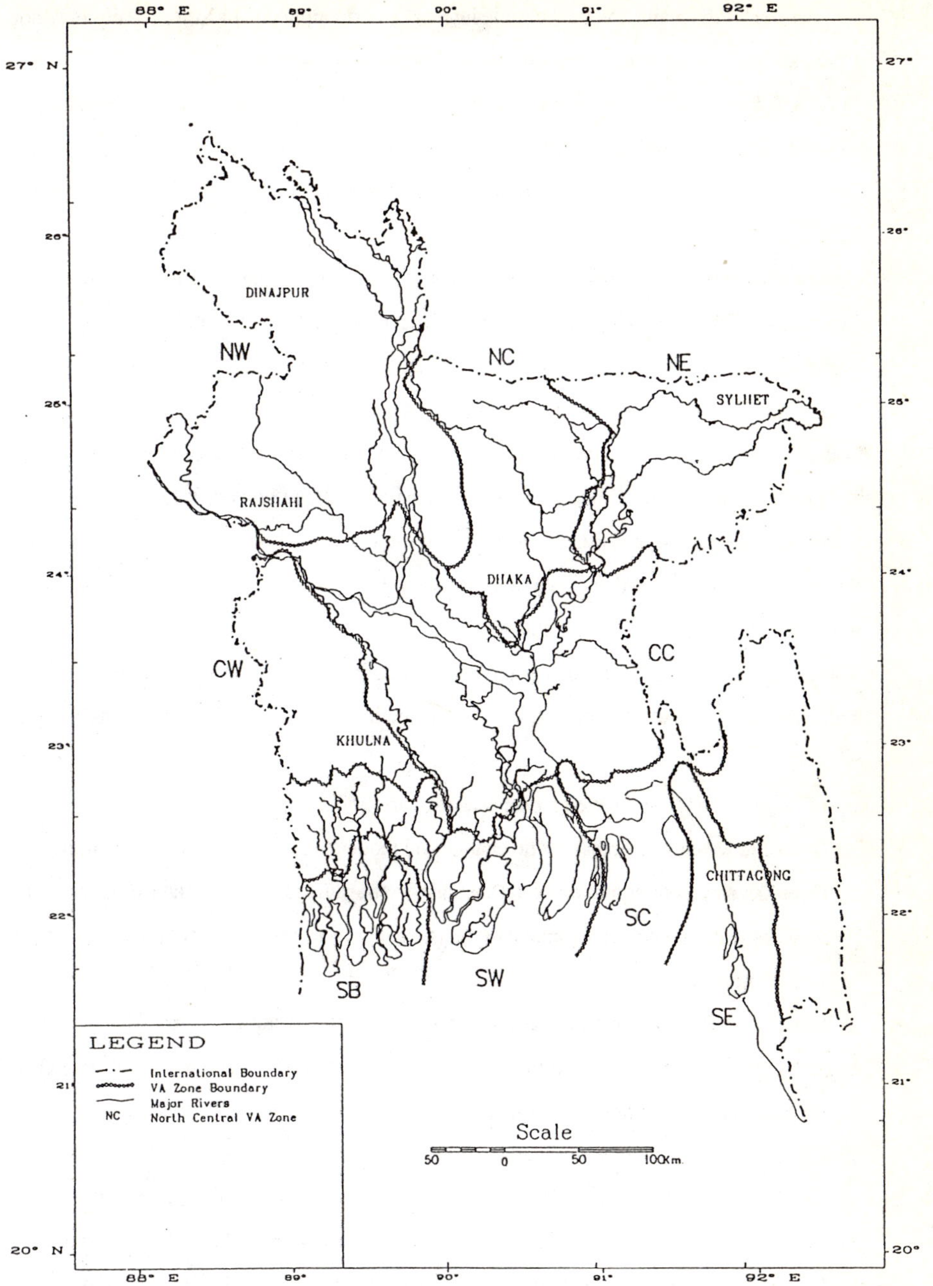

Figure 1 Map of Bangladesh and vulnerability zones. Source: BCAS *et al.* (1994).

mud carried by the major rivers. Possible impacts are: flooding due to SLR, salinity intrusion in soil, surface and groundwater sources and cyclonic storm surges.

- *The South East zone* (SE) runs more or less North/South and forms the eastern boundary of the delta. Impacts: flooding due to cyclonic storm surges.
- *Central West zone* (CW) includes middle and higher floodplains of the Ganges. Impacts: flooding and droughts.
- *Central deltaic and inner zone* (CC) consists of the Ganges, Meghna and Brahmaputra lower floodplains. The rivers are extremely dynamic hydrologically and morphologically. Their hydrologic regimes are characterised by extremely high discharges in the monsoon season, when high local monsoon rainfall adds to large discharges from the Himalayan catchments. In the dry season the river runoff reduces to small values. Impacts: flooding due to ASLR and peak river discharges, drainage congestion leading to flash flood and salinity intrusion.
- *North West zone* (NW) consists of piedmonts of the Himalayan range and middle and higher floodplains of the Bramaputra. Possible impacts: flooding and drainage congestion due to peak river discharges, severe drought.
- *North Central zone* (NC) is middle and higher floodplains of the Meghna river. Possible impacts: flash floods due to drainage congestion and inundations due to peak river discharges.

A range of climate change and development scenarios are evaluated.

- Three climate change scenarios: no change and moderate and severe climate changes.
- Three development conditions for Bangladesh: the situation in 1990 and projections for 2010, assuming two development options: a 'business-as-usual' (BAU) and a 'high-development option' (HDO).
- Three river flow conditions: present, sharing and non-sharing of water. Changes in the discharges of these rivers reflect their upper watershed conditions with respect to afforestation and water consumption as well as agreement with India on sharing the available water, especially during the dry period.

The methodology is based on selected specific *indicators*. The specific indicators are structured in a hierarchy of importance. For example, the indicators flood vulnerable area, flood depth and flood risk represent a higher aspect 'inundation'. Inundation together with low flow and drought belong to the main impact category primary physical effects. Four main categories are considered:

- primary physical effects;
- impacts on natural ecosystems;

- socio-economic stress factors;
- implementation feasibility of response strategies.

Figure 2 presents the main structure of the hierarchical tree of impact categories for each VA-zone. For each of these impact categories, aggregated indices are determined by adding up the individual indicators after a proper scaling and weighing procedure. They represent a level of vulnerability or sensitivity for the considered changes. Further aggregation of these main categories leads to an overall vulnerability index, which is done for the nine VA-zones and for the country as a whole.

The values of such indicators are assessed for the range of scenarios through quantitative analysis and/or expert judgements. In a next step *pairwise comparisons* of scenarios are made to assess the changes in the criteria due to climate change and sea level rise under a specified set of development conditions (Table 2). In total, six pairs of cases have been selected for comparison. Four refer to a high climate change scenario. These represent the *vulnerability* of Bangladesh;

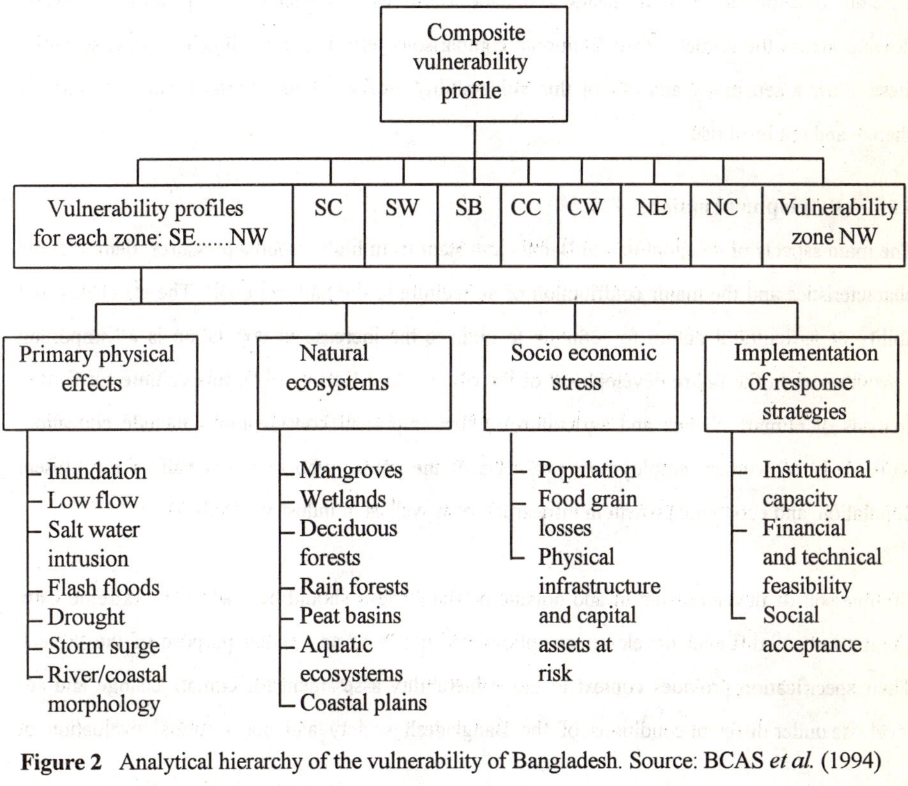

Figure 2 Analytical hierarchy of the vulnerability of Bangladesh. Source: BCAS *et al.* (1994)

Table 2 Main characteristics of the compared cases

Climate change and sea level rise:	Severe/no change				Moderate/no change	
Development:	**Present**	**BAU**	**BAU**	**HDO**	**Present**	**BAU**
Indian rivers:	**(1990)**	**SH**	**NSH**	**NSH**	**(1990)**	**NSH**
• Vulnerability profile (VP)	X	X	X	X		
• Sensitivity of VP to climate change	X		X		X	X
• Sensitivity of VP to development			X	X		
• Sensitivity of VP to water sharing			X			

Notes: BAU, HDO — Business-as-usual and high development options for Bangladesh (in 2010)
SH, NSH — Sharing and non-sharing of waters from Indian rivers.

Source: After BCAS *et al.* (1994).

their comparison gives an indication of how the vulnerability of Bangladesh will change under different development options inside Bangladesh and with respect to the quantities of water flowing across the border. Two additional comparisons refer to a low climate change scenario; these allow a sensitivity analysis of this vulnerability profile for the assessed values of climate change and sea level rise.

3.2 Development options

The main aspects of development in Bangladesh stem from high resource pressures, demographic characteristics and the major contribution of agriculture to the national GDP. The efficiency and ability of agricultural output to continue to outpace the increase in population is an important consideration in the future development of the country (see Karim, 1995, this volume for further analysis of climate change and agriculture). Other important considerations include education, particularly of women, employment prospects of the youth, who represent half of the present population, and economic growth in infrastructure as well as in industry (Table 3).

Comparison of developments in and outside of Bangladesh should be made with extreme care. Comparison of different development options has not been an explicit purpose of the project. Their specification provides context to the vulnerability assessment for climate change and sea level rise under different conditions of the Bangladesh society and not a formal evaluation of alternative development:. Two pathways are compared:

- A *High Development Option* (HDO) in which all the best known development options are implemented. It incorporates improvement in investments, a demographic transition, much improved education, major improvements in agriculture, and a considerable increase in infrastructure.
- A pessimistic *Business-As-Usual* (BAU) development option in which all the present (rather poor) development trends in economic development and population are continued for the next two decades.

Table 3 Characteristics of development options, 2010

Development aspects	**Unit**	**Present (1990)**	**Development scenarios**	
			Business-as-usual	**High development**
Population	% growth/year	1.9	1.9	1.2
	million	109	159	140
Investment/GDP	%	10	12	17
GDP	% growth/year	4	3.4	5.7
Income/capita	US$	210		
	% growth/year		1.5	4.5
Employment	% growth/year	2.8	3.4	4.7
Unemployment	%	28	20	0-2
Literacy	%	40	Low (<50)	High (>80)
Health investment		Low	Low	High
Area under irrigation for Aman	%	30	50	65
Yields: rainfed	% growth/year	-	0.5	0.7
		-	0.7	0.9
Nutritional status	% pop. below minimum	-	Large	Small

Sources: BCAS *et al.*, 1994; Task Force, 1991; World Bank, 1993.

3.3 Upstream development conditions

Assessment of the possible effects of upstream development on river discharges is based on ongoing developments, existing plans and ideas for major infrastructural works. Deforestation is concentrated mainly in the southern hill slopes of the Himalayas in the Ganges catchment and in the Assam region in the Brahmaputra middle reaches and in the eastern rivers. Demographic and economic developments are most pronounced in the densely populated Ganges plain and the catchments of the northwestern rivers. The middle Brahmaputra plain, which is hilly and is one of the most rainy parts of the world, has less potential for intensive development.

The major infrastructure works under discussion are:

- a dam on the Brahmaputra near the Indian-Chinese border which will block all sediments entering India through the river;
- development of a link-canal between the Brahmaputra and the Ganges to augment the waters of the Ganges from the Brahmaputra with a discharge capacity of 3000 m^3/s (about 70 per cent of the Brahmaputra low flow);
- completion of flood embankment projects in the Ganges floodplain in India;
- development of further dam projects on the Ganges and the Meghna in India and Nepal.

Two scenarios for developments in the upstream watershed of the rivers flowing into Bangladesh are defined (Table 4).

- The *non-sharing* (NSH) option is based on a first-come, first-served scenario where the upper riparian users continue to interfere with river flows without considering the concerns of the lower riparian users. Proposed construction, growing water consumption, deforestation, flood protection and lack of agreement on water sharing result in much reduced water flows (principally through the Ganges) and increased high water flows due to increased deforestation upstream.
- The second *sharing* (SH) option assumes a lower level of impact based on an integrated water-sharing agreement between the riparian countries whereby the rights of the Bangladesh share of water is protected. This refers principally to the low flow conditions of the Ganges. This option also considers that the dam on the Brahmaputra and the Indian link-canal are not constructed.

Table 4 Seasonal changes in flows in the Ganges and Jamuna

Scenario	Ganges Increase (%)	Ganges Decrease (%)	Jamuna Increase (%)	Jamuna Decrease (%)
Non-sharing	10	100	10	75
Sharing	5	50	5	25

Source: BCAS *et al.* (1994).

The primary physical effects, impacts on human and natural ecosystems and response strategies are well documented in Bangladesh (see BCAS *et al.*, 1994; BIDS, 1993; BUP *et al.*, 1994; Karim, 1995, this volume).

4 Primary physical effects of climate change and sea level rise

The following sections summarise the ratings used in the vulnerability assessment. Table 5 gives an overview of land types and area subject to flooding. The impact areas are divided into net cultivable areas and settlement areas, and further categorised by the level of likely inundation (F0 to F3). Unprotected areas are prone to greater depth of flooding and more areas may be subject to flooding. The extent to which flood characteristics will change is determined by following a step-wise analysis (Table 6).

Table 5 Land types and area affected by riverine inundation

<table>
<tr><th colspan="12">Total of VA-zones (120.0)</th></tr>
<tr><td rowspan="5">Sund-arbans (5.5)</td><td rowspan="5">Up-land (7.5)</td><td colspan="9">Impact area (105.3)</td><td rowspan="5">Perenially flooded (1.7)</td></tr>
<tr><td colspan="4">Net cultivable area (86.9)</td><td colspan="5">Settlement area total of uncultivable area (18.4)</td></tr>
<tr><td rowspan="2">Dry land (26.4)</td><td colspan="3">Flood vulnerable areas (60.5)</td><td rowspan="2">Dry land (12.9)</td><td colspan="2">FVA</td><td colspan="2" rowspan="2">Not vulnerable</td></tr>
<tr><td colspan="2">Protected (17.2)</td><td>Unprotected (17.2)</td><td>P (2.4)</td><td>UP (3.1)</td></tr>
<tr><td>F0 <30 cm</td><td>F1 30-90 cm</td><td>F2 90-180 cm</td><td>F3 >180 cm</td><td>F0 <30 cm</td><td colspan="2">F1 30-90 cm</td><td>F2 90-180 cm</td><td>F3 >180 cm</td></tr>
</table>

Notes: Areas in parentheses are expressed in thousand km^2: total area of Bangladesh is 144,000 km^2. F0 to F3 are classes of flood inundation. P is protected. UP is unprotected.

Source: After BCAS *et al.* (1994).

The results of the inundation analysis (Figure 3) show that climate change increases the flood vulnerable area considerably, but that in the case of sharing, its impacts are less than under the non-sharing scenario. In the case of flood depth and flood risk, the impacts of climate change are again very considerable, but the impact of high development increases the risk even further. The explanation for this is that the high development option includes construction of embankments, which enhances the vulnerability of non-protected areas for future floods.

Low water flows are important for navigation and for surface water extraction for public water supply, agriculture (irrigation) and other economic activities. Both irrigation and navigation have special significance in the national economy. In the climatic scenarios the low water discharges are reduced, while salt water intrusion further reduces water availability for irrigation.

Table 6 Calculation of inundation changes

Steps	Inputs	Analysis	Technique/tool	Output
1	Scenarios of: Climate change and sea level rise Upstream development Development	Assign boundary conditions for river flows	Manually	Changed boundary conditions for scenarios
2	Changed boundary conditions for river flows	Obtain water levels at 17 selected stations in major rivers	Hydrodynamic model runs using MIKE II	New water levels for 17 selected stations in major rivers
3	New water levels	Correct for bed level changes (17 selected stations in major rivers)	Geomorphological model runs using WENDY	Adjusted water levels (17 selected stations in major rivers)
4	Adjusted water levels	Calculate water level changes in each VA-zone based on 17 selected stations in major rivers	Hydraulic calculations	Average water level changes in floodplain by VA-zone
5	Floodplain water level changes by VA-Zone	Calculate floodplain inundation depths for unprotected	Assumptions on sedimentation in floodplains	Flood depths in unprotected areas by VA-zone Change in flood vulnerable areas by VA-zone
		Calculate flood risk for protected areas	Frequency distribution	Change in flood frequency in protected areas by VA-zone

Source: BCAS *et al.* (1994).

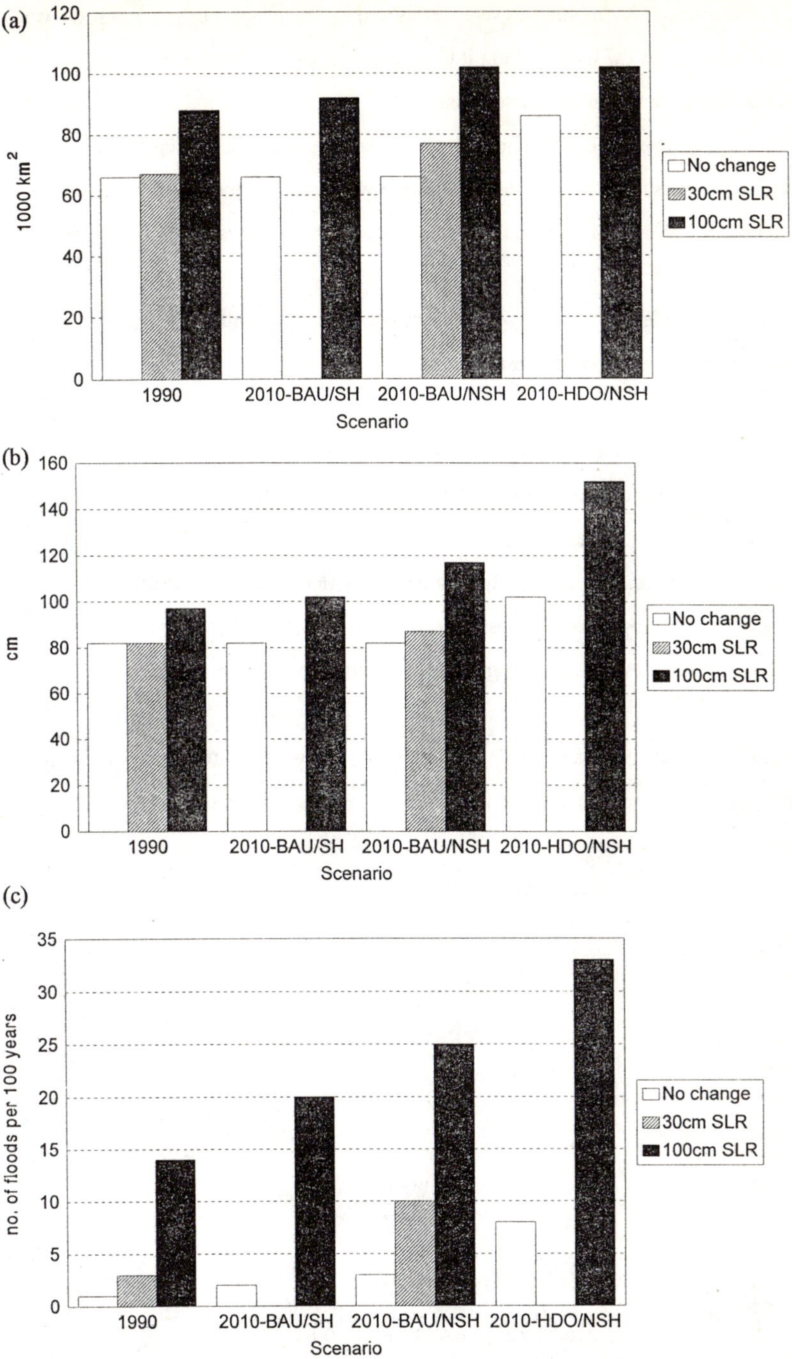

Figure 3 Impacts of inundation: (a) changes in flood vulnerable area; (b) changes in flood depth; (c) changes in flood risk. Source: BCAS *et al.* (1994).

Low flows are characterised by minimum water levels and by the length of the period for which water levels are below a critical level making it unsuitable for specific economic activities. Two indicators have been selected for impact assessment.

- standard low water levels (SLW), defined as the water level at a specific station exceeded on 95 per cent of the days in an average year (i.e., flows were less than the SLW on 18 days);
- low water period, defined for each station as the period for which water levels are below the SLW.

As in the case of the inundation analysis, results of a hydrodynamic simulation model have been used to assess changes in water levels (Figure 4).

At least 1.4 million ha of the coastal and offshore areas of Bangladesh – inhabited by about 15 per cent of the population – are directly affected by the intrusion of salt water into the surface and groundwater systems. High salinity seriously affects agricultural potential of this area: irrigation potential is reduced, crops may be damaged by saline water flooding, and increased soil salinity reduces plant growth and crop yields. It also affects saline-free drinking water for large coastal communities and industries, especially during the dry season.

A rise in sea level increases the area that will be subjected to salt water intrusion into the estuarine surface water system. Lower minimum flows would compound the problems of salt water

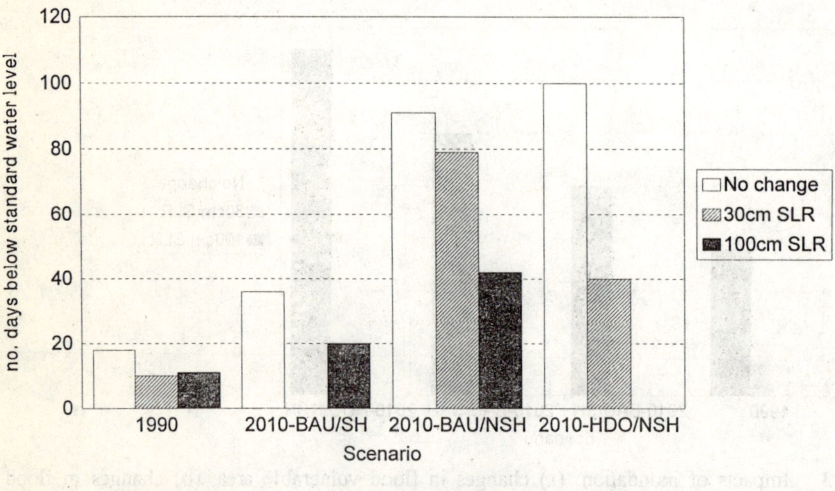

Figure 4 Changes in low water periods. Source: BCAS *et al.* (1994).

intrusion. The effects of sea level rise and changes in river flows may be of the same order of magnitude, but in combination their impacts on the socioeconomic conditions of the whole of Bangladesh would be disastrous.

Flash floods, caused by heavy rainfall in hilly river catchments, occur in the north and east, particularly in the Meghna River basin and the southeastern headwaters of streams in the districts of Chittagong, Noakhali and Chittagong Hill Tracts. A flash flood can be caused by a ten-day period of rainfall of more than 300 mm. Rainfall statistics taken at ten stations in the Surma-Kushiyara-Manu-Khowai River systems in the north for May show that this occurs about once in every two - three years. Flash floods may occur at any time during May and June. Flash floods predominantly damage agriculture and, to a lesser degree, infrastructure. Present agricultural practice in the affected areas is to grow a high-yield rice variety *boro* which is harvested around May 15. If flash floods occur after this date, there is generally no damage to crops; only to the infrastructure. But if flash floods occur before the harvest, crops may be completely destroyed. Under both the moderate and severe climate change scenarios, the incidence of flash floods are expected to increase. The average precipitation is assumed to increase by 18 and 33 per cent, respectively, under monsoon conditions. At present, flash floods and damage assessment in the vulnerable areas are not systematically monitored, so it is impossible to make accurate estimates of likely levels of damage.

Drought refers to moisture stress in the top soil that affects vegetative growth. During the *kharif* season (monsoon, July-September), precipitation exceeds evaporation. Drought in this season is due to shortages of rainfall in the late growing period of transplanted *aman* rice. Such shortages can cause serious yield reductions and delay harvests, thus affecting the cultivation of subsequent crops in the *rabi* and pre-*kharif* season (October-April). During the *rabi* season, evaporation exceeds rainfall and the soil moisture content is heavily dependent on the water remaining after the *kharif* season. Water stress is frequently felt after mid-*rabi*. During the pre-*kharif* period, drought conditions depend on the rainfall pattern, which is erratic.

Drought affects the entire population of the affected areas, forcing the landless and marginal farmers to relocate in search of jobs. Lower food production and unemployment compound poverty, creating social vices.

Under both the moderate and severe climate change scenarios, the intensity of drought will increase. During the *rabi* season, the area severely affected by drought could triple from 4,000 km^2 to about 12,000 km^2 under the severe climate change scenario. During the *kharif* season, the area subject to severe, and moderate drought, could quadruple, affecting both irrigated and rainfed crops (see Karim, 1995, this volume, and Mahtab and Karim, 1992, for further details).

In Bangladesh, cyclonic surges are observed twice a year – in late May and in early November. Since global climate change increases mean sea surface temperature, it may be expected that both the frequency and intensity of cyclones will be affected. An increase in intensity might result in an increase in affected areas. However, there is a lack of data on historic cyclonic processes and their damages. For the vulnerability assessment it is assumed that cyclone intensity will rise only in the presently affected areas for both the climate change scenarios: 10 per cent for moderate climate change and 25 per cent for severe climate change.

The total length of the Bangladesh coastline is about 650 km, of which about 125 km is covered by the Sundarbans and 85 km by Cox's Bazaar beach in the south. The coast can be broadly divided into the east-west oriented deltaic coastline, and a north-south non-deltaic coastline. The deltaic coastline comprises the Ganges tidal plain, which represents abandoned deltas of the Ganges River, and the dynamic Meghna deltaic plain, which is the present conduit of the major rivers. The non-deltaic Chittagong coastal plains is dominated by mud in the north and by sand in the south.

Historically, the coastline has been undergoing erosion and accretion probably in a dynamic equilibrium. The impact of climate change and sea level rise is expected to result in an increase in erosion and a decrease in accretion in the deltaic part of the coast. From an assessment of changes in erosion and accession of the Meghna estuary over the last 20 years using remote sensing data, it was observed that erosion has in fact increased over time. The Chittagong coast is also likely to recede due to increased sea level rise. For a 2 cm per year sea level rise, it is estimated that the coast will recede by 2-3 cm per year or 80-120 m by the year 2030. This means that the Cox's Bazaar beach will completely disappear.

5 Impacts on human and natural ecosystems

Driven by poverty, large populations are already vulnerable to extreme climatic events. It is thus important to evaluate how many people would be at risk in a changed climate. The population impacted by inundation, salinity intrusion, drought, flash floods and cyclones are considered.

Some 71 million people or 65 per cent of the total population of Bangladesh are already affected every year by inundation of slight to moderate intensity. This proportion is likely to increase with climate change and sea level rise (Figure 5). Assuming that the water-sharing option is adopted, under the BAU scenario, 84 per cent of the total population will be affected out of which about 9.3 million people will be affected by severe floods of 90-180 cm. But for the non-sharing option, the flood-affected population increases to 94 per cent. Thus, under the BAU/non-sharing option, the entire country will face floods of varied intensity. At the same time about 20 million people (13 per cent of the total) will be affected by severe floods of 90-180 cm. These results emphasise the importance of sharing water with the upper riparian countries.

Under the high development option (HDO) scenario, due to the presence of large-scale infrastructure installations on river banks, *viz.* embankments etc., the impacts of flooding on

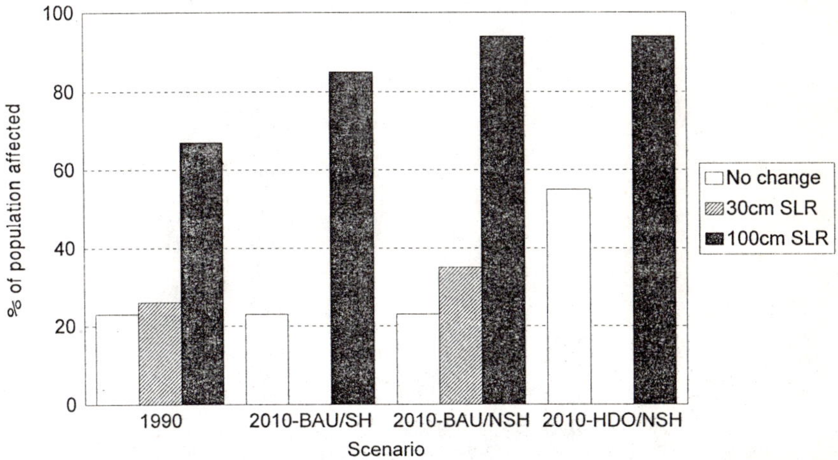

Figure 5 Population affected by inundation with severe climate change. Source: BCAS *et al.* (1994).

the population would be greater: 41 per cent would face floods of 90-180 cm. Under this scenario about four - five million people (three per cent of the total) would be at extremely high risk of flooding.

The total value of capital assets in Bangladesh is estimated at about TK 1,800 billion (1990 values; the exchange rate was about US$1.00 = Taka 40 in 1995). This includes mobile assets, which have been excluded from the pilot study as they are assumed not to be vulnerable to inundations. Immobile physical infrastructure are categorised into five groups: water works, mainly for flood protection and irrigation; houses and settlements; transport network; utilities; and industries. Capital values, based on their 1990 replacement values, are TK 284 billion. This does not include land development, farmhouses and private buildings because of lack of data. Almost 70 per cent of the TK 284 billion is invested in houses and settlements. Most of the investments have been made in the northwest and central zone. Capital values in 2010 have been estimated at TK 410 and 610 billion for the BAU and high development scenarios, respectively.

Through expert judgement, the physical infrastructure has been allocated to different land types (F0 to F3; protected and unprotected) and assessments were made of expected annual damages due to inundation. These assessments were based on a flood which occurs once in every 20 years. Since the change in areas due to inundation for each land type are estimated (above), the total expected damage to infrastructure can be estimated as well (Figure 6).

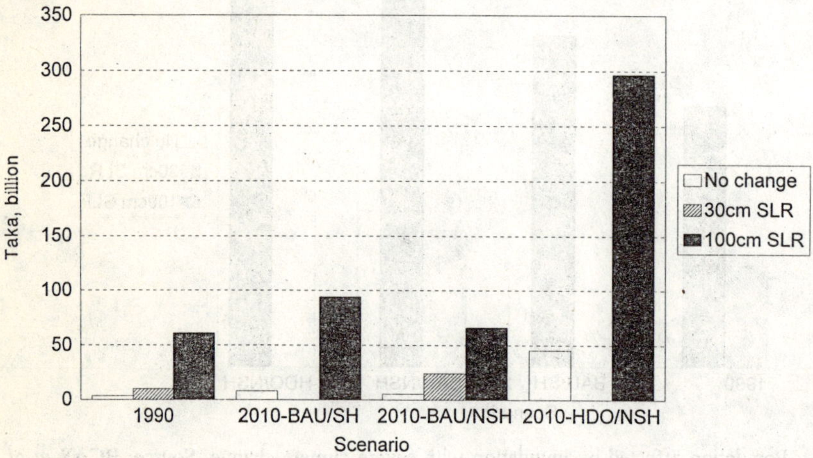

Figure 6 Impact of inundation on infrastructure. Source: BCAS *et al.* (1994).

With climate change and sea level rise, damages would increase by 15 per cent for the 1990 situation and by 21 per cent with BAU development in 2010. For the high development option, the increase in loss is substantially higher; about 41 per cent of the capital value estimated in the year 2010 (this is about 12 per cent of the GDP in 2010). The additional loss of physical infrastructure due to moderate sea level rise (30 cm) is not significant.

Since crop yields vary according to flood vulnerability (land types F0-F4), any change in the flood distribution would change the total production of the main crop, *aman* rice. Under the business-as-usual development scenario, inundation resulting from severe climate change would result in a 13.6 million mt reduction in *aman* rice production when water is shared, compared with a 10.5 mt reduction if water is not shared (Figure 7). Under the high development option, larger areas would be irrigated, flood control embankments would be built, and *aman* rice production would increase. If no climate change is considered, then production would be higher than under present conditions. This increase is attributed mainly to the increase in production of local cultivars, mainly BL *aman* and LT *aman*; although the production of HYV *aman* is reduced because this cannot be grown in F2 areas. Under the severe climate change scenario, production would fall to 70 per cent of current levels (about 19 mt of paddy). Obviously the non-sharing of water would contribute to this huge loss of production – even improvements in irrigation conditions and flood control infrastructures would not compensate for the lost rice production.

With moderate climate change, the area severely affected by drought will increase from 3,600 km^2 to 12,000 km^2 during the *rabi* season. During the *kharif* season, the severely affected area increases foretold with severe climate change.

At present about 73 million people (67 per cent of the population) live in areas that are prone to moderate or severe drought. Of these, about 25 per cent live in areas that are severely and very severely affected during the *kharif* season. The proportion of the population living in severe drought-prone areas would increase from four to nine per cent and 18 per cent under the moderate and severe climate change scenarios, respectively.

Because of the increase in the intensity of drought, there would be more demand for water for irrigation. Irrigation water requirements may increase from 200 cm at present to 310 cm per ha for HYV *boro* in very severely drought-affected areas. The requirement of HYV *aus* will also

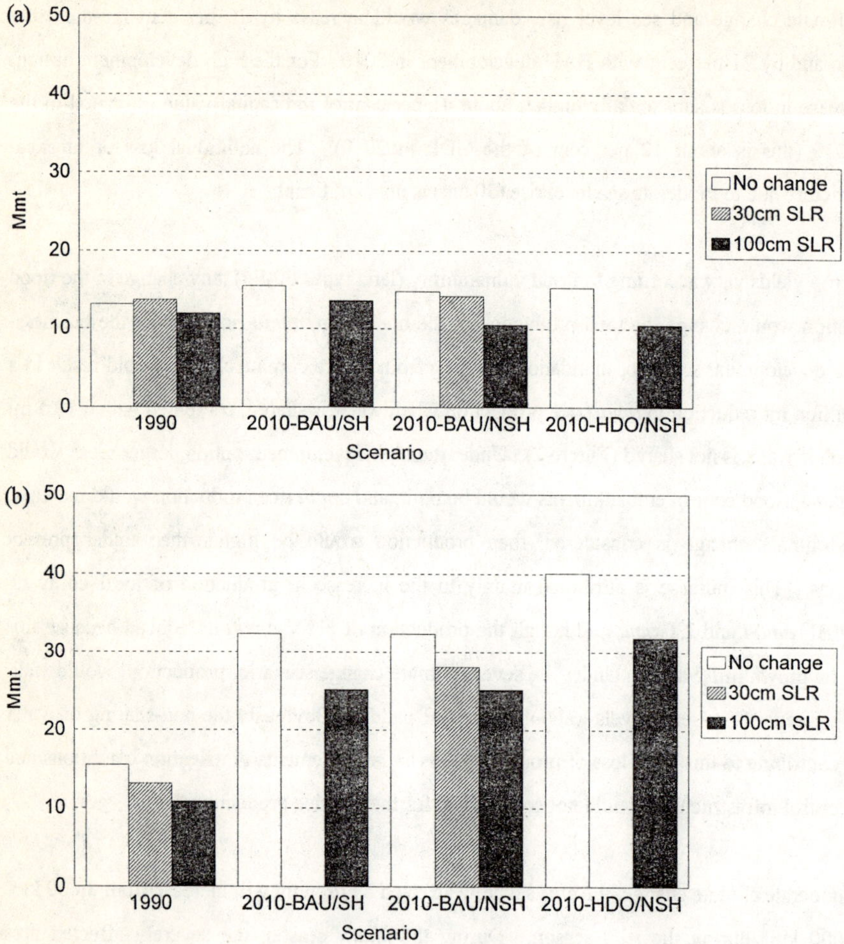

Figure 7 Impact of: (a) inundation on *aman* production; (b) drought on foodgrain production. Source: BCAS *et al*. (1994).

increase by more than 50 per cent from the present level. Under both climate change scenarios, the prospects for growing more irrigated crops will be reduced by the limited availability of irrigation water, leading to lower cropping intensities in many parts of the country.

About 12 million people (10 per cent of the population) now live in areas prone to flash floods. With climate change, these people face more frequent flash floods induced by higher precipitation in the monsoon period.

Completely controlling river bank erosion in Bangladesh presently is beyond the technical, financial and institutional capabilities of the country. Despite huge efforts to reduce erosion, about 110 km^2 of land was lost in 1993-94, making 1.2 million people homeless. Erosion is expected to increase under climate change, putting serious constraints on social and economic development in vulnerable areas.

At present, six per cent of the total population lives in areas affected by salt water intrusion, cyclones, and coastal erosion and accretion. With 100 cm rise in sea level in the severe case, an additional 25,000 km^2 would be affected by surface water salinity intrusion, affecting about 35 million people. The existing balance of coastal accretion and erosion is expected to become negative: instead of an equilibrium situation net erosion is expected.

Important natural ecosystems of international significance include the Sundarbans, which is one of the largest mangrove forests in the world; the rainforests of the Chittagong Hills bordering Myanmar; deciduous forests in the northcentral and northeast zones; peat basins; aquatic ecosystems; coastal plains and the Haor wetlands. Although the possible impacts of climate change on these natural ecosystems are highly uncertain, a qualitative analysis was compiled. Two major ecosystems, Sundarbans mangrove ecosystems and Haor wetland ecosystems, will suffer large impacts.

6 Vulnerability profiles

Vulnerability to the impact of climate change and sea level rise is not an absolute measure. It is a comparative judgement of differences between two conditions of the same development situation in Bangladesh: one with and one without climate change and sea level rise. Examples of such differences are: an 80 cm increase in high water levels; a 20 per cent extension of periods of low water flows in the Bangladesh rivers; a TK18 billion value of infrastructure affected by increase inundations; and a 1.5 per cent decrease in national income. These differences can be compared and judged on their relative importance (weighing). In addition to criteria-related impacts of climate change and sea level rise, indicators referring to the feasibility to implement responsive measures have been included.

The weighted indicators for comparison between scenarios is called a vulnerability profile (VP). Such a profile could be related either to the present or to future development in Bangladesh. Overall vulnerability is expressed in terms of critical, high, medium and low, analogous to the terminology of IPCC-CZMS (1992).

Five development situations have been considered: the 1990 situation; and four different options for 2010 which combine developments within Bangladesh (high and low development) and outside Bangladesh (sharing and non-sharing of water). Six pairwise comparisons were made. Estimating the differences between the cases considered in each pair and weighing these differences resulted in a vulnerability profile for the selected pair. Four out of these six comparisons include vulnerability profiles for a severe climate change scenario (including a 100 cm sea level rise); two present a sensitivity analysis for a moderate climate change scenario. These vulnerability profiles enable a ranking of development conditions in terms of their vulnerability to climate change and sea level rise.

Weighted differences have been combined into aggregated indices. For example, an index for vulnerability due to inundation combines differences in: flood vulnerable area; average flood depths in unprotected areas; risk of flooding in protected areas. Further weighting and aggregation leads to vulnerability indices for the main impact categories – primary physical effects, natural ecosystems, socio-economic stress and implementation of responsibilities. Through a similar procedure the vulnerability indices for the main categories are added to an overall vulnerability index for individual VA-zones. As a final step, these indices are combined into a vulnerability index for the country as a whole. These vulnerability indices are calculated using a multi-criteria analysis technique, the analytical hierarchy process (AHP), which enables both quantitative and qualitative scores to be used. This approach does not yield physically or economically meaningful values but allows ranking and evaluation of different development options with respect to their vulnerability to climate change and sea level rise.

Results are presented graphically in Figure 8, showing the vulnerability indices per VA-zone. These indices are expressed as percentages of maximum scores (the sum of the maximum values for individual indicators). The graph shows a similar vulnerability profile for four situations, the severe climate change scenario for the present, BAU/non-sharing, BAU/sharing, and HDO/non-

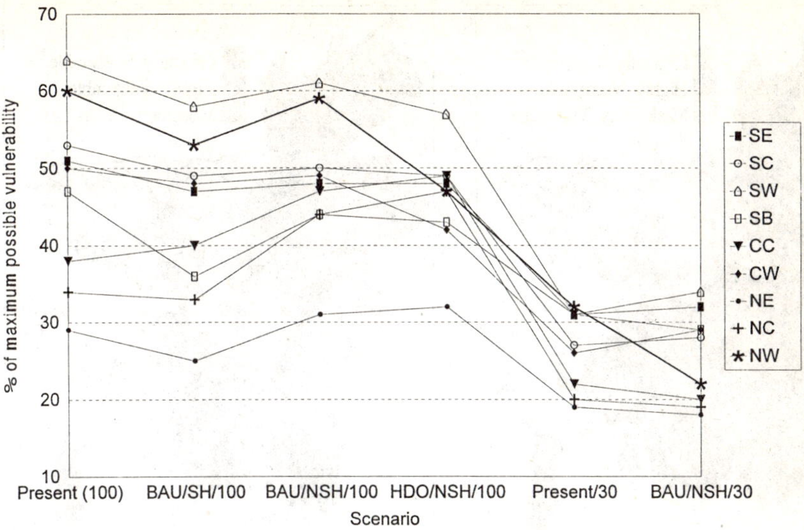

Figure 8 Vulnerability per VA-Zone. Source: BCAS *et al.* (1994).

sharing. This implies that the country remains at the same level of vulnerability as in 1990, irrespective of the expected developments within and outside of Bangladesh.

Figures 9 and 10 show an aggregate vulnerability profile for severe climate change comprised of different development options. In these maps, a score of more than 50 per cent of the maximum possible score has been classified critical, between 25 and 50 per cent is highly vulnerable and below 25 per cent is moderately vulnerable. From these results it can be concluded that a substantial part of the country scores around 50 per cent of the maximum possible vulnerability for most of the considered aspects. The southwest and northwest zones are above average vulnerability, and the northeast zone below average.

It is stressed that these vulnerability indices refer to climate changes relative to the present. The indices therefore refer to relative changes in addition to changes due to development in- and outside of Bangladesh. Major changes and adverse impacts, e.g., on population affected by inundations and the availability of water in the dry period are already expected under the high development and non-sharing option. The climate change impacts add relatively little to these impacts. In some aspects this may result in low vulnerability to climate change and sea level rise. For example, if ecosystems are expected to disappear anyway, they are not vulnerable to further impacts of climate change.

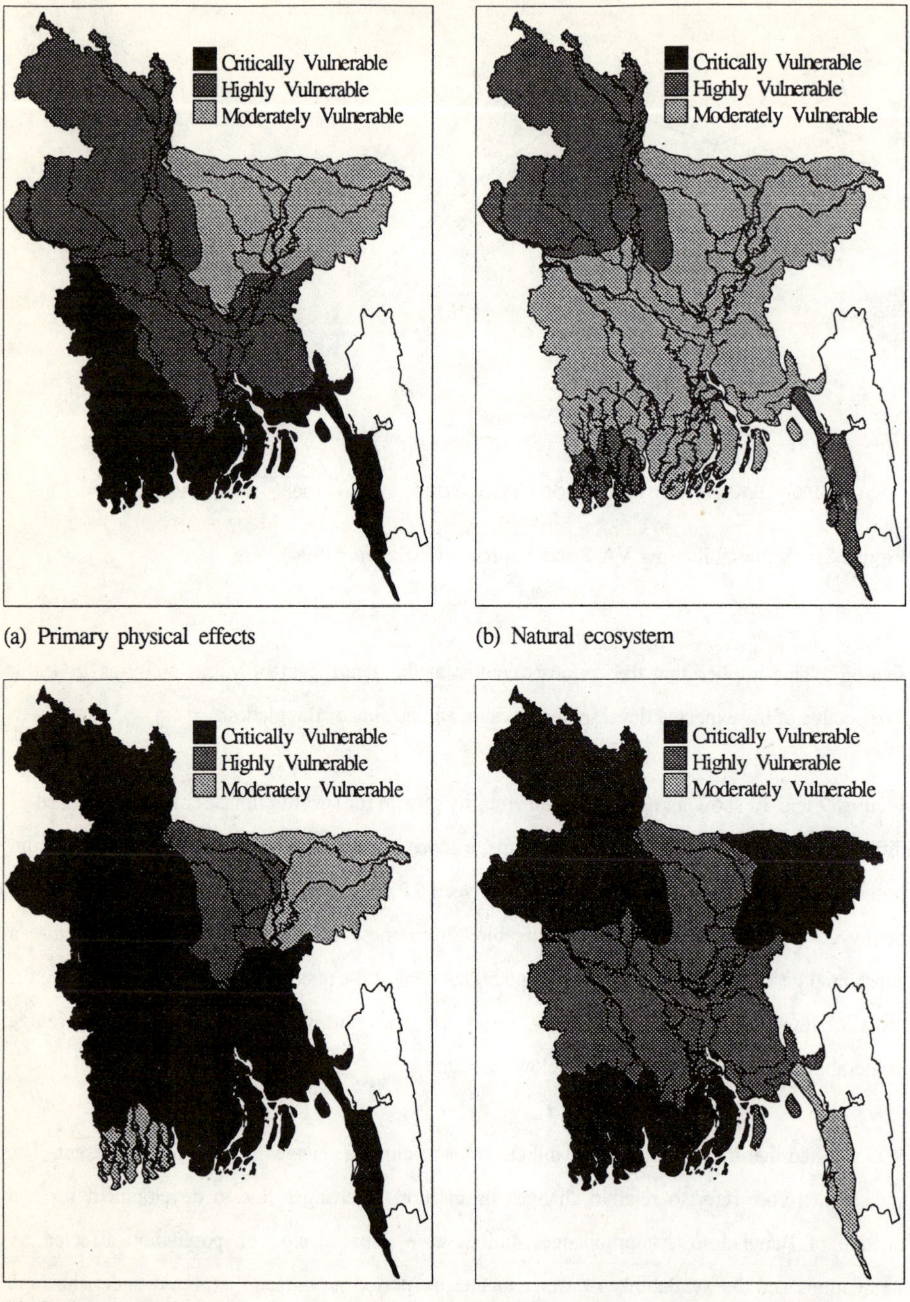

(a) Primary physical effects (b) Natural ecosystem

(c) Socio– economic stress (d) Implementation of response strategies

Figure 9 Vulnerability of Bangladesh to severe climate change. Source: BCAS *et al.* (1994).

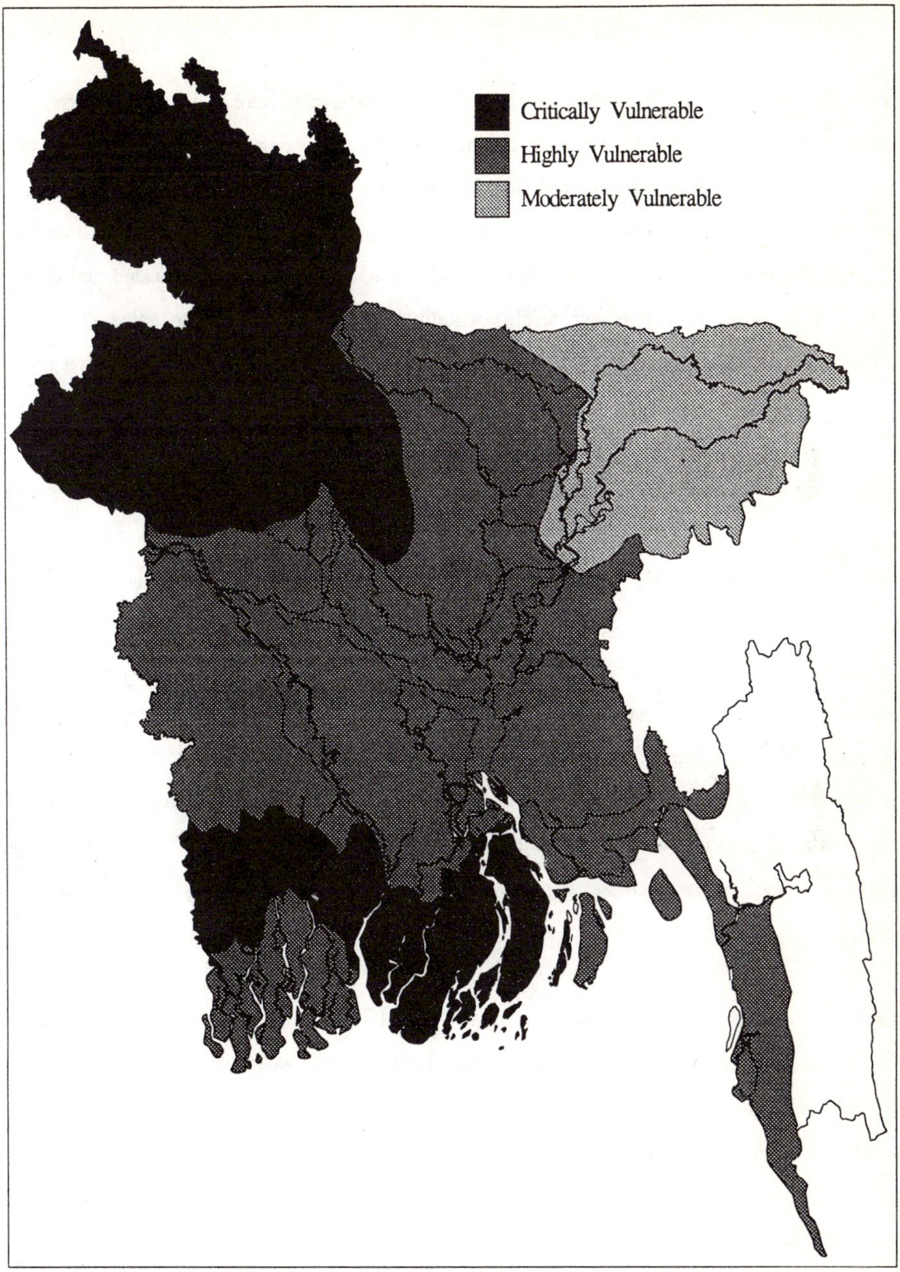

Figure 10 Overall vulnerability of Bangladesh to severe climate change. Source: BCAS *et al.* (1994).

7 Response strategies

The coastal zone of Bangladesh comprises a rich and complicated mixture of different ecosystems, agricultural practices, industrial activities, fishing habitats and navigation and other activities. All these activities and the natural resources on which they are based need to be developed in a way that allows the maximum possible benefit to the largest number of people, without degrading the resources themselves. Available human resources, local traditions and complicated and often inadequate institutional arrangements create possibilities as well as limitations to solve the short range problems of such developments. The impacts of possible climate change and other long term changes on this already complex problem of resource utilisation, makes it even more imperative to develop integrated approaches, technically, economically, socially and institutionally.

As agreed in the declaration of the World Coast Conference in Noordwijk, The Netherlands (IPCC, 1994) and in the Agenda 21 at UNCED in 1992, the need for countries with low lying coastal areas to introduce and implement integrated coastal zone management (ICZM) approaches is imperative to prepare for and to mitigate the impacts of climate change and sea level rise (CZM Centre, 1994). An integrated approach enhances sustainable development. It enables assessment of coastal zone development to reduce vulnerability to the impacts of climate change and sea level rise. Another important aim to place climate change in the proper context of other long range problems – such as the decrease of low water flows in the Ganges – and day-to-day urgent decisions.

Response strategies include reducing greenhouse gas (GHG) emissions and adaptation to changes. Limiting carbon emissions to the level of 1990, which is recommended by the FCCC, not only depends on the government of Bangladesh, but also on other governments. In this case, countries which have high per capita emission of GHGs may act without further delay. In order to limit the emission of GHGs for Bangladesh, a complete GHG inventory is required. Also necessary is an analysis of energy use and energy-related emissions so that the future development of Bangladesh would not be deterred.

Unlike limiting emissions, the adaptive response strategies depend largely on Bangladesh's own capability in terms of finance, technology and human resources. The IPCC/CZM Subgroup (1992) identified three different adaptive measures.

- Retreat: resettle inhabitants, abandon structures in currently developed areas, and require that any new development be set back specific distances from the shore, as appropriate.
- Accommodate: continue to occupy vulnerable areas but accept the greater degree of flooding (e.g. convert farms to fish ponds).
- Protect: defend vulnerable areas, especially population centres' economic activities, and natural resources. This includes building dikes, embankments, sea walls, etc.

Retreat in Bangladesh is not a practical solution, because of the country's high population density. Relocation of people within the country is not acceptable socially. Migration to other countries, especially ones with high resource potential, could only be possible with the consent of the international community.

Accommodation to climate change and sea level rise has been the traditional response to natural hazards, particularly along the coast and rivers. People have already invented many modalities to adapt to hazardous conditions. Based upon this knowledge and participatory planning, appropriate development strategies may be formulated. Possible strategies include developing crop varieties and cropping practices that are adapted to climate change including salinity and drought tolerance. The possibility of brackish water aquaculture may also be experimented with.

In the case of cyclones, the best strategy seems to be to improve forecasting and early warning systems in combination with easily accessible and strong cyclone shelters in sufficient quantities to enable the cyclone-prone population to take shelter. In order to supplement the existing disaster management programme, the institutional capabilities should be further strengthened.

While accommodation strategies need to be pursued, it must be recognised that they only allow the country to live with the problems of climate change and mitigate but not solve or eliminate the problems themselves. Thus it would make sense for Bangladesh to make a major effort in the international arena to ensure that the impacts of future climate change are minimised as much as possible. This will mean the speedy adoption and implementation of international conventions, including the Montreal Protocol and the Framework Convention of Climate Change.

For some parts of the country, such as the Sundarbans mangrove forest, the best coastal protection measure is presently to preserve the protecting mangrove forests as much as possible to allow for future (inland) migration with accelerated sea level rise. In other areas, as in the south central and south east, it would be possible to build embankments or to increase the height of existing embankments to prevent inundation. But it is not possible to prevent intrusion of salt water into the waterways and through capillary action into the groundwater. One possible barrier to such salt water intrusion might be the protection and use of freshwater wetlands as barriers to salt water intrusions, as has been done in other parts of the world.

In the case of cities such as Dhaka, Chittagong and Mongla port, it is worthwhile to build dikes and sea walls to prevent the worst damage from storms and sea water intrusion. Also for the area of poldered agriculture and aquaculture, small-scale embankments may be erected to supplement existing schemes. However, implementing these protective measures would require substantial financial and technical support.

Proper institutional arrangements are at the kernel of ICZM. It is crucial to gather all actors in a proper institutional structure and to submit them to a set of common procedures for the planning and implementation of coastal zone developments. Before starting to improve present management and institutional arrangement in order to facilitate ICZM, it is necessary to analyse the capacity of the present institutions and identify the possibilities and possible bottlenecks for developing and implementing ICZM activities.

The major institutional recommendations are as follows.

Need for a coherent administrative system. Bangladesh is yet to develop systems of government and a people-oriented development/administrative set-up.

Need for a sustained focal institutional framework for development. Bangladesh has not been able to identify and sustain local institutional networks for spearheading and implementing development and policy decisions.

Centralised system of planning. All development schemes are prepared and approved in the capital. Other administrative units such as divisions/districts/thanas lack the power, human resources and the funds to undertake substantial development schemes at the local level.

Need for a legal framework for planning. The Planning Commission and other related agencies have not been created by law but by administrative decisions and orders. Planning is not a mandatory exercise.

Effective mechanism for implementation, coordination, monitoring and evaluation. The bureaucratic bottlenecks and inefficiency of the implementing agencies handicaps implementation of development programmes. Existing programmes should allow for people's evaluation and feedback.

Proper land use planning and effective cost recovery systems. Such plans maximise resource use and efficiency.

Bias in favour of the government sector. Planning and related administrative and financial infrastructures are highly biased in favour of the government sector. It is necessary to identify the development actors from all sectors (including the information sector) and provide them with necessary support to implement development programmes.

The role of the private sector. The private sector's contribution to overall economic development has been marginal. Unless the private sector plays a more dominant and effective role the country may not experience a real change in its current rate of development.

Bias in favour of infrastructure. In most development initiatives the building of roads, culverts, embankments, etc. are given more emphasis than development of human resources, health, education, agriculture and water resources. Resource allocation should be prioritised.

Dependence on donor funding. Such dependence makes development costly, often irrelevant, and unsustainable in many cases. The overall development strategies, therefore often fail to reflect the people's need and priorities.

People's participation. The people at large are bypassed in the overall process of development planning, monitoring and evaluation. Total participation by the people at all stages of development is essential to make it appropriate and efficient.

8 Conclusion and recommendations

The main conclusions are as follows.

- Climate change and sea level rise will affect the whole of Bangladesh and not only the coastal areas. Main impact categories are: inundation; drought; salt water intrusion; and low flow conditions.

- Bangladesh, with 1990 conditions of development, is highly to critically vulnerable to climate change and sea level rise. The main part of the country scores around 50 per cent of the maximum possible vulnerability for most of aspects.
- Bangladesh remains highly vulnerable to climate change and sea level rise, irrespective of future developments in and outside the country. With regard to the primary physical effects, the 1990 situation seems slightly more vulnerable than the three future scenarios while the impacts on the natural ecosystems of non-sharing of water result in slightly higher vulnerability. Differences in aggregated socioeconomic conditions and the feasibility of response strategies are negligible.
- Development in Bangladesh and in the upstream watersheds of the main rivers bring about changes in Bangladesh which are similar to those induced by severe climate changes. The country seems particularly vulnerable to a combination of a high development option in Bangladesh and non-sharing of water from the upstream watersheds.
- The combination of high development, non-sharing of water and severe climate change could bring the country to a critical situation. In particular, the water availability has major consequences on the country's potential for foodgrain production.
- The combination of river flows and water levels is non-linear. For example, the 2010 HDO/NSH and severe climate change scenario results in an average increase of the flood depth by 70 cm whereas the accumulation of individual impacts would amount to only 35 cm.
- Another non-linear combination of effects is for the water balance in the dry period. More irrigation water will be required to compensate for expected changes in evaporation and precipitation, while the availability of water will be substantially reduced due to lower river flows in the dry period and increased salt water intrusion.
- The high development option just compensates for the adverse impacts of climate change on foodgrain production. Improved management practices are expected to bring total production from 32 million mt to about 40 million mt per year. Reduction of production due to drought induced by climate change is expected to be in the same order of magnitude.
- The regional distribution of possible impacts of climate change and sea level rise is dominated by changes in sea level, river discharges and cyclone intensity, resulting in impacts which differ in north-south directions; and temperature, evaporation and precipitation changes, resulting in impacts which differ mainly in a west-east direction.
- The retreat option is only feasible in the southeast zone. Accommodation seems the most appropriate response in the southeast and the coastal zones, where development is highest.

The combination of these effects put the southwest and northwest zones into the highest vulnerability class and the northeast zone in the lowest.

The above conclusions are based on rough assumptions, a limited set of impacts and subjective judgements on vulnerability classifications. Moreover, the flexibility of society to adapt to gradual changes and future technological developments help to mitigate the predicted impacts. On the other hand, important adverse impacts have not yet been studied. These include: salt water intrusion into the surface water and groundwater systems and the influence on soil salinity; and the combination of effects during the dry period of increased water demand for irrigation and the reduced surface water availability during low flow conditions.

In order for Bangladesh to contain and manage vulnerability to the possible impacts of climate change as well as other long term regional and national developments a number of activities have to be undertaken. The framework developed and applied in this study has the potential to become a powerful planning tool to support strategic decision making in Bangladesh. Existing models and available date are sufficient and are able to produce useful information on possible long term impacts of human-induced developments in and out Bangladesh and of changes in climate and sea levels. The Interministerial Government Review Committee on Climate Change – chaired by the Ministry of Environment – played a crucial role in the coordination of the study and the presentation and dissemination of its results and findings. It is strongly recommended to further institutionalise the review committee's role and function and to provide it with operational technical assistance. One of its important functions should be to generate information in support of important planning decisions.

Effective adaptive response to the threat of climate change and sea level rise includes such measures as: the creation of support and extension services to improve or change agricultural practices, negotiations on water-sharing arrangements with India; effective mechanisms for disaster management; development and introduction of desalination techniques; and the plantation of mangrove protection belts. Such measures include many different sectors of the Bangladesh society and should be part of integrated coastal zone management.

Integrated coastal zone management requires operational, institutional arrangements. In particular in training and capacity building, international cooperation and stimulation of transfer of knowledge is urgently required.

Regional cooperation between the riparian countries of the basin is essential to ensure a fair share of freshwater flows, particularly during the dry season. A regional approach to mitigating climate change impacts would be beneficial.

Mitigation strategies in Bangladesh refer to accommodation mainly on a regional scale. Vulnerability studies up to now have concentrated on impacts for the country as a whole or for specified regions. Indicators developed in this study do not differentiate between different stakeholders or social impact categories. Further study is recommended which accounts for the vulnerability of different social groups, e.g. landless people, women and urban poor. A possible approach for such a study could be: (i) identify in a specified area the relevant stakeholders; (ii) formulate their needs with respect to the natural resources, e.g., freshwater, soil conditions and mangrove protection; (iii) establish the possible impacts of climate change and other developments on the availability and accessibility of these resources; (iv) formulate possible measures to enhance and guarantee access to these resources for different stakeholders; and (v) evaluate the feasibility to implement these measures. It is important to relate the vulnerability of different social groups with other types of vulnerability, such as food shortage and poor ambient environmental quality. From the present study, the food security of vulnerable socioeconomic classes is clearly threatened by climate change. If the high development option reduces their vulnerability, for instance to cyclones and erosion, their prospects may dramatically improve. Unfortunately, the business-as-usual option appears more likely, and less optimistic.

Acknowledgements

This paper is based on a summary report by the Department of the Environment (BCAS, RA, CZM-Centre and Approtech, 1994). The report was prepared on behalf of the Department of Environment (DoE) of the Ministry of Environment and Forests (MOEF) by a multidisciplinary team of Bangladeshi and Dutch experts organised by the Bangladesh Centre for Advanced Studies and Resource Analysis from The Netherlands. The study was carried out as input to the Intergovernmental Panel on Climate Change (IPCC) with support from the Government of The Netherlands, Directorate General for International Cooperation (DGIS) of the Ministry of Foreign Affairs, and the Coastal Zone Management Centre of the Ministry of Transport, Public Works and Water Management. Their support, and the editorial assistance of the Environmental Change Unit, are gratefully acknowledged.

References

BCAS (Bangladesh Centre for Advanced Studies), Resource Analysis (RA), Coastal Zone Management (CZM) Centre and Approach Consultants Limited (Approtech) (1994). *Vulnerability of Bangladesh to Climate Change and Sea Level Rise: Concepts and Tools for Calculating Risk in Integrated Coastal Zone Management.* Technical report, Volumes I and II. BCAS, RA, CZM and Approtech, Dhaka, Bangladesh.

BIDS (Bangladesh Institute of Development Studies) (1993). *Regional Study of Global Environmental Issues Project*, Country Study on Bangladesh, BIDS, Dhaka.

BUP (Bangladesh Unnayan Parishad)/Centre for Environmental and Resource Studies / University of East Anglia (1994). *Bangladesh: Greenhouse Effect and Climate Change*, Briefing Documents, No. 1-7, BUP, Dhaka, Bangladesh; CEARS, University of Waikato, Hamilton, and Climate Research Unit, Norwich.

BCAS (Bangladesh Centre for Advanced Studies) (1994). *Wetlands of Bangladesh.* Khan, M.S., Haq, E., Huq, S., Rahman, A.A., Rashid, S.M.A. and Ahmed, H. (eds.). Bangladesh Centre for Advanced Studies and Nature Conservation Movement, Dhaka.

CZM (Coastal Zone Management) Centre (1994). *Common Framework for Integrated Coastal Zone Management: A Communication Paper for Support and Participation.* Coastal Zone Management Centre, Dhaka.

Houghton, J.T., Callander, B.A. and Varney, S.K. (eds.) (1992). *Climate Change 1992: The Supplementary Report to the IPCC Scientific Assessment*, Cambridge University Press, Cambridge.

Houghton, J.T., Jenkins, G.J. and Ephraums, J.J. (eds.) (1990). *Climate Change: The IPCC Scientific Assessment*, Cambridge University Press, Cambridge.

IPCC (Intergovernmental Panel on Climate Change) (1994). *Preparing to Meet the Coastal Challenges of the 21st Century.* Conference Report, World Coast Conference 1993. Ministry of Transport, Public Works and Water Management, National Institute of Coastal and Marine Management/RIKZ and Coastal Zone Management Centre, The Hague.

IPCC (Intergovernmental Panel on Climate Change), WHO (World Meteorological Organisation) and UNEP (United Nations Environment Programme) (1992). *Climate Change: The IPCC 1990 and 1992 Assessments.* IPCC, WHO and UNEP, Geneva.

Intergovernmental Panel on Climate Change - Coastal Zone Management Subgroup (1992). *Global Climate Change and the Rising Challenge of the Sea.* Report of the Coastal Zone Management Subgroup; supporting document for the IPCC-update report 1992. Tidal Waters Division, Rijkswaterstaat, Ministry of Transport, Public Works and Water Management, The Hague.

Karim, Z. (1995). Agricultural vulnerability and poverty alleviation in Bangladesh. In: Downing, T.E. (ed.). *Climate Change and World Food Security*. Springer-Verlag, Heidelberg.

Mahtab, F.U., and Karim, Z. (1992). Population and agricultural land use: towards a sustainable food production system in Bangladesh. *Ambio,* **21:** (1), pp 50-55.

Task Force (1991). *Report of the Task Force on Bangladesh Development Strategies for the 1990s*, Volume I to IV. Government of Bangladesh. United Press Limited, Dhaka.

World Bank (1993). *Bangladesh; Implementing Structural Reform*, South Asia Country Department 1, World Bank, Dhaka.

WWF (World Wide Fund for Nature) (1992). *Climate Change Due to Greenhouse Effect and Its Implications for China.* WWF, Gland, Switzerland

Enhancing Food Security in a Warmer and More Crowded World: Factors and Processes in Fragile Zones

Narpat S. Jodha
Social Policy and Resettlement Division
Environment Department
World Bank
1818 H Street NW
Washington DC 20433, U.S.

1 Introduction

World food security is largely a problem of poor people and poor areas that do not have the resources to ensure dependable access to sufficient quantity and quality of food through production or exchange (Wisner, 1988). This is more so in areas with fragile and marginal natural resources that offer limited and undependable agricultural production and exchange opportunities for people to acquire command over their food requirements (Sen, 1981). The bulk of the mountain/hill regions and dry tropical areas in the developing countries qualify for this status. In fact, if one excludes the political strife-torn areas with high food potential, most charitable donations of food are directed to such areas.

This paper deals with the food security problem in such fragile areas. It illustrates the situation by referring to the mountain/hill areas and the dry tropical areas of south Asia. Instead of presenting a statistical profile of the food situation in the fragile zones, the paper comments on the processes and factors underlying the dynamics of food security/insecurity in such areas. It highlights the emerging scarcity scenarios and indicates the role of population growth and resource degradation therein. By identifying the elements and mechanisms of current scarcity-generating processes and their possible accentuation in the future due to still higher population and potential impacts of warming, the paper indicates the links between the present and future crises influencing food security in fragile resource zones. A dual-purpose approach to the present and future situations is suggested, focusing on measures directed towards enhancing present food security and which may also help strengthen the communities' capacity to withstand future food shortages. This is possible because the factors and processes underlying the two situations have several commonalities. This dual-purpose strategy for enhanced food security must be the integral part of the overall sustainable

NATO ASI Series, Vol. I 37
Climate Change and World Food Security
Edited by Thomas E. Downing

development approach for fragile areas. The basic ingredient for such a strategy can be compiled by looking at the failures of past development interventions that disregarded the imperatives of the resource characteristics of fragile areas; the indigenous production-cum-resource management systems focusing on biomass stability, diversification, resource regenerative processes etc., and the recent successful initiatives that helped identify and harness the fragile areas' niche through equitable inter-regional linkages. The latter has enhanced the range and quality of production/consumption options without undue negative side-effects on the environmental resources of the fragile resource zones. The discussion draws on micro-level studies conducted in India, Nepal, China (West Sichuan and Tibet) and Pakistan with which the author had close association.

2 Focus on fragile resource zones

A fragile resource is one that cannot tolerate the degree of disturbance implied by the intensity of use associated with a specific activity. Thus, strictly speaking, fragility is a relative or context-specific term. Every land resource is fragile, i.e. vulnerable to irreversible damage, when subjected to a degree of use intensity higher than its use capability. Thus the land belonging to use-capability classes IV and above, although good for land-extensive uses, is fragile when assessed with reference to the use intensity implied by the intensive cropping associated with prime agricultural lands (i.e. fertile, well-drained lands with even topography and stable climatic conditions conducive to crop farming).

There may be several other ways to look at the phenomenon of fragility (Turner and Benjamin, 1993; DESFIL, 1988). Besides describing fragility in terms of vulnerability to irreversible damage by higher use intensity (DESFIL, 1988), one can describe it in terms of: low input absorption capacity of the resources; limited scope for resource manipulation; and required high level of biochemical subsidisation of the natural resource to achieve a level of output comparable with that from better land resources. The phenomenon can also be expressed in terms of input:output ratios, where the fragile lands have higher than average ratios. Described in this way, all areas with low potential for crop farming (mountain regions with steep slopes; rain-fed arid and semi-arid tropical areas with undulating topography, low and unstable rainfall; and coastal areas prone to salinity and waterlogging etc.) will fall into the category of fragile resource areas (or fragile areas). Despite their apparent differences, for operational purposes, fragility and its associated attributes impart a broad degree of similarity, if not exact homogeneity, to these areas. This facilitates their consideration as a system in the context of which food security issues can be understood and analysed. In many parts of these areas, under the present pattern of resource use, the threshold limits to maintenance or enhancement of agricultural performance, even by using the inter-regional linkages, seem to have been reached. Further efforts to improve output levels imply over-exploitation of their biophysical resource base and initiation of irreversible processes of resource degradation (Glantz, 1987; Nelson, 1988; and Allan *et al.,* 1988). These areas represent critical zones, where the unsustainability of present patterns of resource use seems to have become an objective reality (Anderson and Jodha, 1994). Their production prospects and output levels, on a per capita basis and in most cases on a per production-unit basis, have declined. Thus, in these habitats, one can observe the visible symptoms of serious food deficit.

Although there are many fragile resource areas, the following discussion relates to: (i) the mountain areas, particularly lower and middle mountains in Himalayas where annual cropping is one of the land-based activities; and (ii) the rain-fed areas in the arid and semi-arid tropical regions (also referred to as dry tropical areas) in India. The mountain regions and the dry tropical regions have several major differences – hence, when necessary, I will refer to the two areas separately. Wherever they share common characteristics or descriptions I will treat them together and refer to them as fragile zones, marginal areas etc. Despite the domination of fragile land resources, these regions also have substantial pockets where fragility is not a problem. The analysis excludes such high-productivity, transformed areas within the mountain and dry tropical regions. Thus the emphasis is more on specific agro-ecosystems rather than geographical or administrative units, where fragile and non-fragile agro-ecosystems coexist.

To the extent that the essence of food security is dependent on production and exchange opportunities, the fragile resource zones constitute a situation with few such opportunities. The biophysical constraints and the side-effects of recent socioeconomic development (partly induced by the former) tend to obstruct the circumstances that could enhance the range of production and exchange options.

The focus of the paper, viz. 'enhancing food security in a warmer and more crowded world', adds to the appropriateness of discussion of fragile areas. Due to fragility and marginality of resources (compared with prime agricultural lands), and their lower carrying capacities, these areas are already overcrowded, and the consequences include emerging food insecurity. Most of these areas already periodically suffer food deficits and account for the bulk of food transfers under charity/relief arrangements. In fact, their current situation can enlighten us about what a 'more crowded world' means and how difficult it is to handle its consequences. This, in turn, may offer some clues to future strategies for enhancing food security in areas with food population imbalances.

Furthermore, most of these areas represent agro-ecological margins (Parry, 1985) where a slight disturbance to their supplies, directly through disruption of their own production systems or indirectly through changes in their external sources of food supplies, due to global warming or other environmental changes, can create a crisis or future food insecurity. Therefore, the fragile resource areas should be prime candidates for the attention of policy-makers. Thus, the low carrying capacity (and overcrowding) and the vulnerability to impacts of future changes (including changes due to global warming) put the fragile areas at the centre of the world food security debate.

Another important factor is the complex invisibility characterising the fragile resource zones. To begin with, due to their inaccessibility and marginal status, their problems do not receive as much – or as early – attention as the poor urban areas with vocal groups and ready media coverage. Even the factors and processes (often emerging as the side-effects of public interventions) accentuating food insecurity in these areas remain invisible (Sanwal, 1989; Jodha, 1991a). The same applies to the off-site impacts of their poverty, misery and food scarcity. For example, poverty-induced degradation of mountain resources affects the downstream areas; degradation of dry tropics adds to the spread and march of desert to neighbouring better-endowed areas. Problems created by environmental refugees from famine-affected areas are well known.

Finally, and most importantly, the conventional green revolution type of approaches to increased food production have not succeeded in most of these areas. However, the components of traditional food security and survival strategies and their rationale constitute a reserve of the traditional wisdom, which could be profitably incorporated into new approaches to the enhancement of food security. This remains invisible or unrecognised because of the ignorance, arrogance and paternalistic approach of the mainstream decision-makers and their biased perceptions (Jodha and Partap, 1993). The enhanced visibility of the constraints and potential of these areas could be a first step in enhancing the food security in the fragile resource zones. However, policy-makers need to reorient their perceptions of the fragile resource areas.

3 Food insecurity in fragile resource zones: the biophysical context

The basic premise of this paper is that food security (both in terms of accessibility and availability of required quantity and quality of food), in the ultimate sense, is a function of production and exchange options available to the people (Sen, 1981; see Bohle, Downing and Watts, 1995, this volume). The range and quality of such options are conditioned by the biophysical attributes of a region's resource base and the pattern of activities (for both production and exchange) it induces or imposes on its inhabitants. The intra-regional differences in the availability of such options (for different groups and individuals) is an equally important aspect of food security (Kates and Millman, 1990). However, this paper is confined to food security in the agro-ecosystem only and hence intra-regional dimensions are not discussed.

The biophysical context of production and exchange options in the fragile resource zones can be understood in terms of the specific characteristics and conditions of these areas, which, in the operational context, separate them from other areas, particularly the prime agricultural areas with highly productive soils, even topography, dependable moisture, and sufficient man-made support facilities to harness their potential. These conditions and characteristics can be listed as inaccessibility (more in mountains than in dry tropical plains), fragility, marginality, diversity (or internal heterogeneity – again more in the mountains than in the dry plains), and niche or conditions/products imparting potential for comparative advantages to these regions. No doubt these characteristics are not specific to fragile resource zones alone, but apply to any region in some measure. However, their degree is incredibly high (compared with the prime agricultural areas) and they have a decisive impact on the production performance of these areas. They are a product of various biophysical factors. Due to the commonality of the biophysical bases and consequent degree of externalities (i.e. spill-over effects of disturbing each other), they are closely interlinked. No doubt not all mountain areas or dry tropical areas have the same degree of any of the above characteristics.

The detailed description of these conditions along with their operational implication is presented elsewhere (Jodha, 1991a; Jodha *et al.*, 1992). In the following discussion, we will focus on them as they obstruct or enhance the circumstances conducive to food security through access to production and exchange opportunities.

3.1 Inaccessibility

Due to slope, altitude, overall terrain conditions and periodical, seasonal hazards (e.g. landslides, snow storms etc.), inaccessibility is the best-known feature of mountain areas (Price, 1981; Hewitt, 1988). In the dry tropical areas, inaccessibility is not of the same order as in the mountains, but it is important when compared with high-productivity, well-watered agricultural areas (Jodha, 1986a). The concrete manifestations of inaccessibility are isolation, distance, poor communications and limited mobility with all their sociocultural and economic implications. The mountains and the dry tropics share these manifestations and implications more significantly than the physical dimensions of inaccessibility. Inaccessibility implies high cost of support systems, impeded exchange linkages, and limited dependability of external support in a semi-closed system. From the food security point of view, these constraints limit the physical production and exchange opportunities. The advantages of scale and technology are restricted. The exchange relations become inequitable. The people have to depend mainly on *in-situ* arrangements for food security, including local resource-centred diversification of the food chain, sharing, recycling etc.

3.2 Fragility

Fragility is the dominant characteristic on the basis of which the mountain areas and the dry tropical areas are chosen for the present discussion. It emanates from the combined operation of slope/altitude or undulating topography as well as geologic, edaphic and biotic factors. Nature's low regenerative capacities and delicate ecological balance are equally important, especially in the dry tropics. Notwithstanding the differences in the relative role of the specific factors in the mountains and the dry tropics, these factors limit the capacities of land resources to withstand even a small degree of disturbance (DESFIL, 1988). Vulnerability to irreversible damage due to overuse or rapid changes extends to physical land surface, vegetative resources, and even the delicate economic life-support systems of the dependent communities (Turner and Benjamin, 1993). Consequently, when resources and environment start deteriorating due to disturbance, they do so rapidly. In most cases, the damage is irreversible or reversible only over a long period (Eckholm, 1975; Hewitt, 1988). Fragility complements inaccessibility and marginality in restricting the range and quality of production, exchange and consumption options by preventing resource-use intensification and high productivity. Fragility, in a way, debars the areas under reference from modern input-intensive technologies to raise food production.

3.3 Marginality

Marginality is another characteristic of these areas that is directly related to fragility. A marginal entity (in any context) is one that counts the least with reference to the mainstream situation. This may apply to physical and biological resources or conditions as well as to the people and their sustenance systems. The basic factors contributing to such a status of any area or community are remoteness and physical isolation, fragile and low-productivity resources, and several man-made handicaps that prevent participation in the mainstream activities (Chambers, 1987; Blaikie and Brookfield, 1987). The mountains and dry tropical plains, being largely marginal areas compared with the prime agricultural regions, share the above attributes of marginal entities and bear the consequences in different ways (Bjonnes, 1983; Jodha *et al.*, 1988). Marginality shares with fragility a number of implications. Besides constraining the range and quality of production options, it reduces the people's capacity to undertake and benefit from high-cost, high-productivity opportunities. Due to marginality, the inter-system linkages make the terms of exchange unfavourable to these areas and their people; the mainstream decision-makers ignore them and treat them as national liabilities.

3.4 Diversity or heterogeneity

The fragile areas in question are internally not homogeneous even in terms of the degree and nature of fragility or marginality. As mentioned earlier, all the characteristics listed here have considerable internal diversity. In the mountain areas one finds immense variations among and within eco-zones, even over short distances. This extreme degree of heterogeneity is a function of interactions of different factors ranging from elevation and altitude to geologic and edaphic conditions (Troll, 1988). In the arid and semi-arid tropical plains, the degree of diversity (although less than in the mountains) is primarily because of topography, soil and precipitation differences (Dixon *et al.*, 1989). Water – a homogenising factor – being limited, the diversity due to other features of land resources persists. The biological adaptations (Dahlberg, 1987) and socioeconomic responses to the above diversities (Price, 1981; Jochim, 1981; Nogaard, 1984) also acquire a measure of heterogeneity of their own. Diversity serves as a source of a complex mixture of constraints and opportunities characterising these areas. As a positive attribute, by supporting interlinked activity patterns, diversity can help enhance the food security in these areas. At the same time, diversity implies a greater degree of location specificity of production activities, which reduces the gains associated with specialisation and scale of activity. It restricts the applicability of generalised (technological and institutional) measures with a record of success elsewhere (Jodha, 1991a).

3.5 Niche or comparative advantage

Because of their specific environmental and resource-related features, both the mountain and the dry tropics provide a niche for specific activities or products (Banskota and Jodha, 1992b; Jodha, 1989a). At the operational level, these areas have comparative advantages over other areas in these activities. In the mountain areas, the examples may include: specific valleys serving as habitats for specific medicinal plants; mountains acting as a source of important high-value products (e.g. fruits, flowers etc.); and mountains serving as the best-known sources for hydroelectric power production. The dry tropical plains, although less productive than the high rainfall mountain areas and having some similarities with other plain areas, may also have comparative advantages for land-extensive activities (e.g. pasture-based animal husbandry), highly moisture- (or humidity-) sensitive cultivars, such as some coarse grains etc. In practice, however, the niche or comparative advantage may remain dormant unless circumstances are created to harness them. The local communities make use of these niche through their diversified activities. Proper harnessing of niche can help in food security through directly usable products or tradeable high-value products. However, their reckless exploitation for the benefit of the mainstream economy can result in the elimination of niche. Harnessing of niche requires infrastructural facilities as well as financial and technological resources, which are lacking in fragile resource areas (Jodha, 1991b).

To sum up, the production options supporting food security in the fragile resource zones are obstructed by the overall limited physical possibilities, the limited relevance and transferability of technologies from outside and difficulties to evolve them internally, and the lack of infrastructural support. All these circumstances in turn are products of the specific biophysical conditions discussed above. The resource characteristics mentioned above also obstruct the exchange-related options that could enhance people's entitlement to food resources. The basic requirements of a successful and equitable exchange activity in these areas are by and large missing. Not only scope for surplus production and processing is limited, but also market and infrastructural support to encourage exchange activities are either missing or very weak as it is too costly to create and maintain them. The roles of inaccessibility, fragility and diversity are shown in Table 1.

Furthermore, because of these conditions, the fragile resource zones are unable to integrate themselves as equals with the mainstream (the prime production) areas and urban areas. Hence, the development programmes and processes bypass them. Due to the marginal status of these regions, *vis-à-vis* the mainstream areas and their dominant decision-makers, fragile areas continue to be neglected or selectively overexploited, when their niche could help the mainstream economies (Banskota and Jodha, 1992b). In fact, irrelevant measures are imposed on these areas, which lead to the disintegration of their production systems and survival strategies (Jodha, 1989a, 1991a, 1991b; Singh, 1992; Whitaker *et al.*, 1991).

4 Food security: the socioeconomic context

Despite limited opportunities (or ones that are not easy to harness) offered by the biophysical conditions, substantial human populations have survived and multiplied in the fragile resource zones. At a relatively low level of demand, they managed their food security through a two-way adaptation process, i.e. adapting the needs and demand (and mechanisms to satisfy them) to the constraints and opportunities offered by the natural resource base and, as far as permitted by their skills and capacities, amending the resource base (e.g. through terracing, water harvesting etc.) to satisfy the communities' growing needs. The focus was on production and enhancement of production options to meet consumption needs. A glance at the traditional farming systems and resource management practices will corroborate this (Brush, 1988; Guillet, 1983; Whiteman, 1988; Jodha, 1990a, 1990b, 1980; Jodha and Partap, 1993; Dixon *et al.*, 1989), as summarised in Table 1 (column 2). Viewed from the food

Table 1 Key elements of traditional food security arrangements in the fragile resource zones

Conditions constraining the production/exchange	Traditional measures to enhance food security opportunities (production/exchange) options
Inaccessability	
Isolation, semi-closedness, poor mobility	Local resource centred, diversified production and consumption activities
Limited access to and dependability of external support (inputs, products)	Local resource regeneration and resource use regulation/rationing (by informal institutional means)
High cost of mobility, infrastructure, support systems and production/exchange activities	Collective sharing, recycling of resources/ products, limited external links
Detrimental to harnessing of niche and gains from trade	Nature and scale of operations as permitted by degree of mobility and local resource availability (e.g. transhumance)
Fragility	
Resources highly vulnerable to rapid degradation, unsuited to high intensity/productivity uses	Resource upgrading; focus on low intensity uses
Limited, low-productivity, high-risk production options; little surplus generation	Resource / product, usage regulation
High overhead cost of resource use, obstacles to infrastructure development, under-investment	Diversification involving a mix of high- and low-intensity land uses; a mix of production and conservation measures
People's low resource capacity preventing use of high-cost, high-productivity/options	Local resource regeneration, recycling, limited external input use; dependence on nature's regenerative processes and collective measures
Marginality	
Low productivity, limited and risky production options; little surplus generation and reinvestment	Focus on security of subsistence needs, low and local input use; integration of production/ consumption activities
Resource scarcities, inability of people to benefit from high-cost, high-productivity technologies/options	Diversification, recycling, collective sharing; greater dependence on nature's processes, self provisioning
Diversity	
Heterogeneity-induced strong location specificity of production options	Small-scale, interlinked diversified production / consumption activities; temporally, spatially differentiated activities for fuller use of environment
Limited use of options involving wider applicability and benefits associated with scale	Location-specific, integrated multiple activities;stability of total system
Niche	
Unique and complex potential opportunities with comparative advantage	Specific opportunities harnessed for petty trade and self-consumption
Opportunities for production and exchange remain under-used for want of resources, infrastructure	Niche-harnessing integral part of diversified resource use

Sources: Whiteman (1988), Sanwal (1989), Pant (1935), Jochim (1981), Guillet (1983), Bjonnes (1983), Jodha (1990a, 1990b, 1991a, 1991b, 1989b, 1980, 1975), Jodha *et al.* (1988), Jodha and Singh (1990), Jodha and Partap (1993), Shrestha and Yadav (1992), Mulk (1992), Songqiao *et al.* (1990), Walker and Ryan (1990), Singh (1992).

security angle, the key dimensions of the traditional arrangements manifested by different practices include the following:

- diversification and flexibility of both food production and consumption activities making fuller use of the temporal and spatial variability of the environment and production resources;
- local resource regeneration and recycling of products and inputs to meet the constraints imposed by fragility, marginality and inaccessibility characterising such areas;
- demand management through informal resource and product rationing and collective-sharing arrangements involving various forms of group action, petty exchange, transhumance etc.; and
- resource updating and new crop introduction as permitted by slowly evolving folk-agronomic knowledge.

The literature on ethno-sciences, traditional farming systems, indigenous knowledge and traditional forms of rural cooperation, which has been receiving increasing attention in recent years, can furnish several examples of how people in the fragile resource zones attempted protection and multiplication of production/consumption options within the constraining biophysical environment. For a ready inventory of such measures, one may glance through different issues of the semi-popular, user-focused journals/newsletters, such as *Indigenous Knowledge and Development Monitor* (the Hague), *ILEIA Newsletter* (Leusden), *IIED Gatekeeper Series* (London), *Honey-Bee* (Ahmedabad), *Forests, Trees and People Newsletter* (Uppsala) and others.

4.1 The changed context

A few important preconditions associated with the traditional food security measures in the fragile areas included the following: (i) they involved adaptations to high-risk, low-productivity environments; (ii) these adaptations required an intimate knowledge and understanding of the local environment; (iii) their implementation, besides the feel of the situation, needed strict adherence to institutional norms to ensure resource-use regulation and group action; and (iv) their feasibility and efficacy were closely associated with low pressure on land resources, which permitted land-extensive production practices and acceptance of stable but low production (as in subsistence-oriented systems). However, of late, the circumstances supporting the effective use of traditional measures for maintenance and multiplications of food security options have radically changed. This has reduced their feasibility and efficacy. The changes are consequences of rapid population growth, increased extent of state interventions, and increased role of market forces (Jodha, 1985a, 1985b, 1986b, 1989a, 1991a, 1991b, 1992b).

4.2 Consequences of population growth

The rapid population growth in the last four to five decades has been a universal feature of most of the fragile resource zones. Since the 1950s, the population of most of these areas has almost doubled. For instance, the present (i.e. year 1991) population density in the hills and mountain areas of Nepal is 126 persons/km^2. The corresponding figures for areas covered by the Hindu Kush Himalayan Region in India and Pakistan are 73 and 56 respectively. The Chinese parts of the Himalayan region had a population density (mainly due to sparsely-populated Tibet) of 12 persons/km^2 for the year 1987 (Sharma and Partap, 1993). In the dry tropical areas of India, the sub-division level data (for 1981) for eight arid or semi-arid states showed a population ranging from 67 persons/km^2 in drier districts to over 300 persons/km^2 in better rainfall districts (Jodha, 1986a). The bulk of this population in these regions directly depends on agriculture. Viewed in the context of the low-carrying capacity of their resources, the present populations already represent an overcrowding of these areas.

The increased population has adversely affected the food security in two ways. Firstly, due to sheer increase in number, the per capita availability of resources has declined. For instance, despite extension of annual cropping to steep slopes and sub-marginal lands in the mountains and the dry tropical areas, the per capita availability of cultivated land has declined. In Nepal, within ten years, it has declined from 0.17 ha (1971) to 0.13 (in 1981). The corresponding figures for Himachal Pradesh (India) are 0.19 to 0.13, and the U.P. hill districts (India) 0.30 to 0.16. The per capita cultivated land (in 1981) was 0.16 ha in the NWFP (Pakistan) and 0.15 ha in West Sichuan (China) (Sharma and Partap, 1993). The per capita cultivated land in the sub-divisions of the dry tropical states of India as mentioned above declined from 0.85 ha to 0.41 ha during the 1950s to 1980s (Jodha, 1986a). Reduced land availability in the absence of productivity growth implied reduced availability of products including food. In fact, if one excludes the high potential pockets within the fragile resource zones, the average yield of several food crops has declined in many parts of the regions under review. For instance, the national statistics show that in the hills and mountains of Nepal the per hectare average yields of paddy, maize, millet and barley have declined by 19 per cent, 21 per cent, 18 per cent and 9 per cent respectively during the period 1980-87 compared with the preceding decade (Banskota, 1992). In Kumaon hills (U.P. India), a decline in the yield of maize, millet and oilseeds has also been reported (Swarup, 1991). According to village-level investigations under International Centre for Integral Mountain Development (ICIMOD)-sponsored studies, most of the mountain villages in Nepal and India showed a decline in the yields of important food crops (Shrestha, 1992; Bajracharya, 1992; Singh, 1992). For the dry tropical region, the areas that have not been able to adopt high-yielding varieties of sorghum, millet etc. have shown a decline in yield (Jodha, 1982, 1992a, 1993a).

Secondly, and more importantly, due to the increased population pressure, the measures and mechanisms directed to option maintenance/enhancement (through diversification, local-resource regeneration, collective sharing, recycling etc.) have become unfeasible or ineffective. Most of the village-level studies conducted by ICIMOD and others indicated that the extent of intercropping, the extent of crop rotations, the number of crop and combinations, and the forms of diversification and linkages have declined over time (Singh, 1992; Shrestha, 1992; Jodha and Singh, 1990; Yadav, 1992; Panday, 1992; Carson, 1992; Jodha, 1980, 1986a, 1986b; Walker *et al.*, 1983). Decline of such practices means resource depletion and reduced carrying capacity finally affecting food security. This is partly because the low man:land ratio, an extremely crucial element of these (land-extensive) practices, is reversed (Jodha, 1991b, 1986b). The increased number is accompanied by qualitative changes in the populations reflected through change in attitudes which are less conducive to group action and adherence to social sanction for collective arrangements (Sharma and Banskota, 1992; Jodha, 1993a). The latter is partly a product of increased state and market interventions, which are directed to modernisation, development etc. The net consequences of these changes are reflected through the changes in land-use practices, cropping patterns, combinations of different land-based activities etc., which in different ways violated the imperatives of resource characteristics in the fragile resource zones. Table 2 gives a broad idea of these changes and their inadequacies.

4.3 Integration and interventions

The changes witnessed by the areas under review also involved their increased physical and market integration with the mainstream economies, the state taking the responsibility for food security in these areas as a part of the welfare-cum-development programmes, and the focused interventions to raise food production and supplies. These measures definitely helped in relaxing constraints generated by resource characteristics such as fragility, inaccessibility, marginality etc. However, most of them, due to their insensitivity to the imperatives of the same characteristics, generated several negative side-effects which adversely affected different aspects of food security (Jodha, 1991a, 1991b, 1986b; Banskota and Jodha, 1992a). The discussion below focuses on public interventions that directly or indirectly affected food security through obstructing or marginalising its underlying processes.

5 The socioeconomic context: resource intensification phase

The rapid erosion of land-extensive, traditional production practices due to demographic changes was followed by a phase where search for food security implied high-use intensity of fragile land resources. The intensification (ignoring resource limitations) took place at both extensive and intensive margins. The initiation of this phase coincided with increased state interventions in the socioeconomic affairs of rural communities (Jodha, 1991b, 1992a). As a part of the nation state's newly-acquired mandate (i.e. extending its activities beyond the law and order affairs in the post-Second World War era), food security also became a part of the state's responsibility. This, in turn, has imparted several unique features to the food security strategies not found in the traditional arrangements (summarised in Table 2). An important feature of public measures directed towards food security issues is that they discard most of the key components of traditional food security measures, and have evolved and been implemented without enough sensitivity to the resource specificities of the fragile resource zones (Table 2, column 3).

The result is that, despite increased efforts, resource allocation and technological input, public interventions have not succeeded in ensuring adequate supplies to the bulk of these areas. On the contrary, the changed resource use practices followed by the farmer and encouraged by development interventions, including subsidisation, have led to the emergence of negative trends, often described as indicators of unsustainability (Jodha, 1991a; Jodha *et al.*, 1992). A

Table 2 The changing complex, orientation and components of food security strategies in the fragile resource zones.

(1) Traditional strategies	(2) Present-day strategies	(3) Implications of (2)
Orientation/context		
Community/household-operated measures, part of subsistence-oriented self-provisioning systems evolved in keeping with the local circumstances; guided by informal institutional arrangements and state of folk knowledge, focus on stability and productivity of total systems	State-administered/patronised/ encouraged activities as a part of formal, generalised development/ welfare programmes: focused on infrastructure, agricultural programmes, distribution network, public sector investment, large-scale projects to raise employment, income, productivity etc., heavy reliance on external experiences	Universalisation of food security issues, relaxation of local situation based food constraints; integration with the mainstream situation Marginalisation of local initiatives and measures; mismatch between imperatives of resource-characteristics and the features of generalised interventions with negative side-effects affecting food security
Resource focus		
Diversified resource use; balancing extensive-intensive land uses, production conservation needs; focus on local resource potential, resource rationing, recycling and regeneration, upgrading; local resource control	Priority to production over conservation, high use-intensity, over-extraction, large-scale resource transformation, e.g. by irrigation projects (ignoring location specificities, diversities)	Over-exploitation, degradation of resource base, sustained and subsidised by the state: bypassing local concerns; emerging long-term unsustainability
Agriculture production front		
Diversified, interlinked multiple land-based activities; practices driven by local demand, local resource, and local knowledge with focus on recycling, flexibility, and stability; productivity of total system; integration of production and consumption activities	Indiscriminate intensification, narrow specialisation, selective commercialisation; sectoral segregation of land-based activities; focus on high-cost, high-productivity, external inputs and technologies, subsidised by the state	Rapid resource depletion, increasing biochemical and economic subsidisation of production systems; emergence of dual (transformed and stagnant) sectors, imposition of inappropriate external options; dependency
Demand management		
Supply-driven demand situation; focus on (diversified) chain rather than food items, informal demand rationing, sharing and recycling practices, transhumance etc.	Focus on supply side mainly, dependence on external supplies and relief, focus on selected foodgrains, increased external demand for selected local items, unregulated market pressures	Rapid growth of pressure on local resources and their degradation, permanent dependency on relief, marginalisation of collective sharing, demand rationing, food chain etc.
Linkages and exchange		
Strong linkages between temporal/ spatially differentiated activities (e.g. farming-forestry-linkages); inter-system (i.e. external linkages) limited; need-based and local capacity driven harnessing of niche for petty trading; transhumance and migration	Improved accessibility-led physical and market integration, better access to external supplies, external market-driven over-extraction of niche and exchange at unequal terms, under-valuation of products, limited plough back of resources	Selective over-extraction of resources with limited local benefits, resource degradation, development of dual-sector economy, people's alienation from resources; domination of mainstream decision-makers, disregard of multiple small niche

Sources: Banskota and Jodha (1992a, 1992b), and Jodha (1993a, 1992a, 1991a, 1991b, 1989a, 1989b, 1986b, 1985b), Jodha *et al.* (1988); and Whitaker *et al.* (1991).

few important features of the food security-oriented interventions (contrasted with the traditional ones in Table 2) are noted below.

5.1 Universalisation of food security issues

Compared with the traditional arrangements, both the orientation and context of food security issues have changed under the new arrangements. Food security is focused through: increased employment and income opportunities (i.e. entitlement enhancement) generated by development projects; agricultural production programmes; and public distribution systems based on mobilisation of food from the surplus-producing areas (Drèze and Sen, 1990). This has definitely reduced the role of local constraints (including food production). Despite some lapses and leakages, the new state-sponsored arrangements have drastically reduced the occurrence of famines, especially in the dry regions of India where the problem was very serious only four to five decades ago (Jodha, 1975; Jodha *et al.*, 1988). However, without belittling these gains, it can be stated that public interventions have generally ignored the location specificities and diversities associated with both demand and supply (production) aspects of food security in the fragile and marginal areas. Food under the new arrangements means a few cereals rather than a chain of seasonal, spatial, ethnic and agro-climatically differentiated diets. Retrospective surveys using the case history method in the hills of Nepal and U.P. (India) recorded more than 20 food items traditionally consumed by the villagers. Now this number is reduced to about five (Singh, 1992). The comparable details of food items recorded in the dry tropics of India ranged between ten and five, excluding several minor items (Jodha, 1985b). Today, the access to food is largely determined by an individual's capacity to acquire it (i.e. entitlement), with little collective-sharing. In the dry tropics, forms of sharing during droughts declined from 12 during 1963-64 to barely three in 1980-81 (Jodha, 1975, 1993a). Furthermore, elements such as informal rationing and recycling etc., have little place under the new market-oriented arrangements. More importantly, the entitlement enhancement approach also had mixed success, as the development interventions (again due to their generalised approach unsuited to the specificities of fragile resource zones) could not significantly alter situations except in selected pockets that were relatively well-endowed, accessible and patronised. Owing to the mismatch between the resource base and the interventions, the state schemes had serious negative side-effects, which eroded the range of production options traditionally available. Most of the development interventions focused on inappropriate resource-use intensification, which encouraged resource degradation and long-term unsustainability (Jodha, 1986a, 1989a, 1991a, 1991b).

5.2 Indiscriminate resource-use intensification

Indiscriminate resource-use intensification under the new arrangements is most clearly visible in agricultural production programmes. The focus of agricultural production programmes had been on (i) food self-sufficiency, (ii) agricultural intensification and narrow specialisation, and (iii) sectoral segregation, as visible in practically all relevant activities ranging from agricultural research and development to foreign-aided development projects.

Focus on food self-sufficiency

Although fully sympathising with the policy-makers' concern to feed the increasing population, one may question the undue focus on food self-sufficiency in fragile areas, especially when food is defined in a narrow foodgrain sense and isolated from the total food chain (a range of seasonally and spatially variable consumables) generated by diversified and interlinked activities. This disregards the historical experience that, even with relatively smaller populations, most of these areas had rarely been self-sufficient in food. The deliberate control of demand, recycling and collective sharing, periodic migration, focus on petty trading involving barter exchange of high-value products for food, strengthening of the totality of the food chain and other strategies were used to adjust to food scarcity in the past. Yet, in the recent past, policy-makers have invariably focused on foodgrain self-sufficiency in the fragile resource areas (Banskota and Jodha, 1992a; Jodha, 1991a, 1991b, 1990b, 1986b; HMG, 1981).

These strategies, in the first place, because of their focus on food grain rather than the total food chain and ignoring the regions' petty but multiple niche (i.e. production with comparative advantage), missed the essence of food security in the context of fragile resources. Furthermore, foodgrain-focused strategies led to the extension of cropping to sub-marginal lands (and steep slopes) not suited for annual cropping, and increased cropping intensity on traditional croplands. From the food security perspective, this reduced the extent of diversification and interlinkages of different activities, reduced local resource regeneration and associated flexibility, led to physical degradation of fragile lands, increased biochemical and economic subsidisation of the agricultural production systems; and reduced average yield of different crops due to inclusion of marginal areas into croplands (Jodha, 1991a, 1991b, 1992a; Yadav, 1992; Singh, 1992).

Agricultural intensification and narrowed focus

The concerns for local food security and rapid agricultural growth were pursued through agricultural intensification. This applied to both food crops as well as high-value cash crops such as vegetables and fruits in the mountains and oilseed crops in dry areas. Agricultural intensification had two interrelated key dimensions. Firstly, programmes focused on a narrow choice of priority crops and varieties, and emphasised only a few attributes (e.g. growth of grain yield rather than stability of total biomass). Secondly, high input-intensive cropping involved application of high-cost external inputs. Both were results of extending agricultural experiences from prime lands, which disregarded the specific circumstances of fragile areas (Jodha, 1986b, 1991a; Jodha and Partap, 1993; Singh, 1992; Whitaker *et al.*, 1991).

Apart from their high dependence on external inputs, this led to the marginalisation of several important local crops and varieties. For instance, in several valleys of the middle mountains of Nepal and the U.P. hills in India, there used to be more than a dozen types of rice. Today, in most cases, one finds hardly two or three varieties developed by agricultural research centres, research stations etc., that occupy the bulk of the rice-growing areas. Horticultural products in many parts of the HKH region, and millet and legume varieties in the dry tropics, are in a similar situation (Singh, 1992; Shrestha, 1992; Jodha, 1986b).

High input-intensive cropping is an unavoidable consequence of the focus on yield levels and the choice of technologies that facilitate it. For multiple cropping and high yields of individual crops, dependence on external (chemical) inputs has rapidly increased. This is partly due to the design of the technologies employed and partly due to the inability of the local resource-regenerative systems (such as nutrition cycling through farming-forestry linkages, biological control of diseases and pests through specific crop rotations etc.) to meet high and temporally concentrated demands of the new systems (Yadav, 1992; Jodha and Singh, 1990).

Sectoral segregation

The required diversity and interlinkages of activities in the fragile resource zones are ignored not only in the production programmes at field level but more so at the policy and planning levels. To suit the pre-existing system of division of labour (or rather division of bureaucratic power), agricultural activities (relating to cropping, animal husbandry, horticulture, forestry etc.) are distributed to various departments and line agencies. Development activities, including project formulation, investment allocation, foreign assistance etc., are separated accordingly. Due to the interlinkages of different resource characteristics (e.g. fragility and diversity) and greater interdependence of different land-based activities (e.g. cropping and animal husbandry), sectoral segregation of decision-making and resource allocation make the development interventions ineffective and often counter-productive. The immediate consequences are narrow specialisation, disregard of interlinkages, and the inability of decision-makers to understand the basis and pattern of the integrated development of agriculture (Sanwal, 1989; Jodha, 1990a, 1991a).

5.3 Missing focus on demand management

Management of demand is not a strong point of the state-patronised food arrangements in the fragile resource zones. In fact, the key emphasis is on supply, be it focused crop production programmes, raising food-purchasing power, or subsidised public distribution. The demand aspect, except during the occasional cases of formal rationing, is rarely emphasised. Not only are elements of traditional arrangements such as diversification of food requirements, recycling, sharing, deliberate consumption curtailment etc. disregarded, but the state policies and programmes deliberately help to raise the pressure of external market demand on resources and selected products of fragile resource areas (Jodha, 1991b). Such products relate to the niche of fragile resource zones such as timber, vegetables and fruits in the mountains and animal products, oilseeds and pulses in the dry tropical areas. Besides the over-exploitation of fragile resources, an over-emphasis on these products has backlash effects on other food crops, which are an important part of local food systems but have little demand in the external market. Displacement of maize and minor millets by high-value vegetable crops in mountain areas and the displacement of millets by oilseeds in the dry tropics are cases in point. Ironically, due to marginality-related attributes, the producers of the high-value crops often do not receive favourable prices. Furthermore, there are no mechanisms to regulate the external demand pressure through fiscal or other means. The consequence is further use intensification and degradation of the natural resource base of fragile areas.

5.4 Linkage and unequal exchange relations

An important dimension of demand management relates to external market linkages and terms of exchange. Improved accessibility and integration of fragile areas with mainstream, urban economies has definitely reduced the vulnerability of these areas to severe food scarcities and famines (Drèze and Sen, 1990). However, the integration of unequals does not necessarily help the weaker and marginal entity. Owing to unequal terms of exchange and under-pricing of their products, in both mountains and dry tropics, integration has more often led to draining of resources. This could partly be a transitional phase where these areas, because of their low absorption capacity, do not directly benefit from new linkages (except for emergency food supplies). But more importantly, because of their marginal status *vis-à-vis* the prime areas and their domination in the decision-making processes, the terms of exchange are usually unequal and unfavourable to these areas. This is reflected through product pricing as well as compensation norms for resource extraction. Examples are minerals, timber, hydroelectric power, animal products, herbs and fruits. The food security implications of such unequal terms of trade are reflected through the local people's loss of control over their resources; inadequate local gains of harnessing the niche (to raise their food-purchasing power); and marginalisation or disappearance of several interlinked production activities with food components (Jodha, 1991a, 1989a; Yadav, 1992; Banskota, 1989).

An extreme case of unequal exchange is provided by the following. The state or its agencies often acquire the whole land tract for harnessing specific products (e.g. minerals) or for preserving biodiversity and wildlife (as under parks and sanctuaries), or for defence-related activities. This phenomenon has a high incidence in fragile resource zones. This is because, despite rapid degradation of their resources, these areas are the only left-over habitats of multiple fauna and flora, are the major areas with relatively sparse and voiceless population, and are the areas that continue to bear the consequences of their marginal status. Dislocations and displacements without adequate compensation and alternatives have significant food-security implications.

Yet another dimension of such unequal inter-system linkages is their side-effects on the organic and economic integrity of the traditional production systems. Nationalisation of forests or state acquisition of common property pastures, adversely affecting farming-forestry livestock linkages that supported the local food systems, is one example (Yadav, 1992; Jodha, 1992a). The direct links of modern dairying with the urban market and substitution of some crucial local inputs by external inputs is another example where the integrity of the total production system has broken down.

5.5 Emerging indicators of unsustainability

The consequences of the breakdown of resource-regenerative, diversified production systems and indiscriminate resource intensification (under the pressure of population growth, enhanced market forces, and development interventions etc.) are manifested through various forms of resource degradation, falling resource availability and resource productivity, and the degree of desperation in people's responses to a deteriorating situation (Bandyopadhyay *et al.*, 1985; Ives and Messerli, 1989; Jodha, 1992a). All directly or indirectly influence food security in fragile zones.

Based on observations and quantified evidence at the micro-level, from different areas within fragile zones, indicators of negative change are summarised in Table 3 (see Jodha, 1991a, 1991b, 1993a; Jodha *et al.*, 1992; Shrestha, 1992; Singh, 1992; Bajracharya, 1992; Hussain and Erenstein, 1992; Shutain and Chunru, 1989). These negative changes, covering a period of four to five decades, manifest the extent of the emerging crisis and scarcities in the fragile resource zones. Viewed differently, these negative changes represent the cumulative type of global environmental change (Jodha, 1992b, 1993b) elaborated below, which can help in relating the current crisis to the future crisis due to the potential impacts of global warming.

It should be noted, however, that the description of negative trends and emerging unsustainability relate to the general situation in the fragile resource zones. There are a few exceptions where the situation has been positively transformed. Our concern in this paper, however, is with the bulk of the areas witnessing negative changes that adversely affect food security.

Table 3: Indicators of emerging crisis and scarcities following the increased pressure and inappropriate resource intensification in fragile resource zones

Mountain areas	Dry tropical areas
Resource degradation	
Soil: Increased landslides, mudslides, collapse of terrain; rapid top soil disappearance from steep slopes	*Soil*: Increased salinity, erosion of top soil; shifting sand covering fertile lands
Water: Reduced (dry season) water flow for irrigation and domestic use; frequent flash floods; increased run-off and soil erosion during rainy season	*Water*: Groundwater salinisation, deepening of water-tables; increased moisture insecurity and frequency/ severity of droughts; decline and unfeasibility of water harvesting systems
Vegetation: Decline of forest/pasture area and productivity; disappearance of minor plant species; biodegradation; reduced biodiversity and biomass availability	*Vegetation*: Emerging plantlessness in forest/pasture, reduced perennial, increased inferior annuals, thorny bushes; reduced carrying capacity and diversity
Resource availability	
Per capita decline in crop-land, irrigated land, grazing land, common lands, reduced access/ availability of forest and its products (fuel, fodder, fibre, food items)	Decline in per capita crop-land, per animal grazing land, per hectare local resource input (e.g. manure); shortage and instability of natural biomass; increased landlessness and reduced access to water facilities
Resource management productivity	
Decline in agricultural diversity, farming-forestry linkages; decline in yield of several crops, unfeasibility of traditional land extensive resource management practices and local resource degeneration, recycling; high biochemical subsidisation of the production systems	Decline in local resource regeneration/recycling, diversity, flexibility in agriculture; decline in crop/ animal productivity, collective resource management and risk sharing; disintegration of traditional farming systems (intercropping, crop rotations etc.)
Impacts/responses	
Desperation and choice for inferior options: Abandonment of old terraces; cropping on more steep slopes; substitution of shallow-rooted crops for deep-rooted crops on eroded slopes; substitution of small ruminants for cattle due to lower carrying capacity of pastures; increased seasonal migration of men and animals, marginalisation of collective sharing; high external resource dependency and indebtedness	*Desperation and choice for inferior options:* Limited abandonment of saline, eroded soils; using land despite unstable and reduced productivity; 'mining the resources', e.g. extension of cropping to sub-marginal lands, premature harvesting/lopping of trees, overgrazing; increased dependency on relief and external resources; decline of collective self-help; substitution of sheep/goats for cattle due to reduced area/yield of pasture; and migration destitutions

Sources: Johda (1990a, 1991a), Shrestha (1992), Singh (1992), Shutain and Chunru (1989), Hussain and Erenstein (1992), Bajracharya (1992). For dry tropics, see Jodha (1986a, 1986b, 1990a, 1991a, 1993a), Jodha *et al.* (1988), Jodha and Singh (1990).

6 Food security in the context of environmental change

The emerging strains on the food situation, caused by increased pressure of demand and the side-effects of measures (e.g. resource intensification) designed to counter them, can be accentuated with the potential impacts of global environmental changes. Global warming-induced climate change and its impacts on the world food situation has consumed considerable attention, time and energy of scientists in addition to scarce material resources. However, despite the statistical sweep and analytical rigour of scientific work, the predicted scenarios, especially in the regional context, continue to be conjectural (Houghton *et al.*, 1992; Schmandt and Clarkson, 1993). In this regard, the fragile resource zones under review share the same level of uncertainties with other regions. In the face of the persistent uncertainties and information (data) gaps, the impact of warming on food security can only be highlighted in terms of potential accentuation of resource degradation and scarcity. Resource degradation and scarcity represent the cumulative type of environmental change and have strong complementarity (in cause and consequence terms) with the systemic type of change represented by distortions caused by the greenhouse gases. An understanding of the complementarities between the two types of environmental change could offer important leads in evolving adaptation strategies against global warming in the regional context. However, this has not received enough attention, for the reasons discussed below (Jodha, 1992b, 1993b).

6.1 Global environmental change: the skewed perspectives

Broadly speaking, systemic change is one that, while happening in one locale, can affect changes in the system elsewhere. The underlying activity need not be widespread or global in scale, but its potential impact is global in that it influences the operation and functioning of the whole system as manifested through subsequent adjustments in the system. Carbon dioxide emissions from limited activities, having impacts on the great geosphere-biosphere systems of the earth and causing global warming, are a prime example. The cumulative type of change refers to localised but widely replicated activities, where a change in one place does not affect change in other places. When accumulated, they may acquire a scale and potential that can influence the total global situation. Widespread deforestation and extractive land use systems and their potential impacts on global environment serve as examples. Both types of change are the products of nature/human interactions and they are linked to each other in several ways (Turner *et al.*, 1990).

In scientific works, discourses and policy advocacy, systemic changes and associated geocentric perspectives are the primary focus. The variables and processes covered are complex and difficult to analyse at the current level of scientific skill and data availability. The uncertainties associated with the change scenarios based on them are, therefore, not unexpected. The cumulative type of change and associated anthropocentric perspectives, however, present a situation that is much simpler and involves fewer uncertainties and unknowns. It is much easier to relate to the human approach to constraints and opportunities presented by the environment. Yet, as they are not sufficiently emphasised in work related to global change, their potential for guiding response strategies remains under-used (Kasperson *et al.*, 1990). Similarly, the complementarities between the two types of change (both in terms of causes and consequences) are not fully understood and analysed. This is largely because of the skewed perspectives of the work and debate on global environmental change, as shown in Table 4 (Jodha, 1993b).

The brief digression into the skewed perspectives on global environmental change emphasises the need for, and the possibility of, using the variables of cumulative change as a basis for: (i) identifying linkages between current crisis and the future warming-induced crisis, and (ii) evolving common strategies to handle the two crises. Effective strategies against the sources and processes of current crisis should therefore be treated as dual-purpose strategies as they will also help the potential impact of warming (Jodha, 1992b, 1993b).

6.2 Warming in fragile resource zones

Although there are no concrete predictions, on the basis of various general circulation models (GCMs) relating to mountains and dry tropics, important details can be summarised (Table 5). If the positive impacts of warming (e.g. the possibility of better cropping prospects through improved moisture and CO_2 fertilisation impacts etc.) are excluded, the key inferences from Table 5 suggest that global warming can influence the production environment and the resource base in mountain areas by accentuating deforestation and changing precipitation (including snow melt). It can dislocate production systems and create more incentives and compulsion to over-extract and degrade the resource base, affecting the food situation. In the dry tropics, a changed precipitation pattern without the certainty of gains in terms of a longer growing season may impart uncertainty to the future. The present marginal areas may be exposed to greater weather

Table 4 Indicators of the skewed perspectives on global environment change

Elements prominently focused	Elements under-emphasised
Systemic type of change	**Cumulative type of change**
Focus on biogeochemical variables and their interaction processes relating to the functions and operation of the geosphere-biosphere systems of the earth	Localised and widely replicated changes in different variables and process of resource (when accumulated) influence the global systems
Geocentric perspective	**Anthropocentric perspective**
Focus on physical dimensions, typically in the natural science framework; concentration on geobiological variables and their complex interaction patterns, with little direct incorporation of the human dimension of changes and change-processes	Primacy of nature-society interactions with focus on their importance to the society; potential mechanism for understanding and handling 'cumulative change' with some possibility of influencing the impacts of 'systemic changes'
Other associated aspects	**Other associated aspects**
Emphasis on long-term horizon (decades/centuries) and inter-generational issues; focus on terminal impacts involving selected variables (e.g. sea level, temperature rise, shift of climatic zones etc.) affecting fundamental equilibrium of world system and atmosphere; analytical methods and material used involve high degree of complexity and sophistication, limited information on several unknowns, limited transparency (for uninitiated ones), multiple uncertainties, and conjuctural nature of predictions	Sensitivity to both intra-generational and inter-generational issues; analytical approaches simpler and oriented towards integration of change processes in the current problem-solving mode; predictions, action/ advocacy focus on short- or medium-planning horizon, greater ease and possibility of associating causes, consequences of and responses to change; greater possibility of integrating geocentric and anthropocentric perspectives
Advocacy and action	**Advocacy and action**
High 'scare and noise' potential of issues covered (e.g. doomsday predictions); approaches to abate/adapt to changes obstructed by uncertainty of change scenarios, induce higher discounting of the potential options, inject vagueness about gains and sacrifices and create more panic and debate than concrete action	Possibility of evolving options within the received (and modified) framework of handling current crises in local contexts; greater scope for clearly associating cost and benefits; greater certainty of potential options and their easy acceptability to decision-makers; possibility of dual-purpose options to handle current and future impacts

Sources: adapted from Jodha (1993b). For various issues and examples that fit into the following grouping of perspectives, see Price (1990), Turner *et al.* (1990), Kates (1990), Flavin (1989), Jodha (1989c, 1992b, 1993b), Glantz *et al.* (1988), Wuebbles and Edmond (1988), Clark (1985), Chen *et al.* (1983), Garcia (1981).

Table 5 Possible impacts of warming in the fragile resource zones

Hindu Kush-Himalayan mountain regions	Dry tropical regions in India
(i) Forests (unmanaged ecosystems)	**(i) Precipitation (mixed picture)**
Qualitative/quantitative changes, disappearance or spatial movement of species; reduced biodiversity, increased deforestation and dislocation of biophysical/ hydrological flows and dependent life-support systems	Temperature increase; increased precipitation from convective rainfall with higher intensity; high rainfall, high run-off, high soil erosion (due to plantlessness and undulating topography), high evaporation with uncertain prospects for longer growing seasons and higher crop production
(ii) Precipitation	**(ii) Areas on margin**
Higher rainfall (convective high-intensity rains), increased run-off, flash floods, soil erosion, mud/ landslides, influencing the biophysical functions of nature and the farming systems	Area currently frequented by prolonged droughts and crop failures may further suffer with greater variability of rainfall, exposure to greater uncertainty and weather risk
(iii) Seasonal snowmelt changes	**(iii) Impact of (i) and (ii)**
Disturbance to hydrological cycles and seasonality of waterflows and well-adapted resource use/production systems, diversity etc.	Disturbance of well-adapted farming systems, reduced groundwater recharge, increased soil salinity, reduced regenerative capacity of nature to support crop and natural biomass-based mixed farming and people's food systems especially in marginal areas
(iv) Resource characteristics	**(iv) Resource characteristics**
Fragility, diversity, niche etc., and their interrelationships to be affected by (i), (ii) and (iii); new contraints (compulsions) and new opportunities (inducements) for higher resource extraction ignoring fragility, marginality etc.	Fragility, marginality etc. may accentuate and adversely affect the resource-use systems; increased dependency on external support, food insecurity

Sources: Data excludes positive impacts of warming which should not adversely affect food security. Adapted from Jodha (1992b), Topping *et al.* (1991) and Houghton *et al.* (1992) for mountains; and Jodha (1993b), Jager (1988) and Houghton *et al.* (1992) for dry tropics.

risks and drought-induced scarcities (Jager, 1988; Houghton *et al.*, 1992; Parry, 1985; Downing, 1992).

Vegetation-related impacts of global warming (although involving significantly different degrees and processes in the two agro-ecosystems) can accentuate trends in deforestation, reduced biodiversity and biomass supplies, and the dependent, diversified farming systems, with direct and indirect consequences on the food situation. Most importantly, in both mountains and dry tropics, warming-induced impacts can reduce the resource-users' sensitivity to the imperatives of features such as fragility, marginality, diversity etc. Materialisation of such possibilities will only accentuate the current consequences of neglecting the above imperatives.

More importantly, and in keeping with the central thread of this paper, resource characteristics such as fragility, marginality and diversity in both the mountains and the dry tropics can be accentuated through the changed biophysical make-up of the regional environment following the impacts of warming (e.g. increased soil salinity due to increased evaporation in dry areas; vulnerability to rapid gully erosion on mountain slopes due to high rain etc.). Furthermore, warming can create circumstances in these regions that could induce or compel people to disregard further the imperatives of resource features such as fragility, marginality, diversity etc., with all the negative consequences already discussed.

Our key purpose in summarising the future possibilities in Table 5 is to see how they relate to the current scenario reflected in Tables 2 and 3. Table 6 presents the broad indicators of a possible worsening of the current crisis with the potential impacts of warming in the context of fragile resource zones. The current negative trends relating to the production base, resource productivity, management practices and vulnerability to these changes have several elements that strongly converge with the components of the negative impacts of global warming through first-, second- and third-order impacts illustrated elsewhere (Jodha, 1989c). For instance, precipitation-based dislocations following warming can accentuate the present problem of soil erosion, and disturb water-harvesting systems and well-adapted farming systems in both mountain areas and the dry tropics.

Table 6 Possible impacts of global warming that could accentuate the current indicators of crisis/ food insecurity in the fragile resource zones

Potential impacts of warming	Indicators of emerging current crisis likely to be affected by warming	
	Mountain areas	Dry tropical areas
Precipitation-based dislocations		
Increased run-off, floods, seasonal changes (M); soil erosion (M, D), landslide/ mudslide (M), increased weather risks (D); disruption of adapted production systems, decline in supplies (M,D)	Increased land/mudslides, flash floods; off-season water scarcity; collapse/ abandonment of terraces; decline of crop-land irrigation facility; reduced cropping diversity, productivity; seasonal food deficit etc.	Increased soil and ground-water salinity, deepening of groundwater, decline of water-harvesting systems, increased frequency and severity of droughts, decline in crop/animal yields, increased external dependency
Forest, biomass-related changes		
Deforestation, reduced biodiversity (M); reduced biomass productivity (M,D); instability of dependent production systems, resource regeneration/stability (M,D)	Deforestation, biodegradation, decline of farming-forestry linkages and resource-regenerative processes; reduced biomass; decline of diversification, reduced resource and product availability	Increasing plantlessness, shifting sands, reduced biomass supplies, increased animal migration, decline of mixed farming systems, reduced resource regeneration, various forms of desertification
Pressure for disregard of resource specificities		
Resource intensification, resource degradation, decline of adapted production systems, practices	Decline of well-adapted production systems; focus on indiscriminate resource intensification, insensitivity to resource limitations; over-exploitation of niche; desperation in resource use	Indiscriminate resource intensification, disregard of resource limitations; decline of diversification, recycling, regeneration; desperation in resource use, poverty resource degradation cycle

M = mountain areas; D = dry tropical areas.

Sources: Jodha (1989c, 1992b, 1993b).

7 Enhancing food security: reversing the negative processes

The above reasoning implies that enhancing food security in the context of increased population and accentuated resource degradation (i.e., the cumulative type of environmental changes) should also serve as a strategy against food insecurity due to the impacts of global warming. Such dual-purpose strategies should focus on arresting and reversing the currently visible negative trends relating to resource availability, resource productivity and access to opportunities based on harnessing regional resources. In practical terms, it would mean avoiding the mistakes of the past and building on the positive experiences of history (Kates and Millman, 1990). In other words, the focus of new strategies must be on reversing the processes that contributed to the currently visible negative trends. To do so, the issues should be addressed at different levels ranging from macro-level policy to field-level operational steps. Previous discussion can be summarised to make the point (Jodha, 1991b, 1989a; Jodha *et al.*, 1992).

7.1 Changing the policy focus and context

At the policy/planning level, the recognition and incorporation of the following interrelated issues is a prerequisite to initiate the processes that may help enhance food security in fragile resource areas:

- Understanding and internalising the imperatives of specific resource characteristics (fragility, diversity, niche etc.) are essential to reorient public interventions to suit the situation of fragile resource zones. This will help amend and evolve interventions in keeping with the specific features of these areas, and reduce the extent and imposition of generalised (less relevant) approaches evolved in other contexts (Sanwal, 1989; Jodha, 1991a, 1991b).
- Recognition and application of the rationale of traditional food supply/ demand management measures to add relevance and effectiveness to present-day steps adopted for food security. This will mean reorienting several production and resource-use strategies and decentralising activities (Sanwal, 1989; Jodha, 1991a; Jodha and Partap, 1993).
- Learning from the past experiences of unsuccessful development interventions and their reorientation to suit the specific circumstances of fragile resource zones. This also

implies amendments in generalised approaches and measures to suit the needs and capabilities of fragile areas (Banskota and Jodha, 1992a, 1992b).

- Learning from the experiences of success stories within the overall environment characterised by unsuccessful initiatives. In fact, most of the successful ventures incorporated elements of the above. In these cases, both production and entitlement options have been enhanced without unduly damaging the resource base and environment (Jodha *et al.*, 1992).

7.2 Recognising the essence of food security

The essence of improved food security is the enhancement of the range and quality of production and exchange options that help people to have command over multiple possibilities *vis-à-vis* food requirements. Option-enhancement strategies should be guided by the following considerations:

- In the changed context of the present, the phenomenon of food security extends far beyond the stability of a self-provisioning system that mainly focused on the production and consumption of food. Now, it should focus more on people's command over options reflected through what are described as entitlements (Sen, 1981). The strategies directed to increased entitlements would be different from those addressing the stability of subsistence supplies dominated by directly consumable items. Thus food security is directly linked to employment and income-generation activities (Sen, 1981; Drèze and Sen, 1990).
- Option enhancement (in the context of currently visible changes in the fragile resource areas) requires: (a) rehabilitation and upgrading of the resource base and dependent production systems using nature's regenerative processes, usage regulation and application of modern science and technology, and (b) diversification of resource use, production activities (or rather income-generation activities), and demand itself. Again, in the changed context, diversification cannot be confined to biomass (including food) production activities but will have to incorporate interlinked activities of the primary, secondary and tertiary sectors, as experiences from most of the transformed areas have shown (Jodha, 1991b). (c) Intensification is an essential precondition for increased production from a resource base. To avoid inappropriate and resource-degrading intensification, the focus has to be on intensifying a total production system (involving diverse and interlinked activities) rather than intensifying selected components (e.g.

single crop within an overall cropping/farming system) that generate imbalances and lead to disintegration of production systems. Science and technology have to play a crucial role in this. (d) Inter-system or inter-regional linkages are another step towards option enhancement. The aforesaid steps can help enhance production, but the gains from them (increased entitlement) can be achieved only with appropriate value-added activities and exchanges. This requires effective and equitable exchange relations and inter-system linkages based on comparative advantage. This is more so when harnessing regional niches. (e) Entitlement enhancement through increased employment and income opportunities is the central purpose of development interventions. Hence, deliberate diversification of income/employment-generating programmes and their adaptation to local circumstances is a key step towards ensuring food security (through entitlement enhancements).

- Demand management is an important aspect of food security (Jodha, 1975). At the micro-level, demand management should cover elements such as diversity of food/food chain, recycling, sharing, consumption-production linkages, occupational diversification-based spatial and seasonal spread of demand pressure etc. These unrecognised aspects of food security need strengthening. Admittedly, the formal food security arrangements cannot enforce such essential informal measures. However, formal interventions through technologies, decentralisation etc. can create objective circumstances to which people can respond and adopt such measures.

Food deficits and food transfers are important aspects of demand management at the macro-level in the fragile resource zones. Despite all the steps suggested above, the physical production of food may have periodical shortages in several parts of these regions. Food transfers from surplus-producing areas have been the government's response to such shortages. This has two problems. Firstly, in remote, less accessible areas, such supplies are undependable. Focus on local production (with traditional features such as diversification, recycling etc.) is still the best strategy there. In more accessible areas, agro-business-oriented activities based on niche can prove to be a more effective means of ensuring food security through entitlement enhancement. The second aspect of food transfer is that, due to their frequent foodgrain deficits, the fragile resource zones are often treated as liabilities for the national economies. This distorts the perception or policy-makers *vis-à-vis* food security issues in the fragile resource zones, implying a charity orientation to food security in such areas (Jodha *et al.*, 1988; Banskota and Jodha, 1992b). However, off-site impacts of decline of fragile resource zones due to inappropriate land use for increased foodgrains and the net transfer of resources from these areas to the mainstream economy (due to under-pricing, inadequate compensations etc.) are considered, liability-dominated perspectives are not justified. A clear understanding of the comparative advantages of the fragile resource zones and their unique contribution to the national economy will help change such prevailing notions (Anderson and Jodha, 1994). The possibilities of food security through increased production and/or entitlement enhancement, using the aforementioned steps, are corroborated by the experiences of agricultural and rural transformation in mountain areas such as Himachal Pradesh (India), Miyi and Ningnan counties (West Sichuan, China), Ilam district (Nepal), as shown by ICIMOD studies. However, because of the high degree of fragility and the limited carrying capacity, fragile zones may not be able to accommodate rising numbers of people (and animals) for ever. Hence, control over internal and external pressure on resources is essential.

7.3 Focal issues at the operational level

Table 7 indicates both the components of strategies for enhancing food security and their operational implications. Accordingly, food-security enhancement simultaneously emphasises the growth of physical production and the opportunities for entitlement. However, their combination will vary according to circumstances such as relative accessibility or inaccessibility of areas, type and extent of niche etc. In the current context, increased physical production possibilities in the fragile resource zones are directly associated with:

- Rehabilitation of degraded resource base by means of resource-use diversification, usage regulation, regeneration etc., where the use of modern science and technology that incorporates the rationale of the traditional systems could be of great help.
- Resource upgrading, involving various conservation and protection measures as well as measures to harness under-used resource potential (through afforestation, water harvesting in a watershed context etc.). Besides technological inputs and initial capital investments, people's participation is a key factor.
- Restoring the intensity of production systems to enhance physical production with resource regeneration/protection measures. This implies greater attention to diversification and regenerative processes that have been marginalised following

Table 7 Steps to enhance food security in the fragile resource zones

Focal areas and component	Implications and action areas
Growth in physical production	
Resource rehabilitation	Diversification, regeneration, recycling, usage regulation; resource-focused reorientation of agricultural R&D incorporating elements of traditional systems
Resource upgrading	Conservation, stabilisation, reforestation, water harvesting, irrigation measures; public investment, people's participation for small-scale, diversified activities; policy reorientation as required
Restoring integrity of production systems	Diversification, interlinked land-based activities, complementarity of: intensive/extensive land uses, annual-perennials; reorientation of agricultural R&D, support to traditional measures, focus on resource regeneration, high productivity, biomass stability
Higher productivity	Intensification of production systems, i.e. all interrelated activities rather than individual components; mixing short-duration annual-perennials; reorientation of agricultural R&D based on understanding of intensification needs; appropriate support systems
Entitlement enhancement and going beyond biomass-centred, self-provisioning	Harnessing niche, high-value products, agro-business orientation; support systems and policy orientation to suit local diversity
Gains from inter-system links	Inter-regional exchange based on comparative advantages and complementarities, equity of exchange, integrating primary, secondary, tertiary sector activities; infrastructural support and policy reorientation
Fair share in harnessing of niche	Equitable product pricing, fair compensation for resource use, required policy change and local participation
Percolation of developmental gains	Reorientation of development interventions, location-specific focus; sensitivity to local needs, decentralisation, local participation

Sources: Jodha (1991a, 1991b, 1990a, 1989a), Jodha and Partap (1993), Banskota (1989).

inappropriate intensification and narrow specialisation (Freudenberger, 1988). Reorientation of agricultural research and development is a primary institutional step. Recognition and use of rural people's traditional knowledge systems can help change scientists' perceptions about the need and utility of diversification, local resource regeneration etc. (Jodha and Partap, 1993).

- Higher productivity through resource-use intensification is an unavoidable step towards raising the food supplies. However, to avoid inappropriate intensification, a practical approach would be to intensify the local production system and its internal linkages rather than individual components of the system (Altieri, 1987).

Issues relating to entitlement enhancement need reorientation of macroeconomic policies, programme designs and implementation:

- Going beyond the biomass-centred approach, where value-added activities through integration of primary, secondary and tertiary sector activities are emphasised (Mellor, 1988). This is more possible in relatively accessible areas with specific niches and better infrastructural support – hence the increased focus on infrastructural development.• Gains from inter-system (e.g. upland-lowland) linkages are another approach for raising income and employment opportunities for enhancing entitlements (Banskota, 1989). However, this is closely associated with accessibility and infrastructural facilities.
- A related aspect is equitable terms of exchange, especially when regional niche is harnessed for the mainstream economy. This calls for important policy changes relating to regional development (Banskota, 1989; Banskota and Jodha, 1992b).
- The perception through gain of development activities can enhance people's purchasing/exchange capabilities. However, this requires diversified and location-specific reorientation of development interventions to suit local circumstances (Sanwal, 1989).

References

Allan, N.J.R., Knapp. G.W. and Stadel, C. (eds.) (1988). *Human Impacts on Mountains*. Rowman & Littlefield, New Jersey.

Altieri, M.A. (1987). *Agroecology: The Scientific Basis of Alternative Agriculture*. Westview Press, Boulder, Colorado.

Anderson, J.R. and Jodha, N.S. (1994). Agricultural research strategy for more enduring productivity in fragile areas. In: Anderson, J.R. (ed.) *Agricultural Technology: Policy Issues for the International Community*. CAB International, Wallingford, U.K.

Bajracharya, B. (1992). *A Review of Literature Focused on Indicators of Unsustainability of Mountain Agriculture in Nepal.* (A report for ICIMOD.) International Centre for Integrated Mountain Development (ICIMOD), Kathmandu.

Bandyopadhyay, J., Jayal, N.D., Schoettli, U. and Singh, C. (eds.) (1985). *India's Environment: Crisis and Responses*. New Delhi.

Banskota, M. (1989). *Hill Agriculture and the Wider Market Economy: Transformation Processes and Experience of Bagmati Zone in Nepal*. ICIMOD Occasional Paper 10. ICIMOD, Kathmandu.

Banskota, K. (1992). Agriculture and sustainable development. In: Asian Development Bank. *Nepal – Economic Policies for Sustainable Development*. Asian Development Bank, Manila.

Banskota, M. and Jodha, N.S. (1992a). Mountain agricultural development strategies: comparative perspectives from the countries of the Hindu Kush-Himalayan region. In: Jodha, N.S., Banskota, M. and Partap, T. (eds.) *Sustainable Mountain Agriculture*. Oxford and IBH Publishing, New Delhi.

Banskota, M. and Jodha, N.S. (1992b). Investment, subsidies and resource transfer dynamics: issues for sustainable mountain agriculture. In: Jodha, N.S., Banskota, M. and Partap, T. (eds.) *Sustainable Mountain Agriculture*. Oxford and IBU Publishing, New Delhi.

Bjonnes, I.M. (1983). External economic dependency and changing human adjustment to marginal environments in high Himalaya, Nepal. *Mountain Research and Development*, **3** (3).

Blaikie, P.M. and Brookfield, H. (eds.) (1987). *Land Degradation and Society*. Methuen, London.

Brush, S.B. (1988). Traditional agricultural strategies in hill lands of tropical America. In: Allan, N.J.R., Knapp, G.W. and Stadel, C. (eds.) *Human Impacts on Mountains*, Rowman & Littlefield, New Jersey.

Carson, B. (1992). *The Land, the Farmer, and the Future: A Soil Fertility Management Strategy for Nepal.* ICIMOD Occasional Paper 21. ICIMOD, Kathmandu.

Chambers, R. (1987). *Sustainable Rural Livelihood: A Strategy for People, Environment and Development.* IDS Discussion Paper 240. Institute of Development Studies, Brighton.

Chen, R.S., Boulding E. and Schneider, S.H. (eds.) (1983). *Social Science Research and Climate Change: An Interdisciplinary Appraisal*. D. Reidel Publishing Co., Dordrecht.

Clark, W.C. (1988). The human dimensions of global environment change. In: Committee on Global Change. *Towards an Understanding of Global Change*. National Academy Press, Washington, DC.

Dahlberg, K.A. (1987). Redefining development priorities: genetic diversity and agro-ecodevelopment. *Conservation Biology*, **1** (4).

DESFIL (Development Strategies for Fragile Lands) (1988). *Development of Fragile Lands: Theory and Practice*. DESFIL, Washington, DC.

Dixon, J.A., James, D.E. and Sherman, P.B. (eds.) (1989). *The Economics of Dry Management*. Earthscan, London.

Downing, T.E. (1992). *Climate Change and Vulnerable Places: Global Food Security and Country Studies in Zimbabwe, Kenya, Senegal and Chile*. Research Report No. 1. Environmental Change Unit, Oxford.

Downing, T.E., Watts, M.J. and Bohle, H.G.(1995). Climate change and social vulnerability: toward a sociology and geography of food security. In: Downing, T.E. (ed..) *Climate Change and World Food Security*. Springer-Verlag, Heidelberg.

Drèze, J. and Sen, A. (1990). *Hunger and Public Action*. Clarendon Press, Oxford.

Eckholm, E.P. (1975). The deterioration of mountain environments. *Science*, **139**: 764-774.

Flavin, C. (1989). *Slowing Global Warming: A Worldwide Strategy*. World Watch Paper 1. World Watch Institute, Washington, DC.

Freudenberger, C.D. (1988). The agricultural agenda for the twenty-first century. *KIDMA (Israel Journal of Development)*, **10** (2).

Garcia, R. (1981). *Drought and Man. Vol 1: Nature Pleads Not Guilty*. Pergamon Press, Oxford.

Glantz, M.H. (ed.) (1987). *Desertification: Environmental Degradation in and around Arid Lands*. Westview Press, Boulder.

Glantz, M.H., Brown, B.G. and Krenz, M.E. (1988). *Societal Responses to Regional Climate Change: Forecasting by Analogy*. National Centre for Atmospheric Research, Boulder.

Guillet, D.G. (1983). Towards a cultural ecology of mountains: The central Andes and the Himalayas compared. *Current Anthropology*, **24**:561-574.

Hewitt, K. (1988). The study of mountain lands and peoples: A critical overview. In: Allan, N.J.R., Knapp, G.W. and Stadel, C. (eds.) *Human Impacts on Mountains*. Rowman & Littlefield, New Jersey.

HMG (His Majesty's Government) (1981). *Nepal's Experiences in Hill Agricultural Development*. (Proceedings of the seminar.) Ministry of Food and Agriculture, HMG, Kathmandu.

Houghton, J.T., Callander, B.A. and Varney, S.K. (eds.) (1992). *Climate Change 1992 : The Supplementary Report to the IPCC Scientific Assessment*. Cambridge University Press, Cambridge.

Hussain, S.S. and Erenstein, O. (1992). *Monitoring Sustainability Issues in Agricultural Development : A Case Study in Swat in North Pakistan*. PATA Working Paper 6. Saidu Sharif, (NWFP, Pakistan). PATA, Integrated Agricultural Development Project.

Ives, J.D., and Messerli, B. (1989). *The Himalayan Dilemma: Reconciling Development and Conservation*. Routledge, London.

Jager, J. (1988). *Developing Policies for Responding to Climate Change*. Summary of Discussions and Recommendations of the Workshops held in Villach and Bellagio, September-November 1987. WCIP-1. WMO/TD-No. 225. World Meteorological Organization and United Nations Environmental Programme, Geneva.

Jochim, M.A. (1981). *Strategies for Survival: Cultural Behaviour in an Ecological Context*. Academic Press, New York.

Jodha, N.S. (1975). Famine and famine policies: some empirical evidence. *Economic and Political Weekly*, **8** (3): 633-639.

Jodha, N.S. (1980). Intercropping in traditional farming systems. *Journal of Development Studies*, **16** (4): 427-442.

Jodha, N.S. (1982). The role of administration in desertification: land tenure as a factor in the historical ecology of Western Rajasthan. In: Spooner, B. and Mann, H.S. (eds.) *Desertification and Development: Dryland Ecology in Social Perspective*. Academic Press, London.

Jodha, N.S. (1985a). Population growth and decline of common property resources in Rajasthan, India. *Population and Development Review*, **11** (2): 247-264.

Jodha, N.S. (1985b). Market forces and erosion of common property resources. In: *Proceedings of the International Workshop on Agricultural Markets in the Semi-arid Tropics*. 24-28 October, 1982. International Crops Research Institute for Semi-Arid Tropics (ICRISAT), Hyderabad.

Jodha, N.S. (1986a). Common property resources and rural poor in dry regions of India. *Economic and Political Weekly*, **21** (26).

Jodha, N.S. (1986b). Research and technology for dry farming in India: some issues for the future strategy. *Indian Journal of Agricultural Economics*, **41** (3).

Jodha, N.S. (ed.) (1989a). *Technology Options and Economic Policy for Dryland Agriculture*. Indian Society of Agriculture Economics, Bombay.

Jodha, N.S. (1989b). Agronomic adjustments to climatic variations. In: International Rice Research Institute, *Climate and Food Security*. (Proceedings of the International Symposium.) International Rice Research Institute, Manila.

Jodha, N.S. (1989c). Potential strategies for adapting to greenhouse warming: perspectives from the developing world. In: Rosenberg, N.J., Easterling, W.E., Crosson, P.R. and Darmstadter, J. (eds.) *Greenhouse Warming: Abatement and Adaptation*. Resources for the Future, Washington, DC.

Jodha, N.S. (1990a). Mountain agriculture: the search for sustainability. *Journal of Farming Systems Research Extension*, **1** (1).

Jodha, N.S. (1990b). *Drought Management: The Farmer's Strategies and their Policy Implications*. Dryland Network Programme Issues Paper 21. International Institute for Environment and Development, London.

Jodha, N.S. (1991a). Sustainable agriculture in fragile resource zones: technological imperatives. *Economic and Political Weekly* (Quarterly Review of Agriculture), **26** (13).

Jodha, N.S. (1991b). Agricultural growth and sustainability: perspectives and experiences from the Himalayas. In: Vosti, S.A. Reardon, T., von Urff, W. and Witcover, J. (eds.) *Agricultural Sustainability, Growth and Poverty Alleviation: Issues and Policies*. Proceedings of the Conference, September 23-27, Food and Agriculture Development Centre, Feldafing, Germany.

Jodha, N.S. (1992a). *Common Property Resources: A Missing Dimension of Development Strategies*. World Bank Discussion Paper 169. World Bank, Washington, DC.

Jodha, N.S. (1992b). *Global Change and Environmental Risks in Mountain Ecosystems*. FS Discussion Paper 23. ICIMOD, Kathmandu.

Jodha, N.S. (1993a). Common property resources in stressed environment: a case from high risk environment, India. Paper presented to IVth annual conference of the International Association for Study of Common Property Resource, June 16-19, 1993. Manila.

Jodha, N.S. (1993b). Understanding and responding to global climate change in fragile resource zones. In: Schmandt, J. and Clarkson, J. (eds.) *The Regions and Global Warming: Impacts and Response Strategies*. Oxford University Press, New York.

Jodha, N.S., Banskota, M., and Partap, T. (1992). *Sustainable Mountain Agriculture*. Oxford and IBH Publishing New Delhi.

Jodha, N.S., and Partap, T. (1993). Folk agronomy in the Himalayas: implications for agricultural research and technology. In: *Rural People's Knowledge, Agricultural Research and Extension Practices*. IIED Research Series, **3** (1). IIED, London.

Jodha, N.S., and Singh, R.P. (1990). Crop rotation in traditional farming systems in selected areas of India. *Economic and Political Weekly* (Quarterly Review of Agriculture), **25** (13): A28 - A35.

Jodha, N.S., Virmani, S.M., Gadgil, S., Huda A.K.S., and Singh, R.P. (1988). The effects of climate variation on agriculture in dry tropical regions of India. In: Parry, M.L., Carter, T.R. and Konijin, N.T. (eds.) *The Impacts of Climatic Variations on Agriculture: Assessment in Semi-arid Regions*. Vol. 2. Kluwer Academic Publishers, Dordrecht.

Kates, R.W. (1990). *Hunger, Poverty and Human Environment.* Distinguished Speaker Series. Centre for Advanced Study of International Development, Michigan State University, East Lansing.

Kates, R.W., and Millman, S. (1990). On ending hunger: the lessons of history. In: Newman, L.F. *et al.* (eds.) *Hunger in History: Food Shortage, Poverty and Deprivation*. Basil Blackwell, Oxford.

Kasperson, R.E., Dow, K., Golding, D., and Kasperson, J.X. (eds.) (1990). *Understanding Global Environmental Change: The Contributions of Risk Analysis and Management*. Clark University, Worcester.

Mellor, J.W. (1988). Global food balances and food security. *World Development*, **16** (9).

Mulk, M. (1992). *The Farmer's Strategies for Sustainable Mountain Agriculture: Chitral District, Pakistan*. MFS Discussion Paper 28. ICIMOD, Kathmandu.

Nelson, R. (1988). *Dryland Management: The Desertification Problem.* Environment Department Working Paper 8. World Bank, Washington, DC.

Nogaard, R.B. (1984). Coevolutionary agricultural development. *Economic Development and Cultural Change*, **60**: 160-173.

Panday, K.K. (1992). *Sustainability of Environmental Resource Base and Development Priorities of a Mountain Community: Bhardeo, Nepal.* ICIMOD Occasional Paper 19. ICIMOD, Kathmandu.

Pant, S.D. (1935). *The Social Economy of the Himalayas: Based on a Survey in the Kumaon Himalayas*. George Allen & Unwin, London.

Parry, M.L. (1985). The impact of climatic variations on agricultural margins. In: Kates, R.W., Ausubel, J.H. and Berberian, M. (eds.) *Climate Impact Assessment*. Wiley, Chichester.

Price, L.W. (1981). *Mountain and Man: A Study of Process and Environment*. University of California, Berkeley.

Price, M.F. (1990). *The Human Aspects of Global Change: Final Report*. National Centre for Atmospheric Research, Boulder.

Sanwal, M. (1989). What we know about mountain development: common property, investment priorities, and institutional arrangements. *Mountain Research and Development*, **9** (1).

Schmandt, J. and Clarkson, J. (1993). *The Regions and Global Warming: Impacts and Response Strategies*. Oxford University Press, New York.

Sen, A. (1981). *Poverty and Famines: An Essay on Entitlement and Deprivation*. Oxford University Press, Oxford.

Sharma, P. and Banskota, M. (1992). Population dynamics and sustainable agricultural development in mountain areas. In: Jodha, N.S., Banskota, M. and Partap, T. (eds.) *Sustainable Mountain Agriculture*. Oxford and IBH Publishing, New Delhi.

Sharma, P. and Partap, T. (1993*). Population, Poverty and Development Issues in the Hindu Kush-Himalayas.* Paper presented at the International Forum on Development of Poor Mountain Areas. March 22-27, Beijing, China.

Singh, V. (1992). *Dynamics of Unsustainability of Mountain Agriculture*. (An unpublished report on U.P. hill areas, India.) ICIMOD, Kathmandu.

Shrestha, S. (1992). *Mountain Agriculture: Indicators of Unsustainability and Options for Reversal.* MFS Discussion Paper 32. ICIMOD, Kathmandu.

Shrestha, S. and Yadav, Y. (1992). *The Farmer's Strategies for Sustainable Mountain Agriculture in the Middle Hills of Nepal.* MFS Discussion Paper 28. ICIMOD, Kathmandu.

Shutain, G. and Chunru, H. (1989). *Problems of the Environment in Chinese Agriculture and a Strategy for Ecological Development (an overview).* Ministry of Agriculture and Beijing Agricultural University, Beijing, China.

Songqiao, Z., Qinye, Y. and Yuancun, S. (1990). *Systems of Vertical Agriculture in Mountain Areas: A Comparative Study of Hengduan and Qilian Mountains of the Tibetan Plateau.* MFS Discussion Paper, 22. ICIMOD, Kathmandu.

Swarup, R. (1991). *Agricultural Economy of Himalayan Region.* G.B. Pant Institute of Himalayan Environment and Development, Almora, India.

Topping, J.C., Qureshi, A. and Sherer, S.A. (1991). *Implications of Climate Change for the Asian and Pacific Region.* Paper presented at the Asian Pacific Seminar on Climate Change. January 23-26, 1991. Nagoya, Japan.

Troll, C. (1988). Comparative geography of high mountains of the world in the view of landscape ecology: a development of three and a half decades of research and organisation. In: Allan, N.J.R., Knapp, G.W. and Stadel, C. (eds.) *Human Impacts on Mountains.* Rowman & Littlefield, New Jersey.

Turner, B.L., II, and Benjamin, P. (1993). Fragile lands and their management. In: Ruttan, V.W. (ed.) *Agriculture, Environment and Health: Towards Sustainable Development into 21st Century.* University of Minnesota Press, Minneapolis.

Turner, B.L. II, Kasperson, R.E., Meyer, W.B., Dow, K.M., Golding, D., Kasperson, J.X., Mitchell, R.C. and Ratick, S.J. (1990). Two types of global environmental change: definitional and spatial-scale issues in their human dimension. *Global Environment Change,* **1** (1): 14-22.

Walker, T.S., Singh, R.P. and Jodha, N.S. (1983). *Dimensions of Farm-Level Diversification in the Semi-arid Tropics of Rural South India.* Economics Programme Progress Report 51. ICRISAT, Patancheru, India.

Walker, T.S. and Ryan, J.G. (1990). *Village and Household Economies in India's Semi-arid Tropics.* Johns Hopkins University Press, Baltimore.

Whiteman, P.T.S. (1988). Mountain agronomy in Ethiopia, Nepal and Pakistan. In: Allan, N.J.R., Knapp, G.W. and Stadel, C. (eds.) *Human Impacts on Mountains.* Rowman & Littlefield, New Jersey.

Whitaker, Meri L., Shenoi, P.V. and Kerr J.M. (1991). Agricultural sustainability, growth and poverty alleviation in the Indian semi-arid tropics. In: Vosti, S.A., Reardon, T., von Urff, W. and Witcover, J. (eds.) *Agricultural Sustainability, Growth and Poverty Alleviation: Issues and Policies.* Proceedings of the conference September 23-27, Food and Agriculture Development Centre, Feldafing, Germany.

Wisner, B. (1988). Of crunching numbers and milling grains: are international and family food security comparable? Background paper, Faculty Workshop on *Issues in International Food Security,* June 13-19. Colgate University, Hamilton, New York.

Wuebbles, D.J. and Edmonds, J. (1988). *A Primer on Greenhouse Gases.* National Technical Information Service, US Department of Commerce, Springfield, Virginia.

Yadav, Y. (1992). Farming-forestry-livestock linkages: a component of mountain farmers' strategies. In: Jodha, N.S., Banskota, M. and Partap, T. (eds.) *Sustainable Mountain Agriculture.* Oxford and IBH Publishing, New Delhi.

Global Climate Change and Agricultural Productivity in Southern Africa : Thought for Food and Food for Thought

Roland E. Schulze, Gregory A. Kiker and Richard P. Kunz
Department of Agricultural Engineering
University of Natal
Pietermaritzburg 3201
South Africa

1 Background

The population of southern Africa (defined in the context of this paper as the Republic of South Africa, plus the Kingdoms of Lesotho and Swaziland) is projected to increase from its present 40 million to between 70 and 90 million by the year 2035. To meet the food demands of this growing population, crop production will have to expand at three per cent per annum (Arbuthnot, 1992). This will not be an easy task, however, as the southern African subcontinent is largely semi-arid and sub-humid, has a diversity of soils, physiography, agricultural crops grown, and management levels at which they are grown. Above all these factors, however, is the wide range of climates, characterised by a marked intra-seasonal and inter-annual variability of rainfall. In terms of resource management, this is a high-risk environment which, in the agricultural industry, be it in the commercial or subsistence sector, implies in most areas uncertain production, frequent crop failures and consequently a drain on state finances through subsidies and drought relief.

Of the total land area in southern Africa most is used for agriculture. In South Africa, for example, this proportion is 89 per cent of an area of 1.22 million km^2. However, 17 000 ha of agricultural land are lost per annum to industrial expansion, mining and formal as well as informal urban expansion. A further 18 000 ha of agricultural land per annum are currently being converted to commercial afforestation (Arbuthnot, 1992). This shrinking area of agricultural land is, *inter alia*, forcing agriculture into climatically more marginal areas associated with even more delicate sensitivities to climate and to economic risks (Human, 1992), implying that crop production per unit area in the higher agricultural production areas will need to be increased substantially.

Superimposed on to this already high-risk environment are the further considerable uncertainties and possible threats linked with as-yet-unknown regional impacts of an anticipated global climate

NATO ASI Series, Vol. I 37
Climate Change and World Food Security
Edited by Thomas E. Downing

change associated with the augmented greenhouse effect. With the manifold questions about the magnitudes and directions of climate change, CO_2 'fertilisation' effects, possible system feedbacks, spatial shifts in agricultural belts, sensitivities to crop yields and changes in vulnerability to crop failure, global climate change has ramifications into food production, food supply, its distribution subsystems, inter-regional food dependence and international trade.

Any national or regional agricultural plan, food security strategy or resource management programme requires, as a prerequisite, a basic knowledge and interpretation of the intrinsic environmental production capacity of the region, its regional distribution patterns and its natural variability. This is necessary in general agricultural terms as well as in regard to the production potential of the region's staple food crop. If, in addition, climate change poses a threat to food security and a region's economic viability, then there is a strong motivation to assess impacts of climate change over and above those of natural variability, particularly if it is surmised that the influences of a climate change on the inter-annual variability and thus risks of yields is likely to be more problematic than its consequences on average yields.

2 Objectives

To address the above problems related to food security, a first step is to set up the infrastructure for, and to then apply, appropriate simulation models of production at an appropriate level of regional disaggregation of this physiographically and climatically diverse region. Climatic, edaphic, plant growth rate and agronomic data/information sets, to be used as inputs to these models, also have to be obtained.

This paper outlines the scientific foundations of a food security study by assessing, mapping and interpreting, for the southern African region, the distributions and inter-annual variabilities for:

- Net above-ground primary production, selected as an expression of the intrinsic environmental production capability of a region.
- Maize yields (for both commercial and subsistence management conditions), maize being selected because it occupies 58 per cent of the cropland and is the staple grain crop of southern Africa.
- Distributions mapped for present and future climate scenarios.
- A daily agrohydrological model (*ACRU*, in which is embedded the CERES crop models).

To preface the results and their possible implications to regional food security, the system infrastructure (i.e. regionalisation, databases), the models that were selected, *viz. ACRU* (Schulze, 1989) and CERES (International Benchmark Sites Network for Agrotechnology Transfer, 1993) and their interlinkages, the concepts of primary production, the assumptions made in the maize yield model and the proposed climatic perturbations associated with an enhanced greenhouse effect are first discussed.

3 System infrastructure

3.1 Regional disaggregation

Because of southern Africa's altitudinal range (0-3000 m) and diverse rainfall climate (mean annual precipitation, MAP, less than 50 mm to over 3000 mm), the region has been delineated into 712 relatively homogeneous response zones of differing area according to local variations of, *inter alia*, MAP, altitude, relative relief, aspect and the intensity of agricultural production as criteria (Dent *et al.*, 1988). The 712 relatively homogeneous response zones were mapped on a geographic information system (GIS). The GIS was then linked with appropriate geocoded vegetation, soils and climate information bases.

3.2 Climate data and information bases

The climate data and information bases for each zone consists of the following.

Rainfall. A representative rainfall station for each zone was selected from the national rainfall databases. The main criterion for station selection was a minimum record length of 40 years of reliable daily data, with any missing data filled synthetically (Dent *et al.*, 1988) using recognised techniques for southern Africa developed by Zucchini and Adamson (1984).

Air temperature. A representative temperature station was assigned to each zone so that reliable monthly means of daily maximum and minimum temperature could be obtained. An altitudinal correction factor, based on mean regionalised adiabatic lapse rates for maximum and minimum temperatures as determined by Schulze and Maharaj (1993), was used to adjust the mean monthly temperature data for differences between the average elevation of the zone and the elevation of the temperature station. Daily values of maximum and minimum temperature are derived internally within the *ACRU* model from the monthly means using Fourier analysis (Schulze, 1989).

Reference potential evaporation. To assess primary production, the Class A evaporation pan was selected for determining reference potential evaporation, E_r (Schulze, 1989). Because of the lack of a uniformly dense network of A-pans from which to interpolate E_r, monthly A-pan equivalent evaporation estimates were derived from monthly air temperature data using the Linacre (1977) E_r equation. The Linacre equation, a temperature-based surrogate of the Penman (1948) technique, has been modified for southern African conditions to improve E_r estimates by applying a day-length correction factor to its radiation-related term and by making regional and seasonal adjustments to the wind term (Dent *et al.*, 1988). Daily values of E_r are determined within *ACRU* from the monthly totals using Fourier analysis. Each daily E_r amount is further perturbed by enhancing it by five per cent on non-rain-days and suppressing evaporation by 20 per cent on each rainday with less than 5 mm rainfall (Dent *et al.*, 1988).

Solar radiation. The CERES model uses a version of the Priestley and Taylor (1972) equation to estimate a reference potential evapotranspiration. This equation requires daily input of incoming solar radiation, which has been measured for long durations at only a few meteorological stations in southern Africa (Reid, 1981). However, solar radiation can be estimated from sunshine duration using the Ångström (1924) equation, and sunshine duration is measured reliably at several hundred sites in southern Africa. Reid (1981) determined regional solar radiation:sunshine duration relationships for southern Africa, from which he produced 12 monthly isoline maps of incoming solar radiation. These isolines were geocoded into a GIS to obtain representative monthly values for each of the 712 climate zones. *ACRU* converts the monthly incoming solar radiation to daily values for each zone by Fourier analysis.

3.3 Vegetation

To determine the vegetation water-use in each of the 712 zones, vegetation parameters such as crop coefficients, rooting distributions and rainfall interception values are required as input into the *ACRU* modelling system. Crop coefficients are used to determine the maximum soil water evaporation and transpiration loss by vegetation. Vegetation interception rates are input to determine, *inter alia*, effective rainfall amounts; rooting distributions within the various soil horizons are used in *ACRU* for estimating soil water uptake rates by plants.

Table 1 Generalised vegetation input parameters for the *ACRU* modelling system, derived from minimum temperatures (Schulze and Lynch, 1992)

Monthly mean of daily minimum temperature (°C)	Crop coefficient	Fraction of roots in topsoil horizon	Interception (mm per rainday)
< 5	0.20	1.00	0.40
5-6	0.30	1.00	0.60
6-7	0.40	0.95	0.80
7-8	0.50	0.90	1.00
8-9	0.65	0.85	1.30
9-10	0.80	0.80	1.60
> 10	0.85	0.75	1.70

As there is at present no generalised land cover map (other than for natural vegetation) for southern Africa that can be used in the primary production component of this study, these vegetation parameters were generated using minimum temperatures. Thus a new season's potential growth commenced, proceeded and senesced according to a set of minimum temperature thresholds (Table 1). A natural vegetation of grassland and deciduous shrub was assumed and values were based on experience of the authors, of ecologists and grassland scientists. The temperature thresholds of the vegetation parameters applicable to the *ACRU* model are summarised in Table 1 (Schulze and Lynch, 1992).

3.4 Soils

Soils information pertaining to southern Africa was obtained from the Institute for Soil, Climate and Water, which has identified 84 'broad natural homogeneous' soil regions. The soils were classified, using the South African binomial system, into 41 soil forms, each made up of a vertical sequence of diagnostic horizons easily identifiable in the field, which in turn were further subdivided into a total of 501 soil series according to a variety of soil physical and chemical criteria (McVicar *et al.*, 1977). Each soil form and its range of soil series has been correlated with the United States Department of Agriculture (USDA) Comprehensive System (Soil Survey Staff, 1960) and the Food and Agriculture Organisation (FAO) World Map Legend (Dudal, 1968) to allow for international intercomparison. A computerised decision-support system was developed to interpret the soil classification system hydrologically (Schulze, 1989) so as to derive necessary information such as soil water retention constants, drainage rates and the respective soil horizon depths for each soil form and series. The information from the soils map was again geocoded into a GIS to obtain relevant soil parameter values for each of the 712 climate zones.

4 Climate change scenarios

To develop a perturbed climate input for southern Africa to simulate impacts of the enhanced greenhouse effect, a number of climate change scenarios were derived from an extensive literature review (Schulze and Kunz, 1993). The climatic change scenarios for southern Africa, which are summarised below, and which are based on GCM output, interpretation of past analogues, expert opinion and adaptions of overseas studies, have been adopted by the South African Global Change Committee and task groups of the Interdepartmental Coordinating Committee on Global Environmental Change (ICC) for use in impact studies in southern Africa.

The 'present' base level of atmospheric CO_2 concentration for modelling purposes is taken to be 330 ppm, and this is perturbed to 555 ppm, a level assumed to be attained between the years 2030 and 2050.[1] This approximates a doubling of the pre-industrial revolution (1850) CO_2 level of 280 ppm and is assumed in the models to change instantaneously when 'doubling of CO_2 scenarios' are used.

The incident solar radiation received at the earth's surface is a function of the extraterrestrial radiation, the atmosphere's ability to absorb, reflect and scatter the incoming solar radiation (i.e. atmospheric transmissivity), the degree of cloud cover, and the portion of the incident shortwave radiation that is reflected back to the atmosphere by the earth's surface.
Extraterrestrial radiation, being mainly a function of the sun-earth distance, season and latitude is considered unaffected by a change in the earth's climate. Atmospheric transmissivity to incoming solar radiation is expected to change negligibly with global warming. As precipitation changes are not considered in this impact assessment, cloudiness is assumed to remain unaffected by climate change. Therefore, for this study, incoming solar radiation has been taken to remain unaltered with climate change.

---

1 Editor's note: current projections for a 'business-as-usual' scenario are for CO_2 concentrations of 480 in 2030 and 590 in 2050 (Viner and Hulme, 1993).

The following basic premises were considered in developing temperature change scenarios for southern Africa.

- Global mean air temperature is predicted from GCMs to increase by 1.5-4.5°C for an effective doubling of CO_2, with a 'best estimate' of about 2.5°C (Wigley and Raper, 1992). For southern Africa, high-resolution GCMs predict a 2°C rise (Mitchell, 1991).
- The regional increase in temperature will depend on latitude (ϕ), increasing at higher latitude (Tyson, 1992).
- Diurnally, minimum temperatures are hypothesised to increase more than maximum temperatures.
- Seasonally, winter warming is likely to be greater than summer warming (Houghton *et al.*, 1992).

Based on the above and assuming a 2°C increase in mean temperature at 30°S latitude, the following temperature change algorithm for southern Africa, and used in this study, was developed by Schulze and Kunz (1993), accounting for latitudinal dependence, diurnal differences between anticipated maximum and minimum temperature changes and seasonal temperature differences such that (see Figure 1):

$$\Delta T = \Delta T_\phi(0.9+F) \quad (1)$$

and $$\Delta t = \Delta T_\phi(1.1+F) \quad (2)$$

where ΔT = temperature change for maximum temperature

Δt = temperature change for minimum temperature

$$\Delta T_\phi = \phi/30\Delta T_{30} \quad (3)$$

where ΔT_ϕ = latitudinally adjusted temperature

ΔT_{30} = temperature increase of 2°C at 30°S

ϕ = latitude south

and F = seasonally adjusted temperature change

with $$F = -0.045*12/S+\cos Z \quad (4)$$

where $S = (12\text{-}I)$ when $(1 \le I \le 6)$

$= I$ when $(6 \le I \le 12)$

I = month of the year (1,..,12)

$Z = I\pi/6$

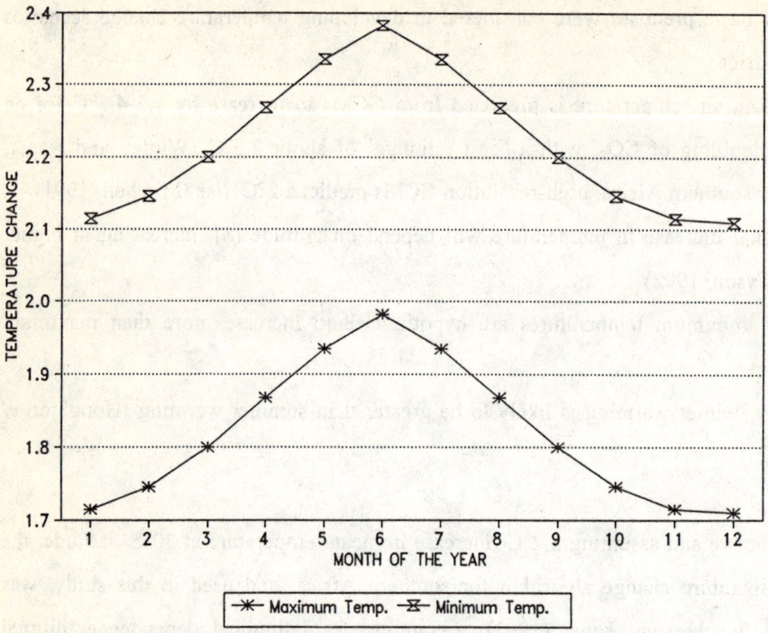

Figure 1 Example of the seasonal scenario changes in minimum and maximum temperatures used in southern Africa for a mean air temperature increase of 2°C at latitude 30°S (Schulze and Kunz, 1993).

A reference potential evaporation change of three per cent per degree C warming was adopted for southern Africa (Schulze and Kunz, 1993), based on suggestions of a similar magnitude for Australia (Pittock, 1991), computations involving the Penman (1948) equation and the fact that the capacity of air for water vapour increases by 5-6 per cent per degree C warming (Rosenberg *et al.*, 1989).

To avoid introducing more uncertainty into this study it was decided, on the basis of the variable and sometimes contradicting GCM predictions of rainfall changes for southern Africa (reviewed by Schulze and Kunz, 1993) and local climatologists' reporting that the present level of understanding regional changes in rainfall over southern Africa could not as yet justify any accurate quantification of rainfall changes, that no changes in rainfall amounts, intensities or temporal/spatial distributions over southern Africa would be input in model runs.

5 Primary production: concepts and estimation

Primary production is a quantitative expression of vegetative matter (e.g. harvestable yield) that can be produced by the natural environment at a location per unit area (e.g. tons per hectare) over a given period of time (e.g. in a season or a year).

It may be conceptualised as a generalised expression of sustainable yield and agricultural production and is a quantification of the dryland (i.e. rain-fed) biomass production expectation of a location. The concept is particularly useful and objective in assessing intrinsic environmental production capability and in comparing the environmental resource potential of one location with others. As such, the concept of primary production can be used as a fundamental tool in overall resource, agricultural and sustainable development planning of a country or, alternatively, a region.

In a major contribution over 25 years ago, Rosenzweig (1968) used total evaporation (i.e. 'actual' evapotranspiration, E) estimates derived from soil water budgeting techniques from 26 mature and undisturbed plant communities in widely ranging environments to derive a generalised equation for net above-ground (i.e. harvestable) production, measured as dry organic matter synthesised per area per unit time:

$$\log_{10}\mathrm{NAAP} = 1.66(\pm 0.27)\log_{10}\mathrm{E} - 1.66(\pm 0.07) \qquad (5)$$

where

NAAP = net annual above-ground production ($g.m^{-2}$)

E = annual total evaporation (mm)

Total evaporation in a natural environment synthesises the two most variable photosynthetic resources, *viz.* water and energy. It is a measure of the simultaneous availability of soil water and atmospheric evaporative demand of a plant community at a given stage of growth in a given period of time and it is the amount of water actually entering the atmosphere through the soil/vegetation complex. Being a quantifiable measure of the energy flow in a plant community, it is a fundamental predictor of production.

6 Simulation models

General features of the *ACRU* and CERES models, used respectively to simulate primary production and maize yields, are presented below.

6.1 The *ACRU* model

ACRU is a physical-conceptual and menu-driven integrated multi-purpose and multi-level modelling system revolving around daily time-step multi-layer soil water budgeting and containing decision-support systems (Schulze, 1989). In addition to run-off components, irrigation demand/supply, reservoir and crop yield options it outputs, *inter alia*, soil water status and total evaporation on a daily basis. The general structure of the model is illustrated schematically in Figure 2. When operated for multiple land uses within a catchment or for multiple sub-regions making up a larger, more complex region *ACRU* operates on an interlinked cell-based structure.

The model partitions and redistributes soil water under saturated and unsaturated conditions and computes evaporation rates from plant intercepted water and the various soil layers according to seasonally varying leaf area index, crop coefficients and root distribution and the relative wetness of respective soil horizons.

For the primary production study, the crop coefficient is generated from minimum temperature according to Schulze and Lynch (1992), implying more rapid vegetative growth after the dormant season in a warmer future temperature scenario (Table 1). In an assumed effective doubling of atmospheric CO_2, transpiration rates are suppressed in the model as a result of increased stomatal resistance associated with enhanced CO_2 concentrations. In *ACRU,* a doubling of CO_2 induces a 33 per cent transpiration suppression for C4 plants, and 22 per cent for C3 plants.

It is vital in primary production modelling to determine at what critical fraction, in the depletion of the plant available soil water reservoir (PAW), plant stress may be assumed to set in, as stress implies a reduction in dry matter production. Although this fraction varies according to daily atmospheric demand and the plant's critical leaf water potential, it was set at 40 per cent of PAW, a value typical for many plants, for the primary production

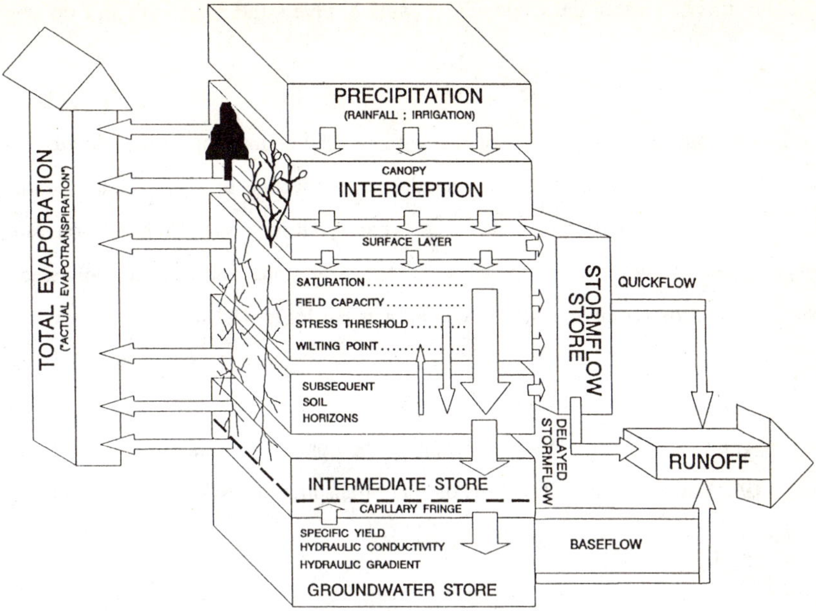

Figure 2 Structure of the *ACRU* agrohydrological modelling system.
Source: After Schulze (1989).

simulations plant stress occurs, the crop coefficient is reduced using an exponential decay function with time from onset of stress (Hughes, 1992). When plant stress is relieved after rainfall, the crop coefficient recovers linearly at a rate dependen on the daily mean air temperature (Hughes, 1992).

For the Rosenzweig (1968) primary production function within *ACRU,* the daily E generated by the model was accumulated over a growing season from July 1 to June 30 (i.e. southern hemisphere mid-winter to mid-winter) to yield annual primary production over the period of climatic record. Various statistics could then be gleaned from the annual series of annual primary production values.

6.2 The CERES model

CERES Version 3.0 is a composite model created from earlier CERES crop models (Versions 2.1 and 2.5), simulating development and yield of maize, wheat, sorghum, millet and barley. The CERES Version 3.0 model is available with the Decision Support System for Agrotechnology Transfer (DSSAT) Version 3.0 (International Benchmark Sites Network for Agrotechnology Transfer Project, 1993). Many new features are available with CERES

Version 3.0, but it is not within the scope of this paper to give a full presentation of the new model.

The CERES model operates on a daily time-step with soil and climate inputs to simulate crop development and yield (Ritchie, 1986). The model is a collection of specialised functions calculating various crop processes such as photosynthesis, evapotranspiration and phenological stages. In addition, the model simulates daily water and nitrogen movement through the plant and soil (Godwin *et al.*, 1984; Godwin and Jones, 1991).

CERES Version 3.0 is useful in climate change research because of its responses to climate change inputs. The CO_2 concentration affects daily canopy photosynthesis by a scaling factor (Rosenzweig, 1988). In addition, CO_2 effects on plant transpiration are simulated by altering stomatal conductance (Peart *et al.*, 1989).

6.3 Linkage of CERES and *ACRU* models

The *ACRU* and CERES models were linked together using *ACRU*'s cell-based structure as a driver program while the CERES model was restructured as a subroutine to execute within the *ACRU* daily loop. The combination of the two models allowed the sharing of rainfall, temperature, solar radiation and soils input information. Users can combine the soils and climate data from either model to simulate either crop growth and/or hydrological related responses.

The *ACRU*/CERES hybrid can be executed in two modes, *viz.* discrete and continuous. The discrete simulation mode starts the crop model on a specific day and ends immediately after harvest. The crop management dates are reset to the next year and *ACRU* continues execution alone until the following growing season's beginning date, when the crop model is executed once again. At the beginning of each discrete simulation, the soil water and nutrient variables are reset to user-prescribed initial conditions.

In continuous simulation mode, the crop model is started on a specific date and simulates all crop functions until harvest. After the crop has reached harvest maturity, the model continues to simulate only the soil water budget, with no crop growth routines executed, resulting in a 'fallow' period until the next seeding date. In this option, none of the soil water or nutrient values are reset to initial conditions at planting, but commence with 'actual' (simulated) values from the continuous simulation.

6.4 Data inputs: management, environment and soils

In southern African maize production, management practices range from highly mechanised to subsistence production. Climate changes in marginal production areas may have a more adverse effect on the smaller farmers (Downing, 1992). One of the most important differences between the two groups is access to fertiliser inputs. In this study, two nitrogen fertilisation schedules were analysed. The first schedule assumed optimum nitrogen availability (i.e. no N stress) to reflect the highly mechanised commercial agricultural practices. The second schedule assumed two nitrogen fertilisations, each of 8.0 kg/ha NO_3, to simulate nitrogen-limited subsistence production. The first fertilisation occurred on October 1, just before the planting date, and the next fertilisation occurred in the first week of January of the next year, midway through the growing season.

Both management practices used the same method for determining a suitable planting date. The planting date was set by an automatic function within CERES (Bowen, 1993) that calculates available soil water in the top 150 mm of soil. The *ACRU*/CERES hybrid was run in the discrete mode, beginning the simulation on September 16 and 'warming up' until the first possible planting date on October 1. The window of opportunity for planting continued until December 21. For continuity in the *ACRU* program, if the desired planting conditions were not met within the window of opportunity, then the crop was assumed to be planted on December 21. To reflect typical southern African cultural practices, the seeding depth was 50 mm with a row spacing of 1.5 m and a plant population of 2.1 plants/m^2 (Sentraoes, 1984). All other management practices were the same in both present and future climate scenarios. Pest effects were not included in the simulations. One cultivar, a medium length of growing season maize variety, was used for all climate zones (Berry, 1993). The genetic coefficients are listed in Table 2.

In using the *ACRU*/CERES hybrid for climate change simulations in southern Africa, all environmental inputs were coordinated through the *ACRU* menu input system. Climate inputs for each response zone included daily rainfall, as well as monthly means of daily maximum and minimum temperature and solar radiation. To compensate for reductions in solar

Table 2 Genetic coefficients for maize used in the *ACRU*/CERES model

Genetic parameter description	Parameter value
Degree days (base 8°C) from emergence to the end of juvenile stage ($\Sigma°C_8$-day)	270
Photoperiod sensitivity constant (dimensionless)	0.75
Degree days (base 8°C) from silking to physiological maturity (Σ $°C_8$-day)	710
Potential kernel number (kernels/plant)	730
Potential kernel growth rate (kernels/day)	7.3

radiation on days when precipitation occurred, a simple perturbation was invoked based on southern African research findings on evaporation fluctuations by Dent *et al.* (1988), *viz.* if the daily rainfall was greater than 5 mm then the mean daily solar radiation value was reduced by 20 per cent, otherwise the mean daily radiation value was enhanced by five per cent.

Daily potential evapotranspiration was calculated within the CERES model using a modified Priestley-Taylor equation (Jones *et al.*, 1986) with daily solar radiation inputs. Rainfall was not altered for future climate change scenarios, as was the case with primary production simulations, and temperatures were changed as described previously.

Soil information from standard *ACRU* files for the 712 climate zones supplied all but three of the necessary CERES soils inputs. The values for the Soil Conservation Service (SCS) curve number, the upper limit of stage one soil water evaporation and the soil albedo were estimated using *ACRU* soils input for the 712 regions in conjunction with the assumptions presented by Ritchie *et al.* (1986). The soil profile in each zone was checked for a minimum depth of 0.75 m. For the maize yield simulations, soil profiles less than this minimum depth were assigned a 0.75 m depth on the premise that only the deeper soils within a zone would be cultivated.

Initial soil water and nitrogen levels were reset each year on the simulation starting date of September 16, two weeks before the first possible planting date, on which date the soil moisture levels were also initialised at 50 per cent available water capacity for all layers. The soil nitrogen levels for ammonium and nitrate were both set at 2.0 g elemental N/mg soil. The two-week period following the simulation starting date allowed the soil water conditions to be adjusted by the local climate conditions occurring for that specific year of simulation.

7 Results from simulations: thought for food

This paper has considered a vegetative response impact analysis that has concentrated directly on estimating the primary effects of environmental variables on crop yields. The approach used in this study has been to quantify the effects of increasing atmospheric CO_2 concentration, and resultant expected increases in air temperature, on crop response and net above-ground primary production. No precipitation changes were considered because of the large uncertainty surrounding the quantification of such changes. The approach assumes an instantaneous change in climate state and estimates the 'before-and-after' yield effects.

7.1 Net above-ground primary production

Results show that net above-ground primary production estimated using the Rosenzweig equation embedded within *ACRU* ranges from less than 2 t/ha/season in the dry and arid western regions to over 14 t/ha/season along the wet and humid eastern seaboard of southern Africa (Figure 3). The increase in primary production from low to high potential is generally in phase with increasing annual rainfall and decreasing reference evaporation from the west to east coast across southern Africa.

The distribution of inter-annual variability of primary production (Figure 4), expressed as the percentage coefficient of variation, shows that the arid west and the northern boundary of southern Africa are characterised as extremely high-risk areas, whereas the eastern and southern regions, although still high risk by most major crop production standards, display lower inter-annual variability.

Figures 3 and 4 illustrate that soil water availability exercises the overriding control on primary production of most South African ecosystems. Net soil water availability is influenced both by supply (rainfall) and demand (evaporation). Therefore, all else being equal, if the rainfall remains the same but temperatures rise, plants should experience lower soil water availability, higher stress frequencies and production should decrease proportionately (Scholes, 1992). More rapid plant development at higher temperatures on the one hand, and the 'fertilisation' effect of increased CO_2 concentrations on the other, could, however, bring into effect feedback.

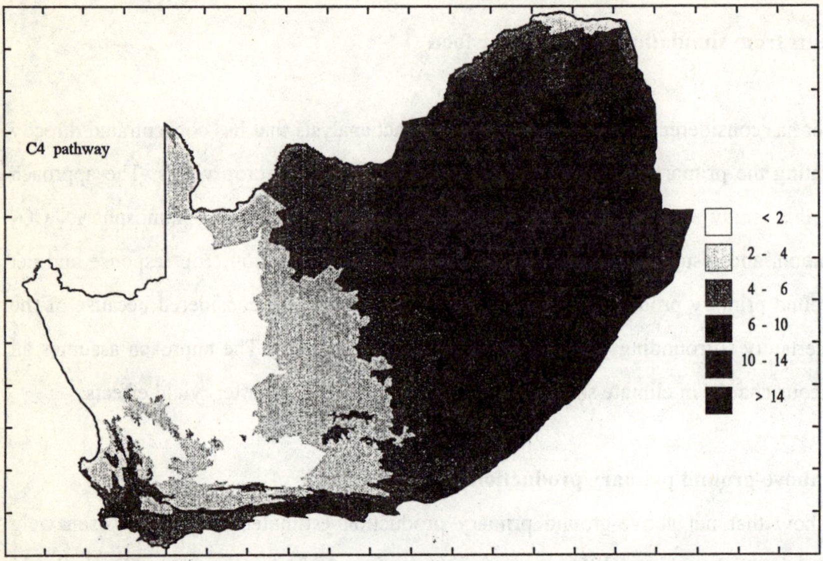

Figure 3 Simulated mean primary production (t/ha/season) over southern Africa for present climatic conditions.

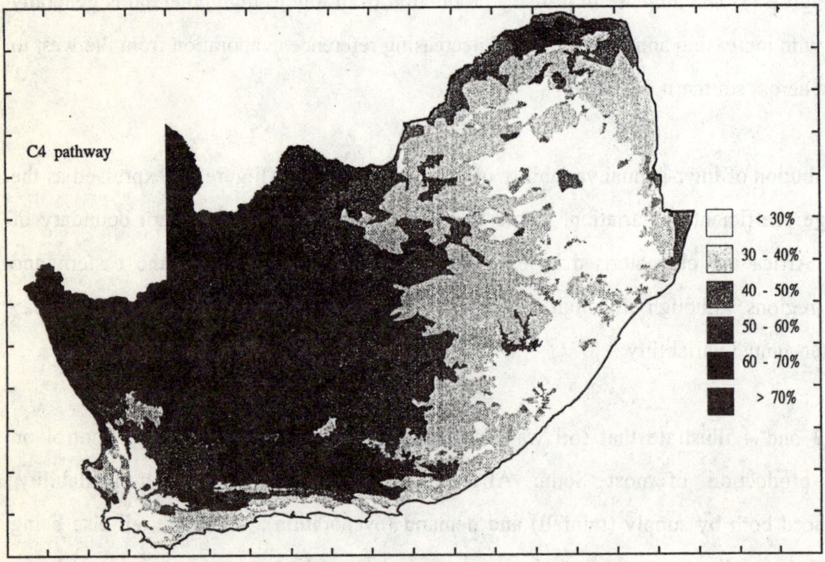

Figure 4 Inter-annual coefficient of variation (%) of primary production over southern Africa.

As 95 per cent of the natural vegetation of southern Africa consists of C4 grasses (Ellis *et al.*, 1980), primary production for a possible future climate with a doubling of CO_2 was subjected to a transpiration suppression of 33 per cent (typical for C4 plants in *ACRU*), with the temperature and reference evaporation perturbations in the model as outlined previously. The results (shown in Figure 5) show similar spatial trends as for the present climate except that primary production is simulated to decrease generally over southern Africa.

A comparison of the changes in primary production from the present conditions to a possible future climate (Figure 6) reveals that the smallest decreases (in t/ha/season) occur in the western regions and the largest decreases occur in the eastern regions.

7.2 Maize yields

Figure 7 displays the mean maize yields simulated for present climate and nitrogen-unlimited conditions. Yields in areas with MAP of less than 300 mm are not shown in this and subsequent figures relating to maize. The areas of highest maize production per ha are found on the eastern coast and north-eastern highland plateau regions where MAP exceeds 600 mm with relatively low variability from year to year. Yields in these regions average

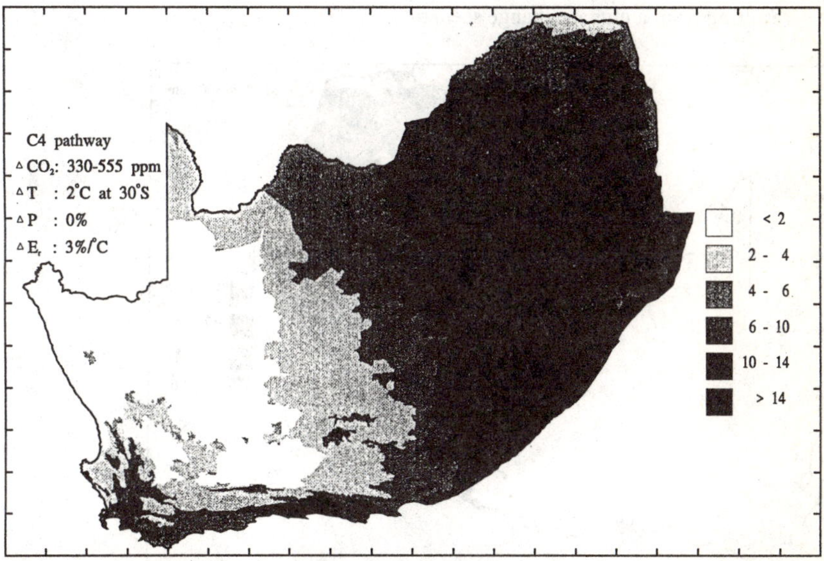

Figure 5 Simulated mean primary production (t/ha/season) over southern Africa for a possible future climate scenario.

between 4.0 and 8.0 t/ha. The more marginal production areas are found to the west of the main production areas. The decrease in MAP and its increased year-to-year variability results in lower mean yields, with higher incidences of crop failure due to severe water stress.

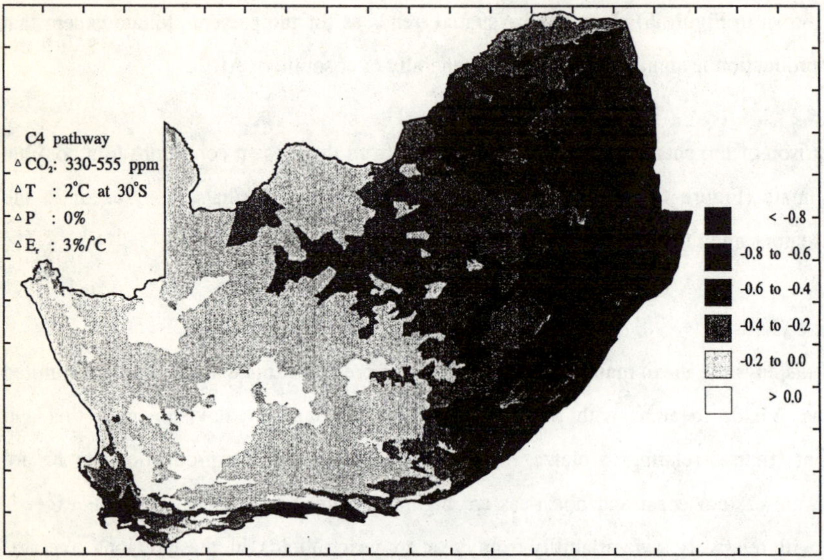

Figure 6 Differences (t/ha/season) in simulated primary production over southern Africa resulting from a climate change scenario.

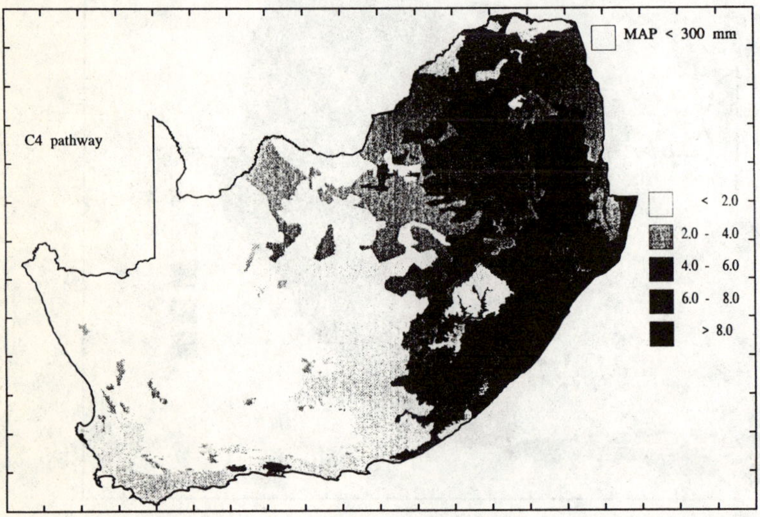

Figure 7 Simulated mean maize yields (t/ha/season) over southern Africa for present climate and nitrogen-unlimited conditions.

In the nitrogen-limited simulations, the spatial patterns of production follow the same geographical trends as those for nitrogen-unlimited production. However, the yields have been reduced significantly by nitrogen stress. In Figure 8, nitrogen deficiency alone has decreased the mean yields to less than one-third of the nitrogen-unlimited yields. This reduction is not uncommon when subsistence yields are compared with those from top commercial enterprises. The yield reduction is observed throughout all maize-producing areas with westward decreases again being due to diminishing rainfall.

Figure 9 depicts the mean maize yields for nitrogen-unlimited simulations in a possible future climate. In areas previously producing mean annual yields of more than 8.0 t/ha, the elevated temperature and CO_2 produces little change in yields, as seen in Figure 9. With no nitrogen stress and abundant rains the crops are, according to the model structure, therefore presently already growing in near optimum conditions.

In more variable production areas (yields between 4.0 and 8.0 t/ha), Figure 9 shows an expansion in the areas yielding up to 8.0 t/ha and a westward expansion of increases into areas previously yielding below 4.0 t/ha. The difference in yield between future and present climate scenarios is shown in Figure 10. The largest change in yields, in t/ha, is found in the north-eastern highveld region. In addition, south of the highveld in Lesotho, over 2.0 t/ha changes are simulated.

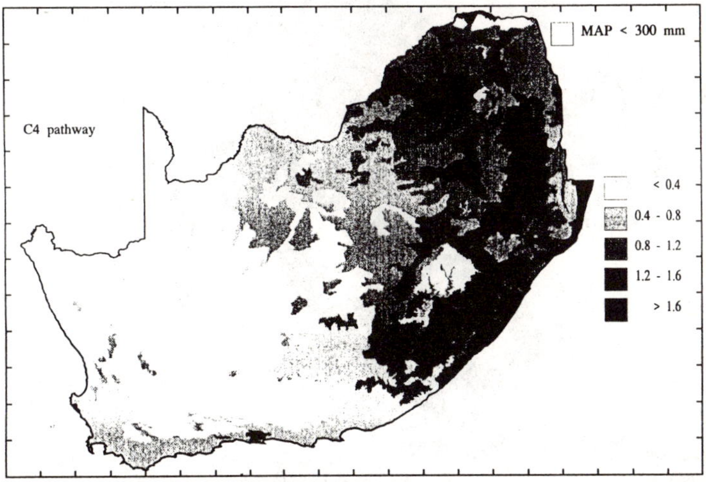

Figure 8 Simulated mean maize yields (t/ha/season) over southern Africa for present climate and nitrogen-limited conditions.

In areas with marginal rainfall for maize production, the climate change scenario has little absolute effect on the already low average yields. Some regions west of the main maize production areas do show slight increases in simulated yields. However, in most low yield

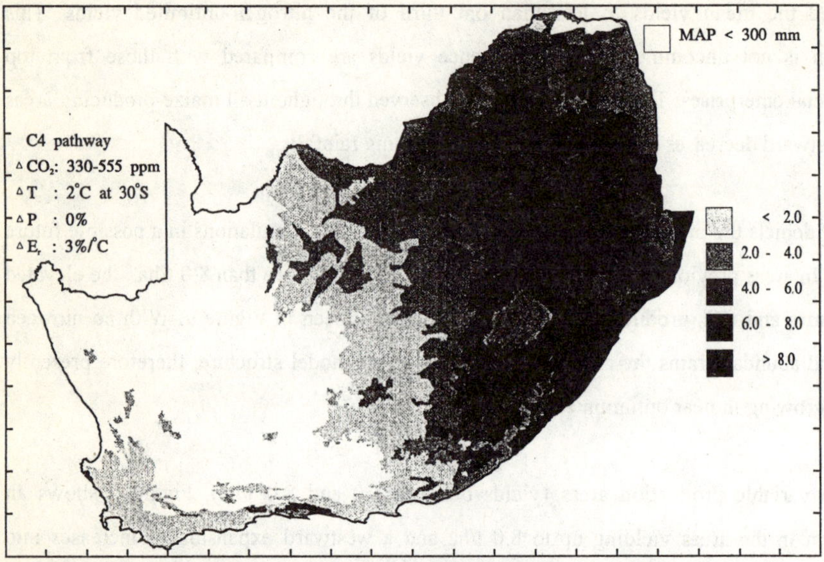

Figure 9 Simulated mean maize yields (t/ha/season) over southern Africa for a possible future climate and nitrogen-unlimited conditions.

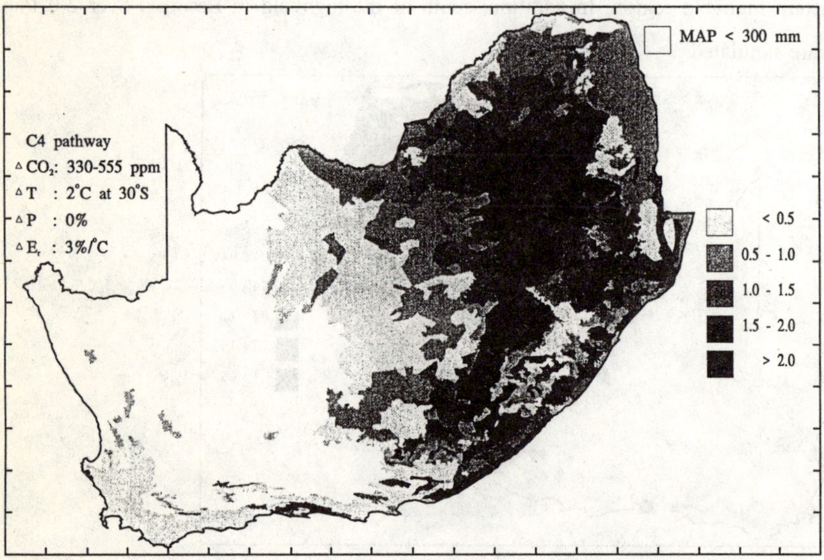

Figure 10 Differences (t/ha/season) in simulated mean maize yields over southern Africa resulting from a climate change scenario.

areas, Figure 10 displays increases of less than 1 t/ha. The same trends are displayed in the nitrogen-limited simulations (distributions not shown in this paper) under possible future climate conditions, again with the increased water-use efficiency leading to larger average yields in high production areas. However, the yield increases are smaller in magnitude due to the significant nitrogen stress.

From an overall food security perspective, therefore, the analysis using a generic primary production function indicates that if negative changes in production occur, they are likely to be relatively minor, whereas the CERES Version 3.0 model in fact points to an overall increase in potential maize production.

8 Points of discussion: food for thought

The results of responses from the *ACRU*/CERES hybridised model to these simple climate change scenarios show a large dependence on rainfall and its intra- and inter-seasonal variation. This is evidenced by nitrogen-unlimited and nitrogen-limited scenarios displaying the same trends in geographical distribution of maize yields, only with the respective values of yields being determined by nitrogen availability.

In these maize yield simulations, the overall effect of temperature increase and CO_2 doubling has been to increase water-use efficiency by reducing plant transpiration. The 2.0°C rise in temperature, reduced in many high production zones by latitudinal and seasonal effects, was more than counteracted by the CO_2 'fertilisation' effect. The primary outcome of this higher efficiency is an increase in mean yields for many areas and thus a reduction in the number of crop failures. This simulated yield increase is seen primarily in regions with MAP between 400 and 800 mm. In the higher rainfall zones, water is not a yield limiting factor to the same degree. As a result, the average yields change very little between simulations of present and future climates. In the lower rainfall zones, the larger number of poor rainfall years coupled with higher average temperatures cancelled any water-use efficiency increases caused by raised CO_2 concentrations. In all climate change scenarios, rainfall variability remains the most important single factor in determining either primary productivity or maize yields. Although changes in average yields do give insights into the climate change question, greater effects on the population and on food security are induced by the impacts of season-to-season variability. This point is illustrated well in Figure 4.

If one major lesson can be learned from the development of the *ACRU*/CERES modelling tool, it would be that fundamental assumptions made in the models play a direct role in the interpretation of simulation results. Two different scales are represented in the models used in the climate change analysis. Primary production is a corporate representation of the environmental capacity through the various climate/soil/vegetative interactions, and as such its distribution is a general index of sustainable agricultural production in southern Africa. The CERES model, on the other hand, simulates individual crops with specific genetic coefficients to delineate cultivar traits. In the primary production model, temperature increases create more rapid early season growth plus an expected elevated potential transpiration, which in turn would be expected to be converted to a higher potential yield by a cumulative transpiration to yield ratio. In suppressing transpiration by as much as 33 per cent, however, to reflect reduced stomatal conductance with higher CO_2 concentrations, the potential yield is reduced because of a lower cumulative transpiration. Assumptions for both CO_2 and temperature change have effects on yield as expected, but the feedforward and feedback mechanisms are relatively simple and their respective magnitudes do not have the same 'sign' as the climate change impacts have with the CERES maize model.

The CERES model circumvents some of the above problems with its more mechanistic physiological basis. Daily canopy photosynthesis is increased by a factor related directly to CO_2 concentration; stomatal conductance is reduced according to crop and CO_2 concentration. The interaction between the two processes is not yet understood fully, i.e. it is not yet known how much CO_2 increase is required to counteract a certain temperature increase and how the water balance is affected. Initial soil moisture and nitrogen conditions are fundamental in simulating environment responses over long periods. The relationship between these two variables is most important, but its significance is lost when initial conditions are reset at each year's beginning simulation date. At this point in the research it seems as though all that can be done is to simulate responses to various accepted climate change scenarios in an effort to improve understanding of the atmosphere/soil/plant continuum, and *not* to interpret the results of climate change scenario impacts at face value.

Clearly, a detailed sensitivity analysis involving both *ACRU* and CERES inputs would help to understand the strengths and the limitations of this powerful integrated tool in the applications that will be demanded of it.

9 Concluding remarks

The natural environment is a complex interactive system. Simultaneous multiple perturbations of this atmosphere/soil/plant system induce further complexities that are compounded by effects of scale and the unknowns of human adaptive responses and of future political developments in southern Africa. As natural and physical scientists many important aspects of food security are not within our control and, for us, the challenges lie primarily in sharpening our modelling skills and predictive capabilities to provide the economic planner and politician with more certainties in answers to questions of climate change.

9.1 What do we not yet know?

A number of important modelling issues have already been addressed in the discussion. However, there remain many other fundamental voids in our knowledge of a future environment and many further constraints in the models which we use to assess its impacts. A few limitations are listed below to illustrate that models are not yet dynamic enough and there is a long road ahead in agricultural impacts studies:

- No changes in farmer perception of climate change and managerial response was incorporated in the models.
- Consequently, technological and cultivation practices affecting crop/climate relationships remained constant in time.
- Soil parameters such as soil water retention constants were assumed to be unaffected by changes in climate variables.
- Root-to-shoot relationships and changing natural fertility through alterations in carbon to nitrogen ratios were also not changed.
- Climate change scenarios at regional scale remain uncertain and speculative at this stage, and precipitation changes may dominate the impacts on agricultural water resources.
- No changes in pests and diseases were assumed, such as likely increases in chilo borer incidence or diplodia root/stem rot in maize (Human, 1992).

Great caution needs to be exercised when interpreting the results of impacts studies.

9.2 What has been achieved?

What this paper has presented is not a series of off-the-shelf answers relating to climate change impacts and food security for southern Africa. The paper has set out to describe the development of a tool with which, given the strengths and constraints of the two models that have been hybridised, scenarios of climate change can be enacted, their responses compared and practical regional issues of food security can now be addressed for southern Africa from a sound scientific platform.

When combining two large modelling systems such as *ACRU* and CERES, the permutations for potential applications are numerous. A major advantage to the linkage of the models is a tool which combines the catchment scale of *ACRU* and field scale of CERES. Hydrological processes through the catchment can thus be linked with detailed field-scale processes and *vice versa*.

The paper has, furthermore, presented a system infrastructure for southern Africa involving detailed climatic, edaphic and vegetative information bases on a scientifically regionalised basis, coupled with the models and integrated with a GIS for purposes of display, comparison and interpretation. This linked modelling system for southern Africa will be used in forthcoming scientifically oriented research to indicate, for example:

- possible thresholds of climate change for significant response to take place;
- sensitivities to variables changing, singly or in combinations;
- regions within southern Africa that may be vulnerable to climate change;
- effects of nitrogen use and efficiency within present and future climate scenarios;
- pest and disease effects with respect to possible climate change.

The foundation has now been laid to couple the model output *via* GIS with issues relating more directly to food security in southern Africa, including:

- analysis of present and possible future production and crop yields and their inter-annual variability based on areas actually under cultivation, district by administrative district;
- analysis of production/variability according to population distribution, accessibility to rural population by road, distance from points of manufacture and export;

- possible management changes to reduce adverse climate change effects on various crops grown in southern Africa;
- analysis of political regionalisation of South Africa into nine provinces and land reform impacts on food production and distribution subsystems.

Acknowledgements

This research was funded by the Foundation for Research and Development, the Water Research Commission and the Fulbright Foundation. Their support, and that of the Computing Centre for Water Research at the University of Natal, is acknowledged gratefully. We thank Mrs C. O'Mahoney for preparation of the document and Mr S.D. Lynch and Ms R.M. Dutlow for their GIS expertise.

This paper is a revised version that first appeared in *Global Environmental Change* Volume 4, Schulze, R.E., Kiker, G.A. and Kunz, R.P., Global Climate Change and Agricultural Productivity in Southern Africa, pp 330-349 (1993). Reprinted with permission from Elsevier Science Ltd, Butterworth-Heinemann Imprint, The Boulevard, Landford Lane, Kidlington OX5 1GB, U.K.

References

Ångström, A. (1924). Solar and terrestrial radiation. *Quarterly Journal of the Royal Meteorological Society,* **50**:121-126.

Arbuthnot, J.D. (1992). The impact of environmental change on southern African agriculture. Report to the Interdepartmental Coordinating Committee for Global Climate Change (ICC) Task Group on the Terrestrial Environment. Department of Environment Affairs, Pretoria, RSA.

Berry, W. (1993). Summer Grain Centre, Cedara Agricultural Research Institute, Cedara, RSA. Personal communication.

Bowen, W. (1993). Decision Support System for Agrotechnology Transfer Version 3.0 (DSSAT V3.0). In: *International Benchmark Sites Network for Agrotechnology Transfer Project (1993).* Department of Agronomy and Soil Science, College of Tropical Agriculture and Human Resources, University of Hawaii, Honolulu.

Dent, M.C., Schulze, R.E. and Angus, G.R. (1988). *Crop Water Requirements, Deficits and Water Yield for Irrigation Planning in southern Africa.* Report 118/1/88. Water Research Commission, Pretoria, RSA. p. 183.

Downing, T.E. (1992). *Climate Change and Vulnerable Places: Global Food Security and Country Studies in Zimbabwe, Kenya, Senegal and Chile.* Research Report No. 1, Environmental Change Unit, University of Oxford, Oxford.

Dudal, R. (1968). *Definitions of Soil Units for the Soil Map of the World.* World Soil Resources Reports No. 33. World Soil Resources Office, FAO, Rome.

Ellis, R.P., Vogel, J.C. and Fuls, A. (1980). Photosynthetic pathways and the geographical distribution of grasses in South West Africa/Namibia. *South African Journal of Science,* **76**:307-314.

Godwin, D.C., Jones, C.A., Ritchie, J.T., Vlek, P.L.G. and Youngdahl, L.J. (1984). The water and nitrogen components of the CERES models. In: *Proceedings of the International Symposium on Minimum Data Sets for Agrotechnology Transfer, Patancheru, India.* International Crops Research Institute for the Semi-Arid Tropics, Andhra Pradesh. pp. 101-106.

Godwin, D.C. and Jones, C.A. (1991). Nitrogen dynamics in soil-plant systems. In: Hanks, J. and Ritchie, J.T. (eds.) *Modelling Plant and Soil Systems.* American Society of Agronomy, Madison. pp. 287-321.

Houghton, J.T., Jenkins, G.J. and Ephraums, J.J. (eds.) (1990). *Climatic Change: The IPCC Scientific Assessment.* Report prepared by Working Group 1, WMO-UNEP. Cambridge University Press, Cambridge.

Houghton, J.T., Callander, B.A. and Varney, S.K. (eds.) (1992). *The Supplementary Report to the IPCC Scientific Assessment.* Cambridge University Press, Cambridge.

Hughes, A.D. (1992). *Sugarcane Yield Simulation with the ACRU Modelling System.* Unpublished M.Sc. dissertation, Department of Agricultural Engineering, Pietermaritzburg.

Hulme, M., Wigley, T., Jiang, T., Zhao, Z., Wang, F., Ding, Y., Leemans, R. and Markham, A. (1992). *Climate Change due to the Greenhouse Effect and its Implications for China.* World Wide Fund for Nature, Gland.

Human, N.B. (1992). *Report to the Interdepartmental Coordinating Committee for Global Climate Change (ICC) Task Group on the Terrestrial Environment.* Department of Environment Affairs, Pretoria.

International Benchmark Sites Network for Agrotechnology Transfer (IBSNAT) Project (1993). *Decision Support System for Agrotechnology Transfer Version 3.0 (DSSAT V3.0).* Department of Agronomy and Soil Science, College of Tropical Agriculture and Human Resources, University of Hawaii, Honolulu.

Jones, C.A., Ritchie, J.T., Kiniry, J.R. and Godwin, D.C. (1986). Subroutine structure. In: Jones, C.A. and Kiniry, J.R. (eds.) *CERES-Maize: A Simulation Model of Maize Growth and Development.* Texas A&M University Press, College Station, pp. 57-58.

Karl, T.R. (1993). A new perspective on global warming. *Transactions of the American Geophysical Union,* **74**:25.

Linacre, E.T. (1977). A simple formula for estimating evaporation rates in various climates using temperature data alone. *Agricultural Meteorology,* **18**:408-424.

McVicar, C.N., De Villiers, J.M., Loxton, R.F., Verster, E., Lambrechts, J.J.N., Merryweather, F.R., Le Roux, J., Van Rooyen, T.H. and Harmse, H.J. von M. (1977). *Soil Classification - A Binomial System for South Africa.* Department of Agricultural Technical Services, Soil and Irrigation Research Institute, Pretoria, p. 150.

Mitchell, J.F.B. (1991). *Simulated Climate Change over Southern Africa in a High Resolution Mixed-Layer Model Experiment.* Hadley Centre for Climate Prediction and Research, Meteorological Office, Bracknell (unpublished paper).

Mühlenbruch-Tegen, A. (1992). Long-term surface temperature variations in South Africa. *South African Journal of Science,* **88**:197-205.

Peart, B., Jones, J.W., Curry, R.B., Boote, K.J. and Allen, L.H., Jr. (1989). In: Smith, J.B. and Tirpak, D.A. (eds.) *The Potential Effects of Global Climate Change on the United States, Appendix C.* US Environmental Protection Agency, Washington, DC.

Penman, H.L. (1948). Natural evaporation from open water, bare soil and grass. *Proceedings of the Royal Society, London,* A **193**:120-146.

Pittock, A.B. (1991). *Developing Regional Climate Change Scenarios: Their Reliability and Seriousness.* Hawkesbury Centenary Conference, University of Western Sydney, Australia. 25-27 November 1991 (mimeo).

Priestley, C.H.B. and Taylor, R.J. (1972). On the assessment of surface heat flux and evaporation using large scale parameters. *Monthly Weather Review,* **100**:81-92.

Reid, P.C.M. (1981). *Energy Aspects of Water-use Efficiency.* Technical Report No. 111. Department of Environment Affairs, Pretoria. p. 217.

Ritchie, J.T. (1986). The CERES-Maize model. In: Jones, C.A. and Kiniry, J.R. (eds.) *CERES-Maize: A Simulation Model of Maize Growth and Development.* Texas A&M University Press, College Station, pp. 3-6.

Ritchie, J.T., Kiniry, J.R., Jones, C.A. and Dyke, P.T. (1986). Model inputs. In: Jones, C.A. and Kiniry, J.R. (eds.) *CERES-Maize: A Simulation Model of Maize Growth and Development.* Texas A&M University Press, College Station. pp. 37-48.

Rosenberg, N.J., McKenney, M.S. and Martin, P. (1989). Evapotranspiration in a greenhouse-warmed world: A review and a simulation. *Agricultural and Forest Meteorology,* **47**:303-320.

Rosenzweig, M.L. (1968). Net primary productivity of terrestrial communities: prediction from climatological data. *American Naturalist,* **102**:67-74.

Rosenzweig, C. (1988). Potential effects of climate change on agricultural production in the Great Plains: a simulation study. In: Smith, J.B. and Tirpak, D.A. (eds.) *The Potential Effects of Global Climate Change on the United States.* Vol. 1:3-1 to 3-30. EPA-230-05-89-050, U.S. EPA, Washington DC.

Scholes, R.J. (1992). *The Impact of Global Change on Terrestrial Ecosystems.* Report to the Interdepartmental Coordinating Committee for Global Climate Change (ICC) Task Group on the Terrestrial Environment. Department of Environment Affairs, Pretoria.

Schulze, R.E. (1989). *ACRU: Background, Concepts and Theory.* Report 154/1/89. Water Research Commission, Pretoria. p. 235.

Schulze, R.E. and Kunz, R.P. (1993). *Climate Change Scenarios for Southern Africa: 1993.* Report to Interdepartmental Coordinating Committee on Global Environmental Change, Pretoria. Department of Agricultural Engineering, University of Natal, Pietermaritzburg. p. 45.

Schulze, R.E. and Lynch, S.D. (1992). Distributions and variability of primary productivity over southern Africa as an index of environmental and agricultural resource determination. In: *ICID International Symposium on Impacts of Climatic Variations and Sustainable Development in Semi-Arid Regions.* Esquel Foundation, Brasilia, p. 16.

Schulze, R.E. and Maharaj, M. (1993). Regional adiabatic lapse rates in southern Africa for maximum and minimum temperatures. Department of Agricultural Engineering, University of Natal, Pietermaritzburg. (Mimeo.)

Sentraoes (1984). *Prosedure vir Potensiaalbepaling van Droëlandmielies.* Report 114. Sentraoes Ko-öperatief Beperk, Ficksburg.

Soil Survey Staff (1960). *Soil Classification - a Comprehensive System (7th Approximation).* US Department of Agriculture, Washington, DC. (Together with supplements to the system dated March 1967 and September 1968.)

Tyson, P.D. (1992). *Recent Developments in the Modelling of the Future Climate of Southern Africa.* 13th Raymond Dart Memorial Lecture, University of the Witwatersrand, Johannesburg. (Mimeo.)

Viner, D. and Hulme, M. (1993). *What Carbon Dioxide Concentrations Should be Used with Climate Change Scenarios?* Technical Note No. 3. CRU/UEA, Norwich.

Wigley, T.M.L. and Raper, S.C.B. (1992). Implications for climate and sea level of revised IPCC emissions scenarios. *Nature,* **357**:293-300.

Zucchini, W. and Adamson, P.T. (1984). The occurrence and severity of droughts in South Africa. Report 91/1/84 and Appendix.Water Research Commission, Pretoria.

Climate Change: Some Likely Multiple Impacts in Southern Africa

Chris H. D. Magadza
University Lake Kariba Research Station
PO Box 48
Kariba
Zimbabwe

1 Introduction

The recent drought in the subcontinent provides an analogue model for gaining some insights into the effects of the projected climate changes. Several water storage facilities failed to satisfy the water demands that they were intended to cater for. Two of the major cities of Zimbabwe were consequently put on strict water rations, at a time when there was an outbreak of cholera. The low levels of both Lake Kariba and Lake Kafue rendered this hydro power complex incapable of meeting the power demands of Zimbabwe and Zambia, which consequently had to import power from neighbouring countries. Strict power rationing to industry, implemented as penalties for exceeding energy quotas, caused significant industrial disruption and job losses. Failure of irrigation dams resulted in production losses in the sugar industry, which also affected the gasohol programme of Zimbabwe. The citrus industry suffered large tree losses which will take several years to replenish. At the height of the drought period mortality amongst wild animals was high, and only artificial feeding and translocation to better watered areas saved many large mammals from mortality. Many tree species also died, including the rare *Androstachys johnsonii.*

The drought episode also revealed some alarming unpreparedness on the part of some Southern African governments to cope with drought situations. Zimbabwe, having disposed of a three year maize back up supply prior to the drought, was forced to import this staple food at some considerable inconvenience to the tax payers, who had to sustain an additional 5 per cent drought levy over and above their normal tax rates. Government had ignored early warnings of impending drought from its technical staff. Similarly government showed a disturbing sense of paralysis on how to deal with the serious water shortage of the two major cities. The drought, though variously described as the most severe in living memory, only represented transient conditions in comparison to the general circulation model (GCM) $2xCO_2$ predictions. The impacts of the GCM equilibrium conditions could thus be much more severe than those elicited by the recent drought.

NATO ASI Series, Vol. I 37
Climate Change and World Food Security
Edited by Thomas E. Downing

Since it is projected that global climate change would probably manifest itself in the accentuation of extreme conditions, i.e. more severe droughts, the incident shows that unless Southern African governments take an informed long term strategy on how to cope with climate change, they are likely to remain unprepared for such extreme conditions.

An impact assessment of projected climate change in Southern Africa is qualified by two major considerations. The first of these is the data base. The sparse distribution of climate stations and consequent paucity of data result in a coarse grid resolution, which negatively affects the construction of climate change scenarios. Furthermore, any analogue modelling of climate change, such as the MINK study of Rosenberg *et al.* (1993), is limited by the historical record. However Downing *et al.* (1989) have shown that useful information can be obtained by an intensive study of a drought episode, such as the 1984 drought in Kenya. This paper does not purport to be such an intensive study, but rather uses the limited data immediately available to sketch out areas of inquiry for major research activities, which would collectively provide a more panoramic view of the likely impacts of climate change in the Southern African subcontinent.

The second qualification is the difficulty of separating impacts of climate change due to greenhouse effects, from impacts of anthropogenic effects that may lead to local climatic changes and ultimately desertification, such as those discussed by Mabutt and Floret (1980). Within Southern Africa similar phenomena have been discussed by Magadza (1986), Whitlow (1988) and Darkoh (1989).

It is difficult to separate anthropogenically-induced climate change, such as in the so called Sonoran discontinuity of Mexico (Balling, 1988) due to albedo effects following deforestation, from greenhouse-driven climate change. This is an important qualification in assessing the impact of climate change in Southern Africa, which is currently undergoing rapid deforestation. Historical and hydrological data indicate a marked alteration in the hydrological performance of the subcontinent, with many streams now showing seasonal flow in the wet season. However, recent drought episodes can give insights into the possible responses of various systems to climate change. It is this experience which will be offered in this paper as a guide to the study of climate change impacts.

2 Background to projected climatic change in southern Africa

The anticipated global climate change has been discussed in several publications and summarised in the Intergovernmental Panel on Climate Change (IPCC) report (Houghton *et al.*, 1990). A more recent assessment, which includes an analysis of the 1991/1992 drought is compiled by Halpert and Ropelewski (1993).

Biological systems, such as vegetation assemblages, have shown an ability to adjust to gradual change. The anticipated climate change due to the man-made greenhouse effect is expected to occur in a few decades. Only small microbial organisms could genetically adapt to changing environmental conditions that rapidly. The majority of life forms, the metazoa, are unable to adapt that rapidly by natural means.

Anticipated changes in precipitation in Africa have been provided by Gleik (1987). The general pattern seems to indicate increased aridity in the arid western parts of Southern Africa, while the eastern part of the subcontinent could receive substantially more rain.

Recent comparisons of GCM control runs (e.g. Geophysical Fluid Dynamics Laboratory (GFDL), Goddard Institute for Space Studies (GISS) and the U.K. Meteorological Office (UKMO)) and observed climate have shown that the models tend to underestimate current temperatures in some regions of Southern Africa while over estimating precipitation (Kalkstein, 1991). There is thus a possibility that the anticipated changes could be more severe than those currently given by the IPCC report (Houghton *et al.*, 1990, 1992).

Although climatologists at the Southerm African Development Community (SADC) conference on climate change have stated that there are no detectable long term aberrations in the climate of Southern Africa, there are certain indications of persistent desiccation in the region. Garangnga (pers. comm.) has analysed rainfall in Zimbabwe and detected diminishing mean annual rainfall and increases in the severity and duration of droughts. Henry Elwell (pers. comm.) also examined the cumulative annual rainfall and compared these with the theoretic trend. His data show that in the past thirty years the mean cumulative annual rainfall has lagged behind expectation by 51 mm. At a regional scale Halpert and Ropelewski (1993) show that 62.5 per cent of the precipitation index percentiles in Southern Africa were below average.

3 The 1991/9192 drought as an analogue for climate change

Southern Africa experienced a severe drought associated with the El Niño Southern Oscillation (ENSO) phenomenon in 1991/1992. Figures 1 and 2 contrast the long term mean climatic variables with those of the drought period. In Figure 1, long term mean temperatures, those obtained during the drought period and the GISS estimates are shown. It can be seen that during the drought period, the seasonal temperatures were consistently higher than the long term means, but were lower than the GISS estimates. In Figure 2 corresponding values for precipitation and evaporation are indicated, as well as the GISS precipitation estimates. During normal seasons precipitation exceeds evaporation during the cropping season from December to March. However, during the drought cycle evaporation exceeded precipitation throughout the season, with the difference between precipitation and evaporation increasing as the growing season progressed. It is evident that if the conditions obtaining during the drought had occurred under the GISS precipitation regime similar moisture deficits would have been obtained. However, as Figure 1 indicates, the GCMs predicted temperatures are higher than those found during the drought, which was described as the worst in living memory.

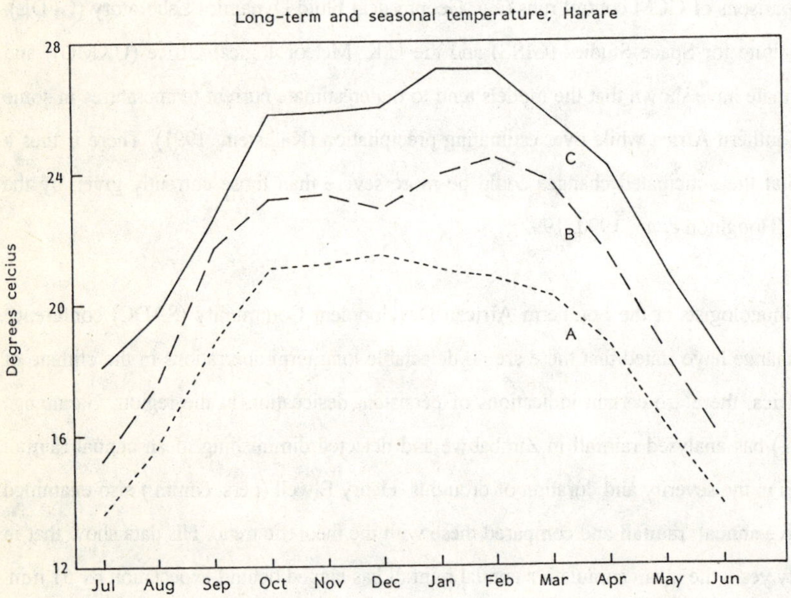

Figure 1 Long term mean, GISS and 1991/92 drought season temperatures. (A) = mean; (B) = 1991/92; (C) = GISS $2xCO_2$.

4 Possible impacts of climate change

4.1 Water resources

The coupling of the temperature model to precipitation is still a matter of discussion. Increased global temperature leads to increased ocean evaporation, and hence greater precipitation. Thus globally it is expected that the impact of global warming would be a global increase in precipitation and run off. However, Gates (1985) notes the lack of agreement between the various GCMs on the magnitude of such global change in precipitation.

Some of the simulation models used to construct possible future climate scenarios predict increases in precipitation in what are currently arid to semi-arid lands. For example, parts of the coastal regions of the Namib desert are likely to experience increases in mean rainfall of up to 1.5 times the current means. However, such precipitation increases also appear to be accompanied by increased mean temperatures, thus raising the evaporation.

Reibsame (1989) has tabulated the relationship between a decrease in run off with a 10 per cent decrease in precipitation at different temperatures. An extrapolation of his data to take into

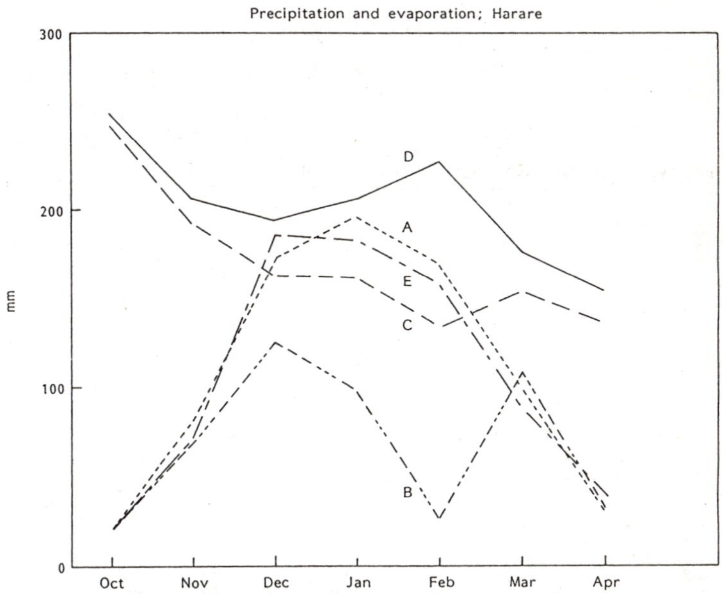

Figure 2 Long term means, GISS and 1991/92 drought season temperatures. (A) = mean precipitation; (B) = 1991/92 precipitation; (C) = mean evaporation; (D) = 1991/92 evapotration; (E) = GISS precipitation

account the high summer temperatures of mid continental Southern Africa, would lead to an estimated reduction in run off in excess of 50 per cent. Figure 3 shows the percentage difference between long term mean evaporation and that experienced during the drought period. These data indicate that during the growing season the decrease in moisture due to high levels of evaporation could be considerable. In Zimbabwe several perennial streams dried up.

These results indicate the effect of increased evaporation at the elevated temperatures. Nemec and Schaake (cited in Reibsame, 1989) have modelled the hydrology of Lake Victoria under various scenarios. Their results indicate that a 10 per cent reduction in precipitation and a 12 per cent increase in evapotranspiration, corresponding to a 3°C temperature rise, would lead to at least a 40 per cent reduction in run off, whilst a similar rise in precipitation for the same rise in temperature would result in no significant increase in run off. It can be surmised therefore that in the continental semi arid to arid tropics, gains in increased precipitation are likely to be nullified by increased evaporation and therefore an overall deterioration in the soil moisture condition as well as stream flow may occur. It is therefore possible that the extent of arid lands in these regions will expand.

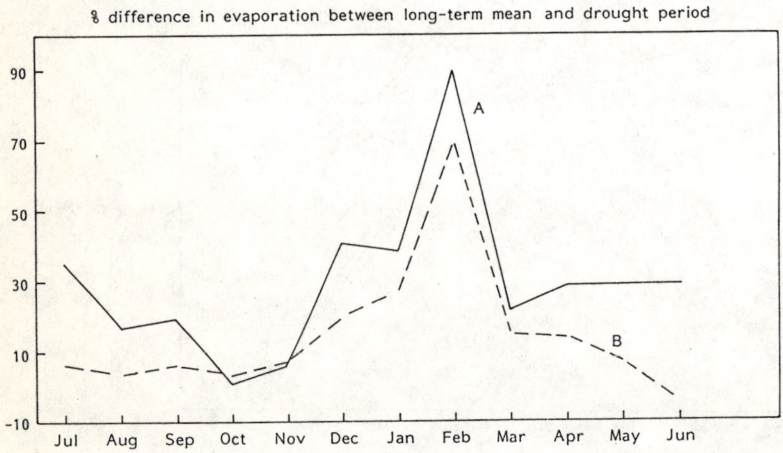

Figure 3 Percentage increase in evaporation at medium (Katsaga) and low (Kariba) elevations during the 1991/92 drought cycle. (A) = Katsaga; (B) = Kariba.

4.2 River flows

Most of Southern African rivers lack historical data of their hydrology such as one finds on the Nile. However, the few hydrological data that does exist have enabled a tentative modelling of the impacts of climate change on river flows in Southern Africa.

Rose (1991) simulated the likely response of the Zambezi River to a doubling of CO_2. The experiment examined the various possible scenarios given by the GCMs, namely an increase of temperature between 2°C and 4°C (Houghton *et al.*, 1990) and a change in precipitation of between -20 per cent (GFDL) and +20 per cent (UKMO).

The experimental results based on an assumption of reduced precipitation gave a much higher reduction in resultant river flow than the increase in flow that would be projected by models assuming an increase in precipitation of equal magnitude.

The UKMO is the only GCM that predicts an increase in precipitation in the region under review. This paper therefore examines the conservative view that as with other middle- to low-latitude semi-arid zones, Southern Africa would experience an overall reduction in precipitation in those areas in which the major river basins lie. It is further assumed that the Zambezi River simulation results would be broadly applicable to neighbouring river basins, such as the Limpopo, and the Save Rivers.

The Limpopo river in particular derives its hydrological input from the semi-arid to arid lands of Botswana and southwestern Zimbabwe. In this area, temperature increases of up to 5°C are anticipated, with a possible 10 per cent precipitation decrease. According to Reibsame (1989), under these conditions evaporative losses would be up to 60 per cent higher than are currently the case. In its upper reaches the river is supplied by seasonal streams which flow only in years of above average rains. It is likely that this condition could extend to the lower reaches of the river, converting the Limpopo into a *wadi* type of stream, such as are found in the Sahel and the arid parts of Somalia.

A further consideration in the implication of climate change are in-stream losses, particularly as they pertain to water transmission in irrigation systems. In Zimbabwe, because of the degraded nature of the so called middle and lowveld areas, which largely comprise the communal lands, the major water storage dams are located on the highveld and transmit water to distant irrigation sites. A combination of stream bed alteration due to siltation and increased temperatures leads to evaporative losses of several hundreds of cubic meters per kilometer (m^3/km) of stream (Magadza, 1993). This situation is likely to compromise irrigated agriculture in the lowveld areas.

4.3 Lakes, dams and wetlands

The likely impact of climate change on the wetlands of Southern Africa can be illustrated by the water balance of the Okovango. Of the 16 billion cubic meters inflow to the swamp 96.25 per cent (15.4 billion cubic meters) is lost by evaporation. Under a 2xCO_2 scenario evaporation would increase by up to 40 per cent, i.e. in excess of 20 billion m^3/yr. Similar considerations would apply to Lake Bangweulu where 90 per cent of the hydrological income is also lost by evaporation.

Kapetsky (1987) examined water levels in a number of reservoirs in northeast Zimbabwe using satellite imagery. His data indicate that under the current climatic regime there was a 49 per cent reduction in surface area between the wet season full supply and the dry season, due to evaporation. Alternately, only 29 per cent of reservoirs detectable at the end of the rainy season were evident at the end of the dry season. Given the projected increase in evaporative losses noted under a 2xCO_2 regime above, and a likely reduction in run off, it would appear that at least in years of drought many of the small to medium impoundments would either dry up or hold water insufficient for the needs for which they were intended. Such a situation would compromise agricultural concerns dependent on stored water.

An analogous situation to this scenario are the water levels in Lake Mtirikwe (formerly Kyle), which dried up during the 1991/92 drought episode following several years of persistent low water levels. Irrigation water supplies to the lowveld sugar estates were severely reduced to safe guard water supply to Masvingo urban area. This estate is the largest irrigation enterprise in Zimbabwe, consisting of over 5 000 ha and, besides supplying the nation's sugar needs and export quotas, it also supplements the nation's petroleum fuels by blending gasoline with alcohol to up to 20 per cent. While the recent droughts are responsible for the state of this reservoir it is also known that the capacity of the reservoir is such that even in years of average rainfall the probability of filling the reservoir is low.

This case provides some insight into what could be a general situation with a number of reservoirs in the subregion. Of considerable interest in this respect would be those water supply reservoirs to urban centres in the semi-arid to arid areas, such as Gaberone dam which supplies Gaberone, the capital of Botswana, as well as the water supply dams to the city of Bulawayo, Zimbabwe, which also failed during the 1991/992 drought period.

4.4 Regional water supply policy

It is estimated that by year 2035 Zimbabwe will have reached its limit in water resources development (Grizic, 1980). There is no doubt that the development of Southern Africa is intimately linked to its water resources development, and more so to its ability to conserve these water resources. The problems is further compounded by high environmental deterioration rates leading to siltation (Magadza, 1984), and subsequent in-stream transmission losses. There is no doubt that the solution to this problem lies more in the political than scientific domain, as overgrazing, deforestation and improper cropping methods are the major causes of siltation (Whitlow, 1988; Chakela, 1981).

The water resources of the Zambezi Basin countries currently cater for a population of some 73.5 million people. These countries show varied situations of water supply security. Certain regions of South Africa appear to have reached the limit of their water resources, hence the projects on importation of water from neighbouring countries, such as the Lesotho Highlands Water Project. Botswana's only natural surface water stocks are in the precarious Okovango delta, so its future expansion of water resources lies in the exploitation of ground waters. As the populations increase the subregion's lakes and reservoirs will have to meet a multiple of needs.

Such multiple use of the available water resources can result in conflicts of interests both at local and international level. As individual countries reach the limits of their own water resources they would probably look to friendly neighbouring countries for supplementation of their water needs. An example is the much mooted project of diverting the Zambezi River at Kazungula to supply Botswana and South Africa, or more recently the proposed Bulawayo pipeline from the Zambezi River. While a dispute of riparian propriety between Botswana and Namibia on the Kovango River caused some embarrassing encounters between the two neighbours, the proposed Bulawayo pipeline drew some articulate protests from other Zambezi riparian states. The issue had to be resolved by the adoption of a general protocol on the exploitation of shared resources within the

SADC countries. These examples illustrate the possible sources of conflict as climate change impact on water resources. The post civil war reconstruction of Mozambique agricultural industry will also depend on availability of water resources. The major water supply rivers of Mozambique originate from outside its borders. Water resources developments as well as catchment degradation there could negatively affect the status of Mozambique's water resources. The sort of political conflicts that can arise from sharing of limited water resources have been discussed by Gleick (1992), and Lowi (1992). It is encouraging to note that the SADC countries have taken note of this situation and have adopted a general framework of shared natural resources.

4.5 Agriculture

Parry (1990) has discussed the likely effect of projected climate change on agriculture. Although most of the data he considers come from the mid to high latitude developed countries it can be assumed that in principal the findings presented in his discussion would apply to the data poor semi-arid region of Southern Africa. For example, research in the Natal province of South Africa indicates a decline in crop yields of the order of 5 per cent to 15 per cent, following an increase in temperature of between 2°C to 4°C.

The total population of the Southern African countries amounts to 88.08 million. On average 55.2 per cent of this population is employed in the agricultural sector (FAO, 1989). The percentage varies from as low as 14.6 per cent in South Africa, where the mining industry absorbs a considerable amount of the labour force, to as high as 82.2 per cent in Mozambique. Generally the less industrialised countries tend to show a higher preponderance of employment in the agricultural sector, with the exception of Namibia, which is largely arid.

The climatic suitability for agricultural production varies within the region and within each country. Namibia and Botswana are the most arid of the Southern African countries, while Malawi, Swaziland and the Natal Province of South Africa are less prone to drought. The mid continent countries (Zambia, Zimbabwe and central South Africa) show a graded climatic change from subhumid in the eastern parts to semi-arid and arid in the western sectors.

In assessing the impact of climate change on agriculture using analogue data it is also important to note the impact of other extrinsic factors, such as civil unrest. Although Mozambique and Angola might also have suffered from the effects of drought, war and its disruption on the agricultural infrastructure, as well as its influence on general morbidity among the agricultural community, could also have influenced the agricultural performance of these countries. Furthermore the general trend in deterioration of agricultural land due to such factors as soil erosion and loss of soil fertility and deterioration of rangelands would also be reflected in agricultural performance. These factors are discussed by Magadza (1986a, 1986b), Moyo *et al.* (1990) and Murphree and Cumming (1993). As an illustration Table 1 shows the loss in water storage capacity in Zimbabwe impoundments, while Table 2 shows some trends in food production in some SADC countries. These data show generally a background of declining food production capability due to environmental degradation.

Figure 4 shows the agricultural indices of six of the ten Southern African countries. The six have been selected for their relatively high variance in the agricultural index over the period depicted. Here variance is used as a simple indicator of the sensitivity of the agricultural production system to disturbance. While in Mozambique, Namibia and Angola civil unrest could have been a contributing factor to levels of production, Figure 4 nevertheless indicates lowered agricultural performance in years of drought.

Table 1 Loss of storage capacity by siltation in some Zimbabwe dams

Name of dam	Year built	Estimate of silting (%)
Astral	1972	46
Chepiri	1956	0.5
Chidoda	1955	30
Gizwe	1955	4
Gudi		100
Hanjane	1956	10
Mahope	1955	15
Marah		10
Mtshele		100
Nyatari		100
Nyazvizvi	1955	15
Shumbayawanda		100
Siteze		66
Siya	1976	heavy siltation
Tselele		100
Zuwa	1955	4

The variance of the agricultural index of each of the Southern African countries over the decade 1977 to 1988 is shown in Figure 5. It is clear from the figure that Zimbabwe, Lesotho, South Africa, Namibia and Botswana show the greatest sensitivity to climate disturbance. Figure 6 also shows the variation in yield per unit harvested area for the period 1979 to 1988. From this figure it can be seen that those Southern African countries with low variances in their agricultural indices also tend to show low yields.

Table 2 Food production and demand in some SADC countries (1969-1984)

Country	Food imports (tonnes)	Growth of food production (%)		Imports/ capita
		1969-1973	1973- 1984	
Angola	47.7	-1.8	-2.2	5.7
Botswana	16.3	0.1	-3.6	16.3
Lesotho	36.9	-1.5	-1.9	24.6
Malawi	5.9	2.7	-0.5	0.9
Mozambique	183.6	0.1	-4.7	13.7
Swaziland	3.3	3.1	0.3	4.7
Tanzania	188.1	-2.1	-1.2	8.7
Zambia	102.1	0.3	-1,7	15.7
Zimbabwe	20.0	1.9	-2.9	2.4

Source: Derived from World Bank (1986).

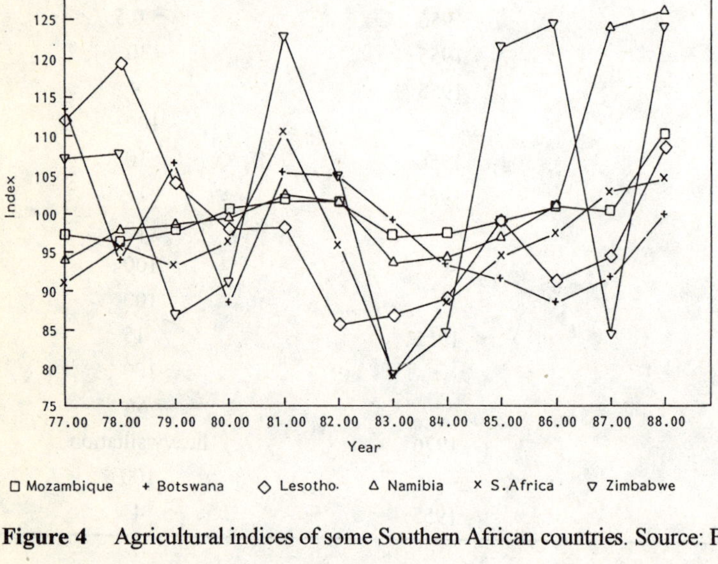

Figure 4 Agricultural indices of some Southern African countries. Source: FAO (1989).

The countries displaying high variances lie on the extremity of the Intertropical Convergence Zone (ITCZ), and depend on the southward aggression of this low pressure system for their rain. Climatic conditions that influence the relative balance between this low pressure system and the South Atlantic high pressure system that dominates the South African weather patterns will influence the variability of rainfall in these mid-subcontinent countries. Periods of relatively high rainfall are currently characterised by a persistence of the ITCZ in the area while drought is associated with the dominance of the so called Botswana High in the mid-subcontinental area. One of the IPCC results (Houghton *et al.*, 1990) is an indication of non symmetrical global warming with the northern hemisphere warming more that the south. Such a situation could lead to abnormalities in the ITCZ oscillation.

In the Southern African region the IPCC scientific assessment of climate change indicates that, though the various GCMs differ in detail, an increase in cropping season temperatures (December/January/February) of up to 4°C, and a precipitation decrease of up to 1 mm per day, corresponding to a seasonal deficit of some 90 mm are projected under a 2xCO_2 atmosphere (Houghton *et al.*, 1990). The assessment also indicates a soil moisture decrease of some 1 mm in the growing season.

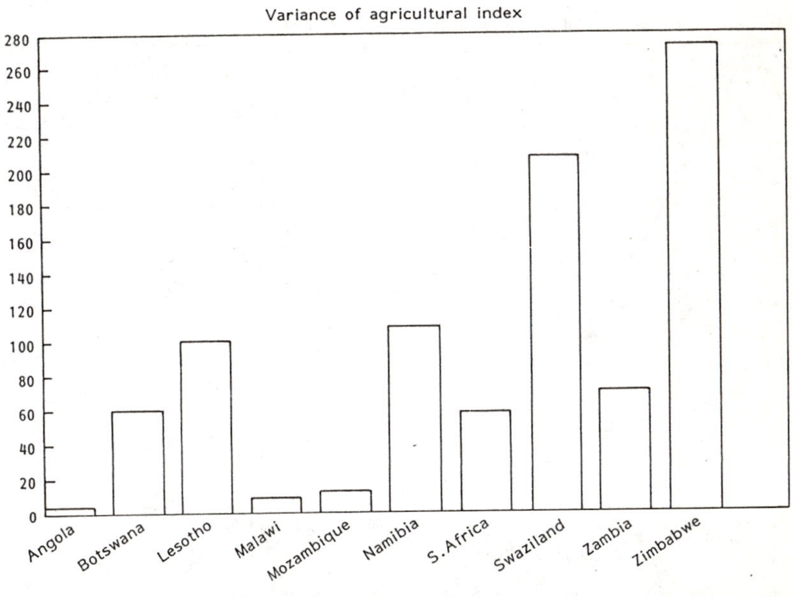

Figure 5 Variance of agricultural indices in Southern Africa. Source: FAO (1989).

The response of various crops to current drought episodes can perhaps be indicated in Figure 6, which shows the yield per hectare during the period covering the drought of the mid 1980s in Zimbabwe. These figures are taken from the commercial farming sector where inputs of agrochemicals would cushion some of the impacts.

Furthermore commercial farms in Zimbabwe, being the former white owned lands, tend to be located in areas where drought would be less severe than the so called communal cropping lands, in which drought relief food supplements often need to be provided. The few points, labelled 'Com. Maize' give some indications of the overall national yield levels, which include communal farm production.

Nevertheless the figure seems to indicate that maize, the staple food of most Southern Africa, is more sensitive to drought, in comparison to sorghum and ground nuts. Furthermore for the period under review the response pattern of both white maize (WMaize) and short season varieties (SsMaize) appears to be similar, with the short season varieties showing consistently lower yields.

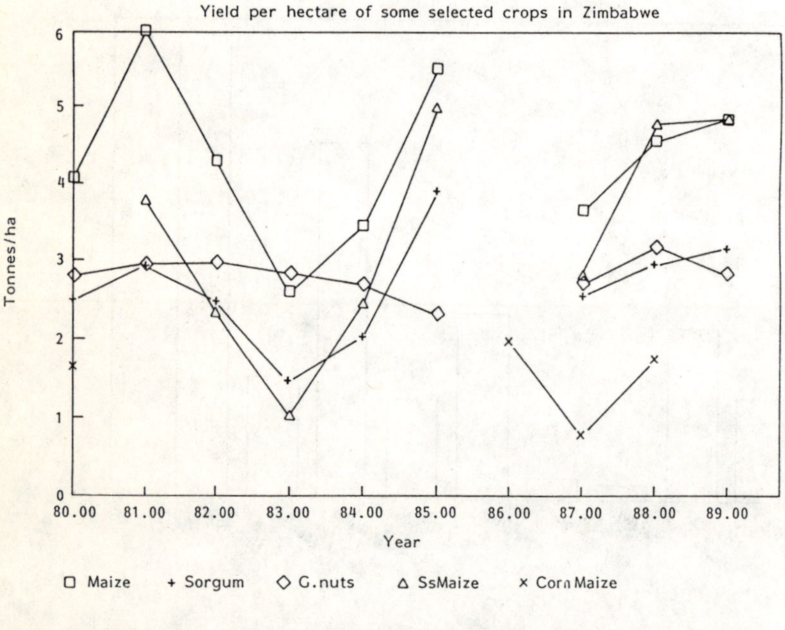

Figure 6 Variances of crop yields in SADC countries.

Table 3 Sensitivity tests for maize yield to climate change in Zimbabwe

	Unit	Banket	Gweru	Chisumbanje
Baseline yield	t/ha	4.7	3.72	2.86
Climate sensitivity				
+2^{0}C	% of baseline	95	92	86
+4^{0}C	% of baseline	80	76	73

Source: After Downing (1992).

Table 3 shows the likely changes in maize yields in selected parts of Zimbabwe. It can be seen that a rise in mean temperature of 4^{0}C could result in substantial reductions in maize production due mainly to soil moisture stress. In the prime maize growing area of Banket to Karoi this reduction could be as much as 20 per cent. In the lowveld the model suggests that maize production would be uneconomic, as production would decrease by as much as 27 per cent.

The projected scenario on the impact of the greenhouse effect is that as the concentration of greenhouse gases further increases, the periodicity and severity of droughts in those parts of the globe which show a proneness to aridity will be enhanced as a response to climate change. Within Southern Africa therefore it appears that those countries which currently show a relatively high variation in crop production in relation to variation in rainfall could exhibit an increase in crop production variability, and hence a reduced certainty in food security. These countries would include Zimbabwe, Botswana, Namibia, Lesotho and parts of South Africa. The inclusion of Lesotho in this group of countries derives from the fact that the bulk of Lesotho's agricultural lands are located in the southwestern lowlands which lie in the Drakensberg's rain shadow.

In those countries which show low variability in crop production because of inherent unsuitability for cropping under current climatic conditions, such as Namibia and Botswana, crop production under enhanced greenhouse effects would respond by reduced crop yields even though the variability could remain low.

Due to the variability of rainfall during the growing season, many crops are irrigated during critical stages of their maturation. Approximately 88 per cent of Zimbabwe's water is consumed by the agricultural industry. Wheat and barley are produced almost entirely under irrigation in all central Southern Africa, as this crop is produced during the dry cool season.

Of the Southern African countries considered here only Zambia and Malawi have substantial inland water resources in the form of large rivers and lakes. The rest of the Southern African countries have had to develop their inland water resources by massive programs of impoundments in order to meet their agricultural irrigation needs (Table 4).

In view of Kapetsky's (1987) findings it is probable that many more of these water impoundments would dry up prematurely in the early dry season, as well as fill and spill less frequently, as is the current case with Lake Mtirikwe.

Reibsame (1989) has indicated the need to increase storage capacity as a safeguard against drought under conditions of reduced runoff. Such a strategy is contingent upon a balancing out of low rainfall periods and normal to above normal rainfall years. However, unless the current trends in siltation in parts of Southern Africa, as shown in Table 1, are remedied, that option will become increasingly not available.

In summary the likely impact of climate change on agriculture in Southern Africa is that in those countries with high variance in the frequency of crop yields due to weather variations there is likely to be an increase in the frequency of crop failures, and water for irrigation could become more scarce. Unless the high current rates of soil erosion are checked the greater storage capacity that would be needed to safeguard against short periods of drought would be compromised by reservoir siltation. The measures needed to safeguard against such eventuality need to be initiated now.

Table 4 Growth of irrigation in Southern Africa (1000 ha).

Country	1972	1977	1982	1987
Angola	nd	nd	nd	nd
Botswana	1	1	2	2
Lesotho	nd	nd	nd	nd
Malawi	7	16	18	18
Mozambique	30	50	74	105
Namibia	4	4	4	4
S. Africa	1017	1060	1128	1128
Swaziland	52	56	60	62
Zambia	13	18	20	20
Zimbabwe	55	90	150	185

Source: FAO (1989). Note: nd = no data

4.6 Fisheries

Magadza (1992) has discussed the likely impact of climate change on the inland fisheries of the Southern African region. Zimbabwe has two types of fisheries at the country level; those which have evolved with the construction of dams, and the natural river fisheries.

Kapetsky and Petr (1984) appraise the status of African Reservoir fisheries. The observations of Kapetsky on seasonal variation in dam levels on the Zimbabwean highveld indicates the difficulty of combining aquaculture and agriculture in a situation of water supply stringency. According to Kapetsky and Petr's (1984) data more than 50 per cent of the dams examined would be unable to support a perennial fishery. Thus aquaculture on the majority of the small to medium impoundments in Zimbabwe would depend on annual stocking, whose viability would be contingent upon the availability of fast growing species which could be harvested in one season.

Some possible benefits from global warming to the fisheries would be the enhanced growth rate of many of the commercially important species. While several of the tilapia species cultured take about four years to reach table size such species would be harvestable at a much shorter culture period. Similarly in crocodile culture it has been shown that supplementary heating can half the time to reach slaughter size. Additionally areas that are suitable for the aquaculture of these organisms would also be expanded, enabling the higher terrain to support a varied aquaculture industry, water resources permitting. Hutton (1984) has suggested that the ecological range of the Nile crocodile on the highveld of Zimbabwe is determined by two effects of temperature:

- an extremely retarded growth rate in which reproductive maturity may not be attained until about thirty years growth; and
- the effect of temperature on sex ratio resulting in the virtual absence of males at the low temperatures of the highveld.

With respect to the natural river fisheries climate change, coupled with river siltation, will only accelerate a process that is already in train; that of habitat reduction and ultimate extinction of local species. Large stretches of the Save River are no longer habitable by fish. The traditional annual fish ceremony of the Hlengwe people of the southeastern part of Zimbabwe is long forgotten due to the siltation of the Runde and lower Save Rivers.

Morris (unpublished data, 1978) noted that an analysis of age groups of *Labeo altivelis* in Lake Chivero revealed missing, or reduced abundance of, age groups following years of below average rainfall. The data indicate failure of the spawning migration during periods of low river flow due to the failure of the migrating fish to surmount obstacles such as weirs, and to find suitable spawning sites.

Magadza (1992) noted the relation between *Linmothrissa* yields on L. Kariba with rainfall. Time series analysis of the *Limnothrissa* fishery on this lake gave the trend equation

$$Y = 1366.43X - 102605 \quad (1)$$

By removing the regressed yields values a detrended time series shown in Figure 7 was obtained. The figure emphasises the relation between depressed pelagic fish yields with low rainfall periods. Magadza *et al.* (1989), have noted that in this lake phosphorus levels in the lake are dependent on annual river flows and that in the dry season periods only trace levels of phosphorus are detected in the pelagium, the rest being tightly held in the biotic system. They inferred that production in L. Kariba was phosphorus limited. It can therefore be inferred that prolonged periods of reduced catchment inflow into L. Kariba would result in depressed fisheries yields in the pelagial fishery.

Figure 8 shows fish landings of the inshore fisheries, which also illustrates the relation between rainfall and inshore fish yields. The nature of droughts on inshore fish yields in large inland lakes may be obscure, but as in the pelagial fishery, nutrients probably play an important role as well. An additional factor is the exposing of breeding sites during water level recession. These considerations, namely nutrient inputs and water levels in inland lakes, have been shown to be operative in other inland fisheries in Southern Africa. (Magadza, 1984, 1990). Since inland fisheries constitute a significant protein source (as much as 70 per cent in Malawi), climate change mediated fisheries yield depression would lead to a substantial reduction in protein supply in the region. For example the total fish harvest in Malawi is estimated at 80 000 tones, 40 per cent of which is contributed by Lake Chilwa. This lake is known to periodically dry up during low inflow period, such as in the 1968 period.

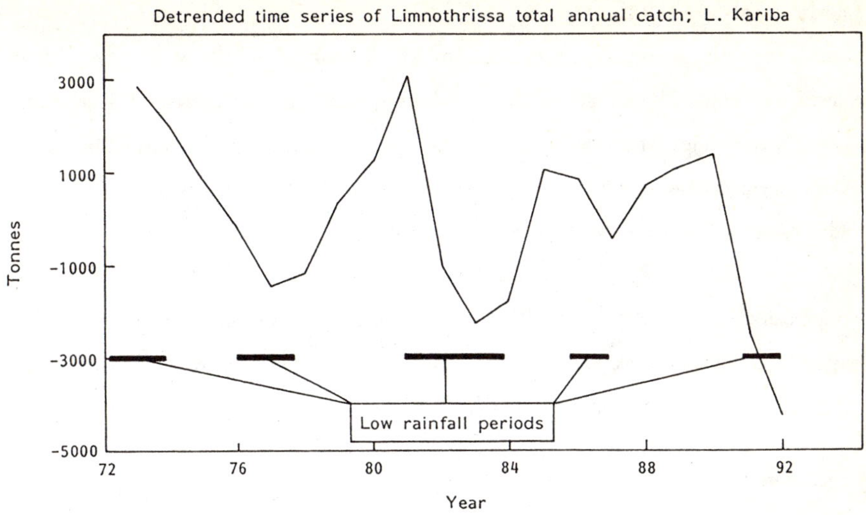

Figure 7 Detrended *Limnothrissa* yields, L. Kariba

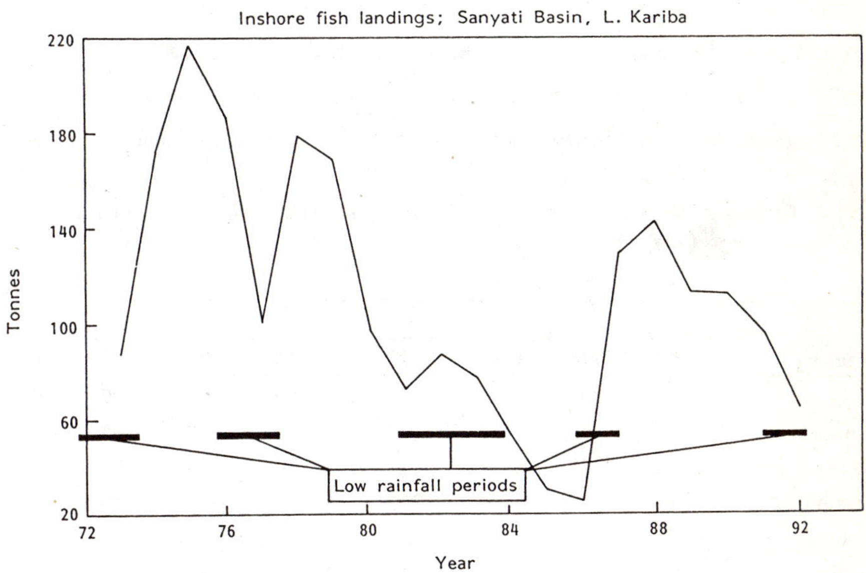

Figure 8 Detrended inshore fish landings, L. Kariba

4.7 Energy supply

The bulk of industrial energy supply to Southern Africa north of the Limpopo rivers is from hydroelectricity, supplied by the power station at Kariba, Kafue Gorge, Liwonde, Kabora Bassa and Kunene. Possibilities for further hydroelectricity developments in the Zambezi basin exist at Batoka Gorge, Devil's Gorge, Mupata Gorge and down stream to the Kafue Gorge. Table 5 shows the relative growth of electrical power generation capacity in some SADC countries.

Table 6 indicates that under all the climate change scenarios L. Kariba would be unable to meet current power production levels. In the extreme case of global warming, reduced precipitation and drought the Zambezi hydro system would fall far short of required power output at present ratings.

4.8 Vegetation

In the limited cases where attempts to characterise the ecological factors controlling productivity in the tropics have been made soil moisture and soil nutrients have emerged as the dominant factors (Cole, 1982, Tinley, 1982 Bell, 1982). In a controlled environment an increase in CO_2 concentration has been shown to elevate photosynthetic activity of crop plants. Most tropical grasses are C_4 plants. An increase in CO_2 has been shown to lead to a higher water use efficiency in these plants. The field application of these results is not clear, but whatever viewpoint we may choose, the relationship between increased temperature on available soil moisture must be

Table 5 Growth of thermal and hydroelectric power installation in some Zambezi Basin countries (kW x 10^3)

	Thermal				Hydro			
Country	**1961**	**1970**	**1980**	**1985**	**1961**	**1970**	**1980**	**1985**
Angola	60	101	200	200	28	211	400	400
Botswana	nd	nd	nd	nd	nd	nd	nd	nd
Malawi	11	21	39	34	1	28	67	176
Mozambique	94	241	280	280	65	114	1520	1523
Namibia	nd	nd	nd	nd	nd	nd	nd	nd
Tanzania	30	94	70	180	20	49	188	260
Zambia	241	184	190	191	43	180	1538	1538
Zimbabwe	502	487	487	906	562	705	705	633
Total	938	1125	1266	1791	719	1287	4418	4470
% Zambezi					93.3	79.8	86.7	85.5

Source: After Pinay, 1988.
Note: nd = no data.

an integral part of any model that aims at predicting the impact of global warming in the sub-humid tropics.

While in the northern hemisphere it is possible to project the redistribution of vegetation types, such as coniferous forest and broad leaf forest, the climate control of vegetation in Southern Africa is not too clear. In the Kalahari vegetation type of Botswana, Skarpe (1986) has shown a rainfall associated vegetation decline. Generally it can be assumed that the so called miombo woodlands (Cesalpinoideae) are associated with humid to subhumid climate with a mean annual rainfall limit of about 500 mm. In drier areas, depending on soil types, the Combretaceae and Mimosoideae tend to dominate. The Zambezi valley and parts of the Limpopo are peculiar in having extensive belts of *Colophospermum mopane*. In broad terms it is suggested here that a possible response of the vegetation type is the retreat of the *Brachystegia* woodlands from parts of Zimbabwe, and its replacement with *Acacia/ Combretum/ Terminalia /Burkea* types. It is also suggested that the semi-arid Karoo vegetation could spread to the Southern and western parts of Zimbabwe.

Table 6 Sensitivity of hydroelectric power generation to climate change scenarios.

Period	System	Base	+4^{0}C	+4^{0}C; +20%	+4^{0}C; -20%
Dry season	Kariba	87	51	91	14
	Kariba (KB)	87	51	91	14
	Batoka (KB)	53	46	50	43
Wet season	Kariba	97	83	98	55
	Kariba (KB)	97	83	98	55
	Batoka (KB)	97	92	97	86
Dry years	Kariba	84	53	89	32
	Kariba (KB)	84	53	89	32
	Batoka (KB)	75	71	75	66
Wet years	Kariba	100	83	100	42
	Kariba (KB)	100	83	100	42
	Batoka (KB)	81	74	79	70

Note: Response is given as percentage of current generation targets of 7208 and 9093 GWH per year respectively for Kariba and Batoka. (KB) denotes when Kariba and Batoka are operated in tandem.

Source: Adapted from Urbeztondo, 1992.

Table 7 depicts the distribution of some of the larger taxa of the woody vegetation of Southern Africa (Coates Palgrave, 1981). At least fifty tree species have a restricted distribution in the Southern subcontinent. Members of the *Leucospermum* and *Leucodendron* genera are particularly conspicuous in this respect.

The table shows that the first five taxa listed in the table have distributions that range north of the Zambezi river. Their distribution south of the Zambezi River follows the 500 mm rainfall isohyet. Of the taxa that have their central distribution in the Southern subcontinent, south of the Zambezi River, the majority abound in the eastern half of the subcontinent, restricted by the 250 mm isohyet. Similarly, the majority of coastal taxa are confined to the wetter Indian Ocean sea front. The distinct mediterranean-type Cape Flora is restricted to the winter rainfall region of the Southern tip of the subcontinent. It is, however, very difficult to forecast future distribution of plant species as a large number of Southern African plant and animal species are highly localised in their distribution. This distribution reflects either localised endemicity or habitat fragmentation due to human land use. Under these conditions species migration would be prevented by ecological barriers, thus leading to local extinctions.

Table 7 Comparison of the geographical range distribution of some major and remnant taxa in the Southern African subcontinent.

	Northern		**Southern**		**Cape**	**Coastal**
	East	**West**	**East**	**West**		
Combretaceae	27	10	1	0	6	0
Caesalpinoidaeae	21	6	0	0	6	2
Papilonoideae	30	13	0	4	14	1
Mimosoideae	33	14	3	-	7	2
Rubiaceae	40	28	1	-	31	0
Compositeae	9	3	1	3	8	1
Liliceae (Aloes)	1	11	3	2	6	2
Arecaceae	3	1	0	0	3	0
Proteaceae	5	12	0	37	1	0
Cupressaceae	1	1	0	2	0	1
Total	170	100	9	48	82	9

Source: Compiled from Coates Palgrave (1981).

In the event of the subcontinent becoming warmer and drier it is likely that the fifty or so species which have localised distributions in the subcontinent will be unable to relocate and therefore may become extinct, especially if the locality happens to be in an area of intense human land use.

Furthermore those species which are thought to have spread south from the more humid central part of the continent are also likely to retreat north as drier conditions develop in the Southern subcontinent.

In the eastern coastal area precipitation is likely to increase, so the flora of this area is unlikely to be affected in the same way as the mid-continental species, but the warmer climate might allow penetration further south into what is now the domain of the Cape Flora.

The net result of these changes at the local level in Zimbabwe could be a reduction in species diversity, since the replacing flora from the drier part of the subcontinent has fewer species. This process will have an influence on the composition of the insect fauna and on the distribution of the animals that feed on them.

4.9 Impact on wildlife

Thackeray (1987) has examined remains of small mammals from the quaternary period in South Africa, dating back to about 50 000 years before present. Using factor analysis and by comparing the present distribution of some of the extent types he was able to characterise the mammalian microfauna of the epoch into warm period and cold period types, with the assemblage of the microfauna corresponding to changes in climate during the epoch. There was also some indication of association of mean body size with the precipitation regime. During this period the highest rate of climate change has been estimated at 1°C per 1000 years, i.e. 0.01° per decade. Thackeray comments that this is the fastest rate of climatic change recorded in the post Pleistocene period. This rate of change is 20 to 50 times slower than that predicted for the effects of our anthropogenically induced climatic change.

The biological result of the climatic change in the Quartenery period discussed by Thackeray (1987) has been a geographical re-distribution of the micro-mammalian fauna of Southern Africa, as the ecologically preferred zones for the various members of the assemblage changed in their geographical ranges.

The relationship between large mammal production and rainfall can be expressed by the equation

$$\text{Biomass} = 0.08\ \text{Rain}^{1.685} \quad \text{(Coe } et\ al.\text{, 1976)} \qquad (2)$$

Subsequent work reviewed by Huntley and Walker (1982) further reveals the role of nutrient availability in modifying the basic biomass rainfall relationship, but it is clear that changes in mean precipitation in these tropical lands will have a significant impact on their ecological productivity. These data lead us to speculate on the continued capability of the Savannas, particularly those of East and Southern Africa, to support large populations of large mammals, such as buffalo, elephant etc., as well as questioning the future of pastoralism in the rangelands of Africa.

Cumming (1982) investigated the relationship between the metabolic biomass of the three most dominant large herbivores and rainfall and noted the decrease in the dominance of large herbivores with decrease rainfall in the precipitation range 200 mm to 1200 mm. Furthermore the percentage dominance of the large herbivores decreased rapidly below the metabolic biomass density of about 5 kg/km^{-2}, a figure corresponding to a mean annual rainfall of 500 mm. These data suggest that in the event of decreased rainfall or soil moisture due to global warming in the African savanna ecosystems, large bodied animals, such as elephant, hippopotamus and the large bovids, are liable to be affected more adversely than smaller species. This result also agrees with that of Thackery (1987) on his study of small mammal fossils in South African Quaternary period.

Flood plains of the Zambezi River system and other inland wetlands are home to a number of wild life species. Figure 9 shows the impact of gradual aridification of the Chobe flood plain on the populations of some of the species associated with the flood plain (Schlettwein *et al.*, 1991). A number of animal and plant specie are associated with smaller wetlands scattered in the region. The data from Chobe wetlands indicate a possible reduction in species diversity as a result of climate change.

In summary, large mammals such as elephant, buffalo and the large antelopes are likely to be threatened. Since most of Southern Africa's wildlife sanctuaries are located in low rainfall areas, which are likely to turn more arid, it may be prudent to consider the relocation of future parks. The finding raises questions regarding the future of the beef industry in Zimbabwe, which is currently based on exotic breeds of cattle from Europe.

4.10 Impact on health and pests

The likely impact on health of climate change would be:

- more rapid multiplication of disease vectors, as well as
- an extended range of disease vectors.

The original location of urban settlements in Southern Africa was determined largely by the distribution of malaria vectors and the occurrence of minerals. The later development of large scale agriculture and cattle ranging was similarly determined by vectors of livestock disease, such as the tsetse fly, *Glossina*.

Population growth and the subsequent vegetation clearing that accompanies it has more or less determined the demise of such vectors as the tsetse fly. However the range of malaria and schistosomiasis vectors is likely to increase both latitudinally and altitudinally. The major cities of Zambia, Zimbabwe and the northern Transvaal could become malaria zones, while schistosomiasis could spread further south as well as into the higher altitudes of Zimbabwe and Malawi.

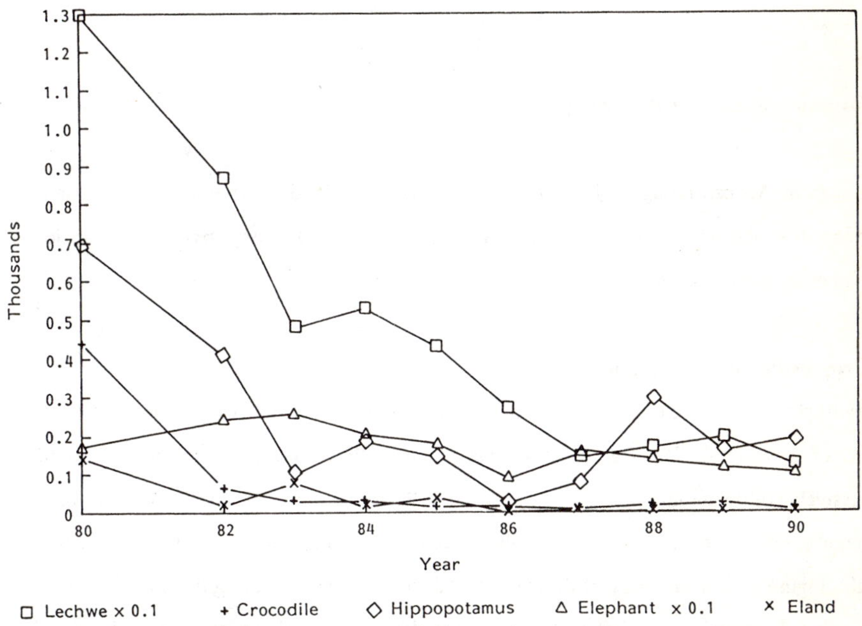

Figure 9 Variation of large mammal populations in the Chobe wetlands

Data on the seasonal incidence of malaria often reveal peaks towards the end of the rainy season. This corresponds to the period when stream flow, and hence the flushing effect of floods is reduced and streams are often dissected into disjointed pools. This condition reduces mosquito larvae mortality resulting in a resurgence of high adult populations. It is postulated here that under climate change incidence of low stream flow and thus high residence period in pools would be more frequent, leading to a longer seasonal incidence of high vector populations.

Furthermore, if drier warmer conditions prevail, such insect pests normally associated with warm arid lands, such as *Locusta*, could become prevalent. *Stomoxys* could spread wider in the woodlands, while *Simulium damnosum*, the river blindness transmitter, could establish itself further south, rendering such tourist attraction as the Victoria falls and its adjacent urban areas of Livingstone and Victoria Falls river blindness zones. The eastern districts of Zimbabwe, with its scenic water falls on the Pungwe and Odzi river systems could also become river Onchocerciasis zones.

During the water shortages of the 1992 drought there were outbreaks of cholera in those urban areas and refugee camps in which the water rationing was severe, leading to several thousand deaths.

5 Overview of the 1991/92 drought

The Southern African drought of the 1991/1992 season was said to have been the worst ever experienced in the subcontinent. A brief overview of its impact may serve as an analogue assessment of likely long term impacts of climate change.

5.1 Progression of the drought

Figure 10 shows the state of Zimbabwe's inland dams over the period covering the wet and dry seasons of 1991/1992. The figure shows that the overall water supply situation deteriorated progressively following the failure of rains in the 1991/1992 season. During this period three of the major cities of Zimbabwe had to install emergency borehole water supplies to cope with the drought. Figure 11 is a summary of Zimbabwe total storage capacity during the drought period. It is seen that the reaction by civil authorities o the development of the drought was somewhat hesitant.

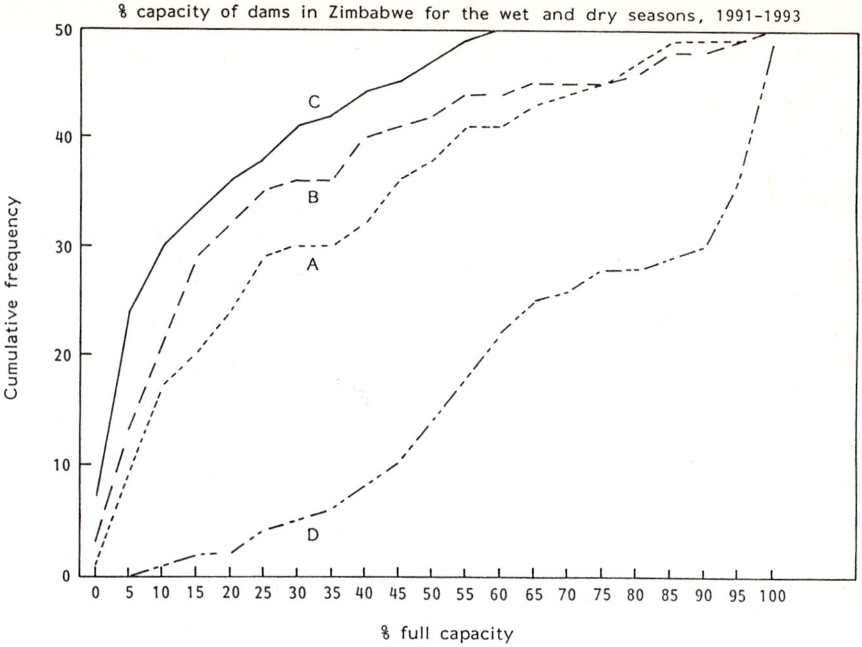

Figure 10 Cumulative frequency of water storage in Zimbabwe during the period 1991 to 1993. (A) = end dry 1991; (B) = begin dry 1992; (C) = end dry 1992; (D) = end wet 1993.

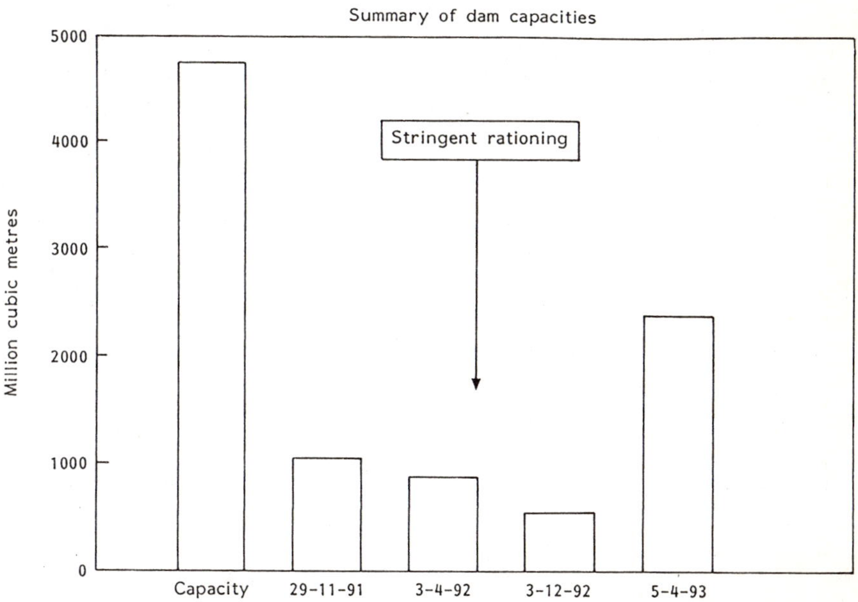

Figure 11 Summary of total water resources in Zimbabwe during the 1991/92 drought

5.2 Energy

Figure 12 shows lake levels in L. Kariba during the drought cycle. Water levels in the dams on the Zambezi reached critical levels, which impaired normal operations both on the Kariba and Kafue plants. The result in Zimbabwe were severe power shortages. The disruption to industry was widespread including labour lay-offs, damage to machinery that was not designed for abrupt power breaks. The retail business suffered large losses caused by the loss of perishable commodities. In order to force power savings the power supply corporation imposed severe punitive tariffs to consumers who exceeded their meagre quotas. Few industries were able to operate within such quotas, thus the power saving programme not only lead to production losses but also uneconomic energy bills.

5.3 Agriculture

The desiccation of dams shown above had a significant impact on irrigated agriculture and horticulture. The entire sugar crop of the year was lost due to non availability of irrigation from Lake Mtirikwe. This loss will be carried over several years as it takes about three years for cane to reach harvestable size. In the Mazoe area hundreds of citrus trees died due to the drying up of Mazoe dam, which has remained virtually empty even after the breaking of the drought.

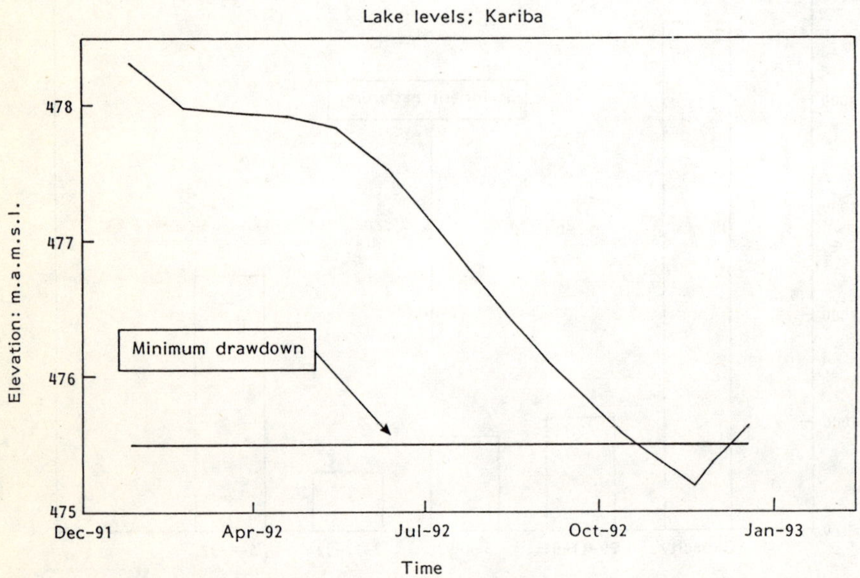

Figure 12 Lake levels and power generation in L. Kariba

The failure of the rains had the severest effect yet recorded on the maize crop, particularly in the communal lands. It is significant to note that the ares most affected are those predicted in the climate model used by Downing (1992). See also Downing *et al.*, 1995, this volume.

5.4 Health

The most notable health impact of the drought in Zimbabwe was the outbreak of cholera on a scale hitherto not experienced. The foci for the outbreak appears to have been the eastern districts of the country, where in Mutare for example piped water supplied only three days in the week.

5.5 Wildlife

The most affected areas in terms of wildlife and ecological impacts were in the eastern lowveld of the country. Woody species such as *Androstachys johnsonii, Brachystegia gloucescens,* and the succulent *Euphorbia cooperi* sustained losses during the drought period. Many small to medium mammals, such as jackals, succumbed but the most notable deaths, estimated in several thousands, were in elephant, buffalo and hippopotami. On a broader scale the impact of long term aridification on wild life populations has been illustrated by the response of large mammal populations to the desiccation of the Caprivi Strip wetlands of Namibia, where there has been a decline of large mammal population, especially those associated with wetlands (Figure 6, after Magadza, 1993).

5.6 Social impacts

The disruption of both traditional and commercial agriculture due to rain failure, and the reduction in industrial production due to energy restrictions led to severe economic and social problems, wherein the most vulnerable social sector were the low wage earners. It is in this sector that layoffs occurred. These impacts have manifested themselves in an unprecedented appearance of the so called 'street kids', juveniles whose parents have simply abandoned them. They lack proper home care, education and all the social amenities that a child growing in normal environment would enjoy.

6 Climate change and desertification

According to the criteria discussed by Barrow (1991) a large portion of Zimbabwe and other Southern African countries are undergoing desertification. A survey by Darkoh (1987) discusses the socioeconomic and institutional factors driving the desertification process in the subregion. Plate 1 shows the emerging mosaic of reflectivity arising from differential loss of surface cover in the farming areas of Zimbabwe. Such land degradation lowers the ecosystem resilience in responding to climate variations, rendering the impact of droughts much more severe than they would otherwise be and thus accelerating the desertification process. Loss of ground cover leads to more rapid loss of soil moisture after rain episodes, a more rapid loss of soil fertility due to irradiative mineralization, increased erosion and thus a more or less permanent reduction of soil depth. These physical changes in soil will result in an impaired ability of the ecosystem to regenerate biomass, resulting in much reduced biodiversity.

7 Adaptations

A stark fact that stands out as a result of recent droughts is the pre-eminence of short term political goals and lack of long term planning strategies to cope with worsening drought. During the 1991/1992 drought Zimbabwe jettisoned its maize stocks and ended up importing large amounts of the same commodity at considerable cost to the tax payer who had to pay an additional drought levy. In the drought recovery programme the Government distributed the high yielding but long growing season SR52 variety of maize to areas of low average rainfall, resulting in the projection of poor harvest in these areas to the following year. Nowhere in Southern Africa have governments undertaken a serious study of climate change impacts and developed a long term strategy such as that produced by Brazil, for example (Magalhães and Glantz, 1992). The climate of the eastern lowveld of Zimbabwe is similar to that of Niamey in the Sahel. Yet development plans in this area assume little water stringency, hoping that dams like Mtirikwe will support large scale irrigated agriculture, and that peasant farming of high yielding grain crops might one day succeed. Government continues to plan large irrigation dams in areas where the current mean annual run off is as low as 40 mm. What the 1991/92 drought demonstrated was that, unlike electrical power grids where failure from one unit can be compensated by increased load from sister units in the system, each catchment system must meet its own water demands independently, which leads to the possibility of multiple but independent failures in the water supply system.

A further factor moulding the response pattern of developing countries such as Zimbabwe is the perception and ranking of risk. Eakin (1993) noted that despite the technical information that went to the Government of Zimbabwe about the impending drought the Executive was more concerned with meeting export targets to satisfy World Bank requirements than the consequences of the drought and 'refused to arrange for imports until January 1992 when there were only a few weeks left of maize stock'.

As in other sub-Saharan countries the political climate in which government can be held accountable for misjudgement is lacking. There is one dominant political party to which the nation demurs. None of the hundreds of thousands of tax payers demanded to know why they should have had to incur the so called drought levy when the country should have had sufficient maize stocks. The news media were content reporting what the white communities were doing to save the wild animals. Had there been famine the majority of the rural folk would have taken it within their stride, and perhaps blamed their poor resource condition as the cause of their inability to cope with the drought. Indeed the strength of such administrations ironically rests in the poverty of the rural folk in that in the event of such catastrophes there would be personal suffering and bereavement perhaps, but little infrastructural investment loss. The former can be accommodated fatalistically but the latter leads to resentment.

7.1 Water resources

The ability of Southern Africa to adapt to climate change will to a large extent depend on its ability to manage the water resources of the region. Therefore effective conservation measures that maximise retention of water within catchments by maintaining high infiltration to recharge ground water reserves, maintain stream flow, and replenish soil moisture would be necessary. Measures that reduce the current rates of soil erosion would be important in preserving high water storage capacity. More water conserving irrigation technologies would be needed. Current overhead spray irrigation methods are wasteful as a good deal of the water is lost as crop foliage intercept as well as aerosol.

7.2 Energy

The extent to which power generation along the Zambezi can be increased is limited. Zimbabwe has thus limited options in its energy policy namely:

- increasing the contribution of thermal power;
- allocating more fiscal resources to power importation from the Zaire basin; and
- opting for low energy consuming industries, such as high technology small mass products, rather than steel mills, i.e. opting out of the raw products economy.

7.3 Agriculture

The scope of possible adaptations in the agricultural sector have been summarised by the IPCC Panel on Response Strategies (WMO/UNEP, 1990). Pertinent to the present discussion is the level of production technology traditionally used in the region. While we have noted the growth of irrigation in the region, the basic current strategy to meet the increasing food demand has nevertheless been the conversion of woodlands to crop lands. Most Southern Africa countries will run out of reserve cultivable land within the first two decades of the next century. Thus the adaptive strategy must have two major components: that of finding suitable drought resistant crop varieties, and that of improving production method to increase yields. The latter imperative requires a drastic revision by policy makers of the so called subsistence or peasant farmer, whose grain yields are normally around 500 kg ha^{-1}. With the extent of climatically cultivable land likely to diminish as result diminished soil moisture conditions under the green house scenario, such low yields would be inadequate to sustain the region's food needs. In Zimbabwe the land issue is highly political and is likely to be used for short term gains. The configuration of environmental issues discussed here make it imperative that whatever the outcome of the land reform issue government's main objective must be to revolutionalise the production technology in order to increase yields and sustainability of production.

A further consideration is the diversification of possibilities of land use in the arid lands. Jawetz (1989) has produced a fascinating discussion of possibilities of desert economies. The narrow view that agricultural production is the paramount land use activity needs to be re-examined with reference to arid zones. Such areas, though agriculturally uneconomic, can generate considerable through what has come to be termed arid economy. Activities such as wild life utilisation have been shown to yield, in monetary terms, a much higher revenue than if they were put to conventional agriculture. A good illustration of this is the Matestsi wildlife management area on the northwestern part of Zimbabwe. Once an impoverished cattle ranching area, this area now has a prosperous wildlife industry. Some of these issues are discussed in Magadza (1986a).

7.4 Ecosystem management

Above all Southern Africa urgently needs a sound ecosystem management strategy. The difference between the Zimbabwe eastern lowveld and the Sahelian region of Niger noted above illustrates the contribution of human activities to the desertification process. While the climatic factors of Zimbabwe's southeastern lowveld are similar to Sahelian conditions, the degradative land use factors of overgrazing and loss of forest cover are yet in their initial stages in Zimbabwe. Conversely, those areas of Zimbabwe that now look like the Sahel due to loss of ground cover could be restored to more resilient ecosystems, as the reforestation around Niamey in Niger attests. If Southern Africa is to avoid catastrophic effects of climate change anthropogenic desertification needs to be combated now.

8 Conclusion

This study has indicated that some Southern African ecosystems are highly sensitive to climate change. This sensitivity imparts a high level of vulnerability to food deficits in the communities, primarily due to crop failures and precarious water resources. The 1991/1992 drought inflicted considerable damage to the food production of Southern Africa as well as its industrial base due to curtailed water resources for plant growth, urban supplies and hydroelectric power generation. However, in comparison to the predicted GCM equilibrium conditions the drought represented only transient conditions, implying that equilibrium conditions could result in more severe impacts.

Acknowledgement

This paper is a revised version that first appeared in *Food Policy* Volume 19, Magadza, C.H.D., Climate change: Some likely multiple impacts in Southern Africa, pp 165-191 (1994). Reprinted with permission from Elsevier Science Ltd, Butterworth-Heinemann Imprint, The Boulevard, Landford Lane, Kidlington OX5 1GB, U.K.

References

Balling, R.C. (1988). The climatic impact of a Sonoran Vegetation discontinuity. *Climate Change,* **13**: 99-109.

Barrow, C.J. (1991). *Land degradation.* Cambridge University Press, Cambridge.

Bell, R.H.V. (1982). The influence of soil nutrient availability structure in African savanna ecosystems. In: Huntley, B.J. and Walker, B.H.(eds.) *Ecology of Tropical Savannas.* Springer-Verlag, Berlin.

Chakela (1981). *Soil Erosion and Reservoir Sedimentation in Lesotho.* Ph.D. Thesis, University of Uppsala, Uppsala. 150 pp.

Coats Palgrave, K. (1981). *Trees of Southern Africa.* Struik, Cape Town. 959 pp.

Coe, M.J., Cumming, D. H. M. and Phillipson, J. (1976). Biomass production of African herbivores in relation to rainfall and primary production. *Oecologia (Berlin),* **22**:341-354.

Cole, M.M. (1982). The influence of soils, geomorphology and geology on the distribution of plant communities in Savanna ecosystems. In: Huntley, B.J. and Walker, B.H. (eds.) 1982. *Ecology of Tropical Savannas.* Springer-Verlag, Berlin.

Cumming, D.H.M. (1982). The influence of large herbivores on savanna structure in Africa. In: Huntley, B.J. and Walker, B.H. (eds.) *Ecology of Tropical Savannas*. Springer-Verlag, Berlin.

Darkoh, M.B.K. (1987). Socio-economic and institutional factors behind desertification in Southern Africa. *Area,* **19**(1): 25-33.

Darkoh, M. B. K. (1989). *Combating Desertification in the Southern African Region: An Updated Assessment.* United Nations Environment Programme, Nairobi. 303 pp.

Downing. T. E., Gitu, K. W. and C. Kimau. (1989). *Coping with Drought in Kenya; National and Local Strategies.* Lynne Reinner Publishers, Boulder. 411 pp.

Downing. T.E. (1992). *Climate Change and Vulnerable Places.* Environmental Change Unit Research Report No. 1. Environmental Change Unit, University of Oxford, Oxford.

Elwell, H. (pers. comm). Institute of Agricultural Engineering, Box BW330, Borrowdale, Harare, Zimbabwe.

FAO (Food and Agriculture Organization) (1989). *Yearbook.* Vol 42. FAO, Rome.

Garanganga (pers. comm). Department of Meteorological Services, P. O. Box BE 150, Belvedere, Harare, Zimbabwe.

Gates, W.L. (1985). The use of general circulation models in the analysis of ecosystem impacts of climate change. *Climate Change,* 7: 267-284.

Gleick, P.H. (1987). Regional hydrologic consequences of increases in atmospheric CO_2 and other trace gases. *Climate Change,* **10**: 137-161.

Gleick, P.H. (1992). *Water and Conflict.* Occasional Paper Series on the Project on Environmental Change and Acute Conflict. International Security Studies Programme. American Academy of Arts and Sciences, University of Toronto, Toronto. 62pp.

Grizik, P.M. (1980). Water: The vital resource. *The Zimbabwe Science News* **14** (12): 297-298.

Halpert, M.S. and Ropelewski, C. F. (1993). *Fourth Annual Climate Assessment.* Climate Analysis Center, NOAA, Washington, DC. 90pp.

Houghton, J.T., Callander, B.A. and Varney, S.K. (eds.) (1992). *Climate Change 1992: The Supplementary Report to the IPCC Scientific Assessment.* Cambridge University Press, Cambridge.

Houghton, J.T., Jenkins, G.C. and Ephraums, J.J. (eds.) (1990). *Climate Change: The IPCC Scientific Assessment.* Cambridge University Press, Cambridge.

Huntley, B.J.and Walker, B. H. (eds.) (1982) *Ecology of Tropical Savannas.* Springer-Verlag, Berlin.

Jawetz, P. (1989). *Potential for Desert Management in Malaska: Beyond Famine in Africa.* Club of Rome, Lusaka.

Kalkstein, L.S. (1991). *Global Comparisons of Selected GCM Control Runs and Observed Climate Data.* United States Environment Protection Agency, Washington, DC. 251 pp.

Kapetsky, J.M. (1987). *Satellite Remote Sensing to Locate and Inventory Small Water Bodies for Fisheries Management and Aquaculture Development in Zimbabwe.* FAO, Rome. 11 pp.

Kapetsky, J.M. and Petr, T. (1984). *Status of African Reservoir Fisheries.* FAO, Rome. 326 pp.

Lowi, M.R. (1992). *West bank Water Resources and the Resolution of Conflict in the Middle East.* Occasional Paper Series on the Project on Environmental Change and Acute Conflict. International Security Studies Programme. American Academy of Arts and Sciences, University of Toronto, Toronto. 62pp.

Mabutt, J. A. and Floret, C. (1980). *Case Studies on Desertification.* UNESCO, Paris..

Magadza, C.H.D. (1992). The distribution and economic importance of lakes in Southern Africa. In: Matiza, T. and Chabwela, H.N. (eds.) *Wetlands Conservation Conference for Southern Africa.* International Union for the Conservation of Nature, Gland.

Magadza, C.H.D. (1986a) Conflicts of resource use on Lake Kariba environs. *Nature and Resources,* **XXII** (4).

Magadza, C.H.D. (1993). Water resources conservation. *The Zimbabwe Science News,* **26** (10/12):82-87.

Magadza, C.H.D. (1984). An analysis of siltation rates in Zimbabwe. *The Zimbabwe Science News,* **18** (5&6): 63-64.

Magadza C.H.D, Heinanen, A. and Dhlomo, E. (1990). Some preliminary results on the limnochemistry of Lake Kariba, with special reference to phosphorus and nitrogen. In: Magadza C.H.D. (ed.) *Ecology of Lake Kariba.* University Lake Kariba Research Station (ULKRS) Bulletin 1/89, University Lake Kariba Research Station, Harare.

Magadza, C.H.D. (1986). Conservation problems in Zimbabwe. In: Magadza, C.H.D. and Dangana, L.B. (eds.) *Woodlands and Water Resources in Eastern and Southern Africa.* UNESCO, Nairobi.

Magalhães A.R. and M.H. Glantz (eds.) (1992). *Socioeconomic Impacts of Climate Variation and Policy Responses in Brazil.* Esquel Brazil Foundation, Brasilia. 155 pp.

Moyo, S., Robinson P., Katerere, Y., Stevenson, S. and Gumbo, D. (1990). *Zimbabwe's Environmental Dilemma: Balancing Resource Inequalities.* ZERO, Harare. 165 pp.

Murphree, M.W. and Cumming, D.H.M. (1993). Savanna land use: policy and practice in Zimbabwe. In: Young, M.D. and Solbrig, O.T. (eds.) *The Word's Savannas.* UNESCO (MAB), Paris. 350 pp.

Parry, M. (1990). *Climate Change and World Agriculture*. Earthscan, London. 157 pp.

Pinay, G. (1988). *Hydrobiological Assessment of the Zambezi River Basin: A Review.* IIASA working paper. IIASA, Laxenburg, Austria. 116 pp.

Reibsame, W.E. (1989). *Assessing the Implications of Climate Fluctuations: A Guide to Climate Impact Studies*. UNEP, Nairobi.

Rose, R. (1991). Personal communication based on M. Sc. Thesis. University of Colorado, Boulder.

Rosenberg, N.J. (ed.) (1993). Towards an integrated impact assessment of climate change: the MINK Study. *Climatic Change,* **24**(1-2). (Special Issue).

Schlettwein, C.H.G., Simmons, R.E., McDonald, A. and Grobler, H.J.W. (1991). Flaura, fauna and conservation of East Caprivi wetlands. *Madoqua,* **17**(2): 67-76.

Skarpe, C. (1986). *Vegetation Ecology in the Western Kalahari in Relation to Large Herbivore Grazing*. D. Phil thesis, University of Uppsala, Uppsala.

Thackeray, J.F. (1987). Late quartenery environmental changes inferred from small mammalian fauna, Southern Africa. *Climatic Change,* **10**: 285-305

Tinley, K.L. (1982). The influence of soil moisture balance on ecosystems patterns in Southern Africa. In: Huntley, B.J. and Walker, B. H. (eds.) 1982. *Ecology of Tropical Savannas.* Springer-Verlag, Berlin.

Whitlow R. (1988). Soil erosion and conservation policy in Zimbabwe. *Land Use Policy,* 419-433.

Williams, M.A.J. and Faure, H. (eds.) (1980) *The Sahara and the Nile.*. Balkema, Rotterdam.

World Bank (1986) *Financing Adjustment with Growth in Sub Sahara Africa,1986-1990.* World Bank, Washington DC. 120 pp.

Zimbabwe Government Central Statistics Office. (1987). *Quarterly Digest of Statistics.*. Government Printer, Harare.

Adaptation of Food Production to Drought in the Senegal River Basin

Brad Bass
Environmental Adaptation Research Group
Institute for Environmental Studies, University of Toronto
Toronto, Canada

Henry Venema
Acres International Ltd.
Niagara Falls, Ontario, Canada

Eric Schiller
Department of Civil Engineering, University of Ottawa
Ottawa, Ontario, Canada

1 Introduction

Neither the Framework Convention on Climate Change (FCCC) nor the Vienna (Ozone Layer) Convention seriously considers adaptation. Indeed, such a consideration may be unnecessary – especially with regard to food security. Agriculture has been and is one of the most adaptable of human production systems. Only an extremely adaptable system could have expanded world food production sufficiently to meet most of the needs engendered by world population growth in the last two centuries. Yet the persistence of drought and hunger in the developing world suggests otherwise. However, this failure is not principally due to climate. It is rooted in the separation of the producers from the means of production and the separation of the techniques of production from the local ecology. This is the perspective of ecological (eco)-political economy (Yapa, 1982).

Considered within the framework of eco-political economy, food production is as much social (the control of the means of production) as it is economic (how the final products are distributed). These 'social relations of production' are the emphasis of political economy. However, people engaging in production also define relationships between their society and the natural environment, or a set of 'ecological relations of production' (Yapa, 1982). The social and ecological relations of production can be reproduced or altered by technological innovations. For example, if agricultural development is capital-intensive, then expensive new technology is required that may be accessible to only a wealthy few. Thus the new

NATO ASI Series, Vol. I 37
Climate Change and World Food Security
Edited by Thomas E. Downing

technology assumes a social bias as access to the factors of production is unequal. Similarly, technology can acquire an ecological bias based on its environmental impact. The alleviation of hunger and the promotion of food security will depend on how the social and ecological relations of production are mediated through the technological biases associated with adaptation.

The emergence of global warming compounds the failure to improve food security in parts of the developing world. There are considerable scientific uncertainties about the likely rate and magnitude of global warming, although the earth is now committed to some enhanced greenhouse effect as a result of past and continuing greenhouse gas (GHG) emissions. Perhaps the most important climatic question for farmers and agricultural policy-makers concerns the regional distribution of the changes, for which the available information is most unreliable. The extent to which any agricultural system must adapt will presumably depend on the amount and kind of climate change, the capacity of the system to minimise the adverse impacts, and the capacity to take advantage of new opportunities.[1] The choice of direction for some developing countries also includes the possibility of a return to (or a stronger preservation of) more traditional forms of agriculture that were 'risk-averse' rather than output or profit-maximising.

It is disturbing that there are few linkages between the world trade community and the climate change community, as many of the technological biases in food production in the developing world are a result of colonial-era policies as well as current trade arrangements (but see Fischer *et al.* (1995) and Riley *et al.* (1995), both in this volume). The sparsity of linkages may be a result of the perception that climate change is a long-term problem, beyond immediate reference to trade. Consideration of climate change, as well as other ecological problems, is necessary in evaluating an agricultural trade policy such as import substitution; agricultural adaptations need to be assessed in the context of macroeconomic and trade policy.

---

1 See the other papers in this volume, notably Schneider (1995) on climate change, Downing *et al.* (1995) for a comparable vulnerability framework, Norse (1995) for discussion of multiple threats to food production, and Eele (1995) for a review of household adaptability.

The Senegal River Basin (SRB) (Figure 1) provides an example of the linkages between trade policy, adaptation to a possible change in climate and the technological biases of the adaptation.[2] The countries that share this basin – Senegal, Mauritania and Mali – face severe foreign exchange shortfalls compounded by the need to import rice in order to maintain low food prices in the urban areas. Against this backdrop, the tri-nation river basin development authority is attempting to execute a shift to irrigated rice production (designated here as policy RP), which stresses large state-managed rice plantations and the regularisation of river flows to permit double-cropping.

Although irrigated agriculture is possible throughout the arid valley of the lower SRB (FAO, 1977), an analysis of Senegal River streamflows at Bakel from 1904 to 1990 suggests that the prolonged Sahelian drought may indicate a permanent reduction in water supply. This is reflected in the downward trend of the last 20-30 years inthe annual flow of the Senegal

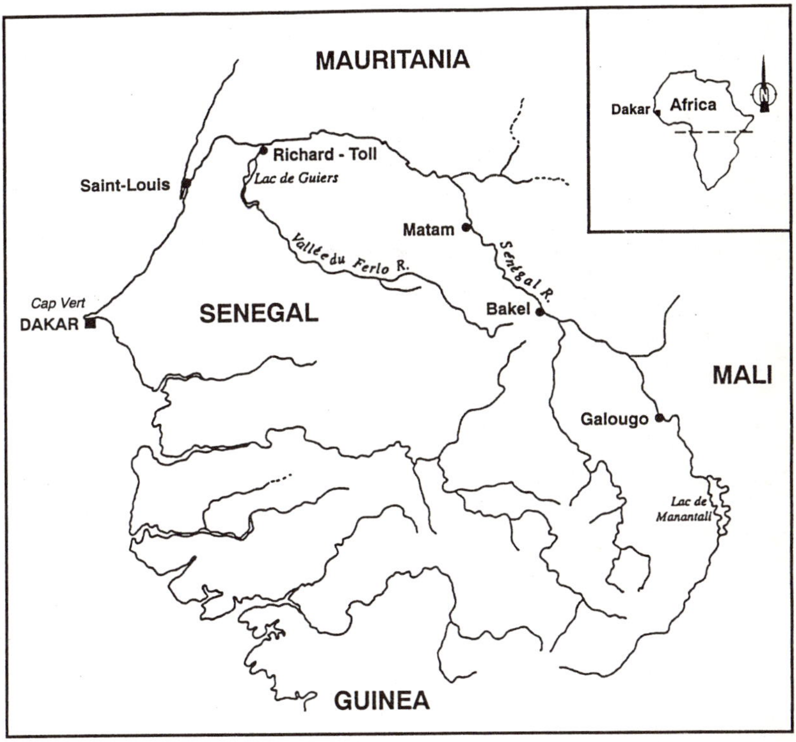

Figure 1 The Senegal River Basin.

[2] See Downing (1992) for a summary of potential impacts of climate change in Senegal.

River at Bakel (Figure 2a). A student's t-test for time series non-homogeneity showed that the average annual flow at Bakel before and after 1960 differed (at a 99.9 per cent significance level), suggesting that the series is not homogeneous and that some intervening factor has altered the milieu and substantially curtailed the availability of water resources. A periodicity of 29 years has been observed in the Senegal River, which is consistent with other rivers in Africa, and as late as 1984 a downward trend in streamflow could not be detected (Lavender and Anderson, 1984). However, the periodic component can only partially account for the relatively recent 30 years of streamflow observations when it is overplayed on the historical and detrended series which includes the data from the remainder of the decade (Figure 2b).

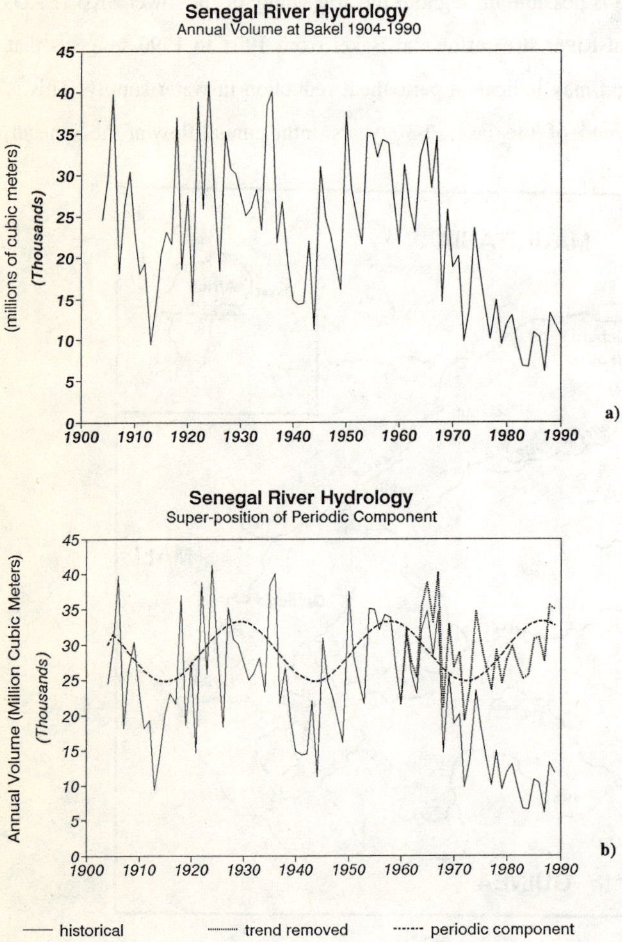

Figure 2 (a) Senegal River annual flow at Bakel, 1904 - 90; (b) superposition of periodic component with Bakel detrended series.

Additional support for the hypothesis that the hydrologic regime of the Senegal River has been permanently altered is derived from Lamb (1985) and Citeau (1990).

Reductions in streamflow restrict the full potential of the policy RP objective to meet urban demands for food and additional demands for water and energy. Under the prevailing drought conditions in the SRB, policy RP will also exacerbate the environmental degradation and the disruption of the village social structure which have been occurring since the colonial era. Within the SRB there are two alternatives to policy RP, constituting a natural resources management initiative of sustainable development, designated here as policy NRM. Policy NRM uses scarce water resources in ways that are more in keeping with prolonged drought, and control is decentralised allowing for input from the individual farmer. In policy NRM, the first alternative is village-scale irrigation projects, or 'Les Perimetres Irriguées Villageois' (PIVs), which are dedicated to low-water-consumption cereal grain crops and are managed by traditional sociopolitical structures. The alternative consists of intensive irrigated agro-forestry projects to redress the primary effects of desertification by using irrigation to re-establish a protective and diverse vegetative cover to maximise biomass production (SYSPRO).

The next section of this paper addresses the link between technological biases in adaptation and the means of production. In Section 3, the SRB experience is presented, followed by some further comments on a protocol for adaptation and some elements of such a protocol. Adaptation needs to go beyond the relationship between GHG and food security. The experience of the SRB illustrates some of the difficulties in formulating a successful adaptation strategy. Specifically, adaptive strategies to cope with problems related to foreign debt may exacerbate existing problems and may be exacerbated by climate change, further jeopardising food security. This is illustrated through a comparison of policies RP and NRM within their historical context.

2 Eco-political economy and factor biases

The problems associated with the adaptation of agriculture to drought in the developing world stem from the limitations on irrigation imposed by the region's hydrologic budget, the connection between ecology and agricultural production, the social impacts of ecological degradation, and the social relations of production which have their roots in the colonial period. An analysis of the political economy focuses only on the social relations of production, and ecological problems are treated as a purely social phenomenon. This type of analysis would not necessarily lead to a recognition of the constraints imposed by climate change reflected in water scarcity.

The primary principle of eco-political economy is that the correlation between technology, economy and ecology is mediated through biases in technology, innovations and adaptations (Griffin, 1974; Yapa, 1982; Ellis, 1988), referred to here as adaptation bias. Adaptation bias can be illustrated through the use of standard microeconomic concepts. In markets where producers face price discrimination in access to inputs, two producers can attain the same production level, represented by isoquant curves Q1-Q3 (Figure 3). Different mixes of inputs or factors of production are represented at points A and B on isoquant Q1 in Figure 4. An adaptation or technological innovation may allow the same output to be produced with a smaller amount of input (isoquant Q1 shifts to position Q2 in Figure 3). If isoquant Q2 is parallel to Q1, then the adaptation is neutral and superior. Likewise, Q3 represents an adaptation that is neutral and inferior. An example of an inferior adaptation is a factor of production that causes soil erosion or a reduction in soil fertility.

If the adaptation is biased in favour of either of the factors of production, then a factor bias is attributed to the technology. In Figure 4, the same output can be achieved by moving Q1 to Q2, but the new innovation is biased towards capital and entails a reduction in labour. If each

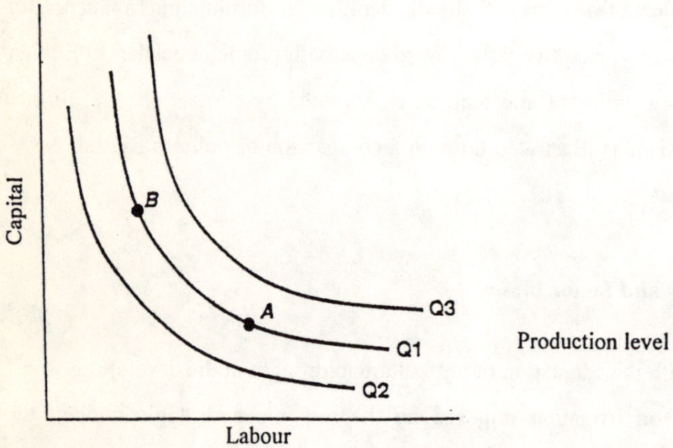

Figure 3 Innovations and technical change.

producer does not have equal access to the factors of production, then, in addition to the factor bias, the innovation acquires a social bias. The adoption of a new technology that carries a social bias will exacerbate the differences between different groups of producers, usually along economic classifications. If access to new technology continually favours wealthier producers, or urban considerations dominate rural area production decisions, these particular social relations of production will control the research and introduction of new innovations.

An adaptation can also acquire a hard or soft ecological bias according to the type and amount of energy required, the amount and nature of waste heat and material generated, the amount of non-renewable resources required and the ecological degradation. There are eight attributes of soft technology (Commoner, 1972) and of particular application to this study are the following: (i) ecologically sound with little disruption of natural cycles; (ii) use of locally available resources; and (iii) labour-intensive and labour-augmenting techniques of production. The high-yielding hybrid grains that were introduced in developing countries as part of the Green Revolution are a good example of a hard path technological innovation. These crops were sold as part of a package that included extended use of nitrogen fertiliser and pesticides in monoculture environment, lack of crop rotation, and an introduction of capital-intensive technologies for planting and harvesting. It has been shown that, even in the Western world, this approach to agriculture has resulted in land degradation (Merrill, 1976; Francis, 1985), eutrophication of lakes and contamination of groundwater (Commoner, 1972), and, in association with the Green Revolution, a loss of genetic variety (Harlan, 1975).

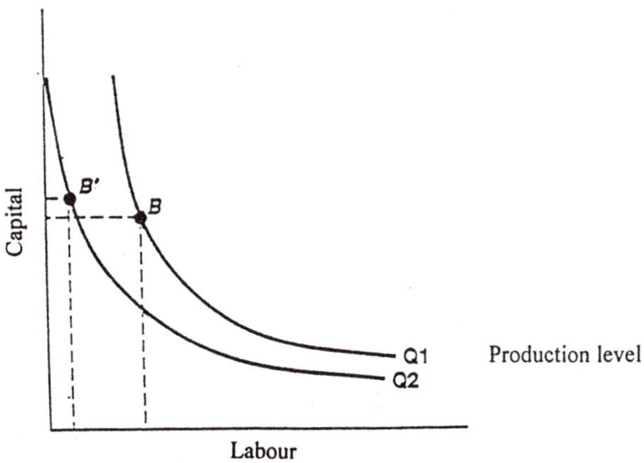

Figure 4 Biased innovations and producer equilibrium.

Throughout the developing world, research has also been biased towards the hard path production of cash crops to the detriment of food production (Lappe and Collins, 1977; Bell, 1986). For example, research on irrigation crop water requirements has focused on monocultures. The water demand is assumed to be a function of crop type and the open-field evapotranspiration potential (Doorenbos and Pruitt, 1977). Little effort is made to account for the positive ecological benefits of shelter-belt development and crop associations, such as the reduction of local evapotranspiration potential and improved soil structure (Lal, 1990). In general, a capital-intensive technique of production corresponds with a hard path technology and Yapa (1982) has also noted a contradiction between the particular social bias of the hard path technologies and ecological stability.

3 Development of the Senegal River Basin

3.1 Historical background

Development of the SRB has been marked by rapid vegetation degradation. The degradation is linked to climatic factors (Ellis, 1987) and anthropogenic factors (Falkenmark *et al.*, 1989). The most important climatic factor is the reduction in average annual rainfall of more than 30 per cent between the periods 1930-69 and 1970-85. Anthropogenic pressures include deforestation, extension of cultivable land and over-grazing. Anthropogenic pressures can be traced to the technological and cultural biases in agricultural policy which altered the ecological relations of production and exacerbated the effects of the colonial-era social relations of production.

The current focus on cash crops continues the social relations of production introduced into the basin during the period of colonisation when the economies in Africa were oriented towards primary production as part of a global division of labour (Bell, 1986; Wallerstein, 1974). The African economies were divided into three geographic zones (Wallerstein, 1976): (i) land owned by European settlers producing for the world market; (ii) production of food crops to supply labour in the export sector and the urban areas; and (iii) the largest zone devoted to subsistence cultivation. The first zone controlled the productivity of the other two zones. With the introduction of taxes and the cash economy, one family member often had to migrate to the urban areas or a portion of land had to be removed from local food. With proximity to the city, an increasing amount of land was devoted to the production of food for the labour market.

The introduction of monoculture cash crop systems south of the Senegal River in 1850 has continued to alter the ecological relations of production through extensive deforestation and a reduction in the amount of cereal crops grown for domestic consumption (Sene, 1992). Deforestation continues presently for groundnut cultivation in Senegal at an annual rate of 60,000 hectares (Seck and Engelhard, 1993). Recent drought years have forced farmers to clear even greater amounts of land to cope with declining yields. The destruction of these natural forests which provide a protective curtain against the desert winds marked the beginning of a domino effect of ecological degradation in Senegal (Sene, 1992). France initiated a rice development policy to ease the large deficit in cereal production and to reduce the vulnerability to the vagaries of the international groundnut market (Engelhard and Ben Abdallah, 1986). The expropriation of land, either directly through settlement or indirectly through taxation, for export crops or urban food supplies forced indigenous farming into more marginal areas.

As the population was forced into more marginal areas and certain other areas were protected, more pressure was brought on a smaller resource base contributing strongly to rural-urban migration (World Bank, 1981). The consumption of rice in African urban areas was, perhaps indirectly, encouraged by government policies aimed at maintaining urban food prices below market levels and the use of land for export crops (World Bank, 1981). Imported wheat and rice were often less expensive than domestically produced foodstuffs, and African governments often resorted to massive imports while local producers in Mali and Senegal were forced to market their rice through parastatal cooperatives that reduced the price to match the imported crops. This reduced agricultural output as farmers have reverted to growing traditional crops outside the system (Lemarquand, 1982), thereby requiring further imports to meet demand.

3.2 Adaptation of rice production to declining streamflow in Senegal

Senegal and Mauritania, the states most dependent on the agricultural component of SRB development, have suffered serious deficits in the production of staple cereal crops. In 1988, Senegal produced only 19 per cent of its rice consumption, and Mauritania fared slightly better, producing 32 per cent of its consumption (CILSS, 1990). Imports of cereals, particularly rice primarily for urban consumers, contribute greatly to foreign debt. The serious production deficits, the prolonged drought and the huge foreign exchange burden of rice importation has led to the development of large state-managed rice plantations to supply as high a measure of rice self-sufficiency as possible (Engelhard, 1991; Seck, 1991).

As in many other African countries (World Bank, 1981), Senegal has tried to implement its top-down rice production policy within large rice perimeters or fields. The rice fields are usually 1000 hectares in size, situated on clay soil flood depressions. The perimeters are diked for flood protection, and include a pump station, a distribution of network canals, a drainage system and usually a drainage sump system. Typically, the pump station consists of two or more pump sets with a total capacity of approximately 5000 litres per second at a design head of 1.5-2 meters. The Food and Agriculture Organisation (FAO) concluded that a total of 376,000 hectares of land could be irrigated throughout the Upper, Middle and Lower valley and in the delta regions of the Lower Basin. The Organisation pour la Mise en Valeur du fleuve Senegal (OMVS) has stated that cultivating this quantity of land at 180 per cent intensity, which includes a second crop on 80 per cent of the land, was in fact a 'perfectly reasonable' objective and may be exceeded in the foreseeable future (Platon, 1981). The regularisation of river flows would permit the annual double-cropping, and the anticipated high productivity of the large perimeters was expected to reduce the national cereal deficit, and provide at least part of the economic justification for the dam construction.

In the Senegal River, regulation strategies in conjunction with agricultural development have been continuously touted as a means of increasing regional prosperity and restoring the seriously degraded basin ecology (M'Baye, 1971). The origins of recent plans date from the 1950s when the French administration drew up plans to exploit the hydroelectric potential in the upper river reaches of Mali and the agricultural potential in Senegal and Mauritania to increase the wealth-generating capacity of the SRB. Senegal, Mauritania and Mali jointly created the OMVS, a supra-national river basin authority charged with the development of the river. The long-term development goals of the OMVS included income reliability, establishment of a more stable ecological equilibrium, reduction of climatic vulnerability and accelerated economic development.

In 1962, based only on the technical feasibility of French engineering studies, the basin states of Mali, Mauritania and Senegal made the political decision to pursue river basin development with concurrent construction of the Diama and Manantali dams. Severe water deficits in the Cap Vert area (the peninsula that environs the city of Dakar and juts into the Atlantic) and salt-water intrusion into the Cap Vert aquifers forced the design of a third infrastructure component to divert water from a natural surface recharge area of the river (Lac de Guiers) via the proposed Canal du Cayor (Potworowski, 1990). The Canal du Cayor project would bring water from the Senegal River to the Cap Vert region of Senegal through intermediate storage at Lac du Guiers, a natural depression located south of Richard Toll. The additional water will also be used to create irrigation plots and to recharge the Cap Vert aquifers. Figure 5 is a schematic diagram of the Senegal River system including irrigation abstractions in the major agricultural development zones and the diversion at Richard Toll into Lac du Guiers to meet urban demands.

CILSS (1990) reports that rice irrigation projects averaged an annual cultivation intensity of 57 per cent, which is well below the design expectations of 180 per cent. Development costs were approximately 5-7.5 million FCFA (Canadian $20,000 - $30,000) per hectare. The policy was characterised by the lack of farmer involvement in design, highly centralised

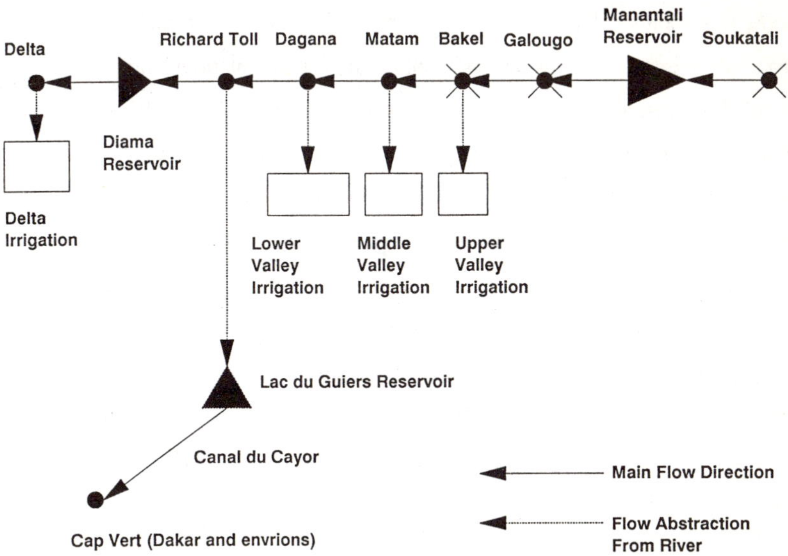

Figure 5 Schematic diagram of the Senegal River Basin development scheme.

management, and the provision of massive subsidies to meet minimum production targets. The farmers regarded themselves as employees; their alienation resulted in poor maintenance and high levels of abandonment and, not surprisingly, the large farms were generally not profitable (Seck, 1991). Policy RP maintains the colonial-era production biases in agricultural development policy and further limits the availability of water resources.

The large volumes of water required for rice cultivation, the evaporation rates common in the basin, and the poor natural drainage of the fields all render the rice perimeters vulnerable to salinisation. In fact, thousands of hectares of former rice production areas have already been abandoned due to salinisation (M'Balo, 1993). M'Balo (1993) also describes the deforestation accompanying rice cultivation, not only to clear the area for cultivation, but also to destroy the habitat of pest birds. Drought has further exacerbated the problem of deforestation by forcing many farmers into supplying charcoal for urban markets as an alternative income-generating activity. Even though the retail unit price for gas is competitive with charcoal, most Senegalese cannot afford the single cash outlay for gas and equipment (Falkenmark *et al.*, 1989).

With the focus on meeting the food, energy and water needs of Dakar and the deforestation of land for large, monoculture farms, the hard technology path of associated large rice plantations has only reinforced the social and ecological relations of production introduced during the colonial period. In addition, this policy ignored the World Bank's recommendations and reproduced many of the policies that failed in other parts of Africa:

> 'Rice ... can be grown efficiently under rain-fed conditions or in small-scale schemes in swamps or riverine valleys while large-scale irrigation schemes are extremely costly both in terms of initial investment and recurrent costs. Therefore, import substitution of these crops *(rice and wheat)*, where technically feasible, will pre-empt a substantial proportion of investible resources. Since urban consumers would be the main outlet for these cereals, ambitious rice and wheat production will reinforce the traditional bias against rural populations.' (World Bank, 1981, p. 58. Italics added by the authors.)

The two alternatives that comprise policy NRM are adaptations to declining streamflow. They are biased towards the needs of the farmers, decentralised decision-making, and an emphasis on labour over capital, appropriate technology and crops with lower water consumption. These biases have been emphasised in many previous studies (Salunke, 1983; Gabel, 1985; Rodale, 1985; WARDA, 1985; Falkenmark *et al.*, 1989; Brusberg, 1990).

The more prevalent alternative, 'Les Perimetres Irriguées Villageois' (PIVs: village-scale irrigation projects), is based on traditional social structures and is the sustainable development success story in the SRB. The PIVs depend directly on external inputs of oil and diesel fuel for the irrigation pumps, but they allow a measure of self-sufficiency. Developed and managed with strong – often sole – local involvement, by 1988, the PIVs had grown rapidly from a total of 1000 ha at 33 sites to 32,900 ha at 1096 sites (Seck, 1991). The relative success of the PIVs is usually attributed to the following factors:

- severe drought that forced a change in traditional agricultural practice;
- availability of the simple motor-pump as an appropriate technology;
- peasant leadership capable of accepting new technologies without upsetting traditional leadership structures;
- social arrangements that vest responsibility for the allocation of water to each smallholder or the political organisation of the village group.

Table 1 summarises the key differences between Large Perimeters (policy RP) and PIVs. This can be contrasted with the high water requirement for rice, the high soil permeability in the region, as well as the large pumping head, due to both the valley relief and the low water levels associated with drought, which make rice cultivation uneconomic and a poor use of scarce water resources. USAID (1984) also describes the exorbitant pumping costs associated with rice cultivation in the Bakel region.

The crop choice on PIVs also has implications for natural resource management. The production costs of some traditional or lower water-consumptive crops, such as sorghum, millet and wheat, are lower than the costs of rice production, and it is possible to obtain yields of more than 4000 kg per hectare under irrigation. Perhaps of greater importance is the failure of state planners to comprehend the role of subsistence behaviour, and this failure lies at the heart of the social alienation and low productivity that plague large and intermediate rice schemes. The prolonged drought has catalysed the rapid spread of PIVs throughout the valley because they are a necessary option for risk minimisation given the present marginality

Table 1 Comparison of large rice projects and village irrigation projects: key features

	Large rice projects	PIV projects
Size (ha)	>1000	10-20
Location	Mostly delta	Mostly valley
Construction costs (FCFA/ha)	4,000,000	400,000 - 900,000
Operating costs (FCFA/ha)	140,000	70,000
Rice yield (t/ha)	4.1	5
Sales to SAED (%)	45	10

Source: Huibers and Speelman (1989).

of rain-fed and flood recession cultivation. Farmers choose PIVs not for the prospect of large surpluses for market, but because it fits their development model of economic diversification and risk aversion under a drought regime.

The PIV offers some hope for ecological and social stability, but the environmental degradation is so far advanced that a combined land-water environmental management plan demands a reorientation of the natural resource base and a maximisation of biomass production by optimal management of available water resources. Environment et Development Action-Tiers Monde (ENDA-TM), an international non-governmental organisation based in Dakar, has attempted a direct response to stop and reverse the environmental degradation, deforestation, soil degradation, desertification and the migration to urban areas. The agricultural division of ENDA, Systemes et Prospective (SYSPRO), has promoted the second alternative to the rice perimeters throughout Senegal. The SYSPRO system uses simple irrigation technology to intensify agricultural systems for year-round biomass production. The key agro-ecological features of the SYSPRO system include inter-cropping, agro-forestry, composting, simple irrigation technology, modest use of external inputs such as fertilisers and pesticides, reliance on labour-intensive cultivation, and the establishment of nurseries for the production of diverse species.

SYSPRO's extensive use of inter-cropping is to a large extent an evolution and intensification of traditional agricultural systems. Like most innovations, inter-cropping also imputes a social bias, but in the opposite direction, because it reduces the need for a costly external input, thereby increasing the accessibility of the SYSPRO system. The agro-forestry aspects of the SYSPRO are an essential production component. Trees act as shelter-belts for irrigated plots. The shelter-belts provide a protective barrier to resist wind and water erosion and also change the local micro-climate by decreasing insolation and evapotranspiration (Lal, 1988). Tree production also serves as a potential fuelwood source. Additional ecological benefits are cited in Lal (1990).

Because of the unique existence of production and water consumption records, a specific project at Sebikotane was selected for an independent investigation of representative crop water requirements of integrated agro-forestry systems. Simulation studies performed with a dynamic programming model (Venema, 1993) confirm that the SYSPRO system has lower overall demand for water resources. In several hydrologic scenarios, which assume cyclical drought or a permanent change in water levels, policy NRM allows greater overall development potential. In addition, policy NRM is maintained at a 65 per cent contre-saison intensity compared with only 50 per cent for policy RP. Thus, operationally, the ecologically important year-round productive capacity of the SYSPRO agro-forestry system can be maintained. Simulation results indicate that, under continued drought conditions, a bias towards the needs of the Cap Vert region through the simultaneous provision of municipal water demands, hydropower production and irrigated rice production could limit the full agricultural development potential to as little as half of what could be developed under an alternative NRM policy.

4 Conclusions

Much of the debate about the role of adaptation in response to global warming has tended to assume that adaptation is only of concern for the future, when climate change has become evident. This perception is subject to the same objection that pertains to mitigative actions. If the world waits for global warming to become manifest beyond any shadow of a doubt, it will be too late to achieve significant mitigation. In the case of adaptation, there are further objections to this 'wait-and-see' attitude. The climate is always changing and, as the above example demonstrates, current climatic variability requires adaptive responses. In addition, as demonstrated in the SRB development, economic or trade policy should be formulated with regard to climate change.

Adaptation requires international cooperation because the underlying cause of some of the poorer nations' vulnerability is connected to macroeconomic concerns and trade policy. To the extent that some action is already under way (to combat the impacts of natural disasters, desertification, deforestation etc.), adaptive responses to climate change can be thought of as a natural extension and expansion of these programs. Indeed many of the potential adaptive responses are similar to or consistent with those actions needed to move towards the goal of sustainable development as proposed by the Bruntland Report, and as elaborated in the Agenda 21 plan adopted at Rio de Janeiro in June 1992.

One of the primary concerns is that adaptation to climate change is consistent with sustainable development. In the developing world, this requires an analysis of the food production systems, from both a social and an ecological perspective, and analysis of the bias in the technology that will be introduced into an area. The example in the SRB illustrates how a change in the adaptation bias can be used to achieve food self-sufficiency in a sustainable manner under a declining water supply.

The scarcity of water resources places a fundamental constraint on irrigation development in the SRB. Irrigation development in the SRB should therefore optimise the allocation of the scarce resource. Government policy (RP) entails following a hard technological path to bring as much rice production into production as possible. The implementation of top-down policies, oriented towards rice production in the SRB, has furthered the process of landscape desiccation and exacerbated poverty by ignoring the effect of technological biases on the social and ecological relations of production of agriculture in the area.

The alternatives aim at providing social and ecological sustainability, satisfying the joint objectives of food self-sufficiency for the existing villages in the SRB, and promoting ecological rehabilitation of the valley through agro-forestry. The effect of the policy NRM projects has been to alter the ecological and social relations of production and break the chain that leads to famine. This has been achieved by introducing a labour-biased production system using simpler, more accessible technology, incorporating the input of local producers, and combining ecological with agricultural goals.

The government's objective in developing the SRB for large-scale rice plantations is driven by a need to reduce a trade imbalance. Although it is not clear that this objective can be met by policy RP, policy NRM will not fulfil the urban areas' demand for inexpensive rice. It is clear that other policies are needed both at the national scale, and within the framework of international agreements, that will support the implementation of sustainable adaptations. At the national scale, governments need to promote policy NRM in order to decrease the rate of rural-to-urban migration and encourage a reversal of this trend. International agreements need to encourage adaptations that are consistent with the goal of sustainability while still recognising the short-term economic burdens that a country such as Senegal faces with respect to the continued need to import rice.

Acknowledgements

The authors would like to acknowledge Lakshman Yapa for contributing Figures 3 and 4.

References

Bell, M. (1986). *Contemporary Africa*. Longman, London.

Brusberg, F.E. (1990). *Analyse des Donnees de Reference et Etude Socio-Economique pour la Period d'Hivernage 1989-1990*. Projet de Gestion de l'Eau et de l'Irrigation, Rapport Intermedeiare, Harza Engineering Company, Technical Assistance Team. Contract USAID no. 685-0280. Devres Inc., Bethesda, Maryland, and Bakel, Senegal.

CILSS (1990). *Problematique des Politiques Riizicoles en Pays Sahelien, Rapport de Synthese*. Comite Permanent Inter-Etats de Lutte contre la Secheresse dans le Sahel, Bruxelles.

Citeau, J. (1990). *Les Satellite Meteorologique, Une Approache Nouvelle de la Connaissance du Climat du Sahel*. These du Doctorat de l'Universite Paul Sabatier, Centre D'Etudes Spatiale des Rayonnements, Toulouse.

Commoner, B. (1972). *The Closing Circle*. Bantam, New York.

Doorenbos, J. and Pruitt, W.O. (1977). *Guidelines for Predicting Crop Water Requirements* – FAO Irrigation and Drainage Paper 24. Food and Agriculture Organisation of the United Nations, Rome.

Downing, T.E. (1992). *Climate Change and Vulnerable Places: Global Food Security and Country Studies in Zimbabwe, Kenya, Senegal and Chile*. Environmental Change Unit Research Report No. 2, Environmental Change Unit, Oxford.

Ellis, F. (1988). *Peasant Economics: Farm Households and Agrarian Development*. Cambridge University Press, Cambridge.

Ellis, W.S. (1987). Africa's stricken Sahel. *National Geographic,* **172**: 140-179.

Engelhard, P. (1991). La vallee 'revistee' ou les 'enjeux de l'apres barrage' cinq ans plus tard. In: Crousse, B., Mathieu, P. and Seck, S.M. (eds.) *La Vallee du Fleuve Senegal*. Karthala, Paris.

Engelhard, P. and Ben Abdallah, T. (eds.) (1986). *Enjeux de l'après-barrage Vallée du Sénégal*. ENDA et République Française – Ministère de la Coopération, Dakar.

Falkenmark, M., Lundqvist, J. and Widstrand, C. (1989). Macro-scale water scarcity requires micro-scale approaches: aspects of vulnerability in semi-arid development. *Natural Resources Forum,* **13**: 258-267.

FAO (Food and Agriculture Organisation) (1977). *Etude Hydro-Agricole du Bassin du Fleuve Senegal.* Rapport de Synthese des Etudes et Travaux, FAO, Rome.

Francis, C.A. (1985). Agricultural policy. *Regenerative Farming Systems.* Proceedings of a workshop, December 10-11, Washington, DC. pp. 80-103.

Gabel, M. (1985). The regeneration of Africa – resources, needs and capacities: inventories for mobilsation. *Regenerative Farming Systems.* Proceedings of a workshop, December 10-11, Washington, DC. pp. 118-132.

Griffin, K. (1974). *The Political Economy of Agrarian Change.* Harvard University Press, Cambridge, Massachusetts.

Harlan, J.R. (1975). Our vanishing genetic resources. In: Abelson, P.H. (ed.) *Food: Politics, Economics Nutrition, and Research.* American Association for the Advancement of Science, Washington, DC.

Huibers, F.P. and Speelman, J.J. (1989). *Water Management in Village Irrigation Schemes of the Senegal River Valley: Technical Aspects.* WARDA/Wagenigen Agricultural University, Water Management Project, St. Louis and Wageningen.

Lal, R. (1988). Soil degradation and the future of agriculture in sub-Saharan Africa. *Journal of Soil Water Conservation,* Nov-Dec: 444-451.

Lal, R. (1990). Low-resource agriculture alternatives in sub-Saharan Africa. *Journal of Soil Water Conservation,* Jul-Aug: 437-445.

Lamb, P. (1985). West African rainfall index. *Climate Diagnostic Bulletin,* **85** (10). NOAA/NWS, Washington, DC.

Lappe, F.M. and Collins, J. (1977). *Food First: Beyond the Myth of Scarcity.* Ballantine Books, New York.

Lavender, S.T. and Anderson, J.E. (1984). *The Recent African Drought Within the Context of Earlier African Hydrological Experience.* Presented at the Sixth Meeting of the Committee of Cooperation of Development in Africa. Energy Group. Rome.

Lemarquand, D.G. (1982). *International Development of the Senegal River.* PhD dissertation, Oxford University, Oxford.

M'Balo, N. (1993). ENDA-SYSPRO, Dakar. (Personal communication.)

M'Baye, K. (1971). Le regime des terres au Senegal. In: Nylander, A. (ed.) *Le Droit de la Terre en Afrique (au Sud du Sahara).* Association Internationale des Sciences Juridiques, Maisonneuve et Larose, Paris.

Merrill, R. (1976). Toward a self-sustaining agriculture. In: Merrill, R. (ed.) *Radical Agriculture.* Harper and Row, New York.

Platon, P. (1981) The development of the Senegal River. *Marches Tropicaux et Mediterraneens.*

Potworowski, A. (1990). A taste of salt. *IDRC Reports,* **18**: 10.

Rodale, R. (1985). Internal resources and external inputs – the two sources of all production needs. *Regenerative Farming Systems.* Proceedings of a workshop, December 10-11, Washington, DC. pp. 1-8.

Salunke, V.B. (1983). *Pani Panchayatt, Dividing Line between Poverty and Prosperity.* Shetkarinagar-Khalad Talukaa-Purandhar District, Pune, Maharahtra, India, Gram Gourav Pratishthan.

Seck, M. (1991). *Programme Intensification Agricole et Nouveaux Systemes de Production.* ENDA-SYSPRO, Dakar.

Seck, M. and Engelhard, P. (1993). *Environnement et Nouveaux Systemes de Production Agricoles: Un Avenir Pour le Sahel?* ENDA-SYSPRO, Dakar.

Sene, I. (1992). Agriculture Segalaise: Diagonstic et Perspectives. *Afrique Espoir,* **9**: 20-22.

USAID (1984). *Bakel Irrigation Project Mid-term Evaluation.* USAID/OMVS, Dakar.

Venema, H.D. (1993). *A Management Planning Model for the Senegal River Basin.* Master's thesis, University of Ottawa, Ottawa.

Wallerstein, I. (1974). *The Modern World-System: Capitalist Agriculture and the Origins of the European World Economy in the Sixteenth Century.* Academic Press, New York.

Wallerstein, I. (1976). The three stages of African involvement in the world economy. In: Gutkind, P.C.W. and Wallerstein, I. (eds.) *The Political Economy of Contemporary Africa.* Sage, New York.

WARDA/Wageningen Agricultural University (1985). *Recherce sur la Gestion de l'Eau dans les Perimetres Irrigues Villageois de la Vallee du Fleuve Senegal.* Phase Preliminaire, Rapport de Synthese, Wageningen Agricultural University, Wageningen.

World Bank (1981). *Accelerated Development in Sub-Saharan Africa.* World Bank, Washington, DC.

Yapa, L.S. (1982). Innovation bias, appropriate technology, and basic goods. *Journal of Asian and African Studies,* **XVII**: 33-44.

Pastoralist Production Systems and Climate Change

Katherine Homewood
Department of Anthropology
University College
Gower Street
London WC1E 6BT, UK

1 Introduction

This paper sketches some fundamental features of pastoral production systems in sub-Saharan Africa, and looks at the special circumstances that make them particularly vulnerable to climate change. Pastoralist systems already have food security problems, and the commonly prescribed solution is to raise their productivity. Before we make suggestions as to what should be done to buffer food security problems in the face of climate change, we should look at the outcome of interventions that have already taken place. Unless we learn from that experience, there is a real risk of making things even worse for the most vulnerable groups. This paper looks at the outcomes of a natural experiment in alternative attempts to raise productivity in one important African case study that typifies some fundamental problems in African pastoralist systems. The observed outcomes have implications for future attempts to intensify production in these systems.

There are, at a conservative estimate, well over 20 million pastoralists in sub-Saharan Africa. Not counting South Africa, they contribute about 25 per cent of all sub-Saharan food production (in grain-equivalents; Jahnke, 1982). They belong to a great diversity of ethnic groups and practise different variations of livestock husbandry as the main source of their subsistence (Sandford, 1983). Pastoralist production systems are not a vestigial hangover from a past era, but a vital and vigorous component of sub-Saharan African cultural, economic and ecological systems.

For all their apparent diversity, these systems display some basic underlying similarities. This paper looks first at the universal ecological and economic constraints under which African pastoralist systems operate. The paper goes on to explore the implications of global

NATO ASI Series, Vol. I 37
Climate Change and World Food Security
Edited by Thomas E. Downing

environmental change for those constraints. Extreme climatic events are predicted to become ever more frequent, but past mechanisms for coping with drought are becoming rapidly less effective. These systems are already vulnerable. For a number of reasons, pastoralist groups tend to be less food-secure than neighbouring groups (e.g. Downing and Parry, 1994). They are also more vulnerable to international conflict, and are often marginalised within their own nations, making the effects of climate-driven vulnerability even worse (Galaty and Bonte, 1992). Constraints on migratory movements and the increasing commoditisation of pastoral produce limit the effectiveness of traditional solutions to problems of environmental stress and food security.

Part of the business of this volume is to identify vulnerable groups and suggest avenues for policy and for research. There is no doubt that African pastoralists as a group are vulnerable. What should be done? It is worth looking first at what has been done, and with what effects. African pastoralist systems have among the lowest indices of productivity per animal in the world. Intensification of the production system has been suggested as the main buffer against adverse change. This is not a new prescription. Since Independence – and, in many cases, since well before it – there have been concerted efforts and considerable resources put into intensifying livestock production systems in sub-Saharan Africa. The paper sets out a quantitative comparison of the outcomes in terms of productivity and welfare of past contrasting attempts to intensify pastoral production systems and improve food security in East Africa. These attempts have achieved little and, in view of future developments, it is important to understand why.

2 Production constraints and subsistence needs

There is a great diversity of pastoralist production systems in arid and semi-arid sub-Saharan rangelands, but they nonetheless all depend on the same crucial variables that broadly define the ecological and economic constraints under which they operate. These variables also define the dilemma of arid-land pastoralism in the face of global environmental change and growing sub-Saharan population numbers.

Firstly, in areas with less than 1000 mm mean annual rainfall, primary production is largely determined by plant-available moisture. Plant-available nutrients are a secondary but nonetheless important factor (Behnke and Scoones, 1992). Secondly, the main determinant of plant-available moisture in drylands is rainfall, which becomes increasingly variable with decreasing mean annual total precipitation. Thirdly, animal biomass and production in the more arid areas are mainly determined by primary production and thus ultimately by climate. Where groundwater and drainage lines affect plant-available moisture, and where there is high enough moisture availability to allow year-round growth and/or woody vegetation cover, other factors such as disease vectors and nutrients become increasingly important. Finally, the extreme variability of rainfall entails extreme variability of primary and therefore secondary production (Behnke *et al.*, 1993).

These basic facts dictate some clear corollaries. Pastoralists at the more arid end of the environmental spectrum either need to be able to move their herds to exploit new areas of plant-available moisture and primary production, or they need extremely high, capital-intensive investment to buffer those changes, coupled with sure mechanisms of rapid destocking and restocking to sustain their production system in the face of rapidly changing resource availability. The first is a strategy that exploits the inherent variability of the environment; the second requires high-cost inputs together with market policies to buffer that environment's vagaries.

The basic ecological facts of pastoralist production also carry economic implications. It is well established that, given the seasonal variability of pastoral environments, pastoralist families cannot generally subsist year-round on the milk and meat their herds produce. Pastoralists may value a milk and meat diet highly, but in practice they must supplement it with wild or cultivated foods. Typically pastoralists sell or exchange pastoral produce for grain or other cultivated foodstuffs that usually make up at least 50 per cent of their dietary energy intake on average, and may account for the whole of the dietary intake during the dry season (Homewood, 1992; Galvin, 1992). The great majority of pastoralist economies operate by virtue of the fact that the calorific and monetary terms of trade broadly favour pastoral rather than crop produce. Weight for weight, milk and meat gain a better price than grain, although grain offers more calories per gram (Swift, 1986; Dahl and Hjort, 1976). Pastoralists can thus ensure a reasonable diet year-round by sale or exchange of milk or slaughter animals for grain.

However, when a dry season stretches into drought, and there are a series of drought years in succession, the animals rapidly lose condition and fetch lower prices, more pastoralists are forced to sell so livestock prices plummet, and livestock mortality rates rise so there are progressively fewer left that any one family can sell to meet its needs. The upshot is that pastoralists face a double jeopardy in drought (White, 1991; Homewood, 1994), with grain prices rising at precisely the time that pastoralist livestock produce less. In these circumstances it is all too common for pastoralist families to lose all their stock and thus their means of livelihood.

Sub-Saharan pastoralist systems have survived and developed over 4000-5000 years in the face of fluctuating environments (Gowlett, 1988). There is nothing new about individual families dropping out of the pastoralist economy, but the buffering mechanisms that may have limited this process, and also provided opportunities to re-enter the system, are progressively disappearing. These coping mechanisms have relied on social networks that share risk and redistribute food and livestock, as well as allowing emergency access to environmental resources not normally used by the group.

3 Climatic change in the sub-Saharan rangelands

Some workers have seen climatic change in African rangelands as primarily the outcome of local and regional land-use practices. For example, Sinclair and Fryxell (1985) attributed the post-1960s downturn in Sahelian rainfall to sedentarisation of previously mobile pastoralist peoples, with attendant overgrazing, decline in ground cover, changing albedo and evaporation and ultimately lower rainfall levels. They saw climatic downturn as part of a vicious downward spiral of environmental degradation and climate feedback set in motion by increased grazing pressure. Probably a greater weight of informed opinion would now link deteriorating climates of African drylands to changing global atmospheric processes, citing the correlation of Sahelian rainfall with the temperature anomalies of the southern oceans (e.g. Pearce, 1991). The consensus emerging among rangeland scientists, contrary to Sinclair and Fryxell's view, is that arid and semi-arid grazing lands are boom-and-bust systems that undergo major climate-driven fluctuations. Pastoralist systems are increasingly seen as operating remarkably well in these systems, with periodic drought losses but without long term environmental degradation commonly ensuing as a result of pastoralist land use (Homewood and Rodgers, 1987; Ellis and Swift, 1988; Behnke *et al.*, 1993).

Although there are considerable differences of interpretation among alternative models of global climate change, there is a strong likelihood of warmer temperatures over much of Africa's drylands, and of an increased frequency of extreme climatic events. Probable humidity changes are less easy to predict. Nonetheless, the implications of global climatic change are fairly clear for these African dryland systems. Most rangelands currently primarily occupied by pastoralists are likely to become more arid, plant-available moisture will probably decrease, levels of primary production are likely to decrease, and the variability and unpredictability of that production will probably rise. For example, in a recent review of the implications of changing climate for Kenya's people and land use, it is clear that the arid pastoralist rangelands are predicted to be particularly adversely affected, with pastoralists (85 per cent food-poor) and agropastoralists (33 per cent food-poor) showing the worst current indices of food security (Downing, 1992). The arid areas of Kenya are predicted to double in size. The very large numbers of smallholders using currently less arid adjacent zones, and their tightly demarcated land use, make it unlikely that pastoral production systems will simply expand into these areas as they become drier, as happened in past climatic downturns in the Rift Valley (Spear and Waller, 1993). Modelling work on the transitions between pastoralism and cultivation suggest instead the counter-intuitive result that pastoralists will themselves be driven to settle and cultivate although conditions for cultivation are progressively deteriorating (Mace, 1993). This theoretical prediction, derived from dynamic optimality modelling, is borne out by empirical observation of historical and contemporary pastoralist societies under stress, such as the Turkana, the Zaghawa, the Chamus and others (Johnson and Anderson, 1988). However, in some semi-arid zones, the trade-off between rising temperatures, rising potential evapotranspiration and rising precipitation may operate to improve conditions for plant growth and production, whether through cultivation or pastoralism (Downing, 1992).

4 Intensification of pastoral production?

One main remedy that is being suggested for the adverse effects of climate change is to intensify production systems where possible so as to guarantee food security (Downing and Parry, 1994). Intensification of pastoralist production systems is not a new idea. For decades, planners have been aware that, although pastoralist systems are of considerable subsistence importance in sub-Saharan Africa, their indices of milk and meat production per animal are among the lowest in the world. Productivity is not an entirely clear-cut issue: per unit of land area, the indices are as good as if not better than those for comparable ecozones in the U.S. or Australia. Nonetheless, other things being equal, if productivity per head could be increased then much of the food security problem could be resolved. Between the 1960s and the 1980s, hundreds of millions of dollars went into attempts to develop the livestock production systems of sub-Saharan Africa (Sandford, 1983). The clear consensus is that these attempts failed to bring about any improvement in productivity, and that the outcome has quite commonly been worse than before in terms of general food security, let alone equitability (e.g. Sandford, 1983; Halderman, 1985; Aronson, 1981; Wyckoff, 1985; Homewood, 1992). This makes it important to look carefully at such attempts to avoid repeating past mistakes.

Maasailand makes a particularly neat case study illustrating the pitfalls of attempts to intensify livestock production systems. This is not only because there are comprehensive data, but also because of the natural experiment provided by the parallel and contrasting interventions in Kenya and Tanzania. Maasailand now comprises some 100,000 km^2 of continuous rangeland cut in two by the Kenya – Tanzania border established in the colonial era. Following their independence in the early 1960s, both Kenya and Tanzania instituted livestock development programmes in their respective parts of Maasailand. Although the Maasai, the rangelands and the production systems were broadly similar on the two sides of the border, the political contexts, level of infrastructure and development policies implemented were in sharp contrast. In both countries, low pastoral productivity was seen as stemming from poor land use and livestock management. In both, land demarcation, control of stocking levels, and investment in water, forage and veterinary inputs were seen as fundamental to intensification and commercialisation of production. However, Kenya was strongly capitalist and livestock development heavily funded, whereas Tanzania was strongly socialist and considerably less well funded.

In Kenya, there was an initial drive to privatise land tenure throughout the Maasai rangelands with individuals staking claim to small plots of a few square kilometres each. This soon ran into trouble: these private holdings were neither ecologically nor economically self-sufficient. Much of the better Maasai dry season grazing land passed rapidly into the hands of cultivators and was lost to the pastoralist system altogether. Kenya then developed a system of group ranches, where groups of cooperating herdowners acquired title to areas of rangeland a few hundred square kilometres in extent. The group ranches – at least in theory – were a means to reduce stocking rates, encourage investment in water, livestock and range improvements, boost turnover and commercial offtakes, and thereby to increase productivity. The whole of Kenya Maasailand is now adjudicated into privately-owned individual, group or company holdings, and the current trend is for renewed subdivision of group ranches.

In Tanzania, a contrasting system of ranching associations was implemented. Each ranching association was to be several thousand square kilometres in extent and to contain a number of socialist livestock villages. Herdowners were intended to cooperate on setting up developments such as veterinary, water and marketing inputs, with state support and United States Agency for International Development (USAID) funds, and were intended to contribute livestock to form the nuclei of communal herds run on intensive lines. In practice, wherever water points, roads and other facilities were built, cultivators rapidly moved in and settled around the new developments, claiming registration as new *ujamaa* villages and diverting the benefits from the target pastoral communities. USAID attempted to resolve the problem by negotiating exclusive land tenure for pastoralists. This ran counter to the Tanzanian *ujamaa* philosophy and, in the late 1970s, the ranching association programme disintegrated.

This paper compares the outcomes of those two decades of contrasting inputs in terms of indices of livestock production – fertility, mortality, milk production and offtake for sale and slaughter; average livestock holdings and wealth distribution; diet, dietary energy adequacy, and nutritional status as indicators of food security for a series of Tanzanian and Kenyan Maasai study sites during 1981-83. Where data are available, they are compared with the figures for 1960s 'pre-development' Maasai.

Given the order of magnitude of difference in the funds invested in Kenya and Tanzania livestock development, and the contrast between the collapse of the Tanzanian programme and the rapid evolution of the Kenyan system, one would expect to find major differences between the two parts of Maasailand. The detailed quantitative comparative review of these outcomes for Kenya and Tanzania Maasailand between the 1960s and 1980s has been carried out elsewhere (Homewood, 1992). Briefly, a retrospective comparison, based on detailed parallel studies during 1980-83 for several study sites in each of Kenya and Tanzania Maasailand, reveals that, far from there being major differences, there are instead striking similarities for all indices of livestock performance, offtake and mean livestock holdings (Tables 1 and 2). Most important as indicators of development status, the dietary energy intake patterns are closely similar, as are the unacceptably high proportions of undernourished and malnourished children, and there are strong indications that food security has not improved since the 1930s (Tables 3 and 4). The only indication of a significant difference between the two parts of Maasailand lies in the larger middle wealth category in Tanzania compared with the larger top wealth category in Kenya (Table 2).

Despite two decades of differing inputs, there is little visible divergence or indeed intensification of the two pastoralist production systems. There are virtually no significant differences between the two parts of Maasailand nor, where data exist to demonstrate it, between the 1960s pre-development and 1980s post-development input figures. Given the similarity of mean livestock holdings, the greater top wealth category in Kenya compared with that in Tanzania, together with other lines of evidence (Graham, 1988), suggests that with the Kenya group ranch system the rich have become richer and the poor poorer to an extent not seen in Tanzanian Maasailand. The poor category has a natural lower cut-off point. Throughout the pastoralist system, people are maintained at nutritional levels verging on the untenably low (see next section; also Galvin, 1992; Homewood, 1992). Individuals and families whose access to land, livestock and pastoral produce falls below that cut-off minimum simply drop out of the system altogether. The information available does not allow any conclusions on the relative rates at which people are being forced out of the pastoralist system in each case. However, although considerable inputs have done little to alter the indices of production, they appear to have crystallised wealth holdings in Kenya and concentrated them into fewer hands than before (Graham, 1988). The working hypothesis that needs testing is that, in the wake of intensive privatisation and commoditisation, there is a

Table 1 Maasai livestock production and offtake

	Tanzania 1981-83	Kenya 1981-83	Kenya 1960s
Calving rates	0.61	0.60	0.55-0.70
no.	153	120	
std. error	0.100		
Cow mortality	0.11	0.06	0.10
no.	342	320	
std. error	0.058		
Milk yield (kg/day)			
Wet season	1.0	1.1	1.0-1.5
no.	35		
std. deviation	0.16		
Dry season	0.7	0.8	0.5
no.	197		
std. deviation	0.141		
no. milkings	> 1000	> 1000	
Offtake (% total herd)	8.1	9.1	8.0

Note: For detail of study sites, methods and sample structure, see Homewood (1992).

Table 2 Maasai livestock holdings

	Tanzania	Kenya
Stock per capita 1980s mean		
Cattle	6.7	9.7
Small stock	8	7.7
Livestock units/reference adult	% households in each wealth category	
<5	34	33
5-13	57	38
>13	9	29
no. households	70	80

$X^2 = 10.89$, df=2, $p<0.001$; Partitioned X^2: cell[2,2] $X^2=0.979$, p=0.32, NS; cell[3,2] $X^2=9.992$, $p<0.002$

Note: For detail of study sites, methods and sample structure, see Homewood (1992).

greater net outflow of destitute and dispossessed people from the Kenya Maasai pastoral system than from the Tanzanian system.

Why is there no change in the majority of indices of production system and food security? A powerful combination of ecological/environmental reasons revolving around the unpredictability of grazing and water resources in this semi-arid environment, together with

Table 3 Maasai diet and nutritional status, 1981-83

	Tanzania	**Kenya**
Contribution to dietary energy (%)	Mean (standard deviation)	Mean (standard deviation)
Milk and milk products	34 (25.3)	48 (32.8)
Meat	10 (7.5)	8 (12.0)
Maize	53 (21.8)	26 (26.3)
Other	4	19
no. household-days	10	127
no. individual-days		1547
Recommended dietary intake (%)	Mean (standard deviation)	Mean (standard deviation)
Protein	141 (25.0)	171 (43.2)
Energy	74 (17.8)	71 (7.6)
no. household-days	10	127
no. individual-days		1547
Nutritional status of children < 5 yrs	%	%
Healthy (>90% wt for ht)	42	47
Undernourished (80% to 90% wt for ht)	35	34
Malnourished (<80% wt for ht)	23	19
no. individuals	232	508

Notes: Maasailand is strongly seasonal, as are the types and amounts of food available as well as workloads for different age/sex classes. Here a single-season Tanzanian sample averaged across three study sites is compared with annual average values for Kenya Maasailand as an initial indication of nutritional conditions. For detail of study sites, methods and sample structure, see Homewood (1992).

related social and socioeconomic factors, operate to buffer changes in productivity and food security indices within the Maasai pastoralist system. These have been analysed in detail elsewhere (Homewood, 1992).

The crucial factor limiting intensification is the ability to respond to seasonal and interannual changes in water, grazing and disease patterns. Historically, that response has relied on movement. Intensification relies on demarcation of land tenure and loss of the mobility that has always underpinned survival. With tightly demarcated land ownership, only individuals who can negotiate reciprocal access to complementary resources can survive within the pastoralist system. In Kenya, many individuals have lost what access rights or land holding they had, and therefore have lost their foothold in the system. The ecological constraints already discussed preclude small private holdings from being self-sufficient. In Kenya, private land tenure has become a means whereby individual pastoralist families may become dispossessed altogether, whereas those who retain private land ownership can negotiate

Table 4 Weights, heights and BMI values for Maasai in the 1980s and 1930s

	Tanzania 1989			Kenya 1930s		
	Adult male	Adult female	Female 18-29yrs	Adult male	Adult female	Female 19-30yrs
Weight	57.6	49.1		60.2	54.2	
no.	88	180		362	333	
std deviation	7.9	6.6				
Height	171.2	159.9		169.3	155.5	
no.	88	180		362	333	
std deviation	6.54	5.5				
BMI	19.7	19.1	19.7	21.0	22.4	22.9
no.	88	180	91	362	333	201
std deviation	2.3	2.3	2.0			

Notes: For detail of study sites, methods and sample structure, see Homewood (1992). Body mass index (BMI: weight/height^2) can give an indication of relative stunting and/or wasting. Stunting refers to a low achieved height attributed to early childhood malnutrition. Wasting indicates more recent nutritional history. The lack of information on the 1930s sample precludes detailed comparison. However, height has increased slightly but not significantly, possibly suggesting marginally better conditions of childhood nutrition. BMI has decreased slightly although not significantly, suggesting rather worse nutritional circumstances at the time of the study than was the case for the 1930s sample. More striking than any difference is the overall similarity between the two samples.

reciprocal access as changing conditions make it necessary to move herds (Grandin and Lembuya, 1987; Graham, 1988). Few can manage the high-cost investment necessary to overcome the major problems of ensuring water and forage availability and disease control on a year-round basis in the face of unpredictable and deteriorating climatic conditions. Other factors mask emergent differences. Aspects of the Maasai social structure still allow conditional reciprocal access to resources such as grazing, and sharing of milk. Wealth is unlikely to be always expressed in better nutritional status, because of the higher value placed on reinvesting milk production in calves that represent the future of the herd, once the most basic human nutritional needs have been met (Grandin, 1988). This does not alter the fact that there has been no obvious improvement in indices of production or in average food security following privatisation of land and investment in the production system.

This is not an isolated example. I know of no other detailed quantitative studies of development outcomes in terms of livestock performance and food security. However, there have been many other evaluations, and there is a widespread consensus that attempts to intensify production in African pastoralist systems have failed, for very good reasons which we cannot afford to ignore.

5 Food security among sub-Saharan pastoralists

Parallel and contrasting development inputs have not improved the vulnerably low dietary energy intakes, nor the alarmingly poor child nutritional status in Kenya and Tanzania Maasai. They have also done little for adult Maasai nutritional status, with 1980s figures as poor as or worse than 1930s samples (Tables 3 and 4; see also Homewood, 1992). The Maasai rank together with the Samburu, the Turkana and the Borana as having low weight for height compared with West African Fulani pastoralists (who tend to be taller and heavier) and various sub-Saharan cultivator populations (who tend to be short and relatively heavy; Galvin, 1992). The figures for the Fulani may be misleading, in that some derive from strongly sedentarised, cultivating Senegalese agropastoralists, and others from pastoral Wodaabe populations in Niger studied before the 1970s and 1980s droughts. Since those data were collected, many communities of Fulani and Tamasheq have become nearly or totally destitute through drought and war (Randall, 1988), barely surviving on gathered foods or in refugee camps. There is no need for anthropometric indices to make the desperate plight of these communities clear. However, as far as communities operating within the pastoralist economy are concerned, it seems clear that East African pastoralist systems maintain large numbers of people at the cost of chronically poor individual and community nutritional status and food security even under normal – let alone drought – conditions.

This chronically poor nutritional status is a product of several factors. Some have already been mentioned, such as the higher priority given to reinvesting milk production in the calves that represent the future of the herd. Others stem from patterns of seasonal production and of market integration. Bimodal climates mean that, in East Africa, milk production is more evenly spread through the year, and transhumant systems are shorter-range. This may make use of milk possible for more people for a longer period through the year, compared with the near-universal pastoralist reliance on cereal for much of the year in Sahelian climates, with their unimodal short rains and a dry season of about nine months. The relatively lower cereal intake is likely to reduce indices of weight for height. There is a long-established contrast between East and West African pastoralists in the type and degree of integration into local, national and international markets. Kerven (1992) traces this difference both to contrasting land-use patterns reinforcing or limiting the opportunities for regular frequent exchange, and to colonial taxation policies encouraging or (in East Africa) discouraging marketing of animals. In general, the West African pastoral Fulani are much more closely integrated into local and national markets than the East African pastoralists (Kerven, 1992); this may be reflected in their greater use of cereals and hence their less vulnerable nutritional states.

6 Implications for pastoral production systems under climate change

The outcomes of attempts at intensification have been set out in detail for the production system and food security of the Maasai, but the general outcome is recognised to have been the same across a wide range of livestock development projects in West as well as East Africa. At the same time, the period from the 1960s to the 1980s also saw the loss of pastoralist access to vital wetlands along many river systems in sub-Saharan Africa (Adams, 1992), with major hydroelectric and irrigation schemes debarring herds from floodplain grazing and riverine watering points, and population growth and permanent cultivation taking up ever more of the highland or groundwater areas formerly used as drought refuges. The large-scale river basin modification schemes have brought little in the way of enhanced production or food security to compensate for the systems they have disrupted. Alongside this progressive loss of central dry season grazing and water resources, there has been an inexorable process of boundary demarcation and closure throughout the rangelands, making it ever more difficult for pastoralists to retain the mobility they need to optimise in the face of fluctuating conditions of climate, primary production and disease (Homewood, 1996). Past flexibility of transhumant and migratory systems is disappearing with the crystallisation of national frontiers, regional administrative boundaries, and local demarcation of private land. Privatisation of land tenure means that increasing numbers of families can lose all access to the basic grazing and water resources necessary for pastoralist production. Growing commoditisation of pastoral produce means they may no longer be able to get help from the wider social network to re-establish themselves in the pastoralist system (Grandin, 1988; Talle, 1988). What are the implications of these trends for pastoral production systems under climate change?

One main response to the threat of climate change is for governments and donors to encourage intensification of food production systems in potentially vulnerable areas such as sub-Saharan Africa. However, the same approach is unlikely to work any better in African rangelands than it did for the 1960s - 1980s programmes. Those inputs were based on the assumption that stock losses and crises of food security in pastoral systems were more the result of poor land use and precarious stocking practices than of periodic drought and environmental shortfall. Environmental rehabilitation, and a more secure production, were to be brought about by demarcation of rangelands, control of stocking levels within individual areas, and investment in water development, dry season forage supplies and veterinary inputs. The result has been rather different. Environmental indices have not changed; droughts and stock losses continue to occur; livestock performance has not changed; food security has not improved. Instead there is a perceptible growth in the proportions of pastoralists who are hired herders rather than herdowners in their own right (Little, 1983), and an as-yet-unmeasured flow of people leaving the pastoral system to become poor cultivators, urban migrants or, at worst, destitute refugees with little hope of a livelihood.

Movement out of a pastoral production system is a long-established response to times of drought or disaster, but the possibilities for accepting refugees from the pastoralist economy have diminished with the spread of permanent cultivation, the relegation of gathering and hunting niches to virtual extinction, and the chronic poverty and unemployment of African urban areas. The more such refugees seeking shelter with farming, urban or other communities, the nearer the system slides to mass migration and famine. Any attempt to minimise the adverse affects of climatic change will have to buffer the production system by safeguarding access to grazing and water, retaining mobility and ensuring restocking, as well as fostering a robust market with realistic terms of trade. Yet interventions ostensibly designed to safeguard access have tended to result in loss of grazing and water resources.

7 Land tenure models and conservation estate in Maasailand

Over the last 100 years, the Maasai have seen the expropriation of a large proportion of their land as settler, cultivator or conservation estate (Homewood, 1996). The contrasting land tenure histories of Kenya and Tanzania Maasailand during 1960-80 have already been discussed. The changing role of conservation estates adds an interesting twist. Unlike land lost to permanent cultivation, a conservation estate retains water and grazing resources in a form that can allow continued pastoralist use. It is also potentially a major source of revenue, although satisfactory revenue-sharing schemes have yet to be established. Conservation estates in different parts of Maasailand have variously excluded the Maasai altogether (Serengeti: Homewood and Rodgers, 1991; Mkomazi: Brockington and Homewood, 1995); established joint conservation/pastoralist development land use (Ngorongoro Conservation Area: Homewood and Rodgers, 1991); attempted community-based conservation management schemes and/or revenue-sharing to make up for potential production forfeited to wildlife (Maasai Mara: Norton-Griffiths, 1993; Amboseli: Grandin, 1986; Western, 1994); and registered communal land tenure for *ujamaa* village grazing lands (e.g. Loliondo GCA: Perkin, personal communication). It is vital to study these alternative land tenure developments, and comparable experiments elsewhere in African rangelands, to establish which – if any – correlate best with medium-term security of livelihoods and, conversely, which are most likely to precipitate dispossession. Only this knowledge will help local and national governments to act before climate change exacerbates the situation.

East Africa has seen past cycles of growth and decline of pastoralist land use (Spear and Waller, 1993). The ecological rationale underlying pastoralism is not made invalid by climate change, but the transitions may become increasingly painful. The crucial resources necessary for the continued survival and operation of East African pastoral systems are already in place: what is in doubt is the continued entitlement of families and communities to gain access to those resources. Current development thinking focuses on making subsistence more secure. In particular, it revolves around translating into a national context the traditional mechanisms that in the past operated to allow flexible negotiation of emergency access to key resources (Behnke, 1993). As the arid areas spread, and refugees from the pastoralist economy accumulate beyond the capacity of other niches to accommodate them, communities may gain a greater say in the management of their local conservation estate. Community-based conservation is currently more catchword than fact but, as it becomes better established, the conservation areas within East African rangelands may come to form the foci for survival not only of wildlife but also of pastoral production systems. There is good evidence that joint land use can work well in savanna ecosystems (Homewood and Rodgers, 1991). Pastoralist land use has preserved both biodiversity and population numbers for savanna wildlife species. National parks and reserves currently deter pastoralist use of their protected dry season water and grazing, but they have also effectively prevented the conversion of these resources to permanent cultivation. As climates deteriorate, the interplay of pastoralist coexistence with conservation values and conservation revenue may come to provide a major source of ecological and economic security in East African rangelands.

8 Conclusion

In this paper, a single special case is used to make a point of much wider general application. In all sub-Saharan pastoralist systems, past fluctuations have been weathered by a range of strategies, from migration through part-time cultivation to short- or long-term movement out of the pastoral system. So what is new, what is contested and what is worth doing about the impact of climate change on pastoral production systems? Surprisingly little is new. There is already a wide array of well-established strategies for survival within arid land production systems. There is also a wide range of familiar suggested policy and development interventions. Currently-predicted climatic changes do not present a qualitatively new problem, but rather a familiar problem with a dwindling and shifting set of potential solutions. What would be new would be a clear understanding of the actual outcomes of those policies and interventions. The solutions that to outside agencies may seem most obviously likely to work, may be those in fact most dangerous to welfare and food security.

What is contested is the best way to improve food security in these vulnerable systems. Many argue that raising productivity *per se* is the key. I would argue that food security would be better achieved by supporting the production system so as to reduce the rate at which individuals and families are dispossessed and forced to move off the land to join the ranks of the urban poor. In many rural production systems in Third World countries, indices of production per head of population are low. But this can be turned around to say that these systems are supporting very large numbers of people in labour-intensive systems that allow the majority to retain their entitlements and their livelihoods. Do we want to raise productivity per head of population at the cost of dispossessing the many?

It is perhaps worth taking this broader view a step further to question the conventional wisdom not only on intensifying production but also on the implications of population and economic growth in African production systems. In their paper in this volume, Fischer *et al.* (1995) use linked models of global climate and economic change to suggest that curbing population increase, boosting economic growth and increasing food production may be the keys to buffering the impact of global climate change on global food security. But the solutions that emerge at a global scale, and which may dictate and reinforce international policy, may paradoxically have the very opposite effect on individual vulnerable populations at a local scale. The present paper documents the way in which interventions geared to increase productivity at best have had no effect, and at worst have compromised food security in African pastoralist populations. This seems a rather common outcome of efforts at intensifying productivity elsewhere in the developing world in general and in Africa in particular (Raikes, 1988). Similar paradoxes apply over population density and economic growth. Empirical evidence suggests that local increases in population density may be vital in establishing food-secure and environmentally-sustainable African rural production systems in formerly vulnerable areas (Tiffen *et al.*, 1994). Economic growth in much of Africa appears to be achieved by the rich getting richer and the poor poorer, actually multiplying rather than reducing the numbers of those vulnerable to hunger.

Past attempts at intensification based on land demarcation have failed to increase production, and at worst have exacerbated food security for many former pastoralist families. Rather than intensifying food production, future improvements may revolve around developing structures of negotiation and arbitration that allow flexible access to key resources in time of need

(Behnke, 1993). This paper's brief review of attempts to manipulate tenure and access to resources in Maasailand shows the rapidly evolving mosaic of different tenure and equity arrangements. There is an urgent need to study the implications of these different patterns of tenure and resource access in pastoralist and other economies, along with the other factors influencing the flow of families dropping out of these systems. The ability to pre-empt environmental change by having in place a set of land use and resource-access strategies that can buffer adverse change will determine the outcomes for peoples adjusting to a warmer, more crowded world. Not climate change alone, but rather resource access in the light of climate change will dictate both the scale of distress, and the burden on national and international bodies left to cope with mass migration, famine and disease.

References

Adams, W. (1992). *Wasting the Rain.* Earthscan, London.

Aronson, D. (1981). Development for nomadic peoples: who benefits? In: Galaty, J., Aaronson, D., Salzman, P. and Chouinard, A. (eds.): *The Future of Pastoral Peoples.* International Development Research Group, Ottawa.

Behnke, R. (1993). Natural resource management in pastoral Africa. Paper presented to the workshop: *Listening to People: Social Aspects of Dryland Management.* UNEP, Nairobi.

Behnke, R. and Scoones, I. (1992). *Rethinking Range Ecology: Implications for Rangeland Management in Africa.* Environment Department. Working paper no. 53., World Bank, Washington DC.

Behnke, R., Scoones, I. and Kerven, C. (eds.) (1993). *Range Ecology at Disequilibrium.* ODI/IIED/Commonwealth Secretariat, London.

Brockington, C. and Homewood, K. (1995). Received wisdom, science, and pastoralists: debates concerning Mkomazi Game Reserve, Tanzania. In: Leach, M. and Mearns, R. (eds.). *The Lie of the Land: Challenging Received Wisdoms in African Environmental Change and Policy.* James Currey Press/International African Institute, London.

Dahl, G. and Hjort, A. (1976). *Having Herds.* Stockholm Studies in Social Anthropology, University of Stockholm, Stockholm.

Downing, T.E. (1992). *Climate Change and Vulnerable Places: Global Food Security and Country Studies in Zimbabwe, Kenya, Senegal and Chile.* Environmental Change Unit Research Report No. 1. University of Oxford, Oxford.

Fischer, G., Froberg, K., Parry, M.L. and Rosenzweig, C. (1995). Impacts of potential climate change on global and regional food production and vulnerability. In: Downing, T.E. *Climate Change and World Food Security.* Springer-Verlag, Heidelberg.

Ellis, J. and Swift, D. (1988). Stability of African pastoral ecosystems *Journal of Range Management,* **41**: 450-459.

Galaty, J. and Bonte, P. (eds.) (1992). *Herders, Warriors, Traders.* Westview Press, Boulder, Colorado.

Galvin, K. (1992). Nutritional ecology of pastoralists in dry tropical Africa. *American Journal of Human Biology,* **4**:209-221.

Gowlett, J. (1988). Human adaptation and long-term climatic change in north-east Africa: an archaeological perspective. In: Johnson, D. and Anderson, D. (eds.) *The Ecology of*

Survival: Case Studies from Northeast African History. Lester Crook Academic and Westview Press, Boulder.

Graham, O. (1988). *Enclosure of the East African Rangelands.* Pastoral Development Network Paper No. 25a. Overseas Development Institute, London.

Grandin, B.E. (1986). Land tenure, subdivision and residential change on a Maasai group ranch. *Bulletin of the Institute for Development Anthropology,* **4** (2), 9-13.

Grandin, B.E. (1988). Wealth and pastoral dairy production. A case study from Maasailand. *Human Ecology,* **16**: 1-21.

Grandin, B.E. and Lembuya, P. (1987). *The Impact of the 1984 Drought at Olkarkar Group Ranch, Kajiado, Kenya.* Pastoral Development Network, Paper No. 23e. Overseas Development Institute, London.

Haldermann, J. (1985). Problems of pastoral development in East Africa. *Agricultural Administration,* **18**: 199-216.

Homewood, K. (1992). Development and the ecology of Maasai food and nutrition. *Ecology of Food and Nutrition,* **29**: 61-81.

Homewood, K. (1994). Pastoralists, environment and development in East African rangelands. In: Zaba, B. and Clarke, J. (eds.) *Environment and Population Change.* Ordina editions.

Homewood, K. (1996) (in press). Ecological outcomes of boundary formation in Maasailand. In: Nugent, P. (ed.) *African Boundaries and Borderlands.* Mansell Press, London.

Homewood, K. and Rodgers, W.A. (1987). Pastoralism, conservation and the overgrazing controversy. In: Anderson, D. and Grove, R. (eds.) *Conservation in Africa: Peoples, Policies and Practice.* Cambridge University Press, Cambridge. pp. 111-128.

Homewood, K. and Rodgers, W.A. (1991). *Maasailand Ecology: Pastoralist Development and Wildlife Conservation.* Cambridge University Press, Cambridge.

Jahnke, H. (1982). *Livestock Production Systems and Livestock Development in Tropical Africa.* Kieler Wissenschaftsverlag vauk.

Johnson, D. and Anderson, D. (eds.) (1988). *The Ecology of Survival: Case Studies from Northeast African History.* Lester Crook Academic and Westview Press, Boulder.

Kerven, C. (1992). *Customary Commerce.* Overseas Development Institute, London.

Little, P. (1983). Absentee herdowners and part-time pastoralists. *Human Ecology,* **13**: 131-151.

Mace, R. (1993). Transitions between pastoralism and cultivation in sub-Saharan Africa. *Current Anthropology,* **34**: 363-381.

Norton-Griffiths, M. (1993). The long-term conservation of wildlife on the rangelands of the Serengeti ecosystem. In: Sinclair, A. and Arcese, P. (eds.) *Serengeti II: Research and Management for Conservation.* University of Chicago Press, Chicago.

ODI/IIED/CS (1993). *New Directions in African Rangeland Management.* Research Workshop, Woburn, 31 May-4 June 1990.

Pearce, F. (1991). Sea change in the Sahel. *New Scientist,* **129**: 31-2.

Raikes, P. (1988). *Modernising Hunger: Famine, Food Surplus and Farm Policy in the EEC and Africa.* James Currey, London and Heinemann, Portsmouth.

Randall, S. (1988). *The Displaced Population in Douentza.* (La population déplacée de Douentza). PIA/ Feru Occasional Report, Save the Children Fund (UK), Bamako, Mali.

Sandford, S. (1983). *Management of Pastoral Development in the Third World.* Overseas Development Institute, London, with John Wiley, Chichester.

Sinclair, A. and Fryxell, J. (1985). The Sahel of Africa: ecology of a disaster. *Canadian Journal of Zoology,* **63**: 987-94.

Smith, A. (1993). *Pastoralism in Africa.* Hurst, London.

Spear, T. and Waller, R. (1993). *Being Maasai.* James Currey, London.

Swift, J. (1986). The economics of production and exchange in West African pastoral societies. In: Adamu, M. and Kirk-Greene, A. (eds.) *Pastoralists of the West African Savanna.* International African Institute, Manchester University Press, Manchester.

Talle, A. (1988). *Women at a Loss.* Stockholm Studies in Social Anthropology, No. 19. University of Stockholm, Stockholm.

Tiffen, M., Mortimore, M. and Gichuki, C. (1994). *More People, Less Erosion: Environmental Recovery in Kenya.* John Wiley, Chichester.

Western, D. (1994). *Conserving Savanna Ecosystems through Community Participation: The Amboseli Case Study.* Community Based Conservation Workshop. Liz Claiborne and Art Ortenberg Foundation, New York.

Western, D. and Ssemakula, J. (1981). The future of savanna ecosystems. *African Journal of Ecology,* **19**: 7-19.

White, C. (1991). Increased vulnerability to food shortages among Fulani nomads in Niger. In: Downs, R., Kerner, D. and Reyna, S. (eds.) *The Political Economy of African Famine.* Gordon and Breach Science, New York.

Wyckoff, J. (1985). Planning arid land development projects. *Nomadic Peoples,* **19**: 59-69.

Agricultural Policy and Climate Change in Mexico

Kirsten Appendini
Centro de Estudios Económicos
El Colegio de México
Camino al Ajusco 20
Mexico DF 01000, Mexico

Diana Liverman
Latin American Studies and Department of Geography
University of Arizona
Tuscon AZ 85721
U.S.

1 Introduction

As Mexico moves towards more open markets in agricultural trade, land and water, with the reprivatisation of land and the initiation of a North American Free Trade Agreement, the Mexican people are facing an uncertain economic future. Food production in Mexico has always been constrained by low rainfall and been at risk from frequent droughts, and is now threatened by climatic changes associated with global warming and regional land use transformation. Yet 20th century Mexico has experienced great increases in agricultural production associated with government policies to develop irrigation, increase the use of chemical inputs, and subsidise prices paid to farmers. At the same time, rapid population growth, urbanisation and export expansion have increased and restructured overall demand for agricultural products such that imports of basic grains have risen significantly in recent years (Sanderson, 1986; Wellhausen, 1976; Yates, 1981). Although Mexico no longer experiences the famines of previous centuries, and several indicators suggest that food security has been attained at a national level, studies show that as much as one- fifth of the population of 85 million is hungry or malnourished (Calva, 1988; Mexico INEGI, 1991).

National and individual food security is very vulnerable to year-to-year variations in climate, domestic agricultural policy and international economic conditions. Domestic agricultural policy is variable and complicated because different population groups have contradictory interests as to the orientation of food policy. The urban poor, rural landless wage-workers, and employers who wish to pay low wages, all demand cheap food. In the absence of

NATO ASI Series, Vol. I 37
Climate Change and World Food Security
Edited by Thomas E. Downing

subsidies, low food prices are a disincentive to agricultural production, hurting peasants and farmers dependent on growing staples. National policy has varied from protectionism and economic self-sufficiency to free trade and export expansion; from free or common access to land and water to private ownership; and from large price subsidies and rampant inflation to withdrawal of government financing and controls on wages and prices (Appendini, 1992; Austin and Esteva, 1987).

This paper describes how variations in agricultural policy and climatic conditions have independently influenced food production and security in 20th century Mexico, and attempts to show how economic policy increases vulnerability to climatic variation and vice versa. We focus on maize (corn), the most significant crop in the Mexican food system and an important basis of food security. Maize is the target of government food policy and is grown by many of the poorest farmers in Mexico on rain-fed, drought-prone land. Our aim is to provide a context for thinking about the implications of two types of global change for Mexico – the internationalisation of economies and the widespread transformation of the environment – through a discussion of how the North American Free Trade Agreement (NAFTA) and global warming may affect Mexican agricultural production.

First we describe the general situation of the Mexican food system today. Then we review how economic policy has moulded food output during the 20th century in decade-long efforts to attain food security based on the goal of national self-sufficiency. This is followed by an examination of the impact of climate variability on agricultural production, and the ways in which agricultural policy has interacted with climate to increase or reduce vulnerability to environmental and social change. Then we consider the implications of global warming for the future of Mexican agriculture. Finally, we discuss the implications of the changing economic environment as Mexico has opened its economy to trade and is seeking its integration into the world economy. This has changed the concept of food security as domestic supply becomes less important because grain can be bought more cheaply on the world market. This policy has differential impacts on different groups because the urban poor may secure access to cheap food, but poor, already malnourished, rural smallholders may be worse off.

2 Mexican food system

Mexican agriculture provides most of the food for a population of about 85 million. During the 20th century, feeding this growing population – which has more than doubled in the last 30 years – has been a great challenge. Food security for the nation has always been a policy priority. Until recently, the government has heavily subsidised food staples through supports for both producers and consumers. For example, high maize prices for farmers have been guaranteed to encourage production, and the price of the consumer food staple made from maize, the tortilla, has also been subsidised to guarantee cheap food, especially for urban dwellers who buy ready-made tortillas. The balance of subsidies has changed over time, sometimes leaning on the production side, most often benefiting consumers. These policies are now being reformed with the current deregulation and liberalisation of the Mexican economy (Appendini, 1992).

In recent decades, importing grain has become a key component of food policy. Mexico became a net food importer in the early 1970s, and imports of maize in particular grew to 20-25 per cent of the total supply during the 1980s. Maize imports are largely distributed to the tortilla-processing industry in the large cities where more than half of Mexico's population live.

Despite, and, in some cases, because of, government food policies, household and individual food insecurity is widespread in Mexico and has accelerated with the economic crisis of the 1980s. Calva (1988) suggests that up to 20 million people in Mexico have caloric or protein deficiency. A 1990 study of children under five found that 21 per cent were malnourished. The hungry include many rural people whose food security is dependent on growing their own food on small plots of drought-prone land, earning a reasonable income from selling crops or working as wage labourers, or obtaining access to government welfare programs or entitlements to food within households. Environmental degradation, mechanisation, economic crisis and reduced government spending are some of the factors that have restricted the entitlements of many poor people to food.

Cropland in Mexico is limited by dry climates and steep topography to about 16 per cent of the total land area. The basic staples, maize and beans, tend to be grown on the 80 per cent of the cropland that is rain-fed. Irrigated land (20 per cent) has largely been devoted to other cash crops ranging from grains and fibres such as wheat and cotton, to high-value export crops such as fruits and vegetables. Irrigated land is spatially concentrated in northwest and central Mexico. The use of chemical fertilisers and, to a lesser extent, hybrid seeds, is fairly widespread in Mexico, but poorer farmers cannot afford to use them in some regions. Average productivity of maize is low at about two tons per hectare, but ranges from more than 10 tons per hectare under irrigation to less than half a ton in drier and less fertile areas.

Seventy per cent of maize is grown on rain-fed land, mostly by the 90 per cent of the maize producers who live on small landholdings of five hectares or less. Although the majority of these small farms consume most of the grain within their households, at an aggregate level they contribute about one- third of the total domestic output which goes to the market. The rest of the maize comes from larger farmers, including about 30 per cent from irrigated areas in 1991 (SARH, 1993). Cropping patterns and yields are highly differentiated and polarised according to the natural endowments of regions as well as by the ability of farmers to obtain capital and purchase agricultural inputs.

3 Mexican food and agricultural policy in the 20th century

The most important policies affecting agricultural production and food security in 20th century Mexico include land reform, infrastructural and technological development, credit availability and subsidies. The early decades of the 20th century were characterized by the social unrest and economic uncertainty of the Mexican Revolution, which was partly driven by the aspirations of poor rural people for land to grow food. Although food production declined in the immediate aftermath of the revolution, the mid-1930s brought massive agrarian reform and increases in food production as peasants on small and collective landholdings gave priority to growing food staples for self-consumption. One important form of land distribution was the *ejido,* land held collectively by a group of farmers which could not be sold or rented under Article 27 of the constitution.

From the 1930s to the mid-1960s, one key source of increased food production was the new cropland distributed to peasants through agrarian reform, and this land contributed food to the market as the peasantry became incorporated into the national economy and began to sell their products. For a time, the main supply of maize and beans was provided by these farmers without requiring large public investment in modernising agriculture in the *ejido* sector. The extent to which the *ejidos* have continued to contribute to local and national food security has been a subject of heated debate. Some authors argue that *ejidos* provide a sustainable livelihood for millions in rural Mexico, others suggest that they are inefficient and contribute little to national agricultural production (Dovring, 1970; Eckstein, 1966; Mueller, 1970; Nguyen, 1979).

Another important factor in increasing agricultural production was prodigious investment in irrigation infrastructure beginning in the late 19th century. Agricultural development was concentrated in the large irrigation districts on which cash and export crops such as cotton and vegetables were being grown. Food staples were grown on rain-fed land with traditional technology and were unable to face the pressing demand of population growth and urbanisation.

By the mid-1950s, it became clear that agrarian reform was tending to redistribute mostly marginal and rain-fed land, and irrigation policy, together with a general agricultural policy favouring private landholdings, was producing a highly unequal agricultural structure. Increasing overall agricultural productivity became an issue as irrigation was not ensuring the security of basic grain production or of the peasant sector. This was even more so, because at the time, Mexico reaffirmed a development strategy of industrialisation through import substitution based on a protected domestic market. Controlling inflation was one of the main factors in guaranteeing stable development, and food price control became a key issue in this policy. Price supports for basic crops were established in 1953, aimed at regulating the grain market and giving stable conditions to farmers, as well as putting a ceiling on food prices.

But from 1963, nominal support prices on basic grains were frozen until 1973. Incentives for growing basic crops were to come from productivity increases rather than prices. This was attained with the Green Revolution which emphasised the use of improved seeds, fertiliser and pesticides to increase yields. The use of inputs was encouraged through state agencies for the distribution of fertiliser, pesticides, seeds and credit, and by the establishment of Conasupo, a state agency that purchased crops at support prices and sold them at lower, subsidised costs to consumers.

By many measures, the Green Revolution was a tremendous success. Maize output grew steadily at an annual rate of 7.6 per cent until 1967 and Mexico became self-sufficient in food crops (Figure 1). The Green Revolution in Mexico was particularly successful for hybrid wheat and maize on irrigated land, but less so for rain-fed maize. Thus, the use of hybrid seeds and chemical inputs was most advantageous on the best land and for the better-off peasantry who could afford the cash expenditures for the new inputs. Peasants continued to grow maize in conditions of low productivity, but commercial farmers in certain regions now found it profitable to grow grains on irrigated and more productive rain-fed lands. Export crops such as cotton were no longer profitable due to international prices, and government was giving strong support to the adoption of Green Revolution technology. This again led to widening the gap between 'modern' agriculture and 'peasant' agriculture (Hewitt, 1976).

But by the end of the 1960s, basic food crops again became unprofitable relative to other cash crops. Maize prices were particularly unfavourable compared with sorghum, which was becoming a competing crop both on irrigated and rain-fed land as demand for animal products increased in Mexico. Maize output in the modern sector of agriculture began to

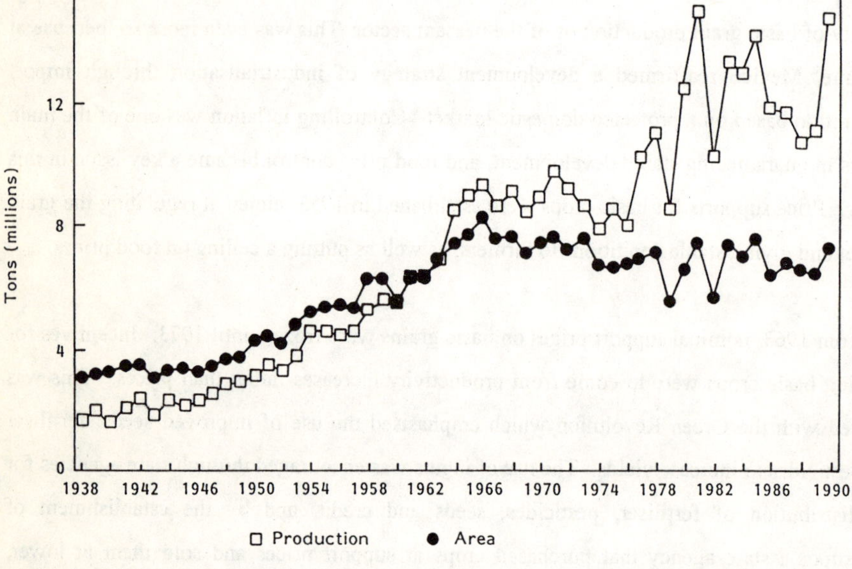

Figure 1 Maize area and production in Mexico, 1983-90. Source: Appendini (1992).

decrease and, by the early 1970s, maize became a peasant crop of the regions of Mexico land that had been largely marginalised from the Green Revolution's technological change and had low productivity. This sector was unable to ensure an increasing supply of this basic food for Mexico's still growing population.

The 1970s witnessed the constraints of Mexico's development policies as inflationary pressures, the balance of payment deficit and external debt began to build up. In the countryside, the contradictions of bimodal development resulted in the stagnation of staple food production due to the marginalisation of peasant agriculture. During the 1970s, state intervention in the countryside expanded in the attempt to increase production of basic food crops. Subsidies increased through the expansion of the rural public credit system (Banrural), the distribution of inputs, the purchase of crops through Conasupo at revised support prices, and investment in rural infrastructure. Crop insurance through a government-sponsored agency was promoted as a response to climatic and other risks.

Domestic supply did not respond to these policy incentives. From 1970 to 1978, foodgrain output grew at an annual rate of only 2.1 per cent while feedgrains grew at 7.4 per cent. Maize output grew at only 0.4 per cent from 1964-66 to 1976-78 and harvested area diminished by 1.5 million hectares (Figure 1). Maize imports began to supplement domestic output from the early 1970s, and have continued to do so ever since. Feedgrains and other commercial crops remained more profitable for larger farms and peasant agriculture remained largely stagnated (Appendini, 1992; Sanderson, 1986).

The use of fertilisers was adopted by a large proportion of peasants during this decade, but it was often to compensate for declining fertility from erosion and continuous, rather than fallow cycle, cultivation, and did not necessarily reflect an increase in yields. Also, subsidies on inputs were not equally accessible to all producers, benefiting those who purchased inputs. The productivity gap between farmers using Green Revolution technology and traditional peasants increased so that, from 1972, support prices did not cover production costs of traditional technology on rain-fed land. If peasants continued to grow and sell maize, it was because they squeezed their income below minimum salary levels.

The supply of maize to the tortilla industry became of increasing concern to the state, especially in the big cities such as Mexico City where cheap and secure food reduced risks of political unrest and kept down demands for higher wages. In 1974, the maize-processing industry for tortillas was decreed 'of public interest' and was taken under complete State control. Those who processed grain and made tortillas (millers and tortillerías) were assigned supply quotas of subsidised grain distributed through Conasupo. This agency now purchased about 20 per cent of domestic supply and was the only agency allowed to import grain (Appendini, 1992). Thus, big cities were increasingly supplied with imported grain - which was disliked by consumers because it produced lower quality tortillas. But real prices of the tortilla diminished during the decade, and the goal of cheaply feeding the urban population was achieved by putting a ceiling on producer prices and subsidising the tortilla industry.

It is important to note that the difficult and expensive task of balancing subsidies, and making decisions on who should benefit and who should pay for the food system, was made much easier by the inflow of Mexican oil revenues in the late 1970s and early 1980s. During 1980-82, Mexico embarked upon the most ambitious food policy scheme in recent history. A special programme called the Sistema Alimentario Mexicano or Mexican Food System (SAM) was launched to increase production of staple crops and attain national food self-sufficiency, while ensuring cheap food for the cities and poor rural population (Austin and Esteva, 1987). Public expenditure increased substantially, channelling funds to support peasants and farmers through subsidies to costs, prices and marketing. To promote new technologies among small farmers, a special risk-sharing programme (Programa de Riesgo Compartido) was introduced in which the state would compensate for any loss due to costs of new technology. Subsidies to consumers also increased to maintain the cheap food policy. Maize output hit a record of 14.0 million tons in 1981.

The policy of resolving the price and food security dilemma by public expenditure was, of course, unsustainable when the debt crisis struck in August 1982. Basic grain production fell dramatically in 1982 – especially maize, which dropped by almost one-third (Figure 2). As we discuss later, this was as much a result of climate variability as economic crisis or government policy, but was blamed on a failure of the SAM. The SAM programme was completely dismantled and food policy was subjected to macroeconomic policy focused on managing the debt crisis and restoring macroeconomic equilibrium.

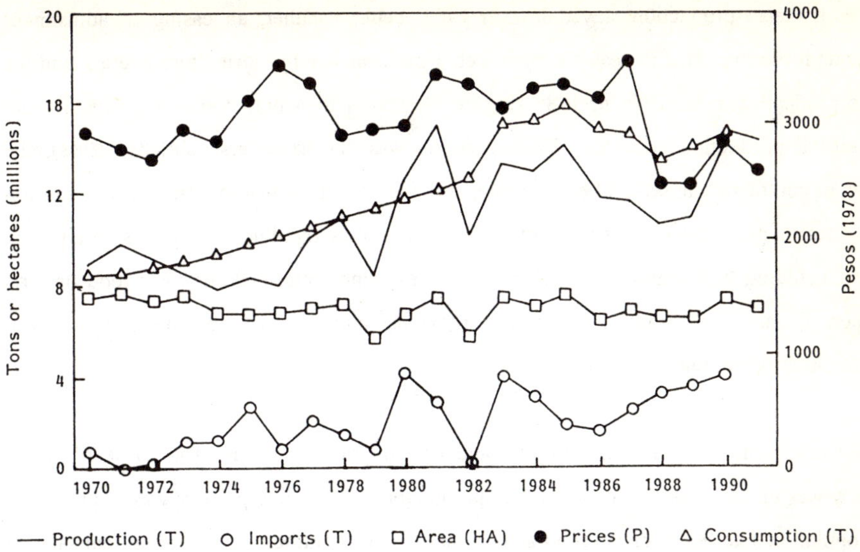

Figure 2 Maize production, area, imports, support prices and consumption. Source: Appendini (1992)

Reducing the public deficit was one of the main policy objectives, and this had direct implications for food policy. During this period, agriculture was not a priority issue on the government agenda and policy was reduced to distributing decreasing public resources. From 1982 to the end of the decade, public expenditure in agriculture decreased at an average yearly rate of eight per cent. This had an impact on the maintenance of infrastructure such as irrigation works and resulted in increasing inefficiency in water use and salinisation. The impact was also felt on credit, where Banrural loans for maize decreased at a yearly rate of 20.2 per cent from 1983 to 1990. Fiscal constraints reduced subsidies on fertilisers and other inputs. Although economic crisis affected Conasupo's capacity to buy domestic grain output at support prices, price policy was perhaps the one instrument that remained as an incentive to farmers, especially for maize which, after a sharp decline in 1983, showed increases in real support prices up to 1987. Maize and beans were both favoured relative to other crops by credit and support prices (Myhre, 1993). Private investment remained stagnant, as it had since the 1970s.

Although maize production stayed low in 1983, better weather, an easing of adjustment policy and support price increases explain better performance for agriculture overall, and for maize production from 1984 to 1986 (Figure 2). From then until the end of the decade, agricultural production declined. This recession was due to adverse climate (1988), the impact of contraction in sectoral expenditure, and to a change in macroeconomic stabilisation policy beginning in 1987. This new anti-inflationary policy fixed nominal prices on all basic goods, including food staples, and controlled the exchange rate, making basic food imports cheaper. From 1989, trade liberalisation removed import tariffs on most agricultural commodities except maize and beans.

During crises of the 1980s, distribution and consumption of staples for household food security was much more of a concern than production. The policy priority was to guarantee the population's access to cheap food as personal income contracted. Minimum real salaries decreased by 50 per cent during the period, making it difficult for many families to purchase adequate food. Subsidies to food staples, particularly in big cities, were not abolished, although they were restructured throughout the decade to diminish the public financial burden of food subsidies. In these years, the gap between the price of imported maize and the domestic support price widened, with domestic prices ending up at double the international price. The state began to depend on cheaper imported maize, rather than domestic grain from Conasupo, to support the tortilla industry. This lowered the cost of the subsidy to this basic food and made it clear that national food self-sufficiency was no longer a goal of food policy.

By the end of the 1980s, the productive capacities of peasants and farmers had been severely eroded by a decade of decreasing government support to agriculture, several years of adverse climate, and decreasing prices. After the new administration of President Salinas took office in December 1988, the agricultural crisis again became an item on the national economic agenda. But, before we summarise the most recent agricultural policy shifts and their implications for food security, we turn to examine the role of one key factor in Mexican food security at national and local scales – climate, and particularly the role of low rainfall in year-to-year variations in agricultural production.

4 Climate variation and food production in Mexico

Almost every year drought endangers agricultural production in some part of Mexico. With 85 per cent of the land classified as arid or semi-arid, Mexican agriculture relies on low, seasonal and variable rainfall. The dry climates, together with steep topography, are the main reason why only 16 per cent of Mexico's land area is considered suitable for crop production. As we noted earlier, only one-fifth of Mexico's cropland is irrigated, and many irrigation districts rely on small reservoirs and wells that deplete rapidly in the dry years.

Throughout Mexican history, drought and climate variation have been associated with migrations, food crises and social unrest. Climate changes have been implicated both in the rise and collapse of powerful pre-Hispanic civilisations such as those in Yucatan and in the valley of Mexico and it is clear that management of scarce water supplies was critical to food security (Sanders *et al.*, 1979). The arrival of the Spanish in 1519 altered climatic risks in important ways. There was a tremendous expansion of irrigation, which increased the range and reliability of crop production while increasing competition for limited water supplies in some regions. However, the Spanish often controlled the best land and water, and their cattle and wheat replaced the more moisture-conserving agriculture of the indigenous people. These changes, together with grain speculation and relentless tribute demands by the colonial powers, have been linked to subsistence crises and to widespread famine in drought years. In 1785, drought is estimated to have triggered the starvation of more than 300,000 people – mostly members of the indigenous Indian population (Florescano, 1969).

As potential evaporation exceeds precipitation in most of Mexico, even slightly below average rainfall can place crops at risk. On average, more than 90 per cent of losses in Mexican agriculture are from drought. In 1990, with favourable weather, about seven per cent of the crop area that was planted was lost to natural hazards. In 1979, with less favourable weather, drought losses alone devastated 19 per cent of the area planted. A large region of central and northern Mexico lost more than 50 per cent of the area planted (Liverman, 1993).

Rain-fed maize production is particularly sensitive to any delay in the start of the summer rainy season (May-June) and to the severity of the midsummer drought called the canicula.

During the 1980s, dry years had significant impacts on national and local food production. Figure 1 shows the dramatic drop in maize production in 1982, which was more a result of low rainfall than of agricultural policy as the critics of SAM claimed. In Puebla, east of Mexico City, rainfall in 1982 was more than 25 per cent below normal (Walsh, 1993). In Oaxaca, to the south, both 1982 and 1983 saw rainfall more than 30 per cent below normal (Dilley, 1993). Yields and production fell correspondingly. Yields fell from more than 3000 kg/ha in 1981, to 1200 kg/ha in 1982 and 1050 kg/ha in 1983, in Puebla. In Oaxaca, maize production dropped from 80,000 tons in 1981 to only 15,000 tons in 1982 and 20,000 tons in 1983. Similar drought-driven production declines at the national level occurred in 1979 and 1988.

When production drops precipitously due to climate, there are local and national impacts on food security. Some peasants do not grow enough to eat and go hungry. Others become temporary migrants and wage labourers. The state usually tries to bring grain into the regions, at least to major cities. When national supplies are inadequate, grain is usually purchased from the United States, but in 1988 it was difficult to make up the shortfall from imports because of weather-related losses in U.S. agricultural production.

Changes in demand for land and water can make some people more vulnerable to climatic variability and drought. For example, population growth has meant more people trying to make a living from the land and has been one of the causes of deforestation, erosion, competition for water resources and increased flood and drought vulnerability. In the valley of Oaxaca, for example, population has quadrupled in the last 30 years, especially around the city of Oaxaca. In some areas of the valley, per capita demand has increased as farmers purchase mechanical pumps and shift from growing maize to alfalfa or vegetables for domestic and international markets. A growing population of middle-class urban dwellers and tourists is increasing per capita water use in the city. As a result, competition for surface water has intensified, and the depth to groundwater has fallen from one to ten meters, especially in low rainfall years (Dilley, 1993).

Another example of urban growth and pressure on water resources is in the northwestern valley of Toluca. In rural districts, underground water has been drained to supply the metropolitan area of Mexico City – which now has a population approaching 20 million –

with water since the early 1970s. Rural communities must now alternate one year with water supplied by wells and one year cultivating on rain-fed land. Peasants switched from rye, wheat and livestock activities – requiring humid soil and suited to the colder climate of the high central plateau – to maize, which can be grown under both irrigated and rain-fed conditions. But they also adopted new technologies to ensure good yields, mainly using chemical inputs. This, together with a credit policy favouring maize, turned the region to almost exclusive maize production, in spite of the fact that maize is more vulnerable to early frost and heavy winds than the other grains. Small landholding peasants were drawn into the market and abandoned the multiple variety of the milpa farming system for self-consumption. The valley has become an important region for supplying Mexico City with maize, but at the same time underground water extraction has made agriculture more vulnerable to climate change (Appendini, 1992).

Both these examples of increasing water demand show potential interactions between agricultural policy – credit, input use, export orientation – and vulnerability to climate. To what extent have agricultural policies described earlier changed vulnerability to drought and reduced climatic threats to food security in Mexico?

5 Interaction of food policy and climate variability

One of the most significant responses to uncertain rainfall has been massive federal investment in irrigation works. This resulted, for the most part, in more reliable and higher volume production, especially in northwest Mexico. In other regions, farmers are still using the traditional irrigation systems built in pre-Hispanic or colonial times. Unfortunately, the ability of irrigation to buffer producers against low rainfall is threatened by growing demand, widespread salinisation, falling groundwater levels and inadequate reservoir supplies, especially when rainfall is low. Although the Comisión Nacional del Agua has formal procedures for allocating water in dry years, and *ejidos* and certain high value crops are supposed to receive priority, in practice it is often the wealthier and more powerful farmers who get the water.

Of the basic grains, wheat is grown mainly on irrigated land. Maize has been grown on irrigated land at certain times, as a response to high support prices, and thus contributed to self-sufficiency. This was the case during the mid-1960s when the Green Revolution and price incentives made the crop profitable on high yielding land. At present, when maize remains the only crop left with a support price and higher profitability than other basic crops, farmers on irrigated land have again switched to maize, accounting for self-sufficiency from 1990 to 1993.

The Green Revolution can be seen as a partial response to climate risk in that some new seeds were bred for drought- and disease-resistance. Wilken (1987) has argued that the new techniques have replaced some traditional hazard-prevention strategies, such as mixed cropping and microclimate modification, and have enabled agriculture to expand into areas of high hazard risk such as deserts, mountains, coastal regions and the disease-susceptible humid tropics. In favourable weather conditions, improved seed and fertiliser give much higher yields of crops such as wheat and corn, especially with irrigation. However, when drought, frost or flood destroys the crop, farmers are often left in debt because of the cost of the technical inputs (Walsh, 1993). Previously, farmers did not purchase inputs, they kept seed from one year to another and perhaps used some green or animal manure as fertiliser. They might lose their crop to drought and be hungry, but would not be taking a major financial risk.

There is a problematic relationship between land reform and drought vulnerability in Mexico. On the one hand, land distribution reduced the vulnerability of significant numbers of landless and poor by providing the opportunity for subsistence food production on their own plot of land. But a large proportion of the land that was distributed was of poor quality and in hazard-prone areas. Irrigated land has tended to stay in the private sector. The vulnerability of *ejidos* is borne out in statistical studies of agricultural losses to natural hazards based on census data. For example, in 1970, the average drought loss on the *ejidos* in the summer growing season was 17 per cent compared with 14.5 per cent on private landholdings over five hectares (Liverman, 1990). In winter, the average *ejido* loss was 11.6 per cent compared with 8.6 per cent in the private sector. Many *ejidatarios* experience this loss in the form of both income from the crops and nutrition from reduced consumption of their produce. These patterns can be explained through a combination of physical, technical and social factors. Explanations of this increased vulnerability are that more biophysically marginal land was given to *ejidos* in the land reform, and that *ejidos* are socially vulnerable because they cannot get access to irrigation, credit, improved seeds or other resources (Liverman, 1993).

Increasingly important for climate vulnerability are the changes brought about by the integration of Mexico into a national and international market and financial system. Subsistence production of corn has been replaced by wheat, dairy, sugar, feedgrain and fibre production for domestic industrial and urban markets. Overall, in latter years, the agricultural system has increasingly been oriented to export production – of fruit, vegetables and livestock – although there have been periods where the state has intervened with subsidies to encourage basic grains.

The increased acreage under fruit, vegetable and feedgrains has increased the demand for water both within and outside the irrigation districts. Alfalfa, tomatoes and lettuce, for example, have a much higher consumptive use of water than maize. As the crop mix becomes more water- demanding, rainfall deficits are more critical. For example, water demand in the irrigation districts of northern Mexico has doubled in the last 20 years. In drought years, reservoir and groundwater resources are increasingly insufficient and water allocations are reduced. Increased production of water-intensive export crops has also increased the potential for international conflict over the waters of the Colorado and Rio Grande, and for competition between agriculture and the industrial and domestic consumers of the rapidly developing border cities (Borza *et al.*, 1993). Mexico's broader economic conditions are also important to rural climate vulnerability. For example, the growth of debt in the 1980s reduced the availability of the state to provide or subsidise rural infrastructure such as dams, credit and inputs.

Despite overall increases in agricultural production, and the ability of irrigation and insurance to reduce climatic risks, it is clear that Mexican agriculture is vulnerable to climatic variations, especially drought. In some years and regions, drought is a real threat to the food security of households, especially when economic conditions raise overall prices and unemployment, and shift government policy away from subsidy and import programmes for basic foods. Thus, any climate change towards warmer, drier conditions could have very serious implications for food security in Mexico.

6 Global warming and Mexican agriculture

Many scientists believe that an increase in atmospheric CO_2 is likely to result in warmer global temperatures and altered precipitation patterns (see Schneider, 1995, this volume). The majority of global warming experiments with climate models have simulated current climate using a CO_2 concentration of about 300 parts per million (ppm) and have compared the results with a simulation based on a doubling of carbon dioxide levels to about 600 ppm. Scientists estimate that this doubling of CO_2, which is likely to occur in the first half of the next century, may raise the global mean temperature by 1.3-5.1°C (IPCC, 1990). The magnitude of temperature changes will vary across regions of the globe and will be accompanied by changes in rainfall and other climatic variables. The poles, for example, are expected to warm more than the equator, and precipitation will increase in some places and decrease in others.

Research on the impacts of global warming in Mexico indicates that climate change may bring warmer and drier conditions to much of Mexico (Liverman and O'Brien, 1991). We have used the results of the major general circulation models to project what may happen to the Mexican climate if concentrations of carbon dioxide in the atmosphere rise to twice pre-industrial levels. There is a considerable debate about the accuracy and validity of these models because of limitations in their ability to simulate some cloud and ocean processes, their coarse spatial scale, disagreements between the models, and their poor ability to reproduce observed climates. The estimates of precipitation are particularly dubious and immediate improvements in the regional projections of climate models are unlikely. We have used the results of five different climate models, as well as extensive sensitivity analyses, in order to try and capture some of the uncertainty in current global warming studies.

On average, the models project that global warming will bring annual temperature increases of 2-5^{0}C in Mexico, depending on the model (Table 1). For rainfall, however, the NCAR model predicts a 23 per cent decrease, and the GISS model a three per cent increase, dramatically illustrating the uncertainty in precipitation forecasts under global warming. Although all parts of Mexico are projected to have temperature increases, and many regions could see rainfall decreases, there are a number of places where rainfall could increase.

In cases where both rainfall and temperature increase, it is not immediately obvious what the effects on soil moisture may be. Will higher rainfall compensate for the increased evaporation brought by higher temperatures? Table 2 shows some estimates for the increased evaporation and water availability in Puebla, an important maize-growing area east of Mexico City. Using relatively simple calculations we estimate that potential evaporation may increase by 7-16 per cent and the annual deficit of water could increase by 18 - 45 per cent. That is, all models project that soil moisture and water availability in the valley of Puebla may decrease. This has serious implications for agricultural production. Our main conclusion, based on climate model projections, is that Mexico is likely to become warmer and drier. All models show increases in potential evaporation, and, in most cases, indicate that moisture availability will decrease, even where the models project an increase in precipitation. Sensitivity analysis of moisture-deficit calculations indicate that water availability could increase only if the models simulated much higher rainfall and relative humidities, or significantly less solar radiation and wind.

Table 1: Climate changes projected by general circulation models for Mexico (averaged change across country)

Model	Temperature (^{0}C)	Precipitation (%)
GFDL	3.11	- 1.75
GISS	3.92	2.74
NCAR	2.38	-23.01
OSU	3.15	- 1.05
UKMO	5.44	- 0.09

Source: Liverman and O'Brien (1991).

Table 2: Climate change and potential evaporation in Puebla, Mexico

Climate	Potential evaporation	Potential evaporation. - precipitation 'water deficit'
Observed	1528mm	694mm
GFDL	+14.2%	+44.7%
GISS	+13.3%	+33.1%
NACRE	+ 7.1%	+22.8%
OSU	+10.0%	+17.5%
UKMO	+16.0%	+29.5%

Source: Liverman and O'Brien (1991)

Table 3: Potential impacts of global warming on Mexican maize yields in Morelos (CERES model)

Model	Rain-fed yield (kg/ha)
Observed	3600
GISS	2900
GFDL	3000
UKMO	2000

Source: Liverman (1992a)

We have also undertaken some analysis of the impacts of global warming on the production of maize – the most important crop in Mexico – using the CERES crop growth model. At Tlatizapan, a site south of Mexico City, current rain-fed corn yields with low fertiliser are 3600 kg/ha. If climate changes and no adaptations are made, yields could decrease to as low as 2000 kg/ha (Table 3). Farmers are, of course, likely to adapt to the projected warming. However, given the environmental and economic constraints on agricultural input use and water resources in Mexico, the ability to adapt to climate change may be limited, especially for poorer farmers (Liverman, 1992).

If global warming occurs as projected, it is likely to interact with some major agricultural and macroeconomic policies which are also likely to dramatically transform Mexican agriculture in the coming decades. We conclude with a discussion of some of these recent and continuing policy changes.

7 New challenges and options for Mexican agriculture

Since 1989, agricultural policy under President Salinas has undergone a complete change as part of the economic restructuring of the Mexican economy towards integration into the world market. The changing policy comprises above all the reorganisation of institutions with the goal of reducing State intervention in the countryside and attracting private investment in agriculture, livestock and forestry. This has led to a complete change in the framework in which peasants and farmers have operated for over two decades. It has meant uncertainty for producers, as access to resources has become more difficult and expensive. Privatisation of land, trade liberalisation and new policy instruments are the core of restructuring.

The main changes in policy that have already had a profound impact on agriculture are:

- the reorganisation of rural credit – reducing heavily subsidised credit for peasants farming at the small and medium scale;
- privatisation of input industries such as fertilisers and seeds;
- abolishing support prices; and
- liberalising trade for all crops except maize and beans.

Small farmers have been excluded from the banking system and may receive a small loan for basic crops – covering about one-third of production costs of maize per hectare – through the poverty assistance programme Pronasol (Programa Nacional de Solidaridad). Credit has become scarce and expensive, and medium–sized farmers have become heavily indebted in the last four years. Access to crop insurance has diminished as subsidies have been abolished. As part of a transition policy, deregulation has been partial. The state still intervenes in the grain market, protecting maize and beans with import licences and by buying up maize and beans at support prices through Conasupo. Other basic products including feedgrains have been liberalised and domestic prices have dropped, equalising international prices.

A contradictory outcome has resulted. Instead of farmers substituting grains for high-value crops such as certain export fruits and vegetables in which Mexico has a comparative advantage, farmers – particularly on irrigated land where it is no longer profitable to grow feedgrains – have turned to maize, taking advantage of the relative high prices and secure market. This seems in the short run to be a safer business decision than to invest in export crops which require large investments and have high transaction costs. Thus, in 1990 and 1991, maize production reached a record harvest of 14 million tons after several years of high imports, and self-sufficiency has been attained for the last three years.

These recent trends in agricultural policy clearly point to integration into the North American Free Trade Agreement (NAFTA) area. It appears that Mexico will become increasingly dependent on grain imports from the U.S. This may mitigate some of the problems referred to in this paper – the constant struggle to increase maize output on poor rain-fed land with a large sector of small and medium-sized peasant producers. At current international prices and with favourable credit extended by the US government for buying grain, Mexico will be able to supply cheap grain and tortillas for its poor consumers. Thus, food security may increase at the national and local level if government maintains some sort of food subsidy for the poor and the world market for grain remains favourable.

Mexico has already negotiated the conditions for maize and beans within NAFTA. For these staples, import licences will be abolished and protection replaced by a tariff which diminishes over a period of 15 years. A tariff-free, increasing import quota of maize will start at 2.5 million tons per year. The issue for Mexican farmers is whether they will be able to increase productivity substantially in the long run and be able to compete with Middle Western US farmers on the Mexican grain market. As yet, there are no signs of a strong government support for developing agriculture among small and medium-sized farmers to reach higher productivity for food crops. A new programme for subsidising all basic crop farmers, Programa para el Campo, falls short of being an incentive to crop production, rather it is a welfare programme handing a small direct payment to farmers according to the hectares they command. Support prices for maize and beans will be abolished after a year's transition period under this new programme.

But these trends also apply to some counter-effects that are already felt in the countryside. Ten per cent of Mexico's labour force are maize producers; they and their families base their livelihoods on small farm households, combining farming with multiple-income generating activities such as wagework and petty commerce. A further deterioration of agriculture and peasant livelihoods as subsidies and supports are withdrawn will put more pressure on the rural and urban labour markets and may increase migration, especially towards and across the border with the United States. Food security at local and household levels will also be more uncertain as poor rural households no longer grow their own crops or sell their land under the new programme to privatise *ejido* land.

Land privatisation and NAFTA could also increase drought vulnerability in northern Mexico. A number of economic models suggest that if NAFTA fully liberalises trade between the U.S. and Mexico, maize production in Mexico will drop dramatically and horticultural production will increase, as basic grains will be increasingly imported from the USA. This would continue the trend towards greater water demand by agriculture in northern Mexico, and the low and variable rainfall of this arid region may eventually provide a constraint to the success of the trade agreement (Borza *et al.*, 1993). If, as seems to be the goal, the privatisation of *ejido* land leads to more intensive agricultural production, this too could increase water demand and drought vulnerability. The rush to private credit and new crops may create a volatile and risky financial situation, especially if extreme weather brings harvest failures in the next few years. Whether water rights and water institutions will change as a result of agrarian reform is as yet unclear, although there is talk of increased water prices, open water markets and the possibility of increased water-use efficiency.

At present it is hard to predict the future of basic food production in Mexico. We are at a crossroads. Changes in land tenure, policy mechanisms and trade liberalisation will certainly have an impact on crop structure and thus food supply. The impact will be differentiated according to regions, crops and the type of farmers and peasants concerned. Poor peasant farmers who remain on the marginal rain-fed land will likely grow maize for self-consumption as long as access to cheaper grain on the market is difficult. Also the degree to which peasants grow food for self-consumption will depend on the alternatives provided by the labour market and on the impacts of shifts in rural property due to privatisation. The future of basic food producers in regions with good rain-fed land who supply the market will depend on specific policy giving them support. Better-off farmers are able to increase productivity and supply part of the food grain on the domestic market because white maize grown in Mexico is still preferred to imported yellow maize as input for the tortilla and other foods.

As for the role of climate in Mexican agriculture, we have tried to show the way in which climatic variation and climatic change influence food production in Mexico. Drought already poses problems for agricultural production and policy in Mexico, especially as increasing water demands place pressure on the irrigation districts. Global warming may present a threat to local and national food security, especially if farmers are unable to adapt to a drier climate or if imports from other regions become more costly. In the long run, Mexican food security will be determined by the complex interaction of macroeconomic policy, social and demographic trends, and environmental change with the risk for individual households sharply differentiated by access to land, water and government support.

Acknowledgement

This paper is a substantially revised and extended version that first appeared in *Food Policy*, Appendini, K. and Liverman, D., Agricultural policy, climate change and food security in Mexico, pp. 149-164. (1994). Reprinted with permission from Elsevier Science Ltd, Butterworth-Heinemann Imprint, The Boulevard, Landford Lane, Kidlington OX5 1GB, U.K.

References

Appendini, K. (1992). *De la Milpa los Tortibonos: La Restructuracion de la Politica Alimentaria en Mexico.* El Colegio de Mexico, Mexico.

Austin, J.E. and Esteva, G. (1987). *Food Policy in Mexico: The Search for Self-Sufficiency.* Cornell University Press, Ithaca.

Borza, K., Leichenko, R. and Liverman, D. (1993). *The Impacts of Climatic and Economic Change on Irrigation Districts in Northern Mexico.* Presentation at the Association of American Geographers Annual Conference, Atlanta, Georgia, April 1993.

Calva, J.L. (1988). *Crisis Agricola y Alimentaria en Mexico 1982-1988.* Fontamara, Mexico.

Dilley, M. (1993). *Climate Change and Agricultural Transformation in the Valley of Oaxaca, Mexico.* Ph.D. dissertation, Department of Geography, Pennsylvania State University, University Park.

Dovring, F. (1970). Land reform and productivity in Mexico. *Land Economics,* **46**(3): 264-274.

Eckstein, S. (1966). *El Ejido Colectivo en Mexico.* Fondo de Cultura Economica, Mexico.

Florescano, E. (1969). *Precios del Maiz y Crisis Agricolas en Mexico 1708-1810.* Ediciones Era, Mexico D.F.

Hewitt, de Alcántara, C. (1976). *Modernizing Mexican Agriculture.* United Nations Institute for Research on Society and Development (UNIRSD), Geneva.

Hewitt, de Alcántara, C. (1992). *Reestructuraciòn Ecómica y Subsistencia Rural.* El maiz y la crisis de los ochenta. El Colegio de Mexico/UNIRSD, Mexico D.F..

Liverman, D.M. (1990). Vulnerability to drought in Mexico: the cases of Sonora and Puebla in 1970. *Annals of the Association of American Geographers,* **80**(1): 49-72.

Liverman, D.M., and O'Brien, K. (1991). Global warming and climate change in Mexico. *Global Environmental Management,* **1**(4): 351-364.

Liverman, D.M. (1992a). The regional impacts of global warming in Mexico: uncertainty, vulnerability and response. In: Schmandt, J. and Clarkson, J. (eds.) *The Regions and Global Warming.* Oxford University Press, Oxford, pp. 44-68.

Liverman, D.M. (1992b). Global warming and Mexican agriculture: some preliminary results. In: Reilly, J.M. and Anderson, M. (eds.) *Economic Issues in Global Climate Change: Agriculture, Forests and Natural Resources.* Westview, Boulder. pp. 332-352.

Liverman, D.M. (1993). *Environmental Change and Economic Transformation: Natural Hazard Vulnerability in Mexico in the Twentieth Century.* Pennsylvania State University, College Station. Unpublished manuscript.

Mexico INEGI (Instituto Nacional de Estadistica, Geografia y Informacion) (1991). *El Sector Alimentaria en Mexico.* INEGI, Mexico D.F.

Myhre, D. (1993). *The Unseen Instrument of Agricultural Restructuring in Mexico: The Growth, Crisis and Erosion of the Official Credit System.* ILET, Mexico D.F. (Preliminary unpublished paper).

Mueller, M.W. (1970). Changing patterns of agricultural output and productivity in the private and land reform sectors in Mexico 1940-1960. *Economic Development and Cultural Change,* **18**(2): 262-266.

Nguyen, D.T. (1979). The effects of land reform on agricultural production, employment and income distribution: a statistical study of Mexican States 1959-1969. *Economic Journal,* **89**: 624-635.

Reilly, J.M. and Anderson, M. (eds.) *Economic Issues in Global Climate Change: Agriculture, Forests and Natural Resources.* Westview Press, Boulder.

Sanders, W.T., Parsons, J.R. and Santley, R.S. (1979). *The Basin of Mexico: Ecological Processes in the Evolution of a Civilization.* Academic Press, New York.

Sanderson, S. (1986). *The Transformation of Mexican Agriculture: International Structure and the Politics of Rural Change.* Princeton University Press, Princeton.

Schneider, S.H. (1995). The future of climate: potential for interaction and surprises. In: Downing, T.E. (eds.) *Climate Change and World Food Security.* Springer-Verlag, Heidelberg.

Walsh, J. (1993). *Climate, Culture and Plan Puebla: Natural and Social Influences on Subsistence Farmers in Mexico.* Honors thesis. Departments of Political Science and Geography, Pennsylvanis State University, College Station.

Wellhausen, E. (1976). The agriculture of Mexico. *Scientific American,* **235**(3): 128-50.

Wilken, G.C. (1987). *The Good Farmers: Traditional Agricultural Resource Management in Mexico and Central America.* University of California Press, Berkeley.

Yates, P.L. (1981). *Mexico's Agricultural Dilemma.* University of Arizona Press, Tucson.

PART V

STRATEGIES TO LIMIT CLIMATE CHANGE AND IMPROVE FOOD SECURITY

Implications of Policies to Prevent Climate Change for Future Food Security

Norman J. Rosenberg
Integrated Earth Sciences
Pacific Northwest Laboratory[1]
901 St. S.W.
Washington, DC 20024, U.S.

Michael J. Scott
Pacific Northwest Laboratory
Richland, WA 99352, U.S.

1 Introduction

There have been a number of comprehensive and detailed studies on the question of what climate change might do to agriculture (Parry *et al.*, 1988; Smith and Tirpak, 1989; Rosenberg, 1993; Council on Agricultural science and Technology (CAST), 1992; Houghton *et al.*, 1990, 1992; National Academy of Sciences, 1992; and others). The potential for adaptation of agriculture to climate change has been evaluated in a number of these studies and in Rosenberg (1992). There has been, however, little if any systematic analysis of how attempts to mitigate or avoid climate change might affect agriculture. Our aim in this paper is to explore this question. Data and analyses on which to draw are limited, so the results of our exploration are preliminary, at best.

Strategies to mitigate or avoid climate change due to greenhouse forcing focus, for the most part, on reducing or eliminating emissions of greenhouse gases into the atmosphere and recapturing them for recycling or sequestration. The gases of greatest concern are carbon dioxide (CO_2), methane (CH_4) and nitrous oxide (N_2O), all of which have both industrial and biogenic, or natural, sources. Fossil fuel combustion and the conversion of forests to other

[1] Pacific Northwest Laboratory is operated for the U.S. Department of Energy by Battelle Memorial Institute under contract DE-AC06-76RLO 1830. The opinions expressed are those of the authors and do not necessarily represent those of Battelle Memorial Institute or the U.S. Department of Energy.

NATO ASI Series, Vol. I 37
Climate Change and World Food Security
Edited by Thomas E. Downing

uses emit about 6 Gt of carbon in the form of CO_2 into the atmosphere. Industrial sources of CH_4 include coal mines and natural gas installations. Methane is emitted from wet soils, including rice paddies and landfills, and is also a by-product of ruminant digestion. Industrial sources of N_2O are relatively minor; this substance is emitted from all soils as the result of both anaerobic and aerobic processes. The use of nitrogen fertilisers has greatly increased the emissions from agricultural soils.

Several strategies have been proposed to mitigate climate change. First and foremost is reduction of fossil fuel combustion. Another strategy is cessation of the conversion of tropical forest conversion to other uses. There are ways that on-farm emissions of CH_4 and N_2O can be reduced. Carbon dioxide can be withdrawn from the atmosphere through the plantation of new forests and the regrowth of those already harvested. Since dry soil is a strong sink for CH_4, the drainage of wetlands is a possible mitigation strategy (e.g. Born *et al*., 1990; Mosier *et al*., 1991). Geoengineering strategies of various kinds have been proposed to facilitate the capture of CO_2 and/or to counteract the greenhouse warming by decreasing the penetration of solar radiation into and through the atmosphere. In this paper, we examine a number of these strategies and speculate on how, if put into effect, they might impact on agriculture. We also attempt to estimate the magnitude of such impacts, although, for reasons that will be obvious, this is somewhat more difficult. Where possible, we attempt to differentiate the impacts of particular climate change mitigation strategies on developing and developed agricultures.

In this paper, we also consider the possibility that mitigation of climate change could have at least one 'negative' impact on agriculture. Because elevation of atmospheric CO_2 concentration increases photosynthetic rate and reduces plant water use, attempts to hold concentrations to current levels will also mean foregoing this offsetting fertilisation effect on global agricultural productivity.

Many examples used in this paper are drawn from the Missouri-Iowa-Nebraska-Kansas (MINK) study (Rosenberg, 1993), an analysis of the possible effects of a 1930s-like 'dustbowl' climate on the agriculture of the Missouri-Iowa-Nebraska-Kansas region, as it is now and as it might be 40 years from now, by which time the greenhouse warming and consequent climate change may actually be observed. The MINK study dealt not only with climate change effects on agriculture but also on forestry, water resources and energy. What happens to these sectors under climate change bears on agriculture, and what happens to all of these sectors bears on the total regional economy.

The authors of this paper will be neither surprised nor offended if our good readers identify other, more interesting or imaginative ways in which climate change mitigation strategies might affect agriculture. Nor will we be surprised if the weights we have assigned to items on our 'laundry list' are disputed.

2 Energy costs for agriculture

Attempts to reduce emissions of CO_2 and other greenhouse gases from energy production should increase the unit costs of energy production from fossil fuels. Initially this would have a detrimental effect on the costs of energy used in agriculture, would likely increase the overall costs of agricultural production, and could in some cases lead to changes in the comparative advantage of some production regions over others. We emphasise 'initially' since adaptation to higher fossil fuel derived energy costs might further decrease energy intensity in agriculture. In such a case it is possible that technological capacity in developed countries would favour rapid adaptation, further disadvantaging the developing countries.

To understand whether these effects are likely to be important in terms of world food security, it is necessary to understand the following:

- how much the cost of energy is likely to increase due to greenhouse-limitation policies;
- how dependent agriculture is on purchased energy;
- how the production of crops will be affected;
- how the changes in food supply will affect world food security.

To address the first question, we assume a $100 per ton tax on carbon contained in fossil fuels.[2] We then calculate the probable changes in fuel prices that would occur as a result of a

---

2 A $100 per metric ton of carbon equivalent (mtce) carbon tax is about the amount necessary to keep carbon emissions at 1990 levels through the year 2030. The cost impact varies by fuel. It would increase the cost of coal to electric utilities by 165 per cent, the cost of gasoline by about 27 per cent (26¢/U.S. gallon, residential natural gas by 25% ($1.50/tcf), and electricity by 27 per cent (1.76¢/kwh). Taxes of $250/mtce or more have also been considered. See Bradley *et al.* (1991).

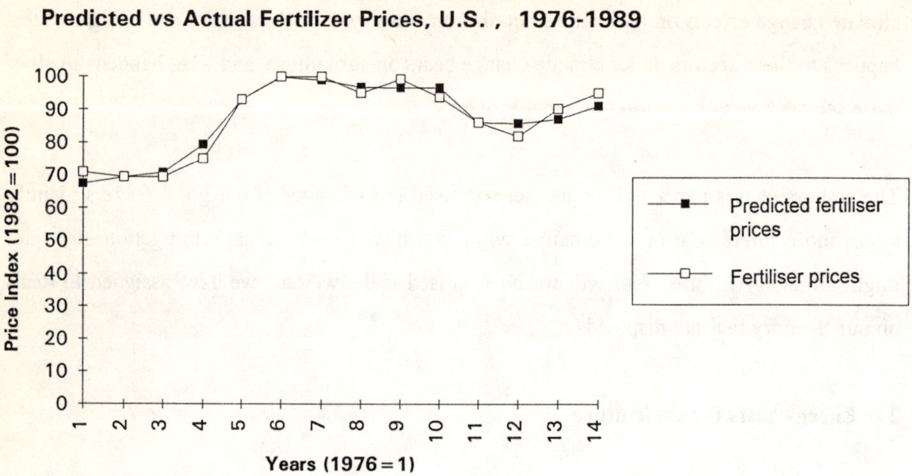

Figure 1 Predicted vs. actual fertiliser prices, United States, 1976 to 1989

U.S. carbon tax assessed on fossil fuels. It is also possible to estimate the impact on prices of energy, fertilisers and other agricultural chemicals. This is illustrated in Figure 1, where a regression equation on fuel prices was used to 'predict' fertiliser prices. We then took the estimate of a carbon emission tax of $100 per metric ton of carbon equivalent (mtce) and estimated its future consequences for fuel and farm chemicals prices. Translating the $100/mtce tax into an increase in oil and natural gas price and extrapolating the regression relationship for energy and chemicals costs into the future yields an estimated increase of about 30 per cent for fuels and about 15 per cent for fertilisers and other farm chemicals.

The dependence of U.S. agriculture on fossil fuels is illustrated with data from the MINK study. The energy costs of the various on-farm uses of energy include direct uses of petroleum, natural gas, and electricity in end uses such as ploughing, planting, trucking and harvesting; fertiliser and pesticides application; lighting and ventilation for animals; irrigation; and crop drying. In addition, the applied fertilisers and pesticides embody a significant amount of energy used in their production. Natural gas, for example, is a major input to nitrogenous fertiliser production. Energy used in the four-state region declined significantly during the 1980s. Table 1 indicates the total amount of energy used and the change for the agricultural census years of 1978-87.

Table 1 Direct and indirect energy used per dollar of farm output in Missouri-Iowa-Nebraska - Kansas, selected years, MINK study

Year	Direct energy		Fertilisers and pesticides		Total energy	
	Trillion BTU	1000 BTU/1982$ Value of crop production	Trillion BTU	1000 BTU/1982$ Value of crop production	Trillion BTU	1000 BTU/1982$ Value of crop production
1978	253.8	10.3	166.7	6.8	420.5	17.1
1981	229.1	9.2	197.6	7.9	426.7	17.1
1987	182.1	6.8	180.8	6.7	384.6	13.5

Sources: Author estimates based on data from Darmstadter (1991) and the 1987 census of agriculture.

Figure 2 provides estimates of the changes in the prices of energy and farm chemicals, and in the intensity of their use in U.S. agriculture that have occurred since the late 1970s. In general, the prices of fuels and agricultural chemicals trended upward, relative to farm output prices, until 1982, and have since declined. Both direct energy use and agricultural chemicals use increased until 1982, declining thereafter in the MINK states and the nation as a whole.

Farmers will respond to higher fuel and chemicals prices by economising on their use. A $100 per ton carbon tax would cause total fuel and chemical costs per acre to rise by less than the initial increase of 30 per cent for fuels and 15 per cent for chemicals. Based on recent U.S. history of energy use in agriculture, it appears that a 1 per cent increase in price of fuels in agriculture results in a reduction of 0.3 per cent in their consumption (in economic jargon, a 'price elasticity' of -0.3), whereas in agricultural fertilisers and chemicals, a 1 per cent increase in price leads to about a 0.6 per cent decrease in use. With a carbon tax of $100/ton, farmers can be expected to reduce fuel use by about 10 per cent and fertiliser use by about 8 per cent. The 'future' situation for total energy costs per acre, given these price increases and substitutions, is shown in Table 2. Year-to-year variability in fuel and fertiliser use has been greater than 10 per cent during the last 20 years. The change likely would have little impact on agricultural output in the U.S.

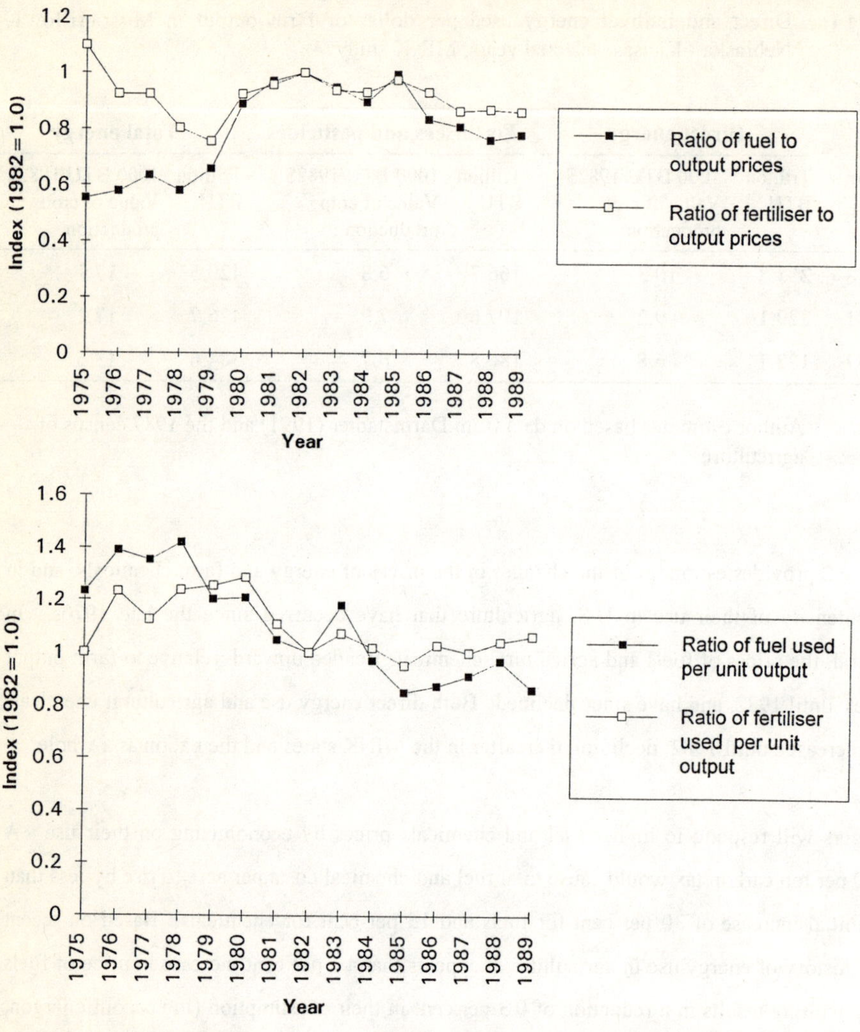

Figure 2 Impact of prices on agricultural use of fuels and chemicals in the United States, 1975 to 1989.

Table 2 estimates 'current' (1989) values for direct and indirect energy expenditures per acre in the MINK states and energy expenditures per acre at future prices with the $100 per ton carbon tax. Because irrigated farms dominate the total in Nebraska and Kansas but are rare in Missouri and Iowa, the values in Table 2 for 'all farms' are close to those for irrigated farms in Nebraska and Kansas, whereas they are close to those for the non-irrigated farms in Missouri and Iowa. Table 2 assumes that prices for direct energy rise by 30 per cent, prices for agricultural chemicals rise by 15 per cent, and farmers react to this price rise by reducing

Table 2 Per-acre direct and indirect costs of energy in agriculture, Missouri-Iowa-Nebraska-Kansas, MINK study (1982), 1989 (author estimate), and with future carbon tax of US$100 per ton (forecast)

Per acre expenses ($)	Missouri			Iowa			Nebraska			Kansas		
	1982	1989	Future	1982	1989	Future	1982	1989	Future	1982	1989	Future
Direct energy												
All farms	14	11	13	23	18	21	19	15	18	12	10	12
Irrigated	69	55	65	83	66	78	50	40	47	51	41	49
Non-irrigated	13	10	12	22	18	21	9	7	8	9	7	8
Commercial fertiliser												
All farms	21	23	24	25	25	27	18	20	21	11	12	13
Irrigated	78	85	90	78	85	90	42	46	49	36	39	42
Non-irrigated	19	21	22	25	27	28	7	8	9	8	9	10
Other agricultural chemicals												
All farms	15	19	20	15	19	20	9	11	12	6	8	9
Irrigated	45	57	61	48	61	65	19	24	26	18	23	24
Non-irrigated	14	18	19	15	19	20	5	6	6	4	5	5
Total energy												
All Farms	50	53	57	63	62	68	46	46	51	29	30	34
Irrigated	192	197	216	209	212	233	111	110	122	105	103	115
Non-irrigated	46	49	53	62	64	69	21	21	23	21	21	23

Sources: Darmstadter, 1991; U.S. Department of Agriculture, 1990, 1991; and author estimates.

energy use according to historical U.S. patterns. As is clear from Table 2, total energy expenditures per acre in this region would likely rise by about 10 - 15 per cent if carbon taxes were assessed, even with the estimated reduction in the use of energy.

The impact of the increased energy expenditures on U.S. agricultural productivity is likely to be slight, however. In the last 15 years, the amount of energy used in agriculture has declined by more than 20 per cent, while the amount of agricultural chemicals used has also declined slightly. During the same period, output of agricultural crops has grown by more than 20 per cent (U.S. Department of Agriculture, 1991).

Although the prospective impact of increased energy prices is small, U.S. agricultural production and competitiveness could be adversely affected, even if technical substitution out of energy continues at the pace experienced in the 1980s.[3] How much any country's agricultural sector is affected, however, depends in part on its comparative advantages. If elsewhere in the world there are regions even more energy intensive, the United States would be rendered more competitive by a general increase in energy prices. Conversely, the United States would be rendered less competitive with international competitors that are less dependent on fossil energy for maintaining their productivity.[3] Unfortunately, existing international data bases on energy consumption published by bodies such as the Organisation for Economic Cooperation and Development and United Nations do not adequately or consistently distinguish energy consumption in agriculture from energy consumption in other sectors. Therefore, only limited international comparisons of energy intensity in agriculture are possible.

We offer one such limited comparison, that of the United States and India. Energy is a significant cost component of U.S. agriculture, and although we think of energy as a key element in the worldwide Green Revolution, recent data indicate that neither in the United States nor in India (one of the Green Revolution countries), is the cost of energy an overwhelming component of total costs (Figure 3). In the United States in 1989, the

---

[3] This assumes that the increase in energy prices does not induce such significant technological changes in the U.S. that overall U.S. costs fall relative to those in other countries *despite* relatively high energy costs. Given the disparity among nations in research and technology capacity, this result cannot be ruled out.

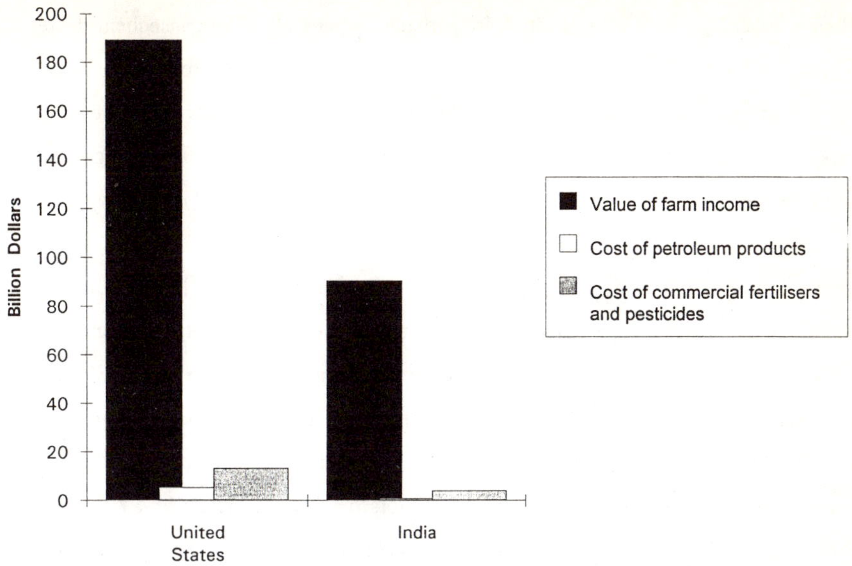

Figure 3 Cost of petroleum fuels and farm chemicals compared with the value of production in U.S. and Indian agriculture, 1989.

combined costs of petroleum, fertilisers and farm chemicals were about 10 per cent of the total value of farm output. In India, these costs were about 5.5 per cent (U.S. Department of Agriculture, 1990; and Centre for Monitoring the Indian Economy, 1992). Thus, a major increase in energy costs to prevent global warming would increase the costs of energy used in agriculture in both countries; other things being equal, this could increase the costs by more in the relatively energy-dependent U.S. production system than in India.[4] However, in neither country is the increase significant when compared with the effects of technological change by which total factor productivity (the ratio of the value of output to the cost of all inputs to production) has been increased in both countries in the last 10 years by 25 per cent or more. Crop production per acre has increased by 20 - 25 per cent (U.S. Department of Agriculture, 1990; Centre for Monitoring the Indian Economy, 1992).

To summarise, the effect of an increase of \$100/mtce to combat global warming would increase the price of energy used in agriculture by about 30 per cent in the U.S. and the

4 This, again, leaves aside the differential research, development and technical deployment capabilities of the two countries. Overall, the U.S. might find it easier to adjust inputs, and India, because of its lower average energy intensity in agriculture, might not find it cost-effective to adopt all of the energy-saving techniques that U.S. agriculture would.

overall costs of energy used in agriculture by perhaps 15 per cent. The consequent decrease in energy use is not expected significantly to affect costs, productivity; or food security in the developed countries or even in some of the agriculturally more-advanced developing countries, such as India. However, when we consider the developing countries as a group, the outcome is less certain.

The Green Revolution has been energy-intensive. In the process of joining the Green Revolution, India began to use much more mechanical energy, fertilisers and agricultural chemicals than it had previously. Between the growing seasons 1974-75 and 1989-90, for example, yields per acre of food crops in India increased by about 60 per cent. Some of this increase was due to improved crop strains; however, consumption of commercial fertilisers per hectare increased 4.8 times, the number of tractors per hectare increased 4.7 times, and the consumption of electric power per hectare increased 5.1 times. Thus, some of the increase in yield undoubtedly was due to increased application of energy in agriculture. We note that these increases took place against a background of steep energy price increases, so increased application of energy in Indian agriculture was profitable, even with high and rising fuel prices. As we noted earlier, the prospective increase in energy prices due to imposition of carbon taxes might not critically increase the costs of Indian agriculture.

However, consider a developing country attempting to follow in India's footsteps. Here the result of future increases in energy prices is less obvious. Such a hypothetical country likely would be using less mechanical energy, fertilisers and agricultural chemicals per hectare than India does now and, theoretically, should be less affected. However, rising energy prices might be more likely to discourage the adoption of energy-intensive techniques at the margin, because farmers in these countries (being less productive, poorer and with less access to borrowed capital than those in India) would be less likely to be able to afford to purchase the necessary energy inputs. Thus, ironically, the poorest and least energy-intensive countries might be the very ones most disadvantaged in attempting to join the Green Revolution.

3 Hydropower

The movement towards reducing fossil fuel usage will increase demand for sources of energy that do not release CO_2 to the atmosphere or that, at the least, add only CO_2 that has recently been removed from the atmosphere by photosynthesis. Candidate energy sources that meet this criterion are nuclear fission, wind, solar, geothermal, biomass and hydroelectricity. Wind, solar and biomass (especially the latter) require large land areas and could, therefore, compete for land with agriculture.[5] This matter is reviewed in a later section. In the remainder of this section we deal with the competition between hydroelectric generation and irrigation for water resources.

Despite the good they have done or were intended to do, dams on large rivers throughout the world have been a mixed blessing. The Aswan High Dam, for example, traps the sediment that from time immemorial replenished the soils of the Nile Valley. Damming the Columbia River in the U.S. Pacific Northwest has interfered with the natural salmon spawning cycle. Many other examples can be given. For these and other reasons, old dams are viewed unfavourably by many in the environmental movement, and because few rivers remain in the 'wild' (uncontrolled) state, new dams are actively opposed. Whether or not pressures to reduce fossil fuel use will cause environmentalists to reconsider their opposition to the construction of new dams to provide hydropower remains to be seen.

Developed river basins are managed for many purposes, of which hydroelectric generation is only one. Other purposes include navigation, recreation, fish and wildlife habitats, the provision of drinking water and cooling water for thermoelectric plants and other industrial processes, the transport and dilution of sewage and industrial wastes, irrigation and the transport of saline drainage waters from irrigated lands. These various uses often are in conflict throughout the entire year or in particular seasons.

Environmental concerns are driving the movement for reduction in fossil fuel use and, hence, are increasing the relative desirability of hydropower. However, environmental uses are often in conflict with hydropower. Frederick (1991) describes one instance of such conflict. The least tern and the piping plover are designated as endangered and threatened species.

---

5 Elliot *et al.* (1991) have estimated the area of lands in the U.S.A. that are windy enough (300 W/m^2) to be useful for wind energy conversion. They allowed for the exclusion of 30 per cent of the agricultural land, 10 per cent of rangeland and 50 per cent of the forested lands with adequate wind resources. Still, the amount of land area available (about 1.040 million square kilometers) equals only 13.5% of the total land area of the U.S. Of this usable area no more than 10 per cent would be occupied by the turbines. Presumably the remainder could be farmed or grazed.

These birds nest on low-lying sandbars and islands downstream of three of the upper Missouri River basin dams. The Corps of Engineers is required at times to alter reservoir releases (and hence, hydropower production) during the May through mid-August nesting season to avoid jeopardising the habitat of the tern and plover.

Surface runoff into streams, rivers and reservoirs is finite, depending on the vagaries of seasonal and annual precipitation and evaporative demands – that is, on the weather and climate. Decades of research into various methods of weather modification have yet to provide evidence of an operational capacity to augment precipitation. Therefore, supply remains finite, if variable, while demands for water for all purposes increase. Will pressures to reduce use of fossil fuels increase the value of water used for power generation? Will such increased value result in increased allocations of water to hydropower despite growth in competing demands?

What impacts do irrigation and hydropower have on one another? Butcher *et al.* (1986) explain the competition in these terms:

- irrigation changes the seasonal pattern of electricity demand, streamflow and hydropower supply;
- irrigation consumes large amounts of electricity for pumping;
- irrigation depletes streamflows and thus reduces downstream hydropower.

The loss of hydropower when water is withdrawn and consumed for irrigation depends on the quantity of water diverted and consumed and the amount of developed head that the water would fall through if left in the stream.

Currently, hydropower is one of the higher-valued uses of water. In many river basins, irrigation is one of the lowest valued, particularly when the water is applied to low-value, extensive crops such as alfalfa and pasture. Electricity generation is a non-consumptive use of water and irrigation is one of the most consumptive uses. In most developed river basins, hydropower is a more valuable use of water at the margin than is irrigation. The difference in relative economic value of water can only grow as the demand for CO_2-clean energy grows and as the costs of fossil fuels rise when anti-climate change policies take effect.

Historically, hydroelectricity has been typically the least expensive form of electricity to produce, and regions receiving a large proportion of their electric power from this source enjoy significantly lower costs for electricity.[6] For example, in 1989, the average cost of electricity provided to residential consumers in the U.S. was $25.67 per million Btu; in gas- and nuclear-dominated California it was $31.71; in hydroelectric-dominated Washington State, it was $14.54 (Darmstadter, 1991). Most of the trans-Mississippi west except for the southwest in the lower 48 states, the Tennessee Valley and Alaska derive more than a quarter of their power from hydroelectricity and, therefore, would be particularly vulnerable to desiccation due to natural droughts and/or climate change (Gleick, 1990). The primary users of water for irrigation in the U.S. are the 17 western states. In the U.S., the west is the region in which the most severe conflicts between hydropower, irrigation and urbanisation are likely to occur.

In many regions, tensions already exist between irrigation and hydroelectric use of water. A case study of the Snake River by Miller (1990) is instructive. Fostered by growing demands for power and by occasional droughts that impair the water security of hydroelectric facilities, the Snake River basin has had to explore institutional changes. Withdrawals for irrigation have reduced the potential for generating low-cost electricity further downstream. Through a series of negotiations and litigation, water security and transferability of water rights have been arranged within the basin. Compacts have been arranged 'to limit future encroachment on in-stream water rights for hydropower generation by irrigation depletions'. A water bank was created that initially permitted single-year transfers. This system now permits purchases of water rights for up to 25 years. While the Snake River case actually may be the exception rather than the rule in the relatively water-abundant Columbia River basin, competition among hydropower, irrigation and water users is common in most western U.S. river basins, with the Colorado and Missouri being prime examples.

The trend of transferring water away from agriculture and to higher value uses is likely to accelerate for many reasons. This is particularly true if demand for CO_2-clean hydropower increases because of public policy aimed at diminishing the likelihood of climate change. If

6 This is true for existing facilities but would not necessarily be true at the margins for new construction. In Alaska in the mid-1980s, for example, gas-fired combined cycle power plants were found less costly than large new hydroelectric facilities.

climate change of the kind that general circulation models predict for North America, i.e. warmer and drier in summer with less runoff and diminished stream flow, were actually to occur despite efforts at mitigation, these transfers would become more problematic.

To give the latter statements some dimension, we refer again to the MINK study. In the water resources portion of that study, Frederick (1991) and Rosenberg (1993) compared current (1951-80) streamflow at 15 gauging stations in the Missouri, Upper Mississippi, and Arkansas-White-Red River basins with flows at the same stations during the dry years of the 1930s. The 15 stations selected were in watersheds in which no man-made changes (for example, diversions, dams) had been made in the intervening years. The ratios of the 1930s to current flows were applied to the sub-basins represented in order to scale up to the entire river basin. The calculations show that if the climate of the 1930s were to recur today, flows would be reduced from the long-term means by 28 per cent, 28 per cent and 7 per cent in the Missouri, Upper Mississippi and Arkansas basins, respectively (Figure 4).

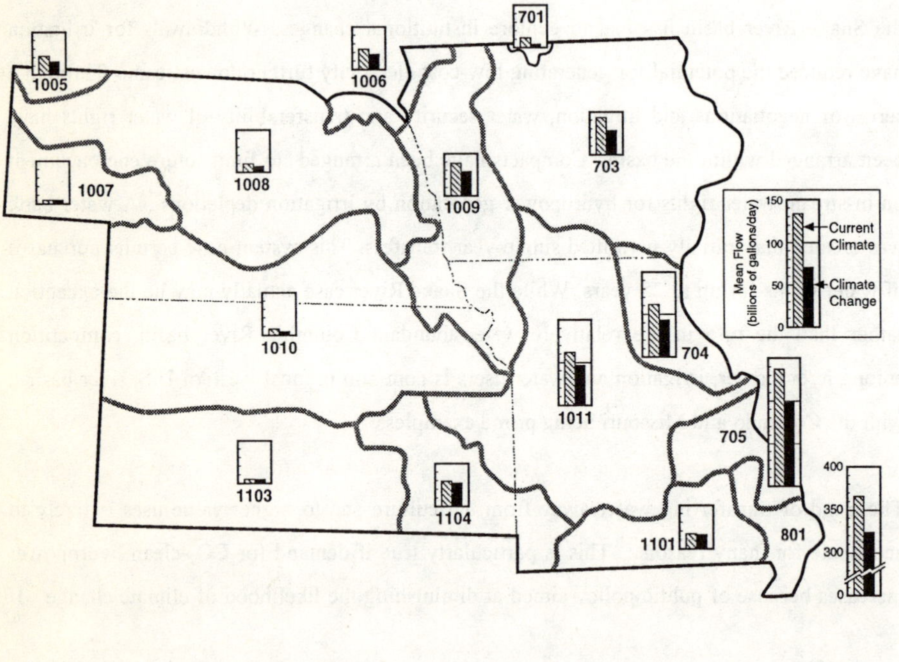

Figure 4 Impact of climate change on mean natural stream flows in selected subbasins of the Missouri, Upper Mississippi and Arkansas-White-Red river systems (billions of gallons/day).

Frederick (1991) also observed that, were the large reservoirs on the upper Missouri to continue operating under current Corps of Engineers assumptions, it would be necessary to curtail hydropower production by about 50 per cent. Of course, the demand for irrigation water would be increasing at the same time: first, because of greater evaporative demand due to higher temperatures and drier air and second, because the profitability of irrigated agriculture would increase as dryland production falls. The assumption of no change in irrigated acreage and irrigation water applied in amounts determined by climatic conditions leads to an increased demand of 39 per cent in Nebraska and 12 per cent in Kansas, the two MINK states with significant irrigated acreage today.

Would climate change exacerbate the competition between irrigation and hydropower more or less and in similar or different ways in developing and developing countries? The answers are not obvious but the questions are worth exploration.

Dam construction leads everywhere to inundation of fertile agricultural lands in low-lying floodplains. In tropical Africa (according to Scudder, 1989) the loss of fertile, productive floodplain land cultivated with traditional techniques is clearly not offset by the small areas gained for formal irrigation as a by-product of dam construction.

In India, for example, it is the loss of such floodplain lands and displacement of the largely agricultural populations settled on them that prompts much of the current opposition to development of the Narmada river (Moore, 1993). This project involves construction of some 30 huge dams, 135 medium-sized dams and 3000 small dams along the river's 800 mile course and its 41 tributaries.

Once constructed, dams are usually operated to provide constant flows for hydropower generation and seasonally adjusted flows for formal irrigation projects. In many of the developing countries, this regulation often interferes with the practices of traditional floodplain irrigation and recession farming downstream of the dams. Of course, these traditional practices can be made less risky and less vulnerable to the vagaries of the weather by controlled releases that permit well-timed flooding. This is unlikely to happen, of course, if such releases disrupt hydropower generation.

Not all regions would necessarily be drier as the result of climate change. But the picture emerging will not be greatly changed by other assumptions. Even without climate change, that picture is one of much less water available for all purposes, a premium placed on power supplies not derived from the combustion of fossil fuels, and increased (but not necessarily beneficial) demand for water to maintain or even enlarge the irrigation enterprise. We venture on the basis of the foregoing to assert that hydropower is likely to out-compete irrigation for increasingly scarce water supplies.

4 Farm emissions of methane and nitrous oxide

Methane (CH_4) and nitrous oxide (N_2O) are greenhouse gases. Although less potent than CO_2 because of their far smaller concentrations in the atmosphere, they nevertheless contribute significantly to current greenhouse warming potential (Houghton *et al.*, 1990, 1992). Methane has industrial and biogenic sources; N_2O is primarily biogenic. Of the many natural sources of CH_4 on the farm, sources of greatest importance are ruminant animals (via enteric fermentation) and animal wastes. Methane is also emitted under anaerobic conditions from wet soils, although emissions vary greatly depending on soil conditions. Paddy rice is one very important source. Soil is also a sink for CH_4 in which it is oxidised under dry conditions. Nitrous oxide is formed in soils under both aerobic and anaerobic conditions. Nitrogenous fertilisers greatly increase the volume of N_2O emitted from agricultural lands.

Climate forcing by any greenhouse gas is a complex function of its infrared absorptive power (radiative forcing per unit mass), its mean lifetime in the atmosphere, and the quantities emitted. According to Houghton *et al.* (1992), the radiative forcings of CH_4 and N_2O per unit mass are, respectively, about 60 and 200 times that of CO_2. The mean lifetime of a CO_2 molecule in air is about 120 years. Methane and N_2O have lifetimes of roughly 10 and 150 years. For the same unit mass, CH_4 and N_2O have, respectively, cumulative lifetime global warming potentials 21 and 290 times that of CO_2. Nonetheless, CO_2 is the predominant greenhouse gas because so much more of it than of CH_4 and N_2O is emitted into the atmosphere: from all sources, over 22 billion tons of CO_2, about 500 million tons of CH_4, and about 20 million tons of N_2O are emitted. Mitigation of greenhouse warming requires that the net emissions of all of these gases be reduced. In this section, we deal only with suppression of CH_4 and N_2O emissions. Management to decrease agricultural emissions of

CO_2 and increase capture and sequestration of carbon are dealt with peripherally in a later section.

More specifics on methods for reducing CH_4 and N_2O emissions are given below. However, before dealing with the means of reducing greenhouse gas emissions from agriculture, it is useful to gain some perspective about how much agriculture contributes to overall global emissions. Figure 5, drawn from a report of CAST (1992), shows that direct CO_2 emissions from U.S. agriculture contribute about 0.4 per cent to the total global climate forcing potential of that gas; U.S. agriculture contributes about 1 per cent to the global CH_4 and N_2O forcing.[7] At least in the United States, major efforts to reduce or eliminate agricultural emissions of the three gases would have little global impact; their total elimination would, of course, require total elimination of U.S. agriculture.

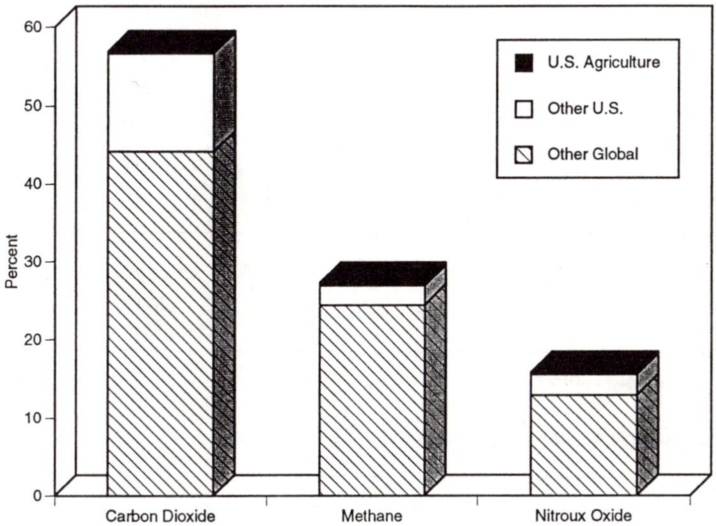

Figure 5 Climate forcing by carbon dioxide, methane and nitrous oxide driven by global emissions, U.S. emissions and U.S. agriculture.
(Source: CAST, 1992, Figure 4.1.1)

---

7 Chen (1990) shows that adding energy used in production of farm equipment, fertilisers and pesticides brings direct and indirect energy consumption in agriculture in the developed world to about 5 per cent of national energy consumption. Adding energy used in food processing, storage and transportation may increase the energy attributable to food production to 17 per cent. Based on these estimates, CO_2 production attributable to the integrated food and fibre industry show the U.S. share of CO_2 emissions could be as high as 3.4 per cent of the world total.

The picture for global agriculture is different. According to Houghton *et al.* (1992), global agriculture (rice paddies, enteric fermentation, and animal wastes) accounts for about 32 per cent of all CH_4 emissions (Table 3). Biomass burning, most of which is attributable to agricultural practices, accounts for an additional 8 per cent. Twenty-two per cent is emitted from wetlands, much of which could, conceivably, be drained for agricultural uses other than paddy rice, although there are strong ecological reasons for not doing so. Table 3 indicates that current best estimates of CH_4 sources and sinks fall within very broad bands of uncertainty.

Table 3 Estimated sources and sinks of methane (Tg CH_4 per year)

	Best estimate	Possible range
Sources		
All sources		
Natural + anthropogenic	515	331-850
Agricultural sources		
Rice paddies	60	20-150
Enteric fermentation	80	65-100
Animal wastes	25	20-30
Biomass burning	40	20-80
Total agricultural	205	125-360
Wetlands	115	100-200
Sinks		
All sinks	500	420-520
Atmospheric increase	32	28-37
Removal by soil	30	15-45

Source: Houghton *et al.*, 1992, Table A1.3.

Table 4 Estimated sources and sinks of nitrous oxide (Tg N per year)

	Range
Sources	
All sources	
Natural + anthropogenic	5.18 - 16.1
Agricultural sources	
Cultivated soils	0.03 - 3.0
Biomass burning	0.2 - 1.0
Sinks	
Photolysis in stratosphere	7 - 13
Atmospheric increase	3.0 - 4.5
Removal by soils	?

Source: Houghton *et al.* 1992, Table A1.5.

No 'best estimates' of N_2O sources and sinks are given in Table 4, only the range of reported estimates. Agricultural sources may be as small as 0.05 per cent of the lowest estimate of all natural plus anthropogenic sources (5.18 Tg N/annum) or as high as 25 per cent of the highest estimate. Biomass burning is again included in this estimate as an agricultural source. Thus, it appears that on the global scale (assuming the highest estimates of agricultural emissions of CH_4 and N_2O hold true), the reduction of CH_4 and N_2O through changes in agricultural practice could significantly reduce the potential for greenhouse warming.

We now turn to the questions of how to reduce agricultural emissions of the subject greenhouse gases. CAST (1992) provides details on ways in which the total emissions of CH_4 from agriculture might be reduced. Net CH_4 emissions can be reduced by improving the feed utilisation efficiency of ruminant animals, by proper treatment of animal manures, by increasing the efficiency of paddy rice production and by draining wetlands that naturally emit CH_4. Methane emission is unavoidable in paddy rice production. The appropriate strategy for meeting growing demands for rice is to increase productivity per unit of land, rather than to put more land into paddy. Deep placement of nitrogen fertilisers stimulates production and decreases the ratio of CH_4 emitted per unit of rice produced. Improvements in biological efficiency of rice to produce higher yields should also reduce the amount of CH_4 per unit of rice. An absolute reduction in CH_4 production per unit land area can be achieved by reducing the amount of organic residue returned to the soil. If the residue were burned, however, the benefits of carbon sequestration would be lost. Residue management that includes rotation of paddy with dryland crops would reduce CH_4 production per unit area, but then rice would have to be produced on more acres to make up for the production shortfall. Another suggested approach is to invest more effort in improving the productivity of upland rice and encouraging the adoption of other cereal grains into the diet of nations in which rice is the staple food.

It is difficult to see where any of the methane-reducing practices described here would greatly strain agricultural economies. The proposed practices that lead to greater productivity of rice production could enable a reduction of area planted to that crop. A reduction in the amount of organic matter incorporated into paddy soils could, over time, lead to a reduction in soil fertility, however. None of the proposed practices appear likely to alter the supply of rice radically or greatly to affect world prices for this commodity.

With respect to CH_4 emissions from ruminant animals, the CAST report concludes that significant reductions are possible through improvements in the biological efficiency of the animals. This can be accomplished through conventional breeding programs, improvements in feed formulations and pharmaceuticals, and the use of additives such as somatotropin to increase feed efficiency and animal gains. CAST suggests, as well, that decreased human consumption of animal products would contribute to reduction of agricultural CH_4 emissions.

Historically, manure and legumes have been used to provide nitrogen for crop production. Use of manufactured nitrogenous fertilisers has been widespread since the 1950s. Nitrogen fertilisers are costly, and farmers have a natural interest in using them efficiently. The desired outcome with respect to climate change is to reduce both the quantity of nitrogen lost as N_2O and its share of the total gaseous nitrogen emissions, of which N_2O is the greatest in quantity. The CAST report provides details on ways to decrease N_2O emissions.

Both the 'natural' and manufactured nitrogen fertilisers undergo chemical processes in the soil that result in the synthesis of N_2O and other nitrogen compounds. The amounts and the rates are determined by management practices, biogenic processes, soil properties and climate. Soil management offers opportunities to improve the effectiveness of applied nitrogen, maximising the amount taken up by the crop, thus minimising both volatilisation and escape to the atmosphere and leaching to depths below the root zone.

Net N_2O emissions can be reduced by improved timing and placement of nitrogen fertilisers and by encapsulating the fertilisers to slow the release of nitrogen. It can also be accomplished through use of nitrification inhibitors. Nitrification is the process whereby fertilisers applied in the relatively immobile forms of NH_3 or NH_4^+ are converted to the NO_3^- ion, which is readily denitrified or leached. Since denitrification occurs primarily under anaerobic conditions, irrigation practices that reduce the number of wet/dry cycles can also aid in decreasing N_2O emissions and potentially improve water-use efficiency since, inevitably, some water is lost in each irrigation.

All of the practices described above as being helpful in reducing N_2O emissions also contribute to improving the efficiency of fertiliser use and decreasing costs of crop production. This is one case in which strategies for mitigating greenhouse warming should be acceptable in the agricultural sector and, in fact, a considerable research effort is already devoted to improving fertiliser-use efficiency.

5 Competition for land: afforestation and herbaceous biomass

It is generally held that the best lands for growing crops are those already growing crops. Much of that land was originally forested and until recent times was cleared for agriculture with great exertion by man and beast. Forest conversion for cropping and grazing continues to this time and at rates in the tropics far exceeding historical experience. Equipment powered by internal combustion engines can remove forests much more rapidly than human and animal power ever could.

Afforestation and the application of management to existing forests, now mostly unmanaged, have been proposed as ways of helping to mitigate greenhouse warming. The aim is to capture through photosynthesis a large portion of the carbon emitted into the atmosphere by fossil fuel combustion, deforestation and biomass burning. The captured carbon either can be sequestered in the roots and boles of trees and in soil organic matter, or the wood can be burned or otherwise converted to provide energy as a substitute for fossil fuels.

Estimates have been made of the amount of land required for afforestation significantly to reduce net carbon emissions to the atmosphere. The estimates vary widely as the result of varying assumptions about the species to be planted, the biological limits of carbon fixation of the various candidate species, climatic and soil conditions, and time required for a planted stand to reach maturity. Three of these estimates are given in Table 5. Two of the estimates take a global view of the problem and one focuses on the United States.

Sedjo and Solomon (1989) calculated that, in order to offset the 2.9 Gt of carbon that accumulate in the atmosphere each year after natural scavenging, some 465 million hectares must be planted with trees having production equivalent to that of plantation forests in the southeast U.S.. Marland (1988) calculated that, depending on the productivity of the forests planted, between 500 and 700 million hectares would be needed for the more ambitious goal of removing 5 Gt of the 6.3 - 8.5 Gt C that are emitted into the atmosphere yearly. A notion of the scale of land diversion being discussed can be derived from Figure 6, which shows how large an area is encompassed in Sedjo and Solomon's estimate of 465 million hectares.

Table 5 Estimates of land required for afforestation to capture specified portions of the annual net CO_2 emissions to the atmosphere

Source	Total C removal (Gt/yr)	Total land area required (million hectares)	Assumptions
Marland, 1988	5.0	500	Production equivalent to best tropical forests
Marland, 1988	5.0	700	Production equivalent to short rotation sycamores in southeastern U.S. (Georgia)
Sedjo & Solomon, 1989	2.9	465	Production equivalent to tree plantations in southeast and northwest U.S.
Moulton & Richards, 1990		28.7[a]	Net U.S. carbon emission = 1.27 Gt/yr 10 per cent reduction

Note: Economically marginal and environmentally sensitive croplands, pasture lands and forest lands on which growth rate of trees could be enhanced. Beyond a 15 per cent reduction in net carbon emissions would require croplands almost exclusively.

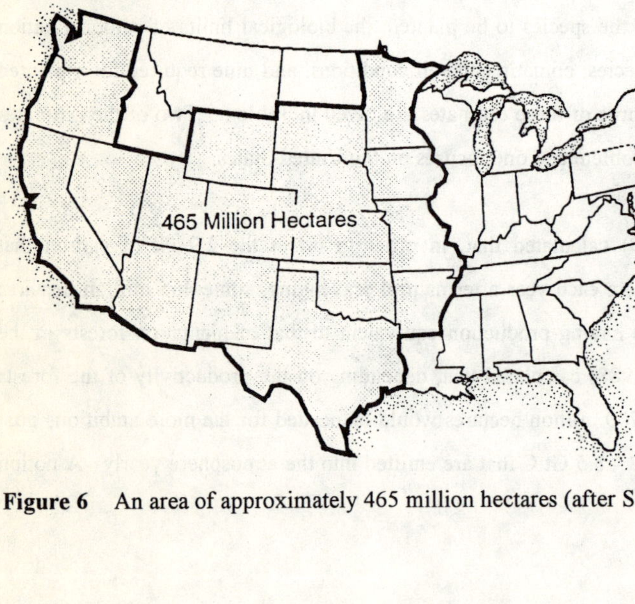

Figure 6 An area of approximately 465 million hectares (after Sedjo and Solomon, 1989).

Afforestation for sequestering carbon easily captures the imagination. Not as much prominence has been given, however, to the potential for recycling carbon through production of herbaceous biomass crops. These crops would be grown as feedstocks for chemical manufacture of substances to replace petroleum-derived liquid fuels and natural gas, or for direct combustion as boiler fuel in energy generation. Although possibilities exist for the production of non-woody biomass on marginal lands, at least in the U.S., the better agricultural lands will be needed to produce biomass competitively (Tyson, 1990). A study of the potential for biomass to provide a significant share of the energy needs of the Netherlands (NOVEM, 1992) concludes that, if all of that nation's land were devoted to biomass, about 22 per cent of its energy needs could be met. However, at the most, 0.5 million hectares of land could be diverted for biomass, which would provide only about 5 per cent of the Netherlands' energy requirements.

In a sense, the production of non-woody biomass is agriculture; the inputs and management required to grow 'energy sorghum' differ little from those for corn. Sylviculture, on the other hand, involves different planting and harvesting methods and much longer rotations. There are two questions to explore: (i) the social effects of carbon-sequestration practices (i.e., what happens to the farmers), and (ii) the effects on food security.

What would be the overall effect of large-scale land diversions away from food production to the sequestering and/or recycling of carbon to mitigate climate change? In calculating the costs of carbon fixation through afforestation, Moulton and Richards (1990) and Adams *et al.* (1993) have shown sharply rising costs with each increment of additional carbon capture. Adams *et al.* calculated, for example, that the first 10 per cent of U.S. emissions can be captured at a cost of about $18 per ton. To sequester 50 per cent, however, would cost $55 at the margin. The steeply rising cost of sequestration reflects rising opportunity costs of agricultural land, as more land of better quality is diverted from cropping. Commodity costs would necessarily rise with reduced production. Of course, U.S. emissions need not be offset by plantings only in the U.S.. Investment in tropical forest plantations may make more sense, but impacts on food-producing potential in the tropics would likely not be small.

Moulton and Richards (1990) focused on opportunities for afforestation in the U.S. to offset portions of that nation's net emissions of carbon to the atmosphere. A goal of offsetting 10 per cent of the net U.S. emissions of 1.27 Gt C (1,270 Tg C) per annum could be accomplished on 28.7 million hectares of economically marginal pasture lands, croplands, and forest lands where productivity could be improved by management. If more than 15 per cent of the net U.S. emissions are to be offset, better agricultural lands must be diverted. To increase carbon sequestration from 15 to about 50 per cent (725 Tg C/yr), for example, would require 224 million acres (90.7 million hectares), or about 48 per cent of all U.S. cropland (Moulton and Richards, 1990).

This discussion has focused on the temperate zone, for which a number of detailed economic analyses are available. However, plantation forestry to offset accumulations in the atmosphere may be more readily accomplished in the tropics. Sedjo and Solomon (1989) believe that large areas of 'degraded lands' in the tropics with few economic uses could be made available for afforestation. Costs of afforestation would be considerably lower in the tropical countries. Recently deforested lands in the tropical rainforest zone are included in the category of 'degraded lands.' Economics will probably cause many such areas to revert, naturally or through human intervention to forest.

Figures on the uses of U.S. cropland suggest that about two-thirds is devoted to the growing of wheat, corn (maize) and soybeans, with each crop occupying about one-third of that area. Assuming that the land removed from agriculture is proportionately distributed among all crops and has average productivity, we can calculate an approximate maximum amount of lost farm production of these three major crops.[8] The loss of 48 per cent of U.S. production would be about 40 million tons of wheat, 70 million tons of corn and 26 million tons of soybeans (7 per cent, 15 per cent, and 25 per cent, respectively, of world production). While clearly an overestimate of the impact, these values are significant, and would be a matter of some concern for world food security.

It is beyond the scope of this paper to attempt a nation-by-nation or region-by-region analysis to identify the best places for afforestation and non-woody biomass production. It is

---

8 This is a maximum for two reasons: i) the land most valuable for agriculture would likely remain in agriculture, meaning that average yield would overstate average loss in production on the average hectare diverted into forests; and ii) farmers would likely farm remaining land more intensively, resulting in higher yields.

sufficient to say here – what was obvious to the reader before – that in any real world efforts at afforestation, lands of marginal value to agriculture (i.e. highly erodible lands, wetlands etc.) will be the first committed to this purpose, but if a significant impact is to be made, highly productive agricultural lands will also be required. Diversion of the highly productive lands will make feeding the world that much more difficult.

6 Geoengineering

Geoengineering consists of several approaches intended to offset climate change, or its impacts, through intentional control of climate or of the concentrations of greenhouse gases. The interventions fall under five general categories: (i) collecting and disposing of CO_2 from flue gas streams; (ii) increasing net uptake of CO_2 into the terrestrial biosphere; (iii) increasing net uptake of CO_2 into the oceans; (iv) changing the Earth's energy balance by altering the albedo (the fraction of incident solar energy scattered or reflected back to space without being absorbed); and (v) altering internal processes in the climate system. Actions that might be taken under categories (ii) and (iv) could, conceivably affect agriculture in a direct way.

A balance between incoming short-wave solar radiation and outgoing long-wave infrared radiation maintains the global heat balance, which energises the climate system. A doubling of atmospheric CO_2 (or its equivalent, with other greenhouse gasses included) would increase the global average radiation flux at the top of the troposphere by 4.4 W/m^2 (Houghton *et al.*, 1990). On average, the Earth absorbs about 240 W/m^2 of the 340 W/m^2 solar energy flux. Thus, a decrease of less than 2 per cent in sunlight penetrating to the lower atmosphere could approximately offset the effects of an equivalent CO_2 doubling. Intentional manipulation of the Earth's albedo could help to limit global warming.

Our goal in this section is to describe a few of the albedo manipulations that have been proposed to mitigate climate change in terms of how they might affect agriculture. The issue of carbon sequestration through afforestation has already been mentioned in Section 5.

Modifications of planetary albedo can be made in space. One plan proposed by NASA involves the use of 55,000 orbiting mirrors of 100 km^2 in near-Earth orbit. If arrayed parallel to the Earth's surface, 1 per cent of the incoming solar radiation could be reflected; if arrayed perpendicular to the incoming radiation, 2 per cent could be reflected. Shadows cast on the surface by these mirrors would be roughly similar to those of an eclipse. Photosynthesis would be affected, but by how much and where would depend on orbiting characteristics of the mirrors.

It has also been proposed that clouds of soot particles be lofted into space to absorb and scatter solar radiation. Backscattering would effectively reflect some of the radiation into space. It has been calculated that 500 million kg of soot in a cloud of 5 million km^2 would be needed to absorb 1 per cent of the incoming radiation. Again, the impact of this type of solar shield would affect photosynthesis, the effect depending on the orbiting characteristics of the cloud and its durability.

Albedo modification can also be done within the atmosphere. It has been suggested that insertion into the stratosphere of small, thin-skinned, helium-filled aluminium balls could be effective in reflecting solar radiation. Corner reflectors could be attached to the spheres to enhance the overall efficiency of the balloons. The balloons would not cast large shadows, so their effect on plant growth would probably be no greater than would be caused by any transitory 2 per cent reduction in insolation.

Injection of various aerosols into the atmosphere provides other ways of reducing absorbed insolation. Current SO_2 emissions and aerosols associated with burning biomass are believed to already exert a cooling effect by increasing reflectivity of the atmosphere (Charlson *et al.*, 1992; Penner *et al.*, 1992). In the stratosphere, injected sulphates would have a reflective effect; injected soot would absorb radiation and prevent it from reaching the lower atmosphere.

Modification of albedo is most easily accomplished at the surface, although it is least effective there, because only about half of the incoming solar radiation reaches the ground. Increasing snow and ice cover (or preventing its loss) provides one way of increasing surface albedo. Turning forests and grasslands into desert would also raise albedo, but it would hardly do much for agriculture. Experiments have been performed to increase the albedo of soils and vegetation. Their aim has not been to protect the planet from climate change but, rather, to decrease the radiant energy load on transpiring surfaces, thereby reducing plant water use. In a series of such experiments in Nebraska from the late 1970s to mid-1980s, soybean plants were maintained with coatings of kaolinite clay and a diatomaceous earth (Baradas *et al.*, 1976a, 1976b). Reflectance in the visible portion of the spectrum was tripled; reflectance increased by 25 per centin the near-infrared. Overall, it was found that the coatings increased crop albedo by about 8 - 10 per cent. Water use was reduced and photosynthesis barely affected, so overall water-use efficiency was improved. In later studies, the Nebraska researchers tested cultivars of soybeans that had been bred to increase their leaf pubescence. Pubescent isogenes had at least triple the number of hairs per unit area than normal. Albedo was greater, water savings occurred, and photosynthesis was essentially unchanged in plots of 'hairy' soybean plants (Baldocchi *et al.*, 1983). Although far-fetched, if prospects for severe climate change warrant, it might be possible to breed and introduce highly reflectant varieties of all cultivated crops and managed forests species. Biotechnology might be used to speed plant breeding if the task becomes urgent enough. The water-saving effects shown in the few field studies described above suggest that the overall impact of breeding for reflectance would be beneficial to agriculture.

To our knowledge, little thought has yet been given to the biological implications of the various geoengineering schemes to reduce insolation from space or from the atmosphere. Direct shading of the surface means that the plants receive only diffuse solar radiation. Diffuse and direct beam solar radiation differ somewhat in spectral composition. It has been shown that the photosynthetic mechanism of many C_3 plants is light-saturated; thus,. these plants are not adversely affected by moderate shading. Additionally, crops with dense canopies may benefit from the greater penetration of diffuse radiation (Rosenberg *et al.*, 1983). On the other hand, most C_4 plants are not light-saturated, even at full sunlight, and do show reduced photosynthesis under shade. However, C_3 rice is often limited by cloudy conditions during rainy seasons in Asia, and this crop might suffer still more if sunshine were further reduced. On the positive side, reduced solar radiation would decrease the energy load on plants, with consequent reductions in evapotranspiration. Would any of the proposed geoengineering fixes alter the quantity or quality of solar radiation enough to make a difference? Possibly so, but the effects would no doubt differ by crop species and by the latitude, altitude and natural cloudiness of the sites affected by the mirrors, dust palls, or whatever the geoengineers somehow convince us to deploy.

7 The CO_2 fertilisation effect foregone

It is a well-known and demonstrable fact that plants exposed to higher-than-normal concentrations of CO_2 respond with an increased rate of photosynthesis. Such increases in photosynthesis normally lead to larger and more vigorous plants, to higher yields of total dry matter (roots, shoots and leaves), and often, to higher yields of harvestable products – fruits, grains, etc. (Rosenberg, 1981, 1992; CAST, 1992). The behaviour described here is demonstrated particularly in plants of the C_3 category which includes most of the world's small grains, legumes, root crops, cool-season grasses and trees. Another category of plants, the C_4 or tropical grasses such as corn, sorghum, millet and sugar cane, are naturally more efficient photosynthesisers than the C_3 plants. They also respond, but less markedly, to increases in atmospheric CO_2. Remarkably, elevating CO_2 concentrations can increase photosynthetic rates in C_3 plants to levels approaching those achieved by C_4 plants at current ambient concentrations. This effect is illustrated schematically in Figure 7.

The magnitude of this CO_2 fertilisation effect is not known precisely because it varies with species, weather conditions, soil fertility and other factors. Kimball (1983a, 1983b, 1986) helped to estimate the effect through extensive reviews of agronomic, greenhouse and growth chamber research in which plants were grown in various concentrations of atmospheric CO_2.

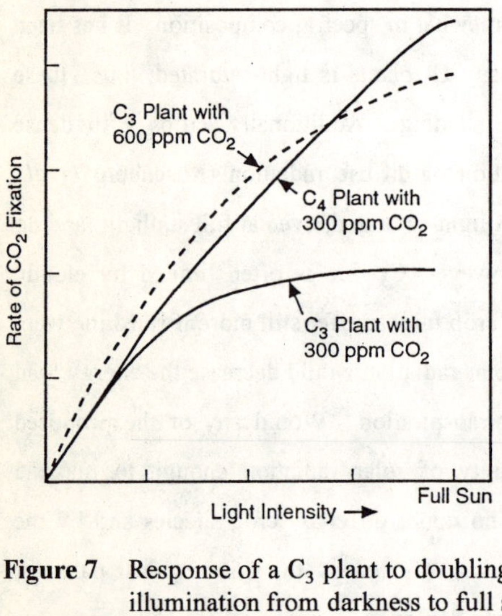

Figure 7 Response of a C_3 plant to doubling of ambient CO_2 concentration over a range of illumination from darkness to full sunlight.

The literature shows that, on average, a doubling of CO_2 from its recent concentration (340 - 356 ppmv), all else constant, will increase growth and yield in C_3 plants by about 34 per cent (+/-6 per cent) and about 14 per cent(+/-11 per cent) in C_4 plants.

Carbon dioxide enrichment also leads to another interesting response in both the C_3 and C_4 plants. Their consumption of water by transpiration is reduced because of partial closure of the leaf stomata (pores) induced by high CO_2 concentration. Reviews of experimental evidence for a wide range of agricultural, woody and weed species have shown that doubling ambient CO_2 from recent levels reduced transpiration by an average of 34 per cent (Kimball and Idso, 1983) and increased stomatal resistance to vapour transport out of the plant by from 51 per cent (Cure, 1985) to 67 per cen t(+/-14per cent) (Morison, 1987). The reduction in transpiration is not accompanied by any significant loss in instantaneous photosynthesis and usually results in greater total photosynthesis where moisture savings occur. Water-use efficiency (production per unit of water consumed) increases because of reduced transpiration and increased photosynthesis.

If the findings from the laboratory, greenhouses and open chambers upon which the above statements are based can be extrapolated to open-field agriculture, we may expect important benefits from the increasing concentration of CO_2 in the atmosphere – increased photosynthesis in many important species and decreased water consumption in most species. We still do not know whether these laboratory-demonstrable CO_2 effects on photosynthesis and transpiration do now or will occur in the future in the field where temperature, moisture and nutrients are the factors that normally limit plant productivity. In developing countries, where limited economic resources and access to technology make it difficult to overcome these limiting factors, it will be more difficult to realise the benefits of CO_2 fertilisation. However, laboratory, controlled environment, and open-top chamber experiments have shown that CO_2 enrichment of the atmosphere can actually reduce the impacts of moisture and salinity stress on plants (Rosenberg *et al.*, 1990). A summary by Allen *et al.* (1990) shows that high temperature stress is also alleviated by CO_2 fertilisation. The effects on nutrient stress remain unclear at this writing.

Carbon dioxide enrichment of the atmosphere could also lead to a number of troublesome effects for agriculture as well as for unmanaged ecosystems. Because of the special benefit that C_3 plants derive from elevated CO_2, infestations of C_3 weeds may become worse where they grow in association with C_4 plants. Additionally, the carbon-to-nitrogen ratio increases in leaves of plants fertilised with CO_2. Some short-term studies show that herbivory increases as insects consume more vegetation to satisfy their nutritional needs (Lincoln *et al.*, 1984). Later studies suggest more complicated outcomes, however. Over longer periods, the population of insects feeding on plants stimulated by higher CO_2 would likely decline in response to the diminished proportion of nitrogen in the plant tissues. As pest populations decline, so would the populations of their predators (Fajer *et al.*, 1989). The long-term ecological changes that may follow from increasing CO_2 concentration in the atmosphere are difficult to predict (Bazzaz and Fajer, 1992). However, the impacts of changes in weed competition and insect activity on agriculture would probably be less profound than they would be in ecosystems that are unmanaged (Rosenberg, 1992).

Because of the many complicating factors induced by the artificiality of experimental environments, it will be some time before we have definitive answers on whether CO_2 fertilisation will actually affect photosynthesis in crops in totally natural open-air environments. Attempts to measure open-air responses began in the 1970s in experiments where CO_2 was released directly into fields of selected crops (see Harper *et al.*, 1973a, 1973b, for cotton; Allen *et al.*, 1974, for corn). These experiments failed to provide definitive results, primarily because of the difficulty of maintaining elevated CO_2 concentrations in the air surrounding plants in the face of normal atmospheric turbulence, which tended to remove it rapidly.

Since these experiments were conducted, technological advances have made open-air CO_2 enrichment research possible. In a program called FACE (which stands for free air carbon dioxide enrichment), sponsored by the U.S. Department of Energy and conducted by the U.S. Department of Agriculture, equipment has been developed to maintain an elevated level of CO_2 in the air within and above crops for the duration of entire growing seasons (Hendrey and Kimball, 1990). The FACE system was originally tested in a Mississippi cotton field using industrial by-product CO_2 and is now working in a field near Phoenix, Arizona. Cotton crops have been grown for two years with CO_2 held at a concentration of about 550 ppmv (Kimball, 1992). Cotton growth and yield has been about 40 per cent greater with CO_2 enrichment. Seasonal water use may not have been affected, although use appears to have been greater early in the season because of more rapid crop growth while the stomatal closure effect reduced use later in the season. Water-use efficiency was improved in direct proportion to the increase in yield.

The evidence strongly supports the notion that CO_2-fertilisation is good for agriculture. But is it good enough to overcome any negative effects of CO_2 emissions on climate? One attempt (perhaps the only one thus far) to estimate the economic impacts of CO_2 fertilisation was made as part of the MINK study. The results of erosion production impact calculator (EPIC) crop model simulations that allow consideration of the direct (non-climatic) effects of elevated CO_2 on crop yield are shown in Table 6. Yields were simulated on some 50 representative farms exposed to a permanent 1930s 'dustbowl' climate under conditions of current CO_2 and elevated CO_2 concentrations (350 and 450 ppmv). The dustbowl years were hotter and drier than the previous and subsequent years. General circulation models forecast that greenhouse warming will lead to hotter, drier conditions in central North America. Yield losses resulting from imposition of the dustbowl climate were aggregated to the state and regional level.

At ambient CO_2 concentrations the dustbowl climate lowered simulated yields of all crops studied. Elevated CO_2 concentration diminished the yield losses for corn, sorghum, and soybeans and produced an increased yield for wheat and hay. The elevated CO_2 concentration reduced the regional loss to regional producers from approximately \$2.7 billion to \$1.3 billion (in 1982 dollars). A regional input-output model IMPLAN (Alward, 1986) was used to track the overall effect on the regional economy. Table 7 shows that adaptations (earlier planting, longer season cultivars, moisture-conserving tillage practices) reduced the

Table 6 Decline in value of feedgrain production (corn, sorghum and soybeans) and of total regional production under the dustbowl climate and 1980s agricultural conditions in the MINK region

Situation	Millions of 1982 \$
(1) Climate change	4374
(2) Adaptation only	3755
(3) Adaptation + CO_2	1847
(4) Reduction in losses attributable to CO_2 effects (2) - (3)	1908

Source: After Bowes and Crosson, 1991, Table 4.5.

loss to the regional economy from \$4.37 to \$3.76 billion. CO_2 enrichment and adaptations together reduced the loss to \$1.85 billion. Thus, in this input-output calculation, the regional economy was spared the loss of nearly \$2 billion by the direct effects of CO_2 and its interactions with adaptations.

These calculations assume that the shortfall of feedgrains is absorbed entirely by the regional animal-feeding industry and by the meat-packing industries, its suppliers, the transporters of its products, etc. In this case the loss to the regional economy would be about 10 per cent. The interacting effects of CO_2 enrichment and on-farm adaptations in response to climate change might reduce this loss to about 6%. These numbers are 'worst case,' because it is likely that the animal-feeding industry would import some feedgrains to make up for the within-region shortfall. Nonetheless, these input-output calculations put credible order-of-magnitude numbers on the benefits that could be foregone if CO_2 enrichment of the atmosphere ceases.

It is difficult to conclude from the scanty analysis described above that we really know how important the economic consequences of the CO_2 fertilisation effect will be. Will the positive benefits of CO_2 fertilisation outweigh the negative effects of greenhouse warming globally or for specific regions? Probably not, but we cannot know for sure. At the least, however, it seems fair to say that, if the climatic consequences of rising atmospheric CO_2 concentration are less severe than we currently think likely (or even benign), attempts to mitigate greenhouse warming may cause us to lose an opportunity for increased agricultural productivity.

Table 7 Changes in crop production and value in the MINK states due to imposition of a dustbowl climate and the ameliorating effects of CO_2 enrichment

	Production (million tons)		**Value (millions of 1982 \$)**	
Crop	Current[a] CO_2	Elevated[b] CO_2	Current[a] CO_2	Elevated[b] CO_2
Corn	-15.03	-9.46	-1644	-1035
Wheat	-0.10	+1.14	-14	+150
Sorghum	-2.15	-1.18	-215	-118
Soybeans	-3.77	-2.09	-789	-438
Hay	-0.80	+1.90	-47	+112
Total			-2709	-1329

[a] Current CO_2 = 350 ppmv. [b] Elevated CO_2 = 450 ppmv.
After Crosson *et al.*, 1991, Table 1.11.

8 Summary and conclusions

Strategies to avoid or mitigate climate change could affect global food security. Increases in the cost of energy will affect all sectors of the economy, with agriculture included. The major effect we see is one in which intensification of developing country agriculture is impeded by rising energy costs. Because it is a carbon-free source of energy, hydropower will be at a premium and will compete more strongly for water supplies at the expense of irrigation.

Strategies that have been proposed for the suppression of on-farm emissions of CH_4 and N_2O need not be very costly. These strategies involve increasing ruminant feed efficiency, increasing the productivity of paddy rice production, and increasing conservation of nitrogen fertilisers – all of which should be environmentally benign and should increase the profitability of farming.

Geoengineering to reduce insolation will be complicated and accomplished only at great cost. Impacts on food security of the various methods proposed would likely be small, although significant effects on particular crops and/or places cannot be ruled out.

Afforestation and production of herbaceous biomass to sequester and/or recycle carbon in significant amounts will require vast areas of land, only a small portion of which can be provided by agriculturally marginal land. Thus, afforestation and biomass will have to compete with agriculture for the use of productive land.

The possibility that potentially significant benefits due to CO_2 fertilisation will not be realised because of limits on fossil fuel use must be considered in any accounting of the possible impacts of climate change mitigation strategies on agriculture and food security. We think that the benefits of CO_2 fertilisation foregone would not be small, but the negative effects of climate change could very possibly outweigh them and for this reason we certainly *do not* urge that attempts to mitigate climate change through reduced emissions and increased sequestration of CO_2 be halted.

Crosson and Rosenberg (1989) argue that the capacity of world agriculture to support a growing population can be accomplished only if the land, water and genetic resources on which agriculture depends are protected. Several strategies to mitigate climate change could make agricultural access to two of these resources (land and water) more difficult. Whether afforestation (or perhaps geoengineering) will negatively alter genetic diversity is a question we have not attempted to examine here, but one that may well warrant study.

Acknowledgement

This paper is a revised version that first appeared in *Global Environmental Change* Volume 4, Rosenberg, N.J. and Scott, M.J. Implications of policies to prevent climate change for future food security, pp 49-62 (1994). Reprinted with permission from Elsevier Science Ltd, Butterworth-Heinemann Imprint, The Boulevard, Landford Lane, Kidlington OX5 1GB, U.K.

References

Adams, D., Adams, R.M., Callaway, J.M., Ching-Cheng, C. and McCarl, B.A. (1993). Sequestering carbon on agricultural land: a preliminary analysis of social cost and impacts on timber markets. *Contemporary Policy Issues*, **11:** pp. 76-87.

Allen, L.H., Jr., Desjardins, R.L. and Lemon., E.R. (1974). Line source carbon dioxide release. I. Field experiment. *Agronomy Journal*, **66:** 609-615.

Allen, S.G., Idso, S.B., Kimball, B.A., Baker, J.T., Allen, L.H. Jr., Mauney, J.R., Radin J.W. and Anderson, M.G. (1990). *Effects of Air Temperature on Atmospheric CO+ Plant Growth Relationships.* Report DOE/ER-0450T (TR048): U.S. Department of Energy, Office of Energy Research, Washington, DC.

Alward, G.S. (1986). *IMPLAN Version 2.0: Methods Used to Construct the 1982 Regional Economic Data Base.* USDA Forest Service General Technical Report RM-000 (draft), Rocky Mountain Forest and Range Experiment Station, Fort Collins.

Baldocchi, D.D., Verma, S.B., Rosenberg, N.J., Blad, B.L., Garay, A. and Specht, J.E.. (1983). Leaf pubescence effects on the mass and energy exchange between soybean canopies and the atmosphere. *Agronomy Journal*, **75**: 537-543.

Baradas, M.W., Blad, B.L. and Rosenberg, N.J. (1976a). Reflectant induced modification of soybean (Glycine max L.) canopy radiation balance. IV. Leaf and canopy temperature. *Agronomy Journal,* **68:** 843-848.

Baradas, M.W., Blad, B.L. and Rosenberg, N.J. (1976b). Reflectant induced modification of soybean (Glycine max L.) canopy radiation balance. V. Longwave Radiation Balance. *Agronomy Journal,* **68**: 848-852.

Bazzaz, F.A. and Fajer, E.D. (1992). Plant life in a CO_2-rich world. *Scientific American*, **266**: 68-74.

Born, M., Door, H. and Levin, I. (1990). Methane concentration in aerated soil in West Germany. *Tellus B*, **43**: 2-8.

Bowes, M.D. and Crosson, P. R. (1991). Consequences of climate change for the MINK economy. Report VI of *Processes for Identifying Regional Influences of and Responses to Increasing Atmosphere CO_2 and Climate Change – The MINK Project.* DOE/RL/0183OT-H12 (TR052H). U.S. Department of Environment, Washington DC.

Bradley, R.A., Watts, E.C. and Williams, E.R. (eds.) *Report to the Congress of the United States: Limiting Net Greenhouse Gas Emissions in the United States.* DOE/PE-0101. U.S. Department of Energy, Washington, DC.

Butcher, W.R., Wandschneider, P.R. with Whittlesey, N.K. (1986). Competition between irrigation and hydropower in the Pacific Northwest. In: Frederick, K.D. (ed.) *Scarce Water and Institutional Change*. Resources for the Future, Washington, DC. pp. 25-66.

CAST (Council for Agricultural Science and Technology) (1992). *Preparing U.S. Agriculture for Global Climate Change.* Task Force Report No. 119. CAST, Ames.

Centre for Monitoring the Indian Economy. (1992). *Basic Statistics Relating to the Indian Economy. Volume 1: All India.* Centre for Monitoring the India Economy, Wori, Bombay.

Charlson, R.J., Schwartz, S.E., Hales, J. M., Cess, R.D., Coakley, Jr., J.A., Hansen, J.E. and Hefmann, D.J. (1992). Climate forcing by anthropogenic aerosols. *Science*, **255**: 423-430.

Chen, R.S. (1990). *Global Agriculture, Environment, and Hunger: Past, Present and Future Links.* R.P.-90-5. The Alan Shawn Feinstein Hunger Program, Brown University, Providence, Rhode Island.

Crosson, P.R., Katz, L.A. and Wingard, J. (1991). *Processes for Identifying Regional Influences of and Responses to Increasing Atmospheric CO_2 and Climate Change – The MINK Project. Report Paper IIA – Agricultural Production and Resource Use in the MINK Region Without and With Climate Change.* DOE/RL/01830T-H7. U.S. Department of Energy, Washington, DC.

Crosson, P.R. and Rosenberg, N.J. (1989). Strategies for agriculture. *Scientific American,* **260:** 128-135.

Cure, J.D. (1985). Carbon dioxide doubling responses: a crop survey. In: Strain, B.R. and Cure, J.D. (eds.) *Direct Effects of Increasing Carbon Dioxide on Vegetation.* DOE/ER-0238, Carbon Dioxide Research Division, U.S. Department of Energy, Washington, DC, pp. 99-116.

Darmstadter, J. (1991). Energy. Report V of *Processes for Identifying Regional Influences of and Responses to Increasing Atmosphere CO_2 and Climate Change – The MINK Project.* DOE/RL/01830T-H11(TR052H). U. S. Department of Energy, Washington, DC.

Elliot, D.L., Wendell, L.L. and Gower, G.L. (1991). An assessment of the available windy land area and wind energy potential in the contiguous United States. PNL-7789/uc-261, Pacific Northwest Laboratory, Richland.

Fajer, E.D., Bowers, M.D. and Bazzaz, F.A. (1989). The effects of enriched carbon dioxide atmospheres on plant-insect herbivore interactions. *Science*, **243**: 1198-1200.

Frederick, K.D. (1991). Water Resources. Report IV of *Processes for Identifying Regional Influences of and Responses to Increasing Atmospheric CO_2 and Climate Change – The MINK Project.* DOE/RL/01830T-H10 (TR052F) 153 pp. U.S. Department of Environment, Washington, DC.

Gleick, P.H. (1990). Vulnerability of water systems. In: Waggoner, P.E. (ed). *Climate Change and U.S. Water Resources.* J. Wiley & Sons, New York.

Harper, L.A., Baker, D.N., Box, Jr. J.E. and Hesketh, J.D. (1973a). Carbon dioxide and the photosynthesis of field crops: a metered carbon dioxide release in cotton under field conditions. *Agronomy Journal*, **65**: 7-11.

Harper, L.A., Box, Jr. J., Baker, D.N. and Hesketh, J.D. (1973b). Carbon Dioxide and the Photosynthesis of Field Crops: A Tracer Examination of Turbulent Transfer Theory. *Agronomy Journal*, **65:** 574-578.

Hendrey, G.R. and Kimball, B. (1990). *FACE: Free-Air Carbon Dioxide Enrichment: Application to Field Grown Cotton.* Report BNL 46155. Brookhaven National Laboratory. Upton.

Houghton, J.T., Callander, B.A. and Varney, S.K. (eds.) (1992). *Climate Change 1992. The Supplementary Report to the IPCC Scientific Assessment.* Cambridge University Press, Cambridge.

Houghton, J.T., Jenkins, G.J. and Ephraums, J.J. (eds.) (1990). *The IPCC Scientific Assessment. Report prepared by Working Group I.* Cambridge University Press, Cambridge.

Kimball, B.A. (1983a). Carbon Dioxide and Agricultural Yield: An Assemblage and Analysis of 430 Prior Observations. *Agronomy Journal*, **75:** 779-788.

Kimball, B.A. (1983b). Carbon dioxide and agricultural yield: an assemblage and analysis of 770 prior observations. *WCL Report* No. 14, U.S. Water Conservation Laboratory, Phoenix.

Kimball, B.A. (1986). Influence of elevated CO_2 on crop yield. In: Enoch, H.Z. and Kimball, B.A. (eds.) *Carbon Dioxide Enrichment of Greenhouse Crops, Vol. II: Physiology, Yield, and Economics.* CRC Press, Boca Raton. pp. 105-115.

Kimball, BA. and Idso, S.B. (1983). Increasing atmospheric CO_2: effects on crop yield, water use, and climate. *Agr. Water Management*, **7:** 55-72.

Lincoln, D.E., Sionit, N. and Strain, B.R. (1984). Growth and feeding response to pseudoplusia includens (*Lepidoptera: Noctuidae*) to host plants grown in controlled carbon dioxide atmospheres. *Environmental Entomology*, **13**: 1527-1530.

Marland, G. (1988). *The Prospect of Solving the CO_2 Problem Through Global Reforestation.* DOE/NBB-0082 (TR039). U.S. Department of Environment, Washington, DC.

Miller, K.A. (1990). Water, electricity and institutional innovation. In: Waggoner, P.E. (ed.) *Climate Change and U.S. Water Resources.* J. Wiley & Son, New York.

Moore, M. (1993). India's lifeline or man-made disaster? *Washington Post* (8/24/93) p. A12.

Morison, J.I.L. (1987). Intercellular CO_2 concentration and stomatal response to CO_2. In: Zeiger, E, Farquhar, G.D. and Cowan, I.R. (eds.) *Stomatal Function.* Stanford University Press, Stanford.

Mosier, A., Schimel, D., Valentine, D., Bronson, K. and Parton, W. (1991). Methane and nitrous oxide fluxes in native, fertilized and cultivated grassland. *Nature* **350**:330-332.

Moulton, R.J. and Richards, K.R. (1990). *Costs of Sequestering Carbon Through Tree Planting and Forest Management in the United States.* Forest Service Report GTR WO-58. US Department of Agriculture, Washington, DC.

National Academy of Sciences (1992). *Policy Implications of Greenhouse Warming. Part Four. Adaptation – Chapter 34 – Farming.* National Academy of Sciences, Washington, DC. pp. 556-566

NOVEM (Netherlands Agency for Energy and the Environment) (1992). *The Feasibility of Biomass Production for the Netherlands Energy Economy.* Report 9210, Apeldoorn, The Netherlands.

Parry, M.L., Carter, T.R. and Konijn, N.T. (eds.) (1988*).* *The Impact of Climate Variations on Agriculture.* Kluwer Academic Publishers, Dordrecht.

Penner, J.E., Dickinson, R.E. and O'Neill, C.A. (1992). Effects of aerosol from biomass burning on global radiation budget. *Science*, **256:** 1432-1434.

Rosenberg, N.J. (1981). The increasing CO_2 concentration in the atmosphere and its implication on agricultural productivity. I. Effects on photosynthesist Transpiration and water use efficiency. *Climatic Change*, **3:** 265-279.

Rosenberg, N.J. (1992). Adaptation of agriculture to climate change. *Climatic Change*, **31:** 385-405.

Rosenberg, N.J. (ed.) (1993). An integrated study of climate change impacts and responses: the MINK project. *Climatic Change* (Special Issue – May-June).

Rosenberg, N.J., Blad, B.L. and Verma, S.B. (1983). *Microclimate: The Biological Environment, 2nd Edition*. Wiley-Interscience, New York.

Rosenberg, N.J., Kimball, B.A., Martin, Ph. and Cooper, C.F. (1990). From climate and CO_2-enrichment to evapotranspiration. In: Waggoner, P.E. (ed.) *Climate Change and U.S. Water Resources*. John Wiley & Sons, New York.

Scudder, T. (1989). Conservation vs. development: river basin projects in Africa. *Environment,* **31**: 4-9; 27-32.

Sedjo, R.A. and Solomon, A.M. (1989). Climate and forests. In: Rosenberg, N.J., Easterling, W.E. III, Crosson, P.R. and Darmstadter, J. (eds.) *Greenhouse Warming: Abatement and Adaption*. Resources for the Future, Washington, DC.

Smith, J. B. and Tirpak, D. (eds.) (1989). The Potential Effects of Global Climate Change on the United States: Appendix C – Agriculture. EPA-230-05-89-053. U.S. Environmental Protection Agency, Office of Policy, Planning and Evaluation, Washington, DC.

Tolbert, N.E. and Zelitch, I. (1983). Carbon metabolism. In: Lemon, E.R. (ed.) *CO_2 and Plants: The Response of Plants to Rising Levels of Atmospheric Carbon Dioxide*. AAAS Selected Symposium 84. Westview Press, Boulder.

Tyson, S. (1990). *Biomass Resource Potential of the United States*. Solar Energy Research Institute, Golden.

U.S. Department of Agriculture. (1990). *Agricultural Statistics, 1990*. U.S. Government Printing Office, Washington, DC.

U.S. Department of Agriculture. (1991). *Economic Indicators of the Farm Sector: Production and Efficiency Statistics, 1989*. ECIFS 9-4. Resources and Technology Division, Economic Research Service, U.S. Department of Agriculture, Washington, DC.

U.S. Department of Energy. (1991). *State Energy Price and Expenditure Report 1989*. DOE/EIA-0376(89). U.S. Department of Energy, Washington, DC.

An End-Use Analysis of Global Food Requirements

William H. Bender
World Hunger Program, Brown University
Providence RI

Food Studies Group
University of Oxford
Oxford, U.K.

1 Introduction

In examining the long-term prospects for global food supply and demand, the balance of research has favoured the examination of global food supply, and not global food demand (Centre for Agricultural and Economic Development, 1967; Chou *et al.*, 1977; Council on Environmental Quality, 1980; Crosson and Anderson, 1992; FAO, 1993; Mitchell and Ingco, 1993; Bongaarts, 1994). Forecasting of global food demand has been limited to an exploration of alternative assumptions about the growth of incomes, prices and population. In particular, the analysis and forecasting of global food demand has used income and price elasticities estimated from historical data, which assume a fixed link between levels of income and levels of food demand. (For expository purposes, this discussion presumes a particular pattern of prices. With changing relative prices, the link between incomes and food demand is more complicated, but this does not change the fundamental argument being made here.)

This paper contributes to the development of long-term scenarios of the food and agriculture sector by challenging the assumption of an exogenous relationship between income and food demand. The technique of end-use analysis is employed to identify four sources of potential change in global food requirements:

- Changes in global calorie requirements resulting from improved nutritional status and changes in physical activity levels.
- Efficiency changes in the storage, handling, transportation and consumption of food which result in changes in the efficiency of the global food system.
- Changes that result from shifts in the structure of diets. Particular attention is focused on the implications for the global demand of starchy staples resulting from shifts in demand in lower

NATO ASI Series, Vol. I 37
Climate Change and World Food Security
Edited by Thomas E. Downing

income countries from a direct demand for starchy staples to demand for livestock and animal products. Increased demand for animal products increases the indirect or derived demand for starchy staples in the form of animal feed.

- Changes to dietary structures due to increasing knowledge and concerns about the health impacts of alternative diets.

This paper argues that there may be significant scope for policy to alter the relationship between income and food demand, and that investments in increasing agricultural supply must be compared with investments in reducing the growth rate of global food demand. A doubling of the world's population by the middle of the next century, along with rising levels of income and living standards, will put substantial pressure on the world agricultural system. The world is no longer land-abundant. The historical expansion of agricultural area slowed down in the 1980s, so most of the increases in future global food supplies must come from an intensification of agriculture and higher yields. However, the ecological side-effects of increasingly intensive agriculture can be substantial, and the sustainability of extremely high yields has not been verified (e.g. Pingali *et al.*, 1990; Bouis, 1990; Crosson and Anderson, 1995, this volume; Norse, 1995, this volume). Many of these side-effects from agriculture are externalities, and are not reflected in the private costs faced by agents in the agricultural sector. As investments in reducing global food demand mitigate the uncosted ecological side-effects of agricultural intensification, comparisons of social returns on investments in global food supply and demand may be significantly different from private returns on investment.

Until the 1970s, it was commonly assumed that the link between GNP on the one hand, and energy supply and environmental pollution on the other hand, was fixed. The constancy of both these links has been successfully challenged by the application of end-use analysis. In both cases, system efficiencies have increased significantly in the past decade in a number of countries, with the use of energy and environmental toxins produced per unit of GNP substantially declining. These increases in system efficiencies result from changes in technical efficiency, as well as from structural changes in the mix of activities. Many of these improvements in system efficiency resulted from end-use analysis, which focuses on the desired outputs of the system (e.g. light, transportation) rather than the components required to provide the outputs (e.g. power plants). This paper provides an end-use analysis of the global food system, and argues that the constancy of the link between income and food demand can be challenged.

The purpose of the food system is not agricultural production, but the satisfaction of nutritional requirements of the world's population, along with providing an enjoyable, healthy and palatable diet. By maintaining a focus on these end uses of the food system, we begin to examine the efficiency with which it satisfies the desired outcomes.

2 End-use analysis

The link between income and food demand has been well documented, and is based on: (i) an increase in expenditure on food with an increase in income; (ii) an increase in calorie consumption with an increase in expenditure on food; and (iii) a shift in diet composition away from income-inelastic goods (e.g. starchy staples) and towards income-elastic goods such as meat, resulting in (iv) a derived demand for feed.

This paper alters the traditional analysis of global food demand. It starts the analysis with global nutritional requirements, and then analyses the individual factors that affect global demand patterns. Most of these factors are correlated with income, resulting in the traditional link between demand and income. Not all these relationships are immutable and independent of policy and investment choices, however. Investments in increasing the agricultural supply (e.g. agricultural research stations) must also be compared with investments in altering global demand patterns. The parallel with the energy sector is again illuminating; investments in energy-saving equipment (e.g. efficient light bulbs) often have much higher returns than investments in energy supply (e.g. power plants), but a variety of incentive structures, policy choices, and the structure of capital markets historically favoured energy supply investments. Similar questions must be asked within the global food system; agricultural supply investments must be compared with agricultural demand investments, and the impact on the long-term structure of global entitlement patterns must be explicitly considered. Furthermore, many environmental side-effects of agricultural production are economic externalities, so that private decisions do not fully reflect social costs. Investments or policy changes that reduce global food demand requirements automatically reduce environmental externalities from agricultural production.

The analysis proceeds in several steps. Consistent with the end-use approach, the analysis focuses on calorie[1] consumption. Although micro-nutrients and other components of food can have large nutritional impacts, under- and over-consumption of calories dominate global nutritional status, and minor shifts in consumption patterns can address most other nutritional problems. Firstly, current global food consumption is estimated from current demographic structures, body weights and activity patterns, and forms the reference base of all subsequent analysis. Secondly, the alternative assumptions regarding nutritional status and activity patterns and their impacts on food requirements are considered. Nutritional status affects food requirements through child malnutrition and adult stature. Changes in activity patterns result from urbanisation, changes in labour force structure and rising incomes. Thirdly, the efficiency of the food system is examined. Finally, the global implications of dietary structure and research into the composition of healthy diets are considered. Results are summarised in Table 1 and Figure 1.

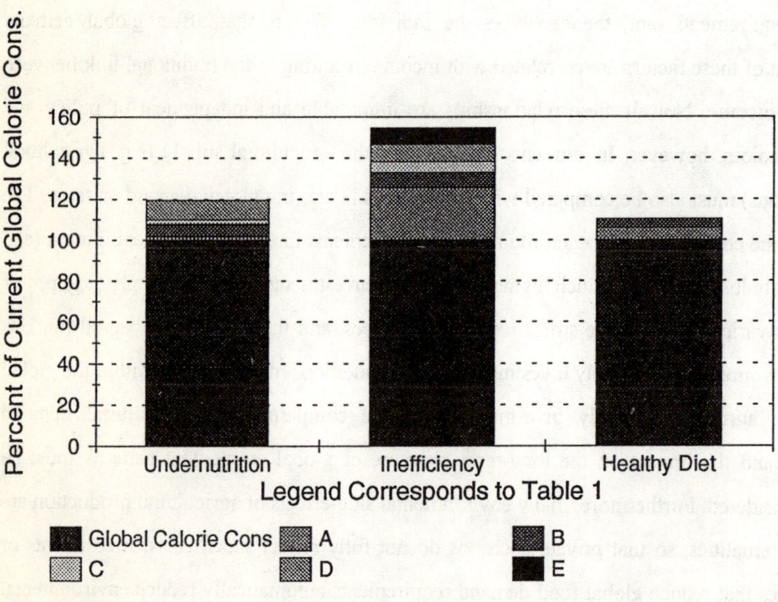

Figure 1 Global summary of end-use calculations, as a percentage of 1990 global calorie consumption.

---

[1] For convenience, the term 'calorie' is used throughout the text, although the actual unit of energy under discussion is a kilocalorie (kcal). This is still commonly used in international data sets rather than the metric (SI) unit, the Joule: 1 kcal = 4.184 Joules.

Table 1 Impact of alternative assumptions on global food requirements

Estimate	Percentage of 1990 calorie consumption
I. Nutritional status and activity patterns (first bar of Figure 1)	
1. Globally attain desired weight for actual height; status quo allowances and activity patterns (A in Figure 1)	1.7
2. No. 1, plus infection allowances and desired activity for children	2.6
3. No. 2, plus desirable activity allowances for adults	7.6
4. Status quo allowances and activity levels, no stunting (relevant only after a generational change) (A+B)	7.7
5. No. 4, with high allowances and activity levels for LDCs (A+B+C+D)	19.3
6. No. 4, with developed country allowances and activity levels globally (A+B+C)	10.0
II. End-use efficiency (second bar of Figure 1)	
7. Cereals marketing and distribution losses (E)	8.0
8. End-use food losses (A+B)	32.9
9. End-use food losses above best practice in developed countries (1.3 times consumption) (B)	7.4
10. Indirect end-use food losses through feed use (C+D)	13.7
11. Indirect end-use food losses above best practice (C)	5.1
12. Direct and indirect end-use food losses (no. 8 plus no. 10) (A+B+C+D)	46.6
13. Direct and indirect end-use food losses above best practice (no. 9 plus no. 11) (B+C)	12.5
III. Healthy diet (third bar of Figure 1)	
14. Reduce fat intake to maximum of 30% of calories from fat; all fat reduction in animal products (A+B)	10.2
15. Reduce fat intake to maximum of 30% of calories from fat; fat reduction proportional to existing fat sources (animal or vegetable) (A)	6.0
IV. Global transfers (for comparison)	
16. Global food aid	0.9

3 Estimating current global food consumption and requirements

It is possible to estimate and project global food requirements on a physiological basis. Following the methodology developed by the Food and Agriculture Organisation (FAO) and James and Schofield (1990), physiological requirements can be estimated in four steps. Firstly, the age and gender structure of the population must be specified. Secondly, the basal metabolic rate of individuals within each age group must be estimated. This is done using the mean heights and weights of each age and gender group, either from survey data, or from estimates using anecdotal data or data from similar countries. Thirdly, the physical activity patterns and resulting energy requirements must be specified. Finally, additional allowances must be made for pregnancy, infections, desirable weights and desirable activity levels. These physiological energy requirements can be converted into food requirements by making assumptions as to diets consumed.

Population is a crucial driving force affecting global food requirements. These requirements will undoubtedly increase as the population doubles in the next century. Population projections affect global food requirements in two ways. Firstly, the size of the population has a direct and obvious impact on food requirements. Secondly, the structure of the population is also important. A rapidly growing population will have a greater proportion of children, who require less food per capita than adults, and will therefore lower the national average per capita food requirement.

Although nutritional status can affect population projections directly through survival and indirectly through fertility decisions, these interactions are not considered here; this paper assumes that population growth is exogenous to the food demand system. For example, compare the food requirements of the Netherlands, with a stable population structure and minimal population growth, and Kenya, with a young population and nearly four per cent population growth. If both populations were perfectly healthy and had the same body heights and weights, and both populations undertook similar activities typical of developed countries, the Netherlands would still require 11.5 per cent more food per capita because of its older population structure. In estimating food consumption and requirements, single-year age groups for 0-17 and aggregated groups for ages 18-29, 30-59 and 60 and over are used, all of which are disaggregated by gender. For most of the calculations in this paper, 1985 population data are used, covering an estimated 92 per cent of the world's population. Besides countries with small populations, the largest missing segments of the global population are the countries of the former Soviet Union, which could create a small bias in the data presented.

Estimates of global food requirements are based on human energy requirements, which in turn are based on estimates of the basal metabolic rate (BMR) for specific population groups. The basal metabolic rate is primarily a function of body size, as measured by weight. However, a healthy weight is dependent on height. Height and weight of a population group depend on past and current levels of nutrition.

Both low and high food intakes can lead to malnutrition – the former typically in less developed countries, and the latter common in the developed countries. A continual low intake of food results in levels of body weight too low for optimal health, whereas excessive food intakes result in body weights too high for optimal health. Childhood malnutrition also affects aggregate food requirements through stunting, or low height for age. Stunting is the result of a complex interaction of food availability, health status and care practices, but unlike wasting is generally permanent. About three-fourths of growth retardation occurs by the age of five, and usually results in a reduction of adult stature. This contributes to a permanent reduction of food requirements throughout the remainder of an individual's life. Reducing childhood stunting will increase per capita food requirements within two to three decades. In a fast-growing population, about 40 per cent of the population and 36 per cent of the food requirements are from individuals aged under 18, and about 60 per cent of the population and 60 per cent of food requirements are from individuals under age 30.

It is well known that chronic energy deficiency and hunger are problems of access to food, rather than problems of food availability (see Downing *et al.*, 1995, this volume). It is nonetheless instructive to examine this proposition quantitatively. Lack of access to food has two main consequences: (i) a decrease in body weight, and (ii) a decrease in activity levels. These two responses together can lower food requirements until a new equilibrium is reached at a lower level of food intake. Data on the activity levels of the malnourished are scarce, but it is straightforward to estimate the change in food requirements necessary to ensure healthy weights for the global population. This can be done using estimates of body weights and heights in different countries. Globally, if all persons received a healthy caloric intake for their current height (as estimated from Frisancho, 1990), the additional required quantity of food would be only 1.7 per cent of current consumption (Table 1, line 1). This is composed of an increase of 2.6 per cent to eliminate low weights for height, but 0.9 per cent of global food consumption would be saved if overnutrition were simultaneously reduced. Developed countries have higher per capita caloric requirements due to their older age structures. Figure 2 shows that these countries all consume more calories than optimum for highest life expectancy (the diagonal line), whereas less developed countries consume below their desired caloric intake.

Similar calculations can be made to determine the impacts of stunting on global food requirements. Although it is theoretically possible to eliminate low weight globally in a few years, a full generational change is required for a reduction of stunting to affect aggregate requirements. Nonetheless, quantifying the impact of stunting on food requirements is useful for long-range scenario development. Hypothetical calculations can reveal the impact of stunting on global food requirements over a period of decades. If the global 1985 population were suddenly increased in height and weight to desirable standards, the global food requirement would increase by 7.7 per cent (Table 1, line 4). Figure 3 shows that stunting can reduce national level food requirements by more than 15 per cent in some countries.

Food is not distributed according to food requirements, however. Future analysis could follow the FAO methodology (FAO, 1985; Statistical Analysis Service, 1992) for estimating the numbers that are chronically energy-deficient. This makes an allowance for inequality in food intakes in order to estimate aggregate food requirements necessary to eliminate malnutrition. The FAO methodology uses a mean national intake and an assumed food intake distribution to calculate the number below a minimum cut-off value representing the minimum household per capita energy need. Food intake distributions are still poorly documented, but the evidence is quickly improving. Recent analysis suggests that the coefficient of variation (CV) of food intakes ranges between a minimum of 0.20 and a maximum of 0.35 (Statistical Analysis Service, 1992, pp. 6, 20). By calculating a cut-off value and assuming a level of inequality, the increase in global food requirements to eliminate chronic energy deficiency can be calculated.

However, an analysis of inequality is bounded by estimates contained here as maximum food intake is physiologically constrained. Measured inequality in household per capita energy intakes can be a result of four factors: (i) inequality in body weights above and below mean population weights; (ii) variation in individual metabolic rates or activity levels; (iii) food waste

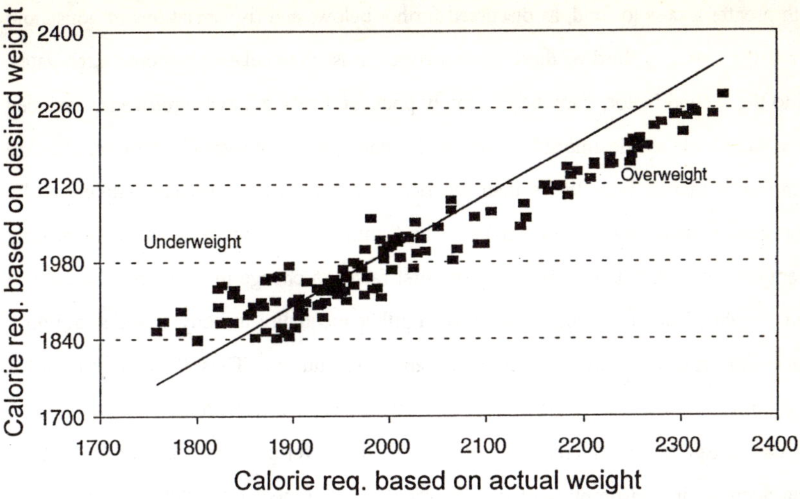

Figure 2 Total calorie requirements: desired weight for height versus actual weight.

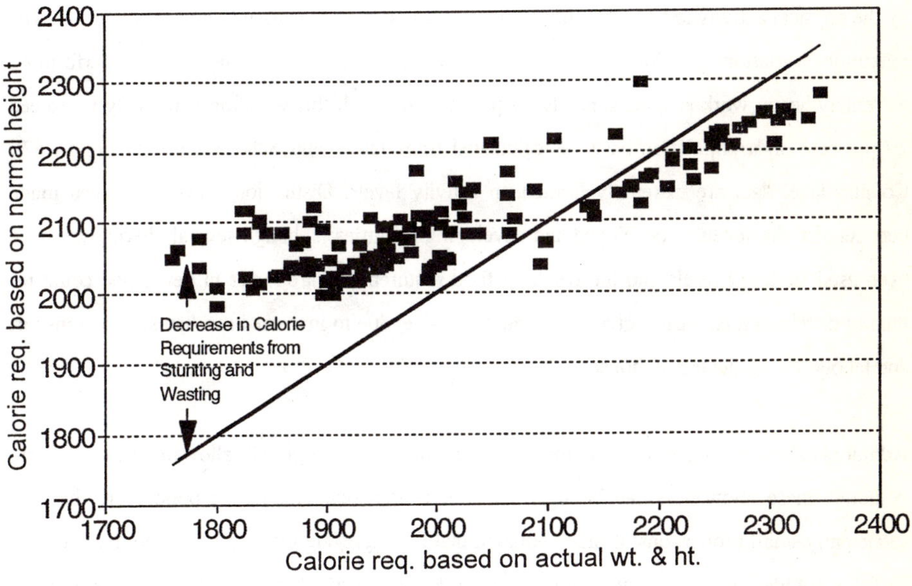

Figure 3 Total calorie requirements: desired weight without stunting versus actual weight and height.

by those with greater access to food, as discussed further below; and (iv) variations in household structure. Only the first and third of these factors concern us in calculating national aggregates. With regard to the second factor, only the extremely poor are likely to have significantly reduced activity levels. Activity levels decline at high levels of income, so the moderately poor are likely to have the same or higher activity levels than the rich. Thus, activity levels are unlikely to be important sources of inequality in food intakes. The fourth factor is unimportant as household variation averages out in national aggregates. The first factor, a change in mean body weight to reach desired weights, has already been discussed. Further inequality in body weights around a desired mean weight is symptomatic of both under- and over-nutrition. Thus, the chief reason for adding an inequality allowance is that the levels of waste are correlated with income (see below). With a constant mean but an increase in inequality, the level of waste is higher, with a compensating decrease in consumption of the poorest. Thus, the inequality allowance is a proxy for a lack of entitlements[2] for the poor, and end-use waste of the rich.

A number of additional factors affect total food requirements. Basal metabolic rates are multiplied by the physical activity level to determine total requirements for a particular population group. The maximum variation in caloric requirements from performing very light work to performing extremely heavy work is approximately 35 per cent, although this variation is unlikely to exceed 20 per cent on a population basis, as calculated from James and Schofield (1990, pp. 54-57). Country-level data are extremely limited on activity levels. Distinctions have only been made between activity levels in developed and developing countries, although several alternatives have been used. Activity levels and the resulting food requirements are lower in developed countries than in developing countries, given the same body size, due to increased mechanisation, transport and labour-saving devices in households.

Additional allowances can be made for pregnancy, infection and growth allowances for children. Although these allowances can be significant for certain segments of the population, they are sufficiently small proportions of the population that the aggregate affect is moderate. Estimate I.2 in Table 1 adds infection allowances for children and has a small impact. (A further food loss is due to morbidity and parasites in all age groups. In this paper, there is no attempt to estimate this loss on a global basis.) Increasing activity levels for adults has a larger impact, as seen in

---

2 Entitlements, a term popularised by Sen, refers to the collection of rights, whether legal, political, economic or social, which allow access to food (Sen, 1976, 1981).

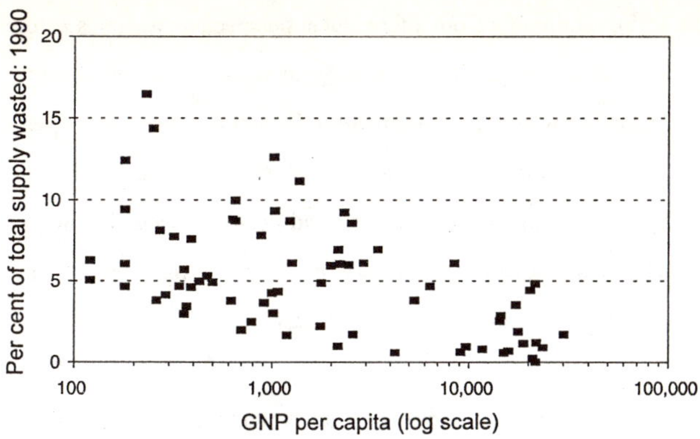

Figure 4 Storage and distribution loss for cereals, in countries with a total supply of more than one million metric tons.

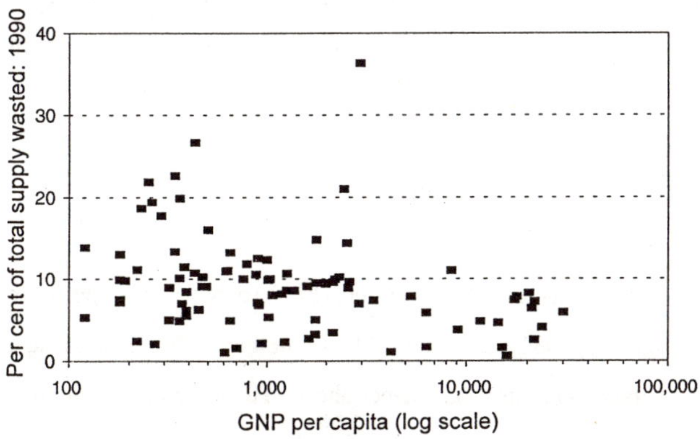

Figure 5 Storage and distribution loss for roots and tubers.

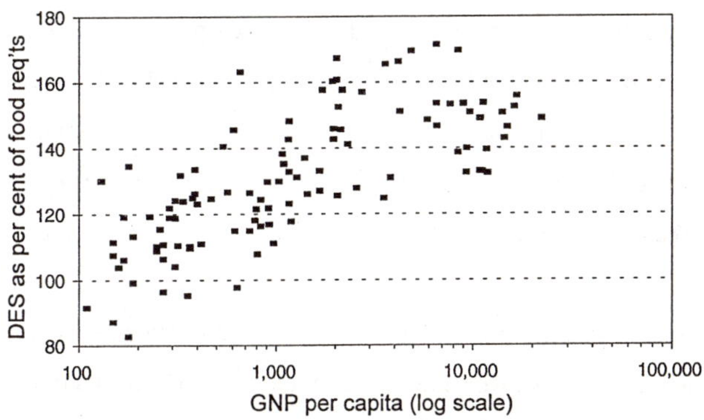

Figure 6 Food system efficiency as a function of GNP per capita.

estimate I. 3. The highest estimate (I. 5), useful only from a theoretical perspective, assumes a new generation that is no longer stunted, but still requires higher allowances in countries currently considered as less developed. The higher allowances are for greater infection rates in children, and for higher activity levels due to a greater amount of manual labour. Estimate I. 6 assumes that the entire global population is (i) not stunted, (ii) maintains a healthy and desirable weight, and (iii) has health status and activity levels comparable with current developed country patterns. This set of assumptions increases global caloric requirements by only ten per cent.

These calculations starkly confirm that the current pattern of malnutrition and hunger is unrelated to food availability, but is instead a function of global entitlements to food. Indeed, the quantity of food required to eliminate underweight status of children and adults is equivalent to the increment required for one year of world population growth, and is insignificant relative to long-term changes in food requirements resulting from population growth, dietary change and food system efficiency.

4 Food system efficiency

This paper examines the efficiency of the post-harvest food system. Pre-harvest losses to pests are estimated to be about 35 per cent (Pimental, 1981). Estimates of losses in the procurement, storage and distribution of food are made by the FAO in its food balance sheets. These are little more than educated guesses, and their accuracy is quite low. Figures 4 and 5 show the percentage of total supply lost in marketing and distribution plotted against GNP per capita. It is not surprising that the degree of loss is correlated with income. Although this form of loss is vitally important in many poorer countries, it is unlikely to be a source of increased global efficiency (Table 1, line 7). If all countries reduced losses to only four per cent of supply, comparable with developed country losses for cereals and with the best developed countries for roots and tubers, the total savings would amount to 1.65 per cent and 5.45 per cent of the global food supply, respectively. Furthermore, these losses will decrease as incomes increase.

On a global basis, one-fourth of food entering the institutional and household distribution system is lost. (This is comparable with Table 1, line 8, but with a denominator of total food demand rather than total consumption.) The end-use efficiency of the food system, as shown in Figure 6, is physiological food requirements compared with nationally-calculated dietary energy supplies

(DES). Unlike most economic goods, the physical quantities of food consumption are physiologically constrained. In a healthy population, the only adaptation to increased levels of food intake are either higher physical activity levels, or increased weight. Empirically, higher physical activity levels are not associated with greater food intake, and in general there is an inverse relationship due to the connection of lower physical activity and higher intakes with higher incomes. Adaptation through increased weight can have only modest effects relative to the inter-country variations in available DES. For example, an increase in intake of 100 calories per day (a modest piece of chocolate and about five per cent of a typical caloric intake), would result in a steady-state increase of four to five kilograms in about one year.[3]

However, estimates of mean body weights are controlled for in country-level estimates of requirements. Therefore, despite some inaccuracies and controversies in the methodologies and data sources, large discrepancies between the dietary energy supply available to households and institutions, as calculated from food balance sheets, and the energy requirements of the population as calculated from body weights, heights and activity levels, must primarily be a result of food losses and waste resulting from food processing activities, spoilage, food preparation waste in households, restaurants and institutions, plate waste, non-food use, or data errors. Not unexpectedly, the level of waste is closely correlated to levels of income, with little end-use food waste at low levels of income, but with 30-60 per cent of food requirements lost in high income countries, where data accuracy is also the highest. With the exception of Belgium, however, the highest levels of loss, equivalent to 60-70 per cent of total food consumption, are found in middle-income countries, including Greece, Ireland, ex-Yugoslavia, Hungary, Bulgaria, Egypt and Libya. In addition to estimates of direct end-use losses (Table 1, line 8), it is possible to calculate the indirect end-use loss if some portion of the loss is animal products, which used feed in their production (Table 1, line 10). There is insufficient evidence to determine what type of food is wasted. This could bias both these estimates and those in the following section. For example,

[3] Multiplying the weight coefficient of the Schofield equations, which estimate BMR from weight in kilograms, by the physical activity level (PAL) coefficient, provides an estimate of the increased calories required for an additional one kilogram of body weight. For adults, the weight coefficient ranges from 8.7 for females aged 30-59, to 15.3 for males aged 18-29. Typical PAL coefficients range from 1.6 to 1.86, implying that an increase of one kilogram in body weight requires a steady-state increase of 14-28 calories in intake.

trimming fat from meat can result in a large caloric waste, but affects fat intakes as discussed in the next section.[4]

Analyses of the sources of this inefficiency have been limited, so it is unclear to what extent increased efficiency is economically feasible. An inter-country comparison could provide useful insights, as countries such as the Netherlands, Finland, Japan and Sweden, at comparable levels of income, produce only about 60 per cent of the loss, on a caloric basis, of countries such as the United States, Belgium, Switzerland and Italy. A reduction in food losses from 50 per cent of requirements to 35 per cent of requirements would directly result in a minimum of a ten per cent decrease in global food supply requirements. This reduction would be greater than ten per cent if a significant portion of the loss included animal products, which would result in a greater than ten per cent reduction in agricultural supply requirements due to the derived demand for feed.

Given the current distribution of food consumption and efficiency, 7.4 per cent (Table 1, line 9) of global food consumption would be saved by reducing the maximum level of loss in any country to 30 per cent of requirements, equivalent to best practice in high income countries. Including indirect losses raises this figure to 12.5 per cent of global food consumption (Table 1, line 13). However, the level of end-use efficiency becomes increasingly important as incomes increase and a greater population has high levels of end-use inefficiency. Despite the minimal examination of these levels of efficiency, it is clear that the social benefits of increased efficiency outweigh private benefits due to the externalities inherent in agricultural production. As with the energy sector, further analysis may determine whether there are policy biases that favour increased production over increased efficiency.

5 Dietary structure

Relationships between the standard of living and diet are predictable and strong. Three main trends occur as incomes increase: firstly, the proportion of income spent on food decreases; secondly, the importance to the diet of starchy staples (cereals, roots and tubers) steadily declines as incomes increase (see Figure 7); and thirdly, the importance of 'luxury' food items steadily increases as

4 For these calculations, it is assumed that the waste of animal products is proportional to their composition in the diet. These calculations also do not take account of trade in processed animal products.

income increases, replacing the caloric value of the starchy staples. This group includes oils and sweeteners, although we are primarily concerned with animal products due to the resulting derived demand for additional cereals in the form of animal feed. Figure 8 shows the current structure of the world's diet, on a caloric basis. The comparison between the structure of the diet in high income developed countries in Europe and the diet in the developing countries of Africa is indicative. Starchy staples comprise 67 per cent of the diet in Africa, but only 31 per cent of the diet in Europe. On the other hand, consumption of animal products increases from six per cent in Africa to 33 per cent in Europe.

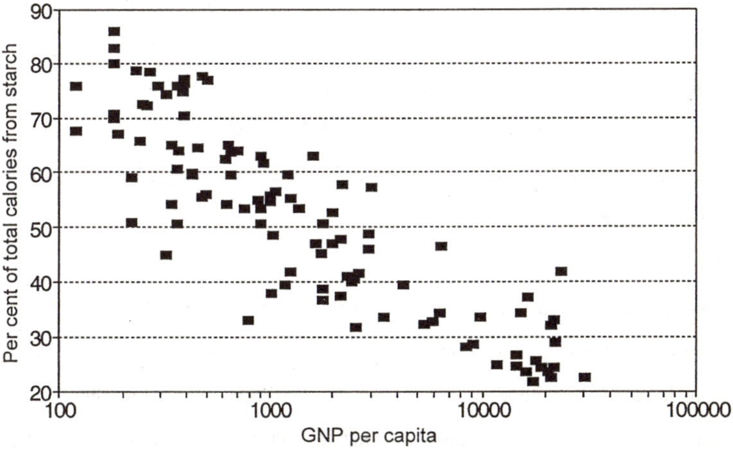

Figure 7 Caloric share of starchy staples in national diets.

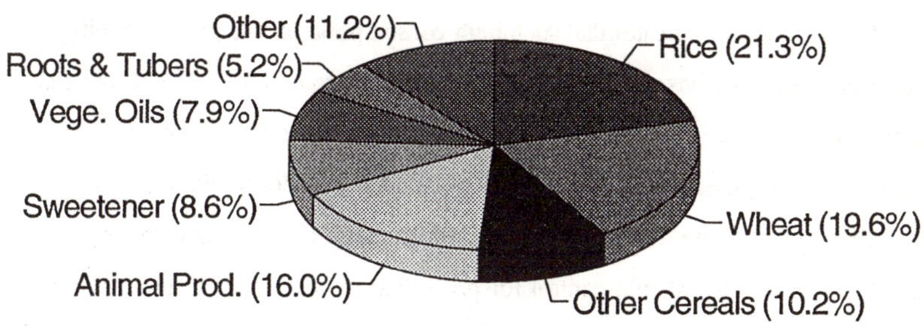

Figure 8 Sources of world calorie consumption: percentage of total.

Thus, a major driving force in increased demand for agricultural production is a switch to increased consumption of animal products. Feeding ratios, the average quantity of cereals and by-products used to produce one unit of livestock output, vary from 0.9 in sub-Saharan Africa to 3.8 in North Africa and the Middle East.[5]

Globally, three-fourths of the energy in livestock feed is provided by forages, most of which is produced on land unsuitable for agriculture, and 6.2 per cent is from by-products (Wheeler *et al.*, 1981). Although there is certainly scope for the improved management of the world's grazing land, substantial increases in animal products will be associated with higher feeding ratios due to the relatively fixed quantity of land suitable for grazing or forage crops. U.S. feeding ratios in 1990 for intensively fed animals were 7.0 for beef (e.g. 7 kilograms of feed for 1 kilogram of beef), 6.5 for pig meat, and 2.7 for poultry (US Congress, 1992). However, these feeding ratios understate the caloric inefficiency of feed conversion, as the caloric density of grain is about 35 per cent higher than that of meat.

The link between income and meat consumption is very strong, and has strong evolutionary roots. (For a variety of viewpoints on the history, evolution and importance of animal products in human diets, see Harris and Ross, 1987; Beitz and Hansen, 1982; Rifkin, 1992; Jensen, 1953.) Nonetheless, satiation on a caloric basis has been reached in total meat consumption in most developed countries, but there are wide differences in per capita consumption of countries at comparable levels of income (see Figure 9). Sweden has only 60 per cent of the meat consumption of the USA and UK, and Japan still consumes less than the world average consumption.

Associated with the increased consumption of animal products is an increase in the consumption of animal fats. Recent research indicates that such high consumption of fats is not a healthy diet, and current recommendations advise limiting dietary intake of fat to no more than 30 per cent of calorie consumption. Of the current world population, 16.8 per cent live in countries already exceeding this average in aggregate (Figure 10). There is already evidence that the mix of meat consumption is changing in some developed countries in response to these concerns (see Figure 11), although additional analysis should control for price. Part of the expected improvements in

5 A livestock output unit, as defined by Sarma (1986), is the sum of the weights of meat (carcass weight) and eggs, plus the weight of milk divided by ten, to reduce to a dry-weight equivalent.

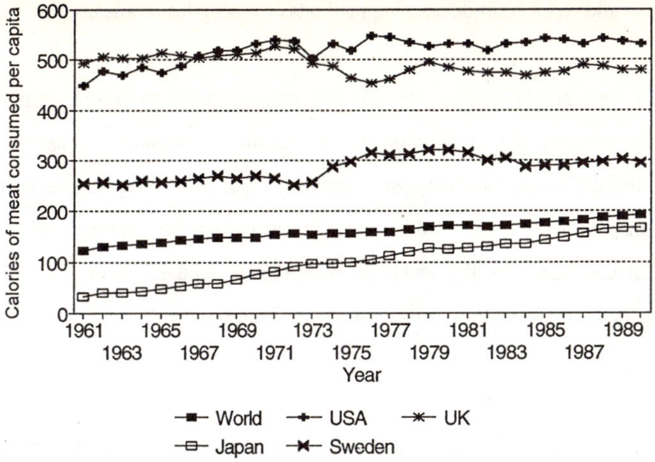

Figure 9 Per capita meat consumption, 1961-90.

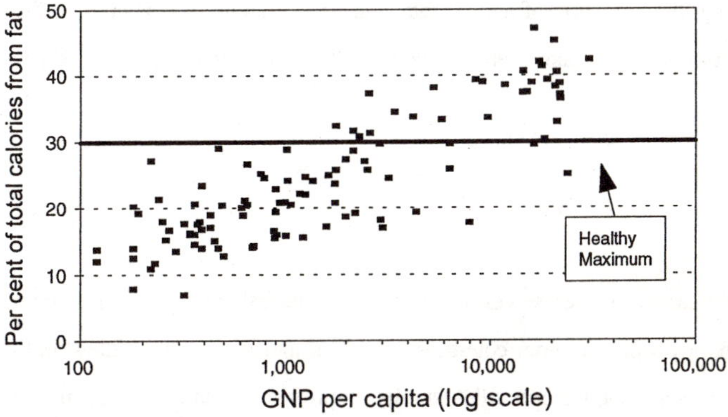

Figure 10 Percentage of calorie consumption from fat, 1989.

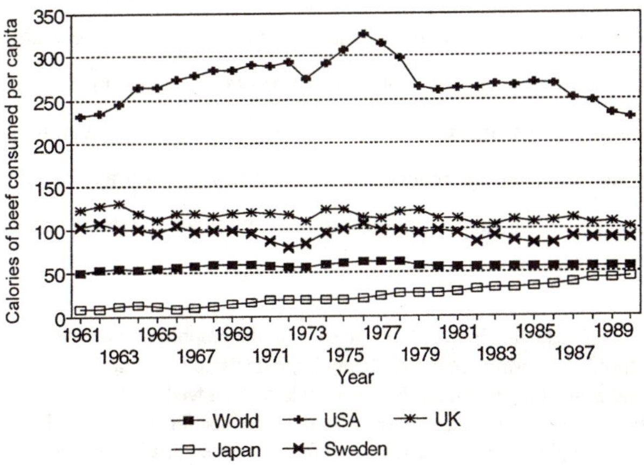

Figure 11 Per capita beef consumption, 1961-90.

feed conversion efficiencies would result from developing methods of raising leaner meat.

If health campaigns on the consumption of animal fats can have impacts similar to those in many countries focusing on cigarette smoking, there is substantial scope for the reduction of feed used in the production of these fats. Indeed, one of the most cost-effective ways of increasing the effective global agricultural supply may be to invest in health campaigns to reduce the consumption of animal products in countries exceeding healthy levels of consumption, and to create a limit, through health awareness, on animal product consumption in countries with rapidly rising incomes.

Section III of Table 1 summarises the possible results of such an investment. Line 14 shows the savings in feed consumption possible if countries above a 30 per cent intake of fat reduced their fat consumption entirely through the reduction of animal fats, and line 15 shows the savings in feed consumption if this reduction in fat intake were made in both animal and vegetable fats, in proportion to current consumption.[6]

6 Regional impact

Improvements in efficiency cannot in themselves solve the major global food problem, namely lack of access to food by poor people and poor countries. Levels of inefficiency are highest in the developed countries of North America, Europe and Australia; improvements in efficiency in these countries would lead to decreased demand, but would do nothing to increase demand in the poorer countries of Africa, Asia, and Central and South America. However, import demand is increasing rapidly in many middle-income countries, particularly in west Asia, north Africa, and some south-east Asian countries. This analysis simply illustrates that there is substantial slack in existing agricultural production and consumption systems that is an alternative source of global food supply if environmental constraints substantially reduce the possibilities or desirability of continued production growth. For countries whose populations are heavily reliant on agriculture as

---

[6] The calculation assumes a proportionate reduction in total feed use equal to the proportion with which total animal fat consumption is reduced. A substitution between animal products to lower fat products could change this calculation, but fat content of animal products is correlated with feed conversion efficiency, so differences are likely to be minor. The calculations also do not adjust for the increased consumption of other foods required to replace the calories in animal fats. However, this is very roughly balanced by the savings in non-cereal feeds, which are also not included in these calculations.

a source of income, particularly in Africa and south Asia, efficiency gains cannot be a source of increased access to food.

7 Conclusion

End-use analysis is effective in raising many questions about the efficiency of the current food system. It confirms the well-known fact that undernutrition results from entitlement patterns, but the absolute quantities of food involved are nonetheless surprisingly minuscule relative to global food consumption. Savings from a reduction in marketing and distribution waste are small, and will naturally accrue as incomes increase. However, end-use food losses are very large in global terms, and will increase rapidly as levels of income increase. These losses will absorb a significant share of increased per capita food production. There are large variations in the level of efficiency among countries at similar income levels, so there is ample scope for the analysis of the determinants of this food system efficiency.

Roughly one-sixth of the world's population lives in countries with low food system efficiency, and unhealthy and high levels of caloric consumption and animal fat consumption. In some countries, aggregate food use could conceivably be reduced by nearly 50 per cent through the promotion of healthy diets, and a systematic attempt to increase end-use efficiency to best practice levels. We have very little knowledge of the reasons for these losses, or of the return on investments in reducing them. Further analysis is required to determine what policies influence the levels of end-use efficiency, and what improvements in this efficiency are feasible. In these countries, per capita food use could significantly decrease. In many of these countries, the existence of surpluses, farmer subsidies and supply management measures make it unlikely in the short run that increases in end-use efficiency will gain political support. However, this analysis does suggest that there is substantial scope for improved efficiency if global production constraints become binding due to increased demand or environmental constraints.

Sixty-two per cent of the world's population already live in countries with over 30 per cent food losses – slightly better than best practice in the developed countries – so future increases in per capita food caloric availability will be entirely increases in inefficiency. Many of these countries are predominately middle-income, and the consumption of animal products will increase with higher incomes, yet still be within recommended dietary guidelines. Some of these countries are approaching recommended maximums, and animal product consumption should be stabilised on health grounds. As a group, these countries should be able to maintain per capita food use as populations increase, and many may be able to reduce per capita consumption if significant end-use efficiency improvements prove feasible.

The remaining 38 per cent of the world's population are in countries where there would be increases in caloric consumption, decreases in system efficiency and increases in consumption of animal products, if incomes increase. Per capita food use will increase in these countries with incomes, although, unfortunately, income growth prospects are poor in a number of these countries.

Although these results are only indicative, they nonetheless raise important questions about the structure of agricultural supply and demand. Demand reductions must be considered as possible solutions to the environmental side-effects of agriculture. As an example, any reduction in food throughput reduces the potentially damaging nutrient flows from agricultural land to a diffuse nutrient sink. There is undoubtedly scope to improve food system efficiency, although substantial work is required to document the sources of inefficiency, and the policy biases that discourage efficiency.

Acknowledgements

Funding for this work was provided by the 2050 Project, a collaborative project of the World Resources Institute, Brookings Institution, and the Santa Fe Institute examining the global transition to sustainability. I wish to thank Robert Chen, formerly of the World Hunger Program at Brown University, and currently at CIESIN, who conceptually originated the application of end-use analysis to the global food system, and encouraged this work. I also wish to thank Günther Fischer, Jock Anderson and John Reilly for comments; remaining errors are mine.

This paper is a revised version that first appeared in *Food Policy* Volume 19, An end use analysis of global food requirements. *Food Policy* 19 (4), pp 1-15 (1994). Reprinted with permission from Elsevier Science Ltd, Butterworth-Heinemann Imprint, The Boulevard, Landford Lane, Kidlington OX5 1GB, U.K.

References

Beitz, D.C. and Hansen, R.G. (eds.) (1982). *Animal Products in Human Nutrition.* Academic Press, New York.

Bongaarts, J. (1994). Can the growing human population feed itself? *Scientific American,* March, pp.36-42.

Bouis, H.E. (1990). *Measuring the Sources of Growth in Rice Yields: Are Growth Rates Declining in South and South-east Asia?.* International Food Policy Research Institute, Washington, DC. (Mimeo).

Center for Agricultural and Economic Development (1967). *Alternatives for Balancing World Food Production and Needs.* Iowa State University Press, Ames, Iowa.

Chou, M., Harmon, D.P., Jr., Kahn, H. and Wittwer, S.H. (1977). *World Food Prospects and Agricultural Potential.* Praeger Publishers, New York.

Council on Environmental Quality (1980). *The Global 2000 Report to the President.* Council on Environmental Quality and the Department of State, Washington, DC.

Crosson, P. and Anderson, J.R. (1992). *Resources and Global Food Prospects: Supply and Demand for Cereals to 2030.* World Bank Technical Paper 184. World Bank, Washington, DC.

Crosson, P. and Anderson, J.R. (1995). Demand and supply: trends in global agriculture. In: Downing, T.E. (ed.) *Climate Change and World Food Security.* Springer-Verlag, Heidelberg.

Downing, T.E., Watts, M.J. and Bohle, H.G. (1995). Climate change and world food security: toward a sociology and geography of vulnerability. In: Downing, T.E. (ed.) *Climate Change and World Food Security.* Springer-Verlag, Heidelberg.

FAO (Food and Agriculture Organization of the United Nations) (1985). *The Fifth World Food Survey.* FAO, Rome.

FAO (Food and Agriculture Organization of the United Nations) (1993). *Agriculture: Towards 2010.* C93/24. FAO, Rome.

Frisancho, A.R. (1990). *Anthropometric Standards for the Assessment of Growth and Nutritional Status.* University of Michigan Press, Ann Arbor.

Harris, M. and Ross, E.B. (eds.) (1987). *Food and Evolution: Toward a Theory of Human Food Habits.* Temple University Press, Philadelphia.

James, W.P.T. and Schofield, E.C. (1990). *Human Energy Requirements.* Oxford University Press, Oxford, by arrangement with the Food and Agriculture Organization of the United Nations.

Jensen, L.B. (1953). *Man's Foods: Nutrition and Environments in Food Gathering Times and Food Producing Times.* Garrard Press, Champaign, Illinois.

Mitchell, D.O. and Ingco, M.D. (1993). *The World Food Outlook.* International Economics Department of the World Bank, Washington, DC.

Norse, D. (1995). The conjunction of threats to regional food production: how serious are environment, economy, population and climate? In: Downing, T.E. (ed.) *Climate Change and World Food Security.* Springer-Verlag, Heidelberg.

Pimental, D. (ed.) (1981). *CRC Handbook of Management in Agriculture.* Vol. 1. CRC Press, Boca Raton, Florida.

Pingali, P.L., Moya, P.F. and Velasco, L.E. (1990). Prospects for rice yield improvement in the post-green revolution Philippines. *Philippine Review of Economics and Business,* **28** (1): 85-106.

Rifkin, J. (1992). *Beyond Beef.* Dutton, New York.

Sarma, L.S. (1986). *Cereal Feed Use in the Third World: Past Trends and Projections to 2000.* International Food Policy Research Institute, Washington, DC. p. 18.

Sen, A.K. (1981a). Famines as failures of exchange entitlements. *Economic and Political Weekly*, **11**.

Sen, A.K. (1981b). *Poverty and Famines*. Clarendon, Oxford.

Statistical Analysis Service (1992). *World Food Supplies and Prevalence of Chronic Undernutrition in Developing Regions as Assessed in 1992*. Statistics Division, Economic and Social Policy Department, FAO, Rome.

U.S. Congress (1992). *A New Technological Era for American Agriculture*. Office of Technology Assessment, OTA-F-474, US Government Printing Office, Washington, DC. p.9.

Wheeler, R.O. *et al.* (1981). *The World Livestock Product, Feedstuff and Food Grain System.* Winrock International, Morrilton, Arkansas.

Policy Lessons from Communities under Pressure

Graham Eele
Food Security Training and Administration Project
Southern Africa Development Community
PO Box 4046
Harare, Zimbabwe

1 Introduction

People living in rural communities throughout the developing world face an uncertain and risky environment. The sources of risk include: economic change and especially policy instability; climatic shocks – both sudden events such as floods and cyclones as well as longer-acting phenomena, especially drought; and more specific incidents such as illness and disablement. In order to survive, people have evolved a number of strategies and mechanisms to deal with these exogenous shocks and to reduce the risk to their survival. Although these strategies and mechanisms have been the subject of study and analysis at the micro-level for many years, especially by social scientists, little attention seems to have been given to their implications for policy at the national and international level. This paper will attempt to demonstrate that rural communities manage risk as part of a continuous process. They develop strategies for production and consumption that help to smooth income flows and that are designed to ensure long-term survival. They adapt, to a greater or lesser extent, to the risks they face on a continuing basis and are aware of the fact that good years are likely to be followed by bad. At the micro-level, the focus is on the protection of livelihoods, of which food consumption is just a part.

To a large extent, however, governments and their international donor partners do not take the same attitude. Although some attention is given to the principle of being prepared for disaster, this is mostly concerned with logistics, and few developing countries seem to take a long-term policy view of risk. Climatic shocks such as drought are seen as unexpected events and are dealt with as emergencies. There has been some success in recent years in helping to avert starvation and death, except in the special cases of war and civil strife, but it is not at all clear whether the long-term capacity of poor countries to manage risk has been increased. At both the macro- and the micro-levels, there is little interaction between emergency relief programmes and longer-term

NATO ASI Series, Vol. I 37
Climate Change and World Food Security
Edited by Thomas E. Downing

development policy. In the short term, the focus is almost exclusively on food security and the protection of levels of food consumption.

Climate change will modify the risks that rural people face. For some, the changes may be beneficial in that the risk of a collapse in their livelihood through climatic shock is reduced. For others, however, climate change may well mean both an overall decline in incomes, through reduced agricultural production, and an increased risk of livelihood failure through climatic shocks, both sudden and more long-lasting. The main concern for policy is where people are unable to adapt to this change – for example, because it happens too quickly for them to adapt satisfactorily, or because they are constrained by poverty and other factors.

This paper is divided into four main sections. The next section provides a framework for analysis based on the concept of exchange entitlements, but extended to include a dynamic element. Section 3 examines the strategies that are adopted by communities to deal with external threats to their livelihoods. Much of the evidence is derived from micro-level studies in sub-Saharan Africa, although some reference is made to other parts of the world. Section 4 looks at the sources of risk facing communities and to what extent these are changing as a result of climate change. In particular, this section focuses on the interaction between types of risk and the extent to which macro-policy can mitigate or exacerbate the effects of climate change. The final section examines the policy lessons from this micro-level analysis. There are three main implications: first, the need for medium- and long-term policy to take into account the likelihood of exogenous climatic shocks and for a much closer integration of relief and development policy; second, the need to move the focus of relief efforts away from food exclusively, towards the protection of livelihoods; and third, the need for policy to support and strengthen micro-level coping strategies.

2 Vulnerability and risk: a framework for analysis

This paper concentrates on the way in which rural communities react to climatic shocks and the policy lessons that can be learnt from this process (see also Downing *et al.*, 1995, this volume). Variation in climatic conditions can lead to two different kinds of threat to rural people. In the first place, unexpected events such as drought or floods can lead to a collapse in livelihoods and, in some circumstances, may lead to starvation and famine. Second, most rural people are subject to some process of seasonality and the phenomenon of regular hungry seasons occurs in many different societies (Chambers *et al.*, 1981). Much of the analysis has been undertaken in the context of transitory climatic shocks that may lead to famine, but the problem of regular hunger and the process whereby repeated shocks can lead to long-term impoverishment are also important. In this paper, we will attempt to bring these ideas together within a framework that concentrates on the sustainability of livelihoods.

Although different communities may face the same risk, in the sense of the likelihood of a climatic shock such as a drought, they will not necessarily be equally vulnerable. It is not the shock itself that causes starvation, destitution and death, it is the combination of the external event and the vulnerability of the people. In a community where people have substantial reserves, where markets work effectively and where external support can be easily mobilised, even a severe drought may have very limited impact. Indeed, this limited impact may well be managed through the operations of capital markets with little or no government intervention. In communities where most people are chronically poor, where there are few reserves and where capital markets are not well developed, which is the case in most rural areas in the developing world, the situation is very different. If external resources, especially relief programmes, cannot be mobilised, then there is a very real possibility of destitution, starvation and death.

The focus of this analysis, then, is on the extent to which people are vulnerable to exogenous shocks. Greatest attention is needed in those places where the probability of occurrence is high or increasing and where vulnerability is greatest. The concept of exchange entitlements developed by Sen (1981) provides an elegant way of expressing these ideas. A household[1] can be considered to be endowed with a set of entitlements including: physical resources, for example, land and capital; human capital and skills; the labour of its members; and access to government or private social security. These entitlements can be exchanged for food at differing exchange rates. The bundle of all possible exchange entitlements available to a household represents the extent to which it can meet its food consumption needs. If there is no possible exchange entitlement that is sufficient to meet the basic nutritional needs of the household, then its members are at risk of starvation.

---

1 In this context, a household is considered to be the basic economic and social unit in which people live, although definitions will vary from one community to another. For convenience, the household is used as the unit of analysis in this paper. It is recognised that this treatment does not capture issues relating to intra-household distribution of food and other resources, nor the vulnerability of specific individuals.

Exogenous shocks can reduce exchange entitlements in two ways. They can reduce the actual set of endowments, for example, by reducing the productive capacity of land or forms of capital such as livestock, or they can reduce the amount of food that can be obtained, for example, by reducing crop yields or by increasing food prices in the market. In practice, vulnerability is highest when these kinds of risk are correlated. For example, when agricultural income is reduced, livestock numbers are reduced, food prices rise and wage rates fall. If the shock is widespread, then entitlements to support from other members of the community will also be reduced.

The exchange entitlement concept provided an important advance in thinking about famines and the impact of climatic shocks. In particular, it focused attention on the fact that 'Starvation is the characteristic of some people not having enough food to eat. It is not the characteristic of there being not enough food to eat' (Sen, 1981). Analysis of the causes and response to famine moved from emphasis on food availability decline (FAD) to giving much greater attention to food entitlement decline (FED). Greater attention was given to the need to devise interventions that helped to protect the entitlements of different population groups affected by exogenous shocks, rather than simply meeting shortfalls in food availability. In particular, the work by Sen led to a much greater emphasis on food security. This term has been defined in many different ways, but a commonly accepted and widely used definition is that proposed by Reutlinger and Selowsky (1975), which focuses on the access by households at all times to enough food to ensure an active and healthy life for all its members.

Although the FED explanation of famines has been widely accepted, some authors have questioned whether it is adequate in all circumstances (for example, Devereux, 1988; Kula, 1988). More generally, the entitlement concept also has some drawbacks as a framework for analysing the impact of exogenous shocks on people (de Waal, 1991). In the first place, it assumes a primacy for food consumption. The original formulation, which focused on the extreme case of starvation and famine, assumed that food is the basic need above all other items and indeed specifically ruled out the possibility of voluntary rationing and hunger.[2] In practice, as is illustrated below, there are many instances of households undergoing voluntary rationing of food consumption and even accepting the possibility of malnutrition because of their desire to protect

[2] In later publications, Sen acknowledged the possibility of voluntary food rationing, both in the situation of less severe food crises and in the context of famine (Drèze and Sen, 1989).

other items of expenditure. Second, the analysis presented by Sen was static; it assumes that food consumption now has no impact on production or consumption in the future. This is clearly not the case; many rural households could continue to meet their food needs in full by selling productive assets such as land, tools and draught animals. In doing so, however, they would jeopardise their future food security and the long-term viability of the household.

What is needed, therefore, is an extension of the exchange entitlement concept to include both present and future consumption. At the same time, the expenditures incurred by households in addition to acquiring food need to be taken into account. Clearly, inadequate food consumption over an extended period of time will lead to illness and eventually death for individuals. The problem is that the definition of what is inadequate varies with individuals, over time, and as the environment changes. In practice, households appear to do everything possible to maintain their future productive base and their future viability. It is necessary, therefore, to average consumption over an extended period of time – certainly one agricultural season and generally more in areas where droughts or other shocks are not uncommon. Households most at risk are those whose exchange entitlements are insufficient to meet their consumption needs (defined to include food and other basic expenditure) on average over a period of several years. Although food aid may be an appropriate way of providing assistance to such households, its effects need to be evaluated against both future as well as present consumption.

3 Response by rural households to climatic shocks

Several recent studies have documented the ways in which rural communities develop and adopt strategies to deal with climatic shocks. (For examples in Africa, see Corbett, 1988; Devereux, 1992; de Waal, 1989; Young and Jaspers, 1992. For examples from Asia, see Jodha, 1981; Torry, 1988; Drèze, 1988. A more detailed review is in Drèze and Sen, 1989, pp. 71-81.) Corbett (1988), in a review of a number of African studies, identifies three main types of coping strategies: insurance mechanisms that may be adopted in anticipation of a food crisis or in the early stages, but which generally do not threaten the long-term productive capacity of the household; the disposal of productive assets; and destitution and distress migration. Devereux (1992), although agreeing that households are likely to adopt a number of different strategies that will change as the intensity and severity of the climatic shock increases, finds that a simple linear progression from one mechanism to another may not describe the actual situation. In the first place, not all strategies are available to all households – for example, some may not receive transfers or may not have access to credit. Second, it is not always easy to distinguish between sales of assets held precisely as a precaution against an exogenous shock and the distress sales of production inputs. For example, a household dependent on crop production may hold assets in the form of cattle that can be sold to help maintain consumption in times of drought. The sale of cattle by this household may represent an early response to the shock, especially if it believes that cattle prices are likely to fall in the future. Sale of cattle by another household that is more dependent on livestock income, on the other hand, may represent a distress sale and could indicate that this household is at serious risk of future destitution.

Both Corbett and Devereux, however, agree that changes in food consumption levels are an early response to climatic shocks and take place even when the affected households have an exchange entitlement that is sufficient to maintain consumption at previous levels. Voluntary food rationing (in the sense that the households or individuals concerned have the resources available to consume larger quantities should they so wish) has also been observed by other researchers. In the Sudan, for example, de Waal (1989) documents how, even at the height of the famine, people were spending as little as one-tenth of their income on food. Although food was available, at a price, they were instead concentrating on maintaining their animals, purchasing seed and hiring labour.

In Northern Ghana, Devereux (1992) documents how nearly 70 per cent of households in the area he studied rationed food consumption in reaction to food deficits in 1988/89. Devereux goes on to examine in some detail the different responses of households both over time and between income groups. He finds that simple models of a linear process whereby households move from one response to another as the severity of the crisis increases do not adequately represent what happens in reality. In the case of regular food problems, such as the hungry season, typically in the period leading up to the harvest, Longhurst (1986) records that the energy expenditure of adults in West Africa (Ghana and Gambia) is greater than their intake in the hungry season, leading to a loss of body weight. A process whereby adults build up body fat after the harvest and then experience an energy deficit in the pre-harvest period seems to be an effective response to a situation where markets do not work effectively or are missing and where many households are dependent on their own production for their food supply.

The literature on responses to food crises and famines illustrates the wide variety of mechanisms and strategies adopted and the complex nature of rural communities. Simple models that postulate a sequence of actions and a number of discrete stages undoubtedly do not reflect this complexity. Asset sales may take place over a period of time reflecting differing returns and the extent to which each type of asset can generate future income. What does seem to be clear from the literature is that voluntary food rationing, through a reduction in the frequency of meals and quantities consumed, is a regular practice for many households. Rural households do adopt a number of *ex ante* responses to expected climatic and other shocks. These include income diversification, asset accumulation, and participation in social arrangements that provide for calls for assistance in hard times. The purpose of the strategies is to help smooth consumption when production (or income) is uncertain. *Ex post*, however, some consumption will be given up if the future viability of the household, in terms of being able to continue to produce and generate income, is jeopardised. The extent of rationing will depend, among other things, on the availability of credit and the access of households to the credit market and on future expectations. The greater the uncertainty about future income, the greater the degree of rationing (Dercon, 1992). An important issue for households and for agencies concerned with trying to protect food security is to have some estimate of the probability of occurrence of events that are so severe that they will threaten the capacity of some people to survive.

4 Policy risk

Rural communities are vulnerable to climatic shocks because a large component of income is derived either directly or indirectly from rain-fed agriculture. The level of vulnerability is affected not only by the degree of variability in climatic conditions, but also by other factors, including the impact of economic and agricultural policy and the operations of factor and product markets. From the point of view of governments and donor agencies, these areas are important because they are amenable to policy. At the very least, governments need to be concerned that they are not increasing the threat of livelihood failure; at best, effective policy can help to reduce vulnerability to climatic shocks and increase the ability of communities to deal with future uncertainty.

The role of markets in times of pressure has been studied by Ravallion (1987) and others. He has shown that, under certain plausible conditions, price instability will worsen the survival prospects of people. This remains true even where there are opportunities for savings and storage. The corollary is that efforts to stabilise prices in times of crisis can have a beneficial effect on food consumption. In the conditions applying in many developing countries, of thin markets, poor and expensive communications and underdeveloped information systems, it is likely that many food markets are 'informationally inefficient' in the sense defined by Ravallion. As a consequence, it is to be expected that markets will over-react to the limited information available about future prices and may well be inherently unstable. Although market liberalisation is expected to lead to lower marketing costs and an economically efficient allocation of resources to storage, processing and other market functions in the long run, in the short to medium term there may well be problems with price instability in situations of climatic variability. Many developing countries, especially those in sub-Saharan Africa, are, or have been, involved in a process of liberalising agricultural product markets. These are the same countries, however, that are most at risk from climatic shocks, and the effective management of inter-annual variation in aggregate food production and availability is still a major concern. There is a need to understand more clearly how to manage the process of market liberalisation in the presence of substantial inter-annual variation in both supply and demand. In particular, a strategy may well be needed to help stabilise prices and to modify expectations if the twin objectives of improving efficiency and strengthening food security are to be achieved.

Agricultural development policy can also contribute to the vulnerability of communities to climatic shocks. Policies designed to maximise the production of basic food crops to achieve national self-sufficiency at all costs are now seen as potentially destabilising. Most African governments have set self-sufficiency in basic food commodities as a major objective for their agricultural policy; the objective has been endorsed by the Lagos Plan of Action, and by international organisations including the Food and Agriculture Organisation (FAO). More recently, the pursuit of self-sufficiency at all costs has been questioned, not only in the academic literature, but in the policy debate in a number of countries, for example, Botswana and Zambia. Reducing a country's reliance on food imports can reduce the risk to aggregate food supplies, through threats to imports, but there is increasing evidence that the level of national self-sufficiency does not necessarily guarantee food security at the local or household level.

An example of the extent to which a policy of national self-sufficiency in basic food can lead to increased vulnerability to climatic change is provided by Zambia. Up to the election of 1991, the government of the time had pursued a policy of promoting the production of maize, which is the staple food of most of the population. Self-sufficiency in maize production was the major goal of agricultural policy. This objective was pursued through a number of policy measures, including: close control of maize marketing, with a pan-territorial, pan-seasonal producer price and legal constraints on private sector trading; centralised processing and storage; control of consumer prices; subsidies on fertiliser prices, transport costs and the consumer price of maize; and the promotion of the cultivation of hybrid maize at the expense of local varieties and other small grains. Over time, this has led to an expansion of the area of maize cultivated, but little change in average yields, with a significant shift in production away from the low-cost areas along the line of rail to the more expensive (in terms of transport and marketing costs) outlying areas. The net effect of the subsidies appears to have been a transfer of income from rural producers to urban consumers at a time when urban-rural income differentials were already very high (World Bank, 1993).

In 1992, Zambia, along with much of the rest of southern Africa, suffered a major rainfall failure in January and February. Although, in aggregate terms, the level of rainfall during the 1991/92 agricultural season was not dramatically less than in previous years, the timing of the period when the rains failed was disastrous for the hybrid maize crop, especially throughout the southern half of the country. Aggregate maize production fell to less than half the previous year's level, and the incomes and food security of many people in the affected areas were threatened. Although widespread famine was prevented by the rapid importation of large amounts of food and the operation of an effective humanitarian relief programme, it has been widely acknowledged throughout Zambia that the impact of the drought was made much worse by the distortions in the agricultural production system induced by the previous policy regime (World Bank, 1993). At both the national and local levels, food supplies and livelihoods had been made more vulnerable because of the over-emphasis on hybrid-maize production. The various subsidies had almost certainly made many rural families worse off and had discouraged investment in local storage and processing.

5 Implications for policy

In order to survive, let alone prosper, most rural people in the developing world necessarily take a long-term view of events. They are well aware of the risks they face, especially those that are climate-induced, and they take a number of steps to reduce their vulnerability. Smoothing consumption is seen as an important objective, especially through strategies that are adopted as insurance against future income failures, but for most people the main objective is to ensure the viability of future livelihoods. To this end, voluntary food rationing is a strategy that is adopted on a wide scale in quite different environments and societies. The evidence from studies of coping strategies from many different countries suggests that the main objective is to preserve productive assets, at least until the point at which lives are actually threatened.

The process of climate change is likely to increase the pressure on rural communities in countries where poverty is widespread. This will be most pronounced in those areas of the world where climate change leads to an increase in the occurrence of extreme events. Clearly, in order to develop national policies that not only promote economic growth but also provide an adequate safety-net in times of crisis, it is important to understand the economic and social environment in which they live and the ways in which they adapt to changing circumstances. This is not to say that policy must always be driven by micro-level considerations, nor that simply because a strategy is adopted by a group of people it must be supported and strengthened. In dealing with the consequences of climate change, however, an understanding of what people do and why can be very useful in identifying an appropriate response. *A priori*, policies that build on and support existing activities are likely to be more effective than those that ignore or directly work against them.

That rural people manage risk in innovative and varied ways has been demonstrated in studies from many different countries, even if these coping strategies are not always sufficient to deal with all eventualities. The process of risk management is continuous and involves the diversification of income sources and the building up of reserves in good times as well as the implementation of contingency measures when pressures increase. In contrast, at the national level, in much of sub-Saharan Africa, the management of risk and variability is usually not an explicit objective of either sectoral or macro-economic policy. Although some attention is given to the principle of disaster preparedness, in practice this is mainly confined to questions of logistics. Droughts, which represent the most serious source of risk for many countries, are managed as one-off events by emergency programmes that are wound down once the emergency is perceived to be over. Development plans and policy frameworks do not take variability into account, even when they cover periods of time that are long enough to make a serious drought reasonably likely and where the level of risk is calculable.

In principle, the probability of drought (or other extreme climatic events) can be identified for many countries, although of course the process of climate change can make this more difficult. Once risk can be quantified, it is possible to identify insurance mechanisms that could be used to help stabilise both national income and consumption (Hay, 1993). The fact that most countries do not attempt to develop long-term policies to deal with drought and instability demonstrates the short-term pressures and the lack of space to take a more measured view. In part, at least, this short-termism has been encouraged by the international community. Aid agencies, and especially the international financial institutions, do not appear to take drought risk into account in their medium-term plans and their influence in setting the policy agenda is significant. The ready availability of emergency food aid has also mitigated against the development of national programmes for risk management. As a result, the swings between the optimism of good rainfall years and the pessimism induced by drought lead to major policy shifts that are costly in both the short and long term.

In theory, if the potential for drought risk insurance exists, then all that is required is an effective and competitive capital market. In practice, however, this is unlikely to occur in most developing countries in Africa for three main reasons. First, capital markets are not yet well developed in most countries. Second, participants in these markets tend to take a short-term view. To a large extent, this represents their view of risk and especially the uncertain policy environment. Governments, however, have a responsibility to take a longer-term view; they represent both future as well as current citizens. Third, the infrastructure is not in place for many people to participate in capital markets. The costs of providing access are likely to exceed the returns to be made.

If a process of management of climatic shocks is to be developed, therefore, governments will need to take the lead role. This is likely to involve a number of different areas. They include: mechanisms to help stabilise product markets, for example, through the holding and release of buffer stocks; investment in infrastructure to help stabilise agricultural production, especially through irrigation; research and development into crops and livestock that are more tolerant of variation in climate; the promotion of drought insurance; and a better integration between relief and development interventions.

There is considerable potential for the promotion of regional trade to promote food security in areas where national risks are uncorrelated (or, even better, are negatively correlated). If the variability in agricultural production within a region is less than that within the individual countries, then trade could be used to help deal with climatic shocks, as long as not all countries are affected at once.

The integration of relief and development activities will require much closer attention to the need to help preserve livelihoods, mirroring the priorities of households and communities. In most African countries, the management of drought relief has focused almost entirely on the availability of and access to food. In general, in recent years, there has been a welcome move away from a concentration on aggregate food availability, although aggregate food import requirements as determined by food balances are still the main tool used by food aid donors to allocate resources between countries. The attention in the 1980s given to food entitlements and food security means that most governments are well aware of the need to take action to protect the food consumption levels of vulnerable groups of the population. The continued emphasis on food, however, has led to other aspects of vulnerability being given rather less attention. The evidence from communities under pressure suggests that their first consideration is to protect productive assets such as breeding livestock, tools and land. Clearly this requires that rural communities remain in their localities and retain access to land and livestock. In much of rural Africa not affected by war, migration in times of drought is often driven by the need for water rather than food.

There is a need, therefore, to expand the focus of humanitarian reliefprogrammes to cover more than simply the provision of food. Although food may be an effective way of providing assistance to people in some circumstances, this may well not be the case in all situations. Where there is an effective private sector distribution and retailing network, for example, it makes little sense for a relief programme to go to the expense of distributing food. Indeed it could even have a negative impact on the development of a private trading and storage network. One of the (implicit) objectives of food aid is to ensure adequate food consumption and to help prevent increased levels of malnutrition but, in practice, the food is seen by recipients mainly as an income transfer. In this case, the use of cash (either in the form of wages or via direct hand-outs) or other mechanisms such as food-stamps may well be more cost-effective.

If relief programmes are to be more directed towards protecting livelihoods, then they need to be evaluated against wider criteria than just food consumption levels and nutritional status. In the longer term, livelihood protection requires investment as well as the action to sustain consumption. Evidence suggests that the investment impact of relief programmes depends on a number of factors, including the form in which the relief is provided, the frequency and size of transfers and the recipient. Although the potential for relief programmes to promote investment appears to be considerable, the extent to which this is considered in their design and implementation is limited (Hay, 1988).

Finally, there is a need for both relief and rural development policy to take greater account of the risk-minimising and coping strategies adopted by rural communities, although their limitations as well as their benefits need to be borne in mind. Perhaps the most important aspects of the different strategies adopted by communities are: the promotion of diversification, of not relying on just one income source; the use of exchange as a way of spreading risk; and the need for flexibility. Relief and development policy is likely to be most effective if it works with and supports these aspects. In particular, it is important to ensure that new policies do not have unforeseen negative impacts on coping strategies.

References

Chambers, R., Longhurst, R. and Pacey, A. (eds.) (1981). *Seasonal Dimensions to Rural Poverty*. Francis Pinter, London.

Corbett, J. (1988). Famine and household coping strategies. *World Development,* **16** (9): 1099-1113.

de Waal, A. (1989). *Famine That Kills.* Clarendon Press, Oxford.

de Waal, A. (1991). A re-assessment of entitlement theory in the light of recent famines in Africa. *Development and Change,* **21**: 469-490.

Dercon, S. (1992). *Household Strategies to Cope with Income Fluctuations: An Analysis of the Effects of Producer Price and Asset Market Interventions on Cotton Producers in Tanzania.* D.Phil. thesis, University of Oxford, Oxford.

Devereux, S. (1988). Entitlements, availability and famine: a revisionist view of Wollo, 1972-1974. *Food Policy,* **13** (3): 270-282.

Devereux, S. (1992). *Household Responses to Food Insecurity in Northeastern Ghana.* D.Phil. thesis, University of Oxford, Oxford.

Downing, T.E., Watts, M.J. and Bohle, H.G. (1995). Climate change and food insecurity: Toward a sociology and geography of vulnerability. In: Downing, T.E. (ed.) *Climate Change and World Food Security*. Springer-Verlag, Heidelberg.

Drèze, J. (1988). *Famine Prevention in India.* Development Economics Paper No. 3, London School of Economics, London.

Drèze, J. and Sen, A. (1989). *Hunger and Public Action.* Clarendon Press, Oxford.

Hay, R. (1988). Food aid and relief development strategies. *Disasters,* **10**.

Hay, R. (1993). *Shocks and Trends: Drought and Public Policy*. Paper presented at a joint NISER-NEPRU Seminar, Windhoek, Namibia. Food Studies Group, Oxford.

Jodha, N.S. (1981). Role of credit in farmers' adjustment to risk. *Economic and Political Weekly,* **13**.

Kula, E. (1988). The inadequacy of the entitlement approach to explain and remedy famines. *Journal of Development Studies,* **25**.

Longhurst, R. (1986). Household food strategies in response to seasonality and famine. *IDS Bulletin,* **17**.

Ravallion, M. (1987). *Markets and Famines.* Clarendon Press, Oxford.

Reutlinger, S. and Selowsky, M. (1975). *Malnutrition and Poverty: Magnitude and Policy Options.* Johns Hopkins University Press, Baltimore.

Sen, A. (1981). *Poverty and Famines: An Essay on Entitlement and Deprivation.* Clarendon Press, Oxford.

Torry, W. (1988). Famine early warning systems: the need for an anthropological approach. *Human Organization,* **47** (3): 273-289.

World Bank (1993). *Food Security Report for Zambia.* World Bank, Washington, DC.

Young, H. and Jaspers, S. (1992*). Nutritional Surveillance: Help or Hindrance in Times of Famine?* Paper presented at an IDS Workshop, Brighton Institute for Development Studies, Brighton.

Climate Change and Food Security: Agriculture, Health and Environmental Research

Vernon W. Ruttan
Department of Agricultural and Applied Economics
University of Minnesota
1994 Buford Avenue
St Paul MN 55108, U.S.

David E. Bell
Harvard School of Public Health
Harvard University
Cambrige MA 02138, U.S.

William C. Clark
Centre for Science and International Affairs
John F. Kennedy School of Government
Harvard University
Cambrige MA 02138, U.S.

1 Introduction

Climate change and food security represent an intersection where agricultural, health and environmental research come together. If one employs a broad concept of food security (Chen and Kates, 1994, 1995 this volume) that goes beyond the conventional notion of an increase in overall food supply to include the food security of households and their individual members, then food security includes the income and resources required to provide or to purchase a household's food and the health required to absorb or use the food provided.

Much of the new technology needed to meet projected food demands into the first decades of the next century is already in the pipeline – in the laboratories and experimental plots of the agricultural research community. It cannot be stressed too strongly, however, that it is not as easy now to visualise the sources of productivity growth that will be necessary to respond to world food demands into the middle of the 21st century as it was three or four decades ago at the beginning of the seed-fertiliser or Green Revolution (Parikh, 1994).

NATO ASI Series, Vol. I 37
Climate Change and World Food Security
Edited by Thomas E. Downing

The purpose of this paper is to outline a vision of the agricultural, health and environmental research systems that must be completed, or put in place, in order to generate the new knowledge and the new technology necessary to meet the challenge posed by Chen and Kates. In developing this vision we draw heavily on a series of consultations held during 1988-90 and an international conference involving leading agricultural, health and environmental scientists held during November 1991[1].

At this stage it is premature to assess the impact of potential climate change on agricultural production at the global level. It is not too early, however, to conclude that there will be very large differential impacts among regions and communities. Some areas will realise substantial gains. Many will be forced to make substantial adjustments in agricultural systems, in settlement patterns, and in living standards. It is important, if these adjustments are to be realised at the least cost and greatest benefit to rural communities, that a fully articulated agricultural, health and environmental research system be fully implemented during the early decade of the 21st century.

Two perspectives emerged from the consultation and the conference. One is that the battle to achieve sustainable growth in agricultural production must be fought out along a broad multidisciplinary front. Poverty undermines health and degrades the environment. Environmental problems such as soil erosion, water logging and salinity, and fertiliser and pesticide residues link the agricultural agenda with issues such as malaria and schistosomiasis control, sanitation, and water and food quality on the health agenda. Environmental changes under way at the global level, such as acid rain, ozone depletion and climate change, will require changes in food production and health practices at the producer and community level. Effective bridges must be built between the 'island empires' of the agricultural, environmental and health sciences.

---

1 The dialogues and recommendations from the initial three consultations, held under the auspices of the "Twenty-First Century Project," with support from the Rockefeller Foundation have been reported in three University of Minnesota Department of Agricultural and Applied Economics Staff Papers (Ruttan, 1989; Ruttan, 1990a; Ruttan, 1990b). Revised versions of the second and third consultation reports have been published by Westview Press (Ruttan, 1992; Ruttan 1994b). The results from the conference, held at the Rockefeller Foundation Bellagio Conference Center are reported in Vernon W. Ruttan (1994a). See Bell *et al.* (1994) for an earlier version of this paper.

A second perspective is the central role of decisions by the family and the community in achieving growth of agricultural production, enhancement of the resource base, and improvements in health. This means that much more effective organisational and institutional links must be built between the suppliers of knowledge and technology and the users. It also means that the institutions must be designed to place the users in a stronger role relative to the suppliers.

During the discussions at the Bellagio conference, a vision of the institutional infrastructure that will be needed to supply the knowledge and technology in the areas of agricultural production, resource management, and health began to take shape. In this paper we draw on the papers and discussion at the Bellagio conference and at the three earlier consultations to outline our vision of the structure of global agricultural, health and environment research systems. We are under no illusion that the process of evolving an effective global research system that will be capable of bridging the island empires of agriculture, environment and health will be easy (North, 1993).

2 Agricultural research

This vision is strongly influenced by the experience of attempts, beginning in the late 1950s, to establish a global agricultural research system (Ruttan, 1986; Baum, 1986). For the architects of the post-World War II set of global institutions, meeting world food needs and the reduction of poverty in rural areas were essential elements in their vision of a world community that could ensure freedom from hunger for all people.

In the immediate post-war years, much of the burden fell on the Food and Agriculture Organisation (FAO). But John Boyd Orr, the first director-general of the FAO, burdened with the memory of the agricultural surpluses of the 1930s, was highly critical of the view that knowledge and technology represented a serious constraint on agricultural production capacity. 'No research was needed to find out that half the people in the world lacked sufficient food for health, or that with modern engineering and agricultural science the world food supply could easily be increased to meet human needs.' (Boyd Orr, 1966, p.160). In the first two post-war decades, assistance for agricultural development in the poor countries was conducted largely in a technology transfer and community development mode. By the late

1950s, it was becoming apparent, however, that the gains in production from simple technology transfer had largely played themselves out.

The inadequacy of policies based on the technology transfer or extension model led, in the early 1960s, to a re-examination of the assumption about the availability of a body of agricultural technology that could be readily diffused from high agricultural productivity to low productivity countries or regions. The result was the emergence of a new perspective that agricultural technology, particularly yield-enhancing biological technology, is highly 'location-specific'. Evidence was also accumulated to the effect that only limited productivity gains could be achieved by the reallocation or more efficient use of the resources available to peasant producers in poor countries.

The new vision that emerged as a guide to the sources of growth in agricultural production was the product of both experience with the improvement in agricultural technology and a reinterpretation of the role of peasant producers in the process of agricultural development.

In the early post-war development literature, peasant producers had been viewed as obstacles to agricultural development. They were viewed as bound by custom and tradition and resistant to change. In an iconoclastic work published in 1964, Schultz advanced a 'poor but efficient' view of peasant producers. They were viewed as making effective use of the resources available to them. But they lived in societies in which productivity enhancing 'high pay-off' inputs were not available to them.

Schultz, drawing on the experience of the Rockefeller Foundation program in Mexico and case studies by anthropologists and agricultural economists, identified three high pay-off investments needed to enhance the productivity of peasant producers. These were: (i) the capacity of the agricultural research system to generate locally relevant knowledge and technology; (ii) the capacity of the industrial sector to develop, produce and market new inputs which embodied the knowledge and technology generated by research; and (iii) the education of rural people to enable them to make effective use of the new knowledge and technology. These insights, from experience and analysis, shaped the response to the food crises of the 1960s and 1970s. The immediate response was the transfer of large resources, including food aid, to the food deficit countries. The longer-term response was the mobilisation of resources to develop a system of international agricultural research institutes and to strengthen national agricultural research systems (Eicher, 1994).

In 1959, the Ford and Rockefeller Foundations collaborated in establishing an International Rice Research Institute (IRRI) in the Philippines. This was followed by the spinning-off of the international activities of the Rockefeller-supported Mexican maize and wheat programs to form an International Center for the Improvement of Maize and Wheat (CIMMYT), and the establishment of an International Institute of Tropical Agriculture (IITA) in Nigeria and the International Center for Tropical Agriculture (CIAT) in Colombia. It became apparent by the late 1960s that the financial requirements to maintain the research and development programs of the four institutes were stretching the capacity of the two foundations. In 1969, consultations were held among the Ford and Rockefeller Foundations, the World Bank, the FAO and the United Nations Development Program (UNDP) that led to the organisation of a Consultative Group on International Agricultural Research (CGIAR). The initial membership consisted of the World Bank, the FAO and the UNDP as sponsors, plus nine national governments, two regional banks and three foundations.

The leadership of the Consultative Group is now centred at the World Bank, which provides a chairperson and a secretariat. Each institute or center is an independent corporate identity governed by its own board of trustees. The CGIAR established a Technical Advisory Committee (TAC) with its secretariat located at FAO in Rome, to provide technical oversight of the research institutes and to advise the CGIAR on priorities and resource allocation among centers. The TAC has been charged with the responsibility of organising comprehensive reviews of the programs of the centers, of evaluating new initiatives, and of overseeing coordination among centers in common program areas such as cropping systems research.

By the early 1990s, the system had expanded from four to 18 centers. The initial centers focused their research on the major food crops grown in developing countries – rice, wheat, maize, potatoes and cassava. These were joined in the 1970s by centers focusing on livestock production, animal disease and genetic resources, on arid and semi-arid areas, food policy, and the capacity of the national research system. At the beginning, a commodity orientation in research and development was adhered to in an effort to ensure that the limited resources available to the system would not be dissipated in unfocused research efforts. In the late

1970s and early 1980s, crop and farming systems research programs were initiated to achieve more effective understanding of soil, water, climate, weed and crop interaction. In the late 1970s, several donors to the CGIAR were instrumental in establishing independent research centers to work on soils, irrigation and agro-forestry.

As the new seed-fertiliser technology generated at the CGIAR centers, particularly for rice and wheat, began to come on stream, some donors assumed that the CG centers could bypass the more difficult and often frustrating efforts to strengthen national agricultural research systems. But experience in the 1960s and the 1970s confirmed the judgement of those who had participated in the organisation of the international centers that strong national research centers were essential if the prototype technology that might be developed at the international centers was to be broadly transferred, adopted and made available to producers.

The location-specific nature of biological technology meant that the prototype technologies developed at the international centers could become available to producers in the wide range of agro-climate regions and social and economic environments in which the commodities were being produced only if the capacity to modify, adapt and re-invent the technology was available. It became clear that the challenge of constructing a global agricultural research system capable of sustaining growth in agricultural production required the development of research capacity for each commodity of economic significance in each agro-climactic region. One response by the CGIAR donor community was the establishment of a new center, the International Service for National Agricultural Research (ISNAR), to provide analytical and technical assistance to national agricultural research systems in strengthening their organisation and management. Another response, particularly during the 1970s, was substantially expanded support for national agricultural research systems.

During 1990-92, five new centers were added to the CG system, increasing the number of centers from 13 to 18. In 1990, the International Irrigation Management Institute (IIMI), the International Center for Research on Agro-Forestry (ICRAF), and the International Network for the Improvements of Banana and Plantain (INIBAP) were brought into the CG system. In 1992, the International Center for Living Aquatic Research Management (ICLARM) was added to the system. This expansion was not accompanied by an expansion of the resources available to the system. By the mid-1990s, a false sense of complacency about food security led a number of donors, particularly the United States, to reduce their commitments to international agricultural research. Support has declined in real terms, producing a 'quiet crisis in the system'.

The crisis has not been only financial. A number of the CGIAR centers are experiencing the difficulties associated with organisational maturity. There is a natural life cycle sequence in the history of research organisations and research programs (Ruttan, 1982, p.132). When they are initially organised they tend to attract vigorous and creative individuals. As these individuals interact across disciplines and problem areas, the organisation often experiences a period of great productivity. As the research organisation matures, however, there is often a tendency for the research program to settle into filling in the gaps in knowledge and technology rather than achieving creative solutions to scientific and technical problems. Since the early 1980s, the managers and boards of several of the CGIAR institutes have been forced to struggle, during a period of budget stringency, with the problem of how to revitalise a mature research organisation.

Efforts to strengthen national research institutes have also been only partially successful. The 1970s witnessed a remarkable expansion of agricultural research capacity in a number of developing countries. The national research systems in India, Brazil, Malaysia and a number of other developing countries began to achieve world-class status in their capacity to make advances in knowledge and technology available to their farmers. A number of other countries, such as the Philippines, Colombia, Kenya and Thailand, achieved substantial capacity to conduct research on their major agricultural commodities. By the mid-1980s the buffeting of a global debt crisis and the weakening of commodity markets had the effect of dampening commitment by a number of aid agencies and national governments to the strengthening of agricultural research. In Africa, many national agriculture research systems that have received generous external support even during the 1980s failed to become productive sources of knowledge and technology (Eicher, 1994).

The role of technical support for farm decision-making by farmers and the capacity to supply producers with the technical inputs in which the new technology is embodied has been a continuing area of controversy. In general, the developing countries have been relatively extension-intensive. The ratio of extension workers to agricultural product has been much higher in developing countries than developed countries (Judd *et al.*, 1987). Weak links between research and extension and between extension and farmers have represented a serious constraint on the diffusion of new technology (Tendler, 1994). During the late 1970s and early 1980s, the World Bank devoted very substantial resources to the support of an intensive training and visit (T&V) system of delivering information about practices and technology to farmers. The system involved a highly regimented schedule in which the field level worker is involved one day each week in intensive training about the information that he or she must convey to farmers (Benor and Harrison, 1977). In retrospect it appears the system erred in placing the extension worker rather than the farmer, or the farm family, at the center of the technology adoption process.

A second constraint on the effectiveness of the transfer of agricultural practices and technology to producers is the weakness of the private sector as a source of both the supply and delivery of knowledge and technology. The emergence of more liberal economic policies since the early 1980s in a number of developing countries is, however, leading to rather rapid growth of private sector suppliers of agricultural technology and to increased research by the suppliers (Evenson *et al.*, 1987; Pray, 1987).

The global agricultural support system is still incomplete. Its deficiencies continue to deprive farm families of the support that they need to meet even current food consumption and income needs (Eicher, 1994; Turner and Benjamin, 1994; Tendler, 1994). Yet the vision of the agricultural support system that will be needed to sustain growth in agricultural production is reasonably clear. During the past several decades, implementations of the vision have been less than adequate in some developed countries and in all but a few developing countries. With the ending of the cold war it may now be possible to extend the vision to farm families in many of the formerly centrally planned economies. One important step will be to place farm families and the farm enterprise in those societies at the center of the agricultural production process. Another important step will be to link the agricultural research systems in the formerly centrally planned economies with a strengthened global agricultural research system.

3 Health research

Division of the gains in health status that can be achieved by even a poor society that devotes significant resources in support of an effective national health policy has been outlined by Gunatilleke (1994). Sri Lanka has achieved health indicators – a life expectancy of about 70 years and infant mortality of below 20 per 1000 live births – comparable with the levels achieved by many societies that are much more affluent. But a vision of the global health research system needed to sustain national health policy has emerged more slowly than the vision of a global agricultural system. Only within the last decade has the health research community begun to articulate the form that such a system might take[2].

For most of the century – since the time of Koch and Pasteur – health research has been thought of principally as laboratory-based biomedical research, seeking 'silver bullets' against specific infections or diseases: new vaccines, new drugs, new surgical techniques. This focus, plus the remarkable improvements in health in recent decades, led to the misperception that all the new knowledge and new technology needed to protect families and communities around the world from debilitation and illness could be generated in the universities, research institutes and pharmaceutical company laboratories of the industrialised countries.

This limited conception was clearly wrong and has been changing rapidly. Three gains in perception are especially important. The first is the recognition that health technologies, to be useful, must be applied in particular social settings. Achieving health improvements requires not only technology but policies, organisations and processes that are adapted to the varied economic, social, cultural and historical circumstances among and within countries. Even vaccines, the simplest of technologies, cannot be applied in Lagos by the same means as in Liverpool.

An effective health research system, capable of conducting the essential national health research (Lucas, 1994), needs epidemiologists, economists, management specialists, and other social and policy analysts in addition to biomedical scientists. Such skills are scarce in

---

2 For an excellent status report and policy analysis, see Commission on Health Research for Development (1990) and World Bank (1993).

industrialised countries. They are grossly deficient in developing countries. But they are essential to identify the precise nature of health problems in different national and local settings, and to design, test and apply appropriate solutions.

A second gain in perception is the recognition that the principal actors in achieving improvements in health are individuals and families, especially mothers. Preventing illnesses and promoting health depends first and most of all on maternal technology – the ability to implement basic knowledge about nutrition, cleanliness, home remedies, and when and how to call on health professionals. Reform involves shifting government spending away from specialised personnel, equipment and facilities at the apex of the health system down the pyramid towards the broad base of a widely accessible community and family core (Mata, 1988; World Bank, 1993).

An effective health research system, therefore, must be organised not simply to serve physicians but to support the flow of health knowledge and technology to communities and families – and to provide for the reverse flow of information from families and communities to researchers about the actual nature of health problems and how they are changing. Such a conception of linking researchers directly to primary actors is customary in agriculture, where research results have long been aimed at farmers as decision-makers. But it is a recent conception in health, particularly in industrialised countries.

A third gain in perception is the recognition that the world's health research efforts are overwhelmingly concentrated in industrialised countries, seeking technologies to address the diseases of the more affluent societies. Only about five per cent of global health research financing is directed to the major diseases and health problems of the less developed countries, where more than 90 per cent of the world's burden of preventable deaths occur (Commission on Health Research for Development, 1990). An effective global health research system must address this huge imbalance, and provide for a large increase in the resources devoted to the health problems of the developing countries.

Combining these three perceptions with the traditional power of biomedical research, one can begin to perceive, dimly, the shape of a global health research system and how to achieve it. Such a system – in health just as in agriculture – will need to be based solidly on national

research systems, capable of supporting decision-makers as they identify and confront health problems. A national health research system requires first of all skills to measure the patterns and determinants of disease, disability and death, and to monitor changes in health status over time. It also requires skills to design, test and evaluate means for applying improved health technologies in local environments, and for making research results available to those who need to use them, from national policy-makers to local families. Every nation needs the capacity to conduct such country-specific research to guide its health activities, and the establishment of such capacity should clearly be given top priority.

Beyond the capacity for essential country-specific research, health scientists in every country will wish to join, as and when they can, in the international effort to advance the world's frontiers of knowledge on the social and biological pathologies of ill-health and disability, and on new technologies to overcome them. In poor countries, the conditions for world-class science are difficult to establish. Nevertheless, a significant number of developing countries – to name just a few: Thailand, India, Egypt, Mexico, Brazil – are beginning to have the capacity to make significant contributions to world knowledge in the health field.

Thus, national health research systems need to begin with the capacity to guide national health activities, and to go on, as conditions permit, to participate in global frontier research. Most developing countries have only rudimentary health research capabilities at present. It is urgent for developing countries, and for the international health assistance community, to commit themselves to building steadily stronger national health research systems. Such systems will need to start small, and to focus initially on the most pressing health problems. But they should be designed with a view to dynamic change over time as financial and personnel resources grow, and as health problems change with the demographic and epidemiologic transitions through which the developing countries will pass over the coming decades.

Thinking about how to achieve an effective global health research system thus begins with the development of strong national systems. But national systems must not be thought of as separate, free-standing entities. On the contrary, it is essential that they be linked together by strong international ties, and draw from the common, growing pool of world-wide health knowledge, with each country adapting advances in health science to its own specific circumstances.

Moreover, it would be a mistake to think of a global system as centred in the industrialised countries, with all scientific advances pioneered there and rippling outward to the developing world. We have already seen major health improvements developed in the Third World, as ambulatory therapy for tuberculosis was pioneered in Madras, and oral rehydration therapy for diarrhoea in Dhaka. As the amount and quality of developing country research steadily rise, a global research system will increasingly be multi-centric – one in which the flows of ideas and new knowledge move in all directions along networks of information and collaboration encompassing scientists from many countries, rich and poor alike.

Thus the guidelines for moving towards a global health research system include: (i) the development as rapidly as feasible of strong national systems, especially in developing countries where they are currently very weak, and (ii) the rapid evolution of international collaborative mechanisms and arrangements. There is much work here for years to come. Two aspects of this overall vision deserve special attention and illumination.

The first is the necessity for building direct relationships between the national health research system and action for health at the community and family level. Kaseje (1994) has described the elements of a community-based health system in Kenya that he helped design and implement, which relies directly on the actions of individual families and communities. The model views the mother as the key health provider. It builds on the strong motivation to carry out her tasks resulting from concern about the current and future well-being of her children and family. Kaseje summarises the concept behind the *Harambee* model as follows.

This model recognises the strengths and resources of the community; seeks to facilitate and enhance these strengths; recognises that communities have always been responsible for their own health, even without the intervention of health professionals; that the mother is the most important and knowledgeable health provider. The mother is not, however, left without resources to carry out her responsibilities. She is reinforced with a strong program of health education, the availability of appropriate technology and materials, and support from NGO and official health programs. The system described by Kaseje does not work perfectly and should not be overly idealised. Kaseje himself expressed considerable concern about breaking the professional and bureaucratic inertia needed to extend and sustain the program (Kaseje, 1994).

It is clear, however, that the resources needed to enable families and communities to provide effective health services to their members are very similar to those identified three decades ago by Schultz to enable peasant producers to become effective suppliers of agricultural commodities. The 'high pay-off' health inputs include:

- The capacity of the health research community to produce the new knowledge and the materials that are appropriate to the resource and cultural endowments of rural communities.
- The capacity of national, regional and local institutions to make the knowledge and the materials available to families.
- The formal schooling and informal education of families, particularly mothers, to make effective use of the knowledge available to them.

The second aspect of the vision of an effective global health system is the nature of the international apparatus needed to sustain a global health research system.

In agriculture, the CGIAR-sponsored set of international research centers serve as leaders of applied science for the Third World and accelerators of linkages between frontier science and Third World problems. There is no comparable set of internationally-supported health research centers in poor countries of the tropics. Lucas (1984) notes that there are only two international centers of significant size in the field of health – the International Centre for Diarrhoeal Disease Research in Bangladesh, and the International Centre for Insect Physiology and Ecology in Kenya (which is concerned with entomology that is relevant to both health and agriculture).

There are strong differences of opinion within the international health community as to whether a system of international health research centers, analogous to the CGIAR centers, would be appropriate or effective:

> On the one hand, internationally organised efforts have the advantage of achieving a critical mass of scientists concentrating on and physically located close to high-priority problems...Internationally organised research efforts can focus on specific problems in a

> multidisciplinary way and demonstrate economies of scale in their operations, making them attractive to external funding. On the other hand, international center salaries are high and their activities, if not carefully targeted, can supersede rather than complement national efforts. (Commission on Health Research for Development, 1950, p.58).

Present constraints on foreign assistance funds suggest that it would be unrealistic to expect that resources could be mobilised in the late 1990s to support an expanded system of international health research centers in the tropics. It seems more likely that the predominant model of international collaboration in the health field will be international networks linking scientists in national institutions (both in industrialised and developing countries) in goal-oriented research programs aimed at specific health problems. A successful example of such collaboration is the Special Programme for Research and Training in Tropical Diseases (TDR), co-sponsored by UNDP, the World Bank and WHO. Started in 1976, TDR focuses on six specific diseases (including malaria, schistosomiasis and leprosy). In addition to supporting research, it invests approximately 25 per cent of its annual budget of $30-35 million in strengthening research capacity in developing countries.

Although international networks of national centers evidently can work effectively in supporting research on particular diseases, there is one extremely important function they cannot perform. The field of health research conspicuously lacks an overview mechanism. In agriculture, the CGIAR (as distinct from the set of centers it sponsors) has built highly valuable methods for surveying the world-wide agricultural research scene in relation to the needs for research results, reviewing continuing research activities (both those of the international centers and of other institutions), and proposing changes in current research priorities and institutional arrangements including, where necessary, the development of new research facilities.

There is no analogous, effective, independent organisation in the health field for assessing progress in research, especially on developing-country health problems, identifying neglected areas, and promoting necessary action. The result is clear. At present, of the three leading infectious disease causes of death in the world (acute respiratory infections, diarrhoeal diseases and tuberculosis), only diarrhoea is addressed by a major, sustained research effort. That is why the Commission on Health Research for Development came to the conclusion that 'a health analogue of the CGIAR assessment and promotion structure could be of great value and should be established' (Commission on Health Research for Development, 1990, p.59).

In response to this challenge, the World Bank has proposed the establishment of a body to coordinate research on health problems of developing countries. Meanwhile, a Geneva-based Council on Health Research, funded by several donors, including Canada and Sweden, has been established to provide support for essential national health research in developing countries. Discussion has been initiated to explore the complementarity between the two initiatives (Aldhous, 1993; World Bank, 1993).

4 Environmental research

If the global research system for agriculture now faces the challenges of maturity and the system for health research confronts those of adolescence, then the global environmental research system still requires pre-natal care.

To be sure, research for environmental conservation has a long and productive history in many parts of the world. Since World War II, this research has been given impetus and direction by at least three waves of concern over the implications of natural resources availability and environmental change for the sustainability of improvements in human well-being. Early work focused on the adequacy and protection of the material base for agricultural and industrial production. By the mid-1970s, increasing attention was also being given to the impact of residuals generated by that production on air and water quality and human health. Rapidly growing awareness of global change in the earth system has provided yet another dimension to our environmental concerns.

Most environmental research to date has been performed in universities, initially with support from major philanthropic organisations such as the Ford Foundation. Prodded by the 1972 Stockholm Conference on the Human Environment, national governments have become increasingly involved as supporters, producers and users of environmental research. Over the last decade, there has also been an explosion in the number and variety of non-governmental organisations active on the world's environmental scene, some of them producing research of the highest calibre and relevance (Livernash, 1992). International programs for environmental research have also expanded dramatically since their 'modern' birth in the International Geophysical Year of 1957. Nonetheless, most important international institutions for environmental research are barely 20 years old – for example, International Council for Scientific Unions' (ICSU) Scientific Committee on Problems of the Environment (SCOPE), United Nations Education, Scientific and Cultural Organisation's (UNESCO) Man and Biosphere (MAB), the International Institute for Applied Systems Analysis (IIASA) and, of course, United Nations Environment Program (UNEP) (Caldwell, 1990). Today's major research programs on global change – the World Climate Research Program (WCRP), the International Geosphere Biosphere Program (IGBP), Human Dimensions of Global Change Program (HDGEC), and the Intergovernmental Panel on Climate Change (IPCC) – are younger still (Jaeger and Ferguson, 1991; Miller and Jacobson, 1992; Perry, 1991).

This impressive and expanding array of activities nonetheless falls far short of the global system of environmental research needed to provide the knowledge base for sustainable development. Still lacking is a coherent institutional structure that can link the world's environmental researchers *both* upward to the international level of policy negotiations and downward to the community-level consumers, producers, health workers and extension agents on whose actions sustainable development must ultimately depend (Bradley, 1994; Lee *et al.*, 1994; Rawlins, 1994). In the wake of the Rio Earth Summit, however, several initiatives are under discussion that could supply important components of such a system and move it substantially closer to reality.

The most ambitious of these is START – a System for Analysis, Research and Training proposed in 1991 by the IGBP in collaboration with the WCRP and HDGEC. START was planned as 'a global system of regional research networks to stimulate research, modelling, and training activities related to global [environmental] change in both the natural and social sciences' (IGBP, 1992, p.5). Its regional focus is based on the realisation that global change wears local faces. The origins, the impacts and the options for managing global environmental change will be different in different parts of the world, and must be understood within their local environmental and social contexts. The initial START planning document divided the world into 13 scientifically coherent regions (Eddy *et al.*, 1991). Within each region, the research network is planned to consist of one or more research centers plus an unspecified number of regional research sites (e.g. university departments, field stations). The networks aim to provide scientists from all parts of the world with the knowledge and infrastructure necessary for them to participate fully in on-going research concerning global environmental change. Initial funding has launched several networks, most notably in south and east Asia.

In addition to the comprehensive plans of START, a number of more focused regional initiatives are also being pursued around the world. For example:

Asia. The Smithsonian-sponsored program on Sustainable Management of Tropical Evergreen Forests has linked leading centers throughout Asia in a unique network for research, training and data collection (Ashton, 1991). Although the tropical forests program has been launched largely through the efforts of private foundations and host-country contributions, Japan and the United States – through their Global Partnership Plan of Action – have promised increased governmental support of environment and conservation research in the region (Lepkowski, 1992).

The Americas. The Inter-American Institute for Global Change Research (IAI) has been established as a regional network of research entities...[that] seeks to achieve the best possible international coordination of scientific and economic research on the extent, causes and consequences of global change in the Americas (IGBP, 1991; Declaration of Montevideo, 1992). It is closely integrated with the START initiative.

Central and Eastern Europe. Several environmental research, development and training institutions have been formed to address the special problems of this region. One notable example with support from a number of western countries is the Regional Environmental Center in Budapest. Since its inception in 1989, the Center has helped to set up environmental surveys, grassroots and non-governmental organisations, new environmental legislation and remediation campaigns (Anonymous, 1992). Its major activity has been building a data base on environmental conditions in the region, coupled with a computer network to disseminate these data to smaller offices for use by local researchers.

Globally. Increasing attention is being given to the need for a permanent international research institution that could tackle environmental problems that transcend individual regions, and link national centers for environmental research into a truly global system. This function is currently performed on a largely ad hoc basis – for example, through studies of ICSU's SCOPE or the IPCC reviews. But the time may well be ripe for complementing such ad hoc efforts with a more permanent home or homes. IIASA, with its focus on problems of global change, has been put forward as one leading candidate (IIASA, 1991; Maddox, 1992).

How these and other initiatives will relate to one another or to existing national research centers is not yet clear. Most of the parties involved are aware of the need for addressing such relations. Early indicators are that their potential complementarities could dominate the inevitable competition for people, programs and funds. The formation of a professional secretariat for START in Washington will, it is hoped, improve the prospects for successful integration of emerging international environmental research efforts.

Against this optimistic assessment, however, it must be noted that in the dialogue leading to recent environmental research initiatives there appears to have been inadequate consideration of appropriate links with agricultural and health research systems. This is a serious omission for two reasons. Firstly, it virtually guarantees that many of the lessons painfully learned in the course of building today's relatively mature network of agricultural and health research systems will be lost on the fledgling environmental effort. Secondly, it perpetuates the 'island empire' problems we referred to at the outset of this essay. There are several possible measures for mitigating these shortcomings:

- The lessons from the development of today's agricultural and health research systems should be incorporated in new environmental efforts. As noted above, all of these reflect a growing appreciation of the central role of family, firm and community-level decisions in shaping sustainable development.
- Means must be designed to ensure that research priorities reflect the environmental problems confronting individual families, farmers, and resource users in the field. The small charmed circle of puzzles that excite laboratory scientists or program administrators should not be allowed to dominate the agenda. The World Bank's report, *Development and the Environment,* is surely correct in its conclusion that the current environmental debate has paid too little attention to the problems of clean water, urban air pollution, indoor air pollution, and severe land degradation that each year kill millions of people, undermine the health of hundreds of millions more, and significantly reduce productivity of people who can least afford it (World Bank, 1992, p. 4; see also Norberg-Bohm *et al.*, 1992).

- We must resist the temptation to search for universal 'silver bullets' that will solve specific environmental problems whenever and wherever they occur. Most causes, impacts and solutions will be intimately associated with particular social circumstances and landscapes. Effective research systems will therefore require significant site-specific components, and must avoid focusing activity in a few elite laboratories of the high-income countries. The need for elite laboratories will remain, in part because of needs for special research and data processing equipment, in part because of the need to bring top scientists from many disciplines together for particular aspects of the necessary research. But specific measures must be implemented to ensure that such regional centers do not bleed talent, funds and equipment from the essential national and local nodes of the research network. A recognition of the need for simultaneous and complementary strengthening of the local, national and regional dimensions of the emerging global environmental research system seems well embodied in the plans for START (Eddy *et al.*, 1991). But a practical vision of essential national environmental research – how it is to be funded and linked to international efforts – has yet to emerge.
- A technology transfer strategy for research and development will be no more successful in dealing with environmental problems than it has been for sustaining improvements in agricultural productivity or human health. This applies not only to conventional north-south transfers, but also to the current spate of enthusiasm for grafting the clean energy systems of advanced OECD nations on to the formerly socialist economies of Europe. Less obviously, but perhaps even more importantly, experience in the agriculture and health sectors warns against the wholesale transfer of institutions as a means of enhancing environmental conservation. This is especially the case in the area of common pool resources, where an uncritical tendency to transfer solutions based on full private property rights or centralised regulation to small-scale, low-income situations has had disastrous consequences. Appropriate alternatives often exist, more finely attuned to local social and environmental conditions (Ostrom, 1990). In general, the need is not to transfer environmental technologies and institutions from 'advanced' to 'developing' regions, but rather to promote more widespread sharing of knowledge, know-how and experience around the world. In particular, in environment as in agriculture and health, the need is to enhance the voice and power of users relative to suppliers of needed research and development.

- An effective global environmental research system must be much more broadly inclusive than is presently the case. The need to incorporate knowledge users in the system in a better way has been stressed in this chapter. The need for an expansion of the capacity to monitor global change is imperative. The environmental research and development potential of the formally centrally planned economies must also be tapped, although this will require institutional innovations to end the traditional exclusion of such societies from the global research system. Finally, the private sector must be encouraged as both a supplier and deliverer of the knowledge needed for environmentally sustainable development. Perhaps no single factor has so inhibited the development of effective global research systems for agriculture and health as the failure to promote incentive and reward structures that can induce constructive private sector involvement. In the environmental field, there is a vast potential for private sector engagement in topics as diverse as energy efficiency to biotechnology. But a number of issues involving intellectual property rights, liabilities and government-industry relations will have to be resolved before the potential can be fully tapped for the benefit of sustainable development (Schmidheiny, 1992).

In summary, an effective global environmental research system will have many of the features of effective agricultural and health research systems. The behaviour of consumers of environmental services and the producers of the residuals – households, farms and factories – that erode or generate environmental amenities will have to be recognised as central to the process of environmental change. The resources that will be needed to place households, farms and factories in a position to respond constructively will depend on: (i) the capacity of the environmental research system to provide the knowledge, including the essential national environmental research, needed by household, farm and factory decision-makers; (ii) the capacity of national, regional and community institutions to transmit the knowledge, technology and incentives to those who make decisions about resource use; and (iii) the depth of understanding possessed by household, farm and factory decision-makers about the consequences of their own actions and the actions of the economic and political institutions in which they participate.

5 Bridging the 'island empires'

We have argued that the 'island empires' of the agricultural, health and environmental sciences can learn from one another as they strive to build global research systems that can support sustainable development. Whether they can, or even should, move beyond passive learning to active cooperation remains to be seen.

There seems little merit in any grand organisational scheme that would attempt to pull the already diverse networks of research in the respective empires under a single roof. And the most dynamic of the existing empires – that dealing with environmental research – simply does not have enough experience in the tough business of running a global network to seem credible as a leader of any major bridging movement. What does seem both feasible and desirable, however, is to begin some modest effort at active bridge-building.

At a minimum, the principals of the three empires might agree to meet regularly – at the international, the national and the local level – in order that they and their senior staff members could get to know one another and exchange information on current activities. An exploration of possible collaboration in global monitoring and other data-gathering activities might be a good early agenda item for such an international meeting. The new UN Commission on Sustainable Development, established at the 1992 Earth Summit, would be one logical convenor for such meetings. But private foundations and NGOs could do a lot to get the ball moving.

At a deeper level, it is essential to realise that the global agricultural, health and environmental research systems outlined in this chapter have important common elements. The global systems outlined in this chapter can be effective only as the underlying sciences – particularly the biological and the social sciences – advance. Advances in the biological and the social sciences are necessary to enlarge the world's understanding of the natural and social phenomena in global change. They are also needed to expand the capacity to apply advances in knowledge to the national and human dimensions of development in the poor countries where most of the world's people live.

The need to enlarge scientific capacity in the poorer countries of the world should not be viewed as a burden on either the developed or developing countries. Rather it is an opportunity to multiply the intellectual talent necessary to advance knowledge relevant to the achievement of sustainable development. Completion of the development of global research systems in agriculture, health and environment is a necessary component of a global effort to establish and mobilise the intellectual capacity and energy that will be needed to sustain development.

Acknowledgement

This chapter is a revised version which first appeared in *Global Environmental Change* Volume 4, Ruttan, V.W., Bell, D.E. and Clark, W.C. Climate change and food security: agriculture, health and environmental research. pp: 63-77. (1994). Reprinted with permission from Elsevier Science Ltd, Butterworth-Heinemann Imprint, The Boulevard, Landford Lane, Kidlington OX5 1GB, U.K.

References

Aldhous, P. (1993). World Bank report calls for network to bolster research. *Science,* **261**: 135.

Anonymous (1992). West's gift becomes a model. *Nature,* **355**: 672.

Ashton, P. (1991). *Sustainable Management of Tropical Evergreen Forests: A Program Proposal.* Harvard Institute for International Development, Harvard University, Cambridge.

Baum, W. (1986). *Partners Against Hunger: The Consultative Group on International Agricultural Research.* World Bank, Washington, DC.

Bell, D.E., Clark, W.C. and Ruttan, V.W. (1994). Global research systems for sustainable development: agriculture, health and environment. In: Ruttan, V.W. (ed.) *Agriculture, Environment and Health: Sustainable Development in the 21st Century.* University of Minnesota Press, Minneapolis. pp. 358-79.

Benor, D. and Harrison, J.Q. (1977). *Agricultural Extension: The Training and Visit System.* World Bank, Washington, DC.

Boyd Orr, J. (1966). *As I Recall.* MacGib and Kee, London.

Bradley, D. (1994). Institutional capacity to monitor the interactions of agriculture and health change. In: Ruttan, V.W. (ed.) *Agriculture, Environment and Health: Sustainable Development in the 21st Century*. University of Minnesota Press, Minneapolis: 308-338.

Caldwell, L. (1990). *International Environmental Policy: Emergence and Dimensions.* 2nd edn. Duke University Press, Durham, North Carolina.

Chen, R. S. and Kates, R.W. (1994). World food security: prospects and trends. *Food Policy* **19** (2): 192 - 208.

Chen, R. S. and Kates, R.W. (1995). Towards a food secure world: prospects and trends. In: Downing. T.E. (ed.). *Climate Change and World Food Security.* Springer-Verlag, Heidelberg.

Commission on Health Research for Development (1990). *Health Research; Essential Link to Equity in Development.* Oxford University Press, Oxford.

Declaration of Montevideo (1992). *Declaration of Montevideo on an Inter-American Institute for Global Change Research.* International America Institute, Washington, D.C.

Eddy, J. A., Malone, T.F., McCarthy, J.J. and Rosswall, T. (1991). *Global Change System for Analysis, Research and Training (START).* Report No. 15, International Geosphere-Biosphere Programme, International Council of Scientific Unions. UCAR Office of Interdisciplinary Earth Studies, Boulder.

Eicher, C. (1994). Building productive national and international agricultural research systems. In: Ruttan, V.W. (ed.) *Agriculture, Environment and Health: Sustainable Development in the 21st Century.* University of Minnesota Press, Minneapolis: 77-103.

Evenson, R.E., Evenson, D.D. and Putnam, J.D. (1987). Private sector agricultural inventions in developing countries. In: Ruttan, V.W. and Pray, C.E. (eds.) *Policy for Agricultural Research.* Westview Press, Boulder.

Gunatelleke, G. (1994). Health policy for rural areas: Sri Lanka. In: Ruttan, V.W. (ed.) *Agriculture, Environment and Health: Sustainable Development in the 21st Century.* University of Minnesota Press, Minneapolis. pp. 209-235.

IGBP (International Geosphere Biosphere Programme) (1991). The Inter-American Institute for Global Change Research. *Global Change Newsletter,* December: 4-5.

IGBP (International Geosphere Biosphere Programme) (1992). A quick start for START: guidelines for regional research networks and centers. *Global Change Newsletter,* **9**: 5-7.

IIASA (International Institute for Applied Systems Analysis) (1991). *Agenda for the Third Decade.* IIASA, Laxenburg.

Jacobson, H.K. and Price, M.F. (1990). *A Framework on the Human Dimensions of Global Environmental Change.* International Social Science, Council/United Nations Economic and Social Council Series 3. New York.

Jaeger, J. and Ferguson, H.L. (eds.) (1991). *Climate Change: Science, Impacts and Policy.* Proceedings of the Second World Climate Conference. Cambridge University Press, Cambridge.

Judd, M. A., Boyce, J.K. and Evenson, R.E. (1987). Investment in Agricultural Research and Extension. In: Ruttan, V.W. and Pray, C.E. (eds.) *Policy for Agricultural Research.* Westview Press, Boulder. pp. 2-38.

Kaseje, D.C.O. (1994). Health systems for rural areas: Kenya. In: Ruttan, V.W. (ed.) *Agriculture, Environment and Health: Sustainable Development in the 21st Century.* University of Minnesota Press, Minneapolis. pp. 236-256.

Lee, J. A., Norse, D. and Parry, M. (1994). Monitoring the effects of acid deposition and climate change on agricultural potential. In: Ruttan, V.W. (ed.) *Agriculture, Environment and Health: Sustainable Development in the 21st Century.* University of Minnesota Press, Minneapolis. pp. 284-307.

Lepkowski, W. (1992). US-Japan global partnership action plan revived. *Chemical and Engineering News,* **70**(29): 16-18.

Livernash, R. (1992). The growing influence of NGOs in the developing world. *Environment,* **34**(5): 12ff.

Lucas, A.O. (1994). The institutional infrastructure for health research in developing countries. In: Ruttan, V.W. (ed.) *Agriculture, Environment and Health: Sustainable Development in the 21st Century.* University of Minnesota Press, Minneapolis: 187-208.

Maddox, J. (1992). Dangers of disappointment at Rio. *Nature,* **357**: 265-266.

Mata, L. (1988). A public health approach to the "food-malnutrition-economic recession" complex. In: Bell, D.E. and Reich, M.R. (eds.) *Health, Nutrition, and Economic Crises: Approaches to Policy in the Third World.* Auburn House, Dover. pp. 265-275.

Miller, R.B. and Jacobson, H.K. (1992). *Research on the Human Components of Global Change: Next Steps.* Discussion Paper 1, Human Dimensions of Global Environmental Change Programme, International Social Science Council, Paris.

Norberg-Bohm, V. *et al.* (1992). *International Comparisons of Environmental Hazards.* CSIA Discussion Paper 92x. John F. Kennedy School of Government, Harvard University, Cambridge.

North, D.C. (1994). Constraints on institutional innovation: transaction costs, incentive compatibility and historical consideration. In: Ruttan, V.W. (ed.) *Agriculture, Environment and Health: Sustainable Development in the 21st Century.* University of Minnesota Press, Minneapolis. pp. 48-70.

Ostrom, E. (1990). *Governing the Commons: The Evolution of Institutions for Collective Action.* Cambridge University Press, Cambridge.

Parikh, K. (1994). Agriculture and food system scenarios for the 21st century. In: Ruttan, V.W. (ed.) *Agriculture, Environment and Health: Sustainable Development in the 21st Century.* University of Minnesota Press, Minneapolis. pp. 26-47.

Perry, J.S. (1991). Global change: from rhetoric to reality. *Review of Geophysics,* (Supplement/April) pp. 39-45.

Pray, C.E. (1987). Private sector agricultural research in Asia. In: Ruttan, V.W. and Pray, C.E. (eds.) *Policy for Agricultural Research.* Westview Press, Boulder. pp. 411-431.

Rawlins, S.L. (1994). Institutional capacity to monitor the services and effects of environmental change in agriculture. In: Ruttan, V.W. (ed.) *Agriculture, Environment and Health: Sustainable Development in the 21st Century.* University of Minnesota Press, Minneapolis. pp. 261-283.

Ruttan, V.W. (1982). *Agricultural Research Policy.* University of Minnesota Press, Minneapolis.

Ruttan, V.W. (1986). Toward a global agricultural research system: a personal view. *Research Policy,* **15**: 307-327.

Ruttan, V.W. (ed.) (1989). *Biological and Technical Constraints on Crop and Animal Productivity: Report on a Dialogue.* University of Minnesota Department of Agricultural and Applied Economics, St Paul.

Ruttan, V.W. (ed.) (1990a). *Resource and Environmental Constraints on Sustainable Growth on Agricultural Production: Report on a Dialogue.* University of Minnesota Department of Agricultural and Applied Economics, St Paul.

Ruttan, V.W. (ed.) (1990b). *Health Constraints in Agricultural Development.* University of Minnesota Department of Agricultural and Applied Economics, St Paul.

Ruttan, V.W. (ed.) (1992). *Sustainable Agriculture and the Environment: Perspectives on Growth and Constraints.* Westview Press, Boulder.

Ruttan, V.W. (ed.) (1994a). *Health and Sustainable Agricultural Development.* Westview Press, Boulder.

Ruttan, V.W. (ed.) (1994b). *Agriculture, Environment and Health: Sustainable Development in the 21st Century.* University of Minnesota Press, Minneapolis.

Ruttan, V.W., Bell, D.E. and Clark, W.C. (1994). Climate change and food security: agriculture, health and environmental research. *Global Environmental Change,* **4**:63-77.

Schmidheiny, S. (1992). *Changing Course: A Global Business Perspective on Development and the Environment.* MIT Press, Cambridge.

Schultz, T.W. (1964). *Transforming Traditional Agriculture.* Yale University Press, New Haven.

Tendler, J. (1994). Tales of dissemination in agriculture. In: Ruttan, V.W. (ed.) *Agriculture, Environment and Health: Toward Sustainable Development in the 21st Century.* University of Minnesota Press, Minneapolis. pp. 146-180.

Turner, B. I., II, and Benjamin, P.A. (1994). Fragile lands: identification and use for agriculture. In: Ruttan, V.W. (ed.) *Agriculture, Environment and Health: Toward Sustainable Development in the 21st Century.* University of Minnesota Press, Minneapolis. pp. 104-145.

World Bank (1992). *Development and the Environment: World Development Report 1992.* Oxford University Press, Oxford.

World Bank (1993). *Investing in Health: World Development Report, 1993.* Oxford University Press, Oxford.

List of Contributors

Ahsan U. Ahmed, Bangladesh Centre for Advanced Studies, 620 Road 10 A (New) Dhanmondi, GPO Box 3971, Dhaka 1209, BANGLADESH

Jock Anderson, World Bank, 1818 H St., NW, Washington DC 20433, U.S.

Kirsten Appendini, Centro de Estudios Económicos, El Colegio de México, Camino al Ajusco 20 , México DF 01000, MEXICO

Brad Bass, Environmental and Adaptation Research Group, Institute for Environmental Studies, University of Toronto, Haultain Building, 170 College Street, Toronto, M5S 1A4, (Formerly, Atmospheric Environment Service, Environment Canada, Downsview, Ontario), CANADA

David E. Bell, Harvard School of Public Health, Harvard University, Cambridge MA 02138, U.S.

William H. Bender, World Hunger Program, Brown University, Providence RI and Food Studies Group, Oxford; correspondence to P.O. Box 991, Groton MA 01450, U.S.

Hans-Georg Bohle, South Asian Institute, Im Neuenheimer Feld 330, 69120 Heidelberg, (Formerly, Albert-Ludwigs-Universitat, Institut fur Kulturgeographie, D-7800 Freiburg, Werderring 4), GERMANY

Robert S. Chen, Manager, Socioeconomic Data and Applications Center, Consortium for International Earth Science Information Network, 2250 Pierce Road, University Center MI 48710, U.S.

William C. Clark, Center for Science and International Affairs, John F. Kennedy School of Government, Harvard University, Cambridge MA 02138, U.S.

Robert Coppock, German-American Academic Council, 1055 Thomas Jefferson, Suite 2020, Washington DC 20007, (Formerly, World Resources Institute, 1709 New York Avenue NW, Washington DC 20006), U.S.

Bruce Crawshaw, 9 Muirfield Park, Gullane, East Lothian EH31 2DS, SCOTLAND

Pierre Crosson, Resources for the Future, 1616 P Street NW, Washington DC 20036, U.S.

Thomas E. Downing, Environmental Change Unit, University of Oxford, 1a Mansfield Road, Oxford OX1 3TB, U.K.

Graham Eele, Food Security Training and Administration Project, Southern Africa Development Community, P.O. Box 4046, Harare, ZIMBABWE, (Formerly, Food Studies Group, 21 St Giles, Oxford, U.K.)

Günther Fischer, Food and Agriculture Project, International Institute for Applied Systems Analysis, A-2361 Laxenburg, AUSTRIA

Klaus Frohberg, Institut für Agrarpolitik, Rheinische Friedrich-Wilhelms-Universitat, 5300 Bonn 1, GERMANY

Katherine Homewood, Department of Anthropology, University College, Gower Street, London WC1 E6BT; correspondence to: 103 Old Road, Wheatley, Oxon OX33 1NX, U.K.

Neil Hohmann, Department of Economics, University of Chicago, Chicago IL 60637, (Formerly, U.S. Department of Agriculture, Economic Research Service, Room 408, 1301 New York Ave., Washington DC 20005), U.S.

Saleemul Huq, Bangladesh Centre for Advanced Studies, 620 Road 10 A (New) Dhanmondi, GPO Box 3971, Dhaka 1209, BANGLADESH

Narpat S. Jodha, Social Policy and Resettlement Division, Environment Department, World Bank, 1818 H Street, NW, Washington DC 20433, U.S., (Formerly, ICIMOD, GPO Box 3226, Kathmandu, Nepal)

Frank Judd (Lord Judd), 21 Mill Lane, Marston, Oxford, U.K.

Sally Kane, Council of Economic Advisers, 317 Old Executive Office Building, Washington, DC 20500, (Formerly, National Oceanic and Atmospheric Administration, 1825 Connecticut Avenue NW, Suite 625, Washington DC 20235), U.S.

Z. Karim, Bangladesh Agricultural Research Council, Farmgate, New Airport Road, Farmgate, Dhaka 1215, BANGLADESH

Robert W. Kates, P.O. Box 8075, Trenton ME 04605, U.S.

Gregory A. Kiker, Department of Agricultural and Biological Engineering, Cornell University, Riley-Robb Hall, Ithaca NY 14853-5701, U.S.

Rob Koudstaal, Resource Analysis, Zuiderstraat 110, 2611 SJ Delft, THE NETHERLANDS

Richard P. Kunz, Department of Agricultural Engineering, University of Natal, Box 375, 3201 Pietermaritzburg, SOUTH AFRICA

Diana Liverman, Latin American Studies and Department of Geography, University of Arizona, Tucson AZ 85721, (Formerly, Department of Geography, Pennsylvania State University, 302 Walker, University Park PA 16802), U.S.

JoAnn McGregor, St. Antony's College, Oxford, OX2 6JS, U.K. (Formerly, Refugees Studies Programme, University of Oxford), U.K.

Chris H.D. Magadza, Lake Kariba Research Station, University of Zimbabwe, P.O. Box 48, Kariba , ZIMBABWE

David Norse, Department of Geography, University College London, 26 Bedford Way, London WC1H 0AP, (Formerly, Environmental Change Unit, University of Oxford, Oxford OX1 3TB), U.K.

Martin Parry, Department of Geography, University College London, 26 Bedford Way, London WC1H 0AP, (Formerly, Environmental Change Unit, University of Oxford, Oxford OX1 3TB), U.K.

John Reilly, Resources and Technology Division, Economic Research Service, U.S. Department of Agriculture, 1301 New York Avenue, NW, Washington DC 20005-4788, U.S.

Norman Rosenberg, Integrated Earth Sciences, Pacific Northwest Laboratory, 901D Street SW, Suite 900, Washington DC 20024, (Formerly, Resources for the Future, 1616 P Street NW, Washington DC 20036), U.S.

Cynthia Rosenzweig, Goddard Institute for Space Studies, 2880 Broadway, New York NY 10025, U.S.

Dale S. Rothman, Sustainable Development Research Institute, University of British Columbia, Vancouver, British Columbia, Canada V6T 1W5, (Formerly, World Resources Institute, 1709 New York Avenue. N.W)., Washington DC 20006, U.S.)

Vernon W. Ruttan, Department of Agricultural and Applied Economics, University of Minnesota, 231 Classroom Office Building., 1994 Buford Avenue, St. Paul MN 55108, U.S.

Eric Schiller, Department of Civil Engineering, University of Ottawa, Ottawa, CANADA

Stephen H. Schneider, Department of Biological Sciences, Stanford University, Stanford CA 94305-5020, U.S.

Roland E. Schulze, Department of Agricultural Engineering, University of Natal, Box 375, 3201 Pietermaritzburg, SOUTH AFRICA

Michael J. Scott, Pacific Northwest Laboratory, Richland WA 99352, U.S.

John Shaw, 112 Kenwood Drive, Beckenham, Kent BR3 2RB, U.K. (Formerly, World Food Programme, via Cristoforo Colombo 426, Rome, Italy)

Henry Venema, Acres International Ltd., Niagra Falls, Ontario, CANADA

Futang Wang, Chinese Academy of Meteorological Sciences, Baishiqiao Road, No. 4b, Western Suburb, Beijing 100081, P.R. CHINA

Michael Watts, Department of Geography, University of California, Berkeley CA 94720, U.S.

Zong-ci Zhao, Chinese Academy of Meteorological Sciences, Baishiqiao Road, No. 4b, Western Suburb, Beijing 100081, P.R. CHINA

Note: Former affiliations refer to positions at the time of the NATO Advanced Research Workshop on *Climate Change and World Food Security*.

INDEX

A

B

C

D

E

F

G

H

I

J

K

L

M

S

T

The ASI Series Books Published as a Result of Activities of the Special Programme on Global Environmental Change

This book contains the proceedings of a NATO Advanced Research Workshop held within the activities of the NATO Special Programme on Global Environmental Change, which started in 1991 under the auspices of the NATO Science Committee.

The volumes published as a result of the activities of the Special Programme are:

Vol. 1: **Global Environmental Change.** Edited by R. W. Corell and P. A. Anderson. 1991.
Vol. 2: **The Last Deglaciation: Absolute and Radiocarbon Chronologies.** Edited by E. Bard and W. S. Broecker. 1992.
Vol. 3: **Start of a Glacial.** Edited by G. J. Kukla and E. Went. 1992.
Vol. 4: **Interactions of C, N, P and S Biogeochemical Cycles and Global Change.** Edited by R. Wollast, F. T. Mackenzie and L. Chou. 1993.
Vol. 5: **Energy and Water Cycles in the Climate System.** Edited by E. Raschke and D. Jacob. 1993.
Vol. 6: **Prediction of Interannual Climate Variations.** Edited by J. Shukla. 1993.
Vol. 7: **The Tropospheric Chemistry of Ozone in the Polar Regions.** Edited by H. Niki and K. H. Becker. 1993.
Vol. 8: **The Role of the Stratosphere in Global Change.** Edited by M.-L. Chanin. 1993.
Vol. 9: **High Spectral Resolution Infrared Remote Sensing for Earth's Weather and Climate Studies.** Edited by A. Chedin, M.T. Chahine and N.A. Scott. 1993.
Vol. 10: **Towards a Model of Ocean Biogeochemical Processes.** Edited by G. T. Evans and M.J.R. Fasham. 1993.
Vol. 11: **Modelling Oceanic Climate Interactions.** Edited by J. Willebrand and D.L.T. Anderson. 1993.
Vol. 12: **Ice in the Climate System.** Edited by W. Richard Peltier. 1993.
Vol. 13: **Atmospheric Methane: Sources, Sinks, and Role in Global Change.** Edited by M. A. K. Khalil. 1993.
Vol. 14: **The Role of Regional Organizations in the Context of Climate Change.** Edited by M. H. Glantz. 1993.
Vol. 15: **The Global Carbon Cycle.** Edited by M. Heimann. 1993.
Vol. 16: **Interacting Stresses on Plants in a Changing Climate.** Edited by M. B. Jackson and C. R. Black. 1993.
Vol. 17: **Carbon Cycling in the Glacial Ocean: Constraints on the Ocean's Role in Global Change.** Edited by R. Zahn, T. F. Pedersen, M. A. Kaminski and L. Labeyrie. 1994.
Vol. 18: **Stratospheric Ozone Depletion/UV-B Radiation in the Biosphere.** Edited by R. H. Biggs and M. E. B. Joyner. 1994.
Vol. 19: **Data Assimilation: Tools for Modelling the Ocean in a Global Change Perspective.** Edited by P. O. Brasseur and J. Nihoul. 1994.

Vol. 20: **Biodiversity, Temperate Ecosystems, and Global Change.**
Edited by T. J. B. Boyle and C. E. B. Boyle. 1994.

Vol. 21: **Low-Temperature Chemistry of the Atmosphere.**
Edited by G. K. Moortgat, A. J. Barnes, G. Le Bras and J. R. Sodeau. 1994.

Vol. 22: **Long-Term Climatic Variations – Data and Modelling.**
Edited by J.-C. Duplessy and M.-T. Spyridakis. 1994.

Vol. 23: **Soil Responses to Climate Change.**
Edited by M. D. A. Rounsevell and P. J. Loveland. 1994.

Vol. 24: **Remote Sensing and Global Climate Change.**
Edited by R. A. Vaughan and A. P. Cracknell. 1994.

Vol. 25: **The Solar Engine and Its Influence on Terrestrial Atmosphere and Climate.**
Edited by E. Nesme-Ribes. 1994.

Vol. 26: **Global Precipitations and Climate Change.**
Edited by M. Desbois and F. Désalmand. 1994.

Vol. 27: **Cenozoic Plants and Climates of the Arctic.**
Edited by M. C. Boulter and H. C. Fisher. 1994.

Vol. 28: **Evaluating and Monitoring the Health of Large-Scale Ecosystems.**
Edited by D. J. Rapport, C. L. Gaudet and P. Calow. 1995.

Vol. 29: **Global Environmental Change Science: Education and Training.**
Edited by D. J. Waddington. 1995.

Vol. 30: **Ice Core Studies of Global Biogeochemical Cycles.**
Edited by R. J. Delmas. 1995.

Vol. 31: **The Role of Water and the Hydrological Cycle in Global Change.**
Edited by H. R. Oliver and S. A. Oliver. 1995.

Vol. 32: **Atmospheric Ozone as a Climate Gas.**
Edited by W.-C. Wang and I. S. A. Isaksen. 1995.

Vol. 33: **Carbon Sequestration in the Biosphere.**
Edited by M. A. Beran. 1995.

Vol. 34: **Climate Sensitivity to Radiative Perturbations: Physical Mechanisms and Their Validation.**
Edited by H. Le Treut. 1996

Vol. 35: **Clouds, Chemistry and Climate.**
Edited by P. J. Crutzen and V. Ramanathan. 1996.

Vol. 36: **Diachronic Climatic Impacts on Water Resources.**
Edited by A. N. Angelakis and A. S. Issar. 1996.

Vol. 37: **Climate Change and World Food Security.**
Edited by T. E. Downing. 1996.

Vol. 38: **Evaluation of Soil Organic Matter Models.**
Edited by D. S. Powlson, P. Smith and J. U. Smith. 1996.

Vol. 39: **Microbiology of Atmospheric Trace Gases.**
Edited by J. C. Murrell and D. P. Kelly. 1996.

Vol. 40: **Forest Ecosystems, Forest Management and the Global Carbon Cycle.**
Edited by M. J. Apps and D. T. Price. 1996.

Springer-Verlag and the Environment

We at Springer-Verlag firmly believe that an international science publisher has a special obligation to the environment, and our corporate policies consistently reflect this conviction.

We also expect our business partners – paper mills, printers, packaging manufacturers, etc. – to commit themselves to using environmentally friendly materials and production processes.

The paper in this book is made from low- or no-chlorine pulp and is acid free, in conformance with international standards for paper permanency.